Molecular Interactions
in Bioseparations

Molecular Interactions in Bioseparations

Edited by

That T. Ngo

Department of Developmental and Cell Biology
University of California
Irvine, California

Plenum Press • New York and London

Library of Congress Cataloging-in-Publication Data

Molecular interactions in bioseparations / edited by That T. Ngo.
     p.   cm.
   Includes bibliographical references and index.
   ISBN 0-306-44435-6
   1. Affinity chromatography. 2. Biomolecules--Separation.
I. Ngo, T. T. (That Tjien), 1944-   .
   QP519.9.A35M65   1993
   574.19'285--dc20   93-34419
                                                            CIP

ISBN 0-306-44435-6

©1993 Plenum Press, New York
A Division of Plenum Publishing Corporation
233 Spring Street, New York, N.Y. 10013

Contributors

Stig Allenmark ● Department of Organic Chemistry, University of Gothenburg, S-41296 Gothenburg, Sweden

Francisco M. Alvarez ● Schering-Plough Research Institute, Kenilworth, New Jersey 07033. *Present address*: Mylan Pharmaceuticals Inc., Morgantown, West Virginia, 26504

Lars I. Andersson ● Department of Pure and Applied Biochemistry, University of Lund, S-221 00 Lund, Sweden

Shalini Andersson ● IFM/Department of Chemistry, University of Linköping, S-58183 Linköping, Sweden

Pascal Bailon ● Protein Biochemistry, Molecular Sciences Department, Hoffmann-La Roche Inc., Nutley, New Jersey 07110

Milan J. Beneš ● Institute of Macromolecular Chemistry, Academy of Sciences of the Czech Republic, Prague, Czech Republic

Arvid Berge ● SINTEF, Applied Chemistry, N-7034 Trondheim, Norway

Gerd Birkenmeier ● Institute of Biochemistry, University of Leipzig, D–04103 Leipzig, Germany

Carey B. Bottom ● Schering-Plough Research Institute, Kenilworth, New Jersey 07033. *Present address*: Chase Laboratory Co., Newark, New Jersey 07105

Michael D. P. Boyle ● Department of Microbiology, Medical College of Ohio, Toledo, Ohio 43699

Irwin Chaiken ● SmithKline Beecham, King of Prussia, Pennsylvania 19406

Prashant Chikhale ● Department of Pharmaceutical Chemistry, The University of Kansas, Lawrence, Kansas 66045

Pedro Cuatrecasas ● Parke Davis Pharmaceutical Division, Warner Lambert Company, Ann Arbor, Michigan 48106

Matthijs Dekker ● Unilever Research Laboratorium, P.O. Box 114, 3130 AC Vlaardingen, The Netherlands

Edith Dellacherie • Laboratoire de Chimie-Physique Macromoléculaire, URA CNRS 494, ENSIC, 54001 Nancy Cédex, France

Björn Ekberg • Bio-Swede AB, IDEON, S-223 70 Lund, Sweden

Ervin L. Faulmann • Department of Microbiology, Medical College of Ohio, Toledo, Ohio 43699

Steinar Funderud • The Norwegian Radium Hospital, Montebello, N-0310 Oslo 3, Norway

Jun Hirabayashi • Department of Biological Chemistry, Faculty of Pharmaceutical Sciences, Teikyo University, Sagamiko, Kanagawa 199–01, Japan

Erik Hornes • Department of Microbiology and Immunology, Norwegian College of Veterinary Medicine, Oslo 1, Norway, and Dynal AS, 0212 Oslo, Norway

Patrick Hubert • Laboratoire de Chimie-Physique Macromoléculaire, URA, CNRS 494, ENSIC, 54001 Nancy Cédex, France

T. William Hutchens • Protein Structure Laboratory, USDA/ARS Children's Nutrition Research Center, Department of Pediatrics, Baylor College of Medicine, Houston, Texas 77030

Gabor L. Igloi • Institut für Biologie III der Universität Freiburg, D–79104 Freiburg, Germany

Shin-ichi Ishii • Department of Biochemistry, Faculty of Pharmaceutical Sciences, Hokkaido University, Sapporo, Hokkaido 060, Japan

Ken-ichi Kasai • Department of Biological Chemistry, Faculty of Pharmaceutical Sciences, Teikyo University, Sagamiko, Kanagawa 199–01, Japan

Shigeo Katoh • Chemical Engineering Department, Kyoto University, Kyoto 606, Japan

Rajni Kaul • Department of Biotechnology, Chemical Center, Lund University, S-221 00 Lund, Sweden

Neeta Khatter • BioProbe International Inc., Tustin, California 92680

Gerhard Kopperschläger • Institute of Biochemistry, University of Leipzig, D–04103 Leipzig, Germany

Takashi Kumazaki • Department of Biochemistry, Faculty of Pharmaceutical Sciences, Hokkaido University, Sapporo, Hokkaido 060, Japan

Eva Linné • Department of Biotechnology, Chemical Center, Lund University, S-221 00 Lund, Sweden

J. Paul Luzio ● Department of Clinical Biochemistry, University of Cambridge, Addenbrooke's Hospital, Cambridge CB2 2QR, England

Shu-Ching Ma ● College of Pharmacy, The University of Michigan, Ann Arbor, Michigan 48109-1065

Leon Marcus ● Pediatric Clinic, Kupat Holim Clalit Health Fund, Kiryat Moshe, Rehovot, Israel

Shlomo Margel ● The Department of Chemistry, Bar Ilan University, Ramat Gan, Israel

Bo Mattiasson ● Department of Biotechnology, Chemical Center, Lund University, S-221 00 Lund, Sweden

Dennis W. Metzger ● Department of Microbiology, Medical College of Ohio, Toledo, Ohio 43699

Satoshi Minobe ● Research Laboratory of Applied Biochemistry, Tanabe Seiyaku Co., Ltd., Osaka 532, Japan

Klaus Mosbach ● Department of Pure and Applied Biochemistry, University of Lund, S-221 00 Lund, Sweden

Michele Nachman-Clewner ● Department of Physiology, Cornell Medical School, New York, New York 10021

Dyer Narinesingh ● Department of Chemistry, The University of the West Indies, St. Augustine, Trinidad, West Indies

That T. Ngo ● BioProbe International Inc., Tustin, California 92680. *All correspondence should be addressed to*: Department of Developmental and Cell Biology, University of California, Irvine, California 92717

Kjell Nustad ● The Norwegian Radium Hospital, Montebello, N-0310 Oslo 3, Norway

Sten Ohlson ● HyClone Laboratories, Inc., Logan, Utah 84321. *Present address*: Department of Natural Sciences, University of Kalmar, S-39129 Kalmar, Sweden

Ørjan Olsvik ● Enteric Diseases Branch, Division of Bacterial and Mycotic Diseases, Centers for Disease Control, Atlanta, Georgia 30333, and Department of Microbiology and Immunology, Norwegian College of Veterinary Medicine, Oslo 1, Norway

Toshiaki Osawa ● Division of Chemical Toxicology and Immunochemistry, Faculty of Pharmaceutical Sciences, The University of Tokyo, Tokyo 113, Japan

S. Oscarsson ● Biochemical Separation Centre, Uppsala University, S-751 23 Uppsala, Sweden

Indu Parikh ● Research Triangle Pharmaceuticals Ltd., Durham, North Carolina 27713

Charles Pidgeon ● Department of Medical Chemistry, Purdue University, West Lafayette, Indiana 47907

J. Porath ● Department of Biochemistry and the Division of Biotechnology, University of Arizona, Tucson, Arizona 85721

Peter J. Richardson ● Department of Pharmacology, University of Cambridge, Cambridge CB2 1QJ, England

Eizo Sada ● Chemical Engineering Department, Kyoto University, Kyoto 606, Japan

Tadashi Sato ● Analytical Chemistry Research Laboratory, Tanabe Seiyaku Co., Ltd., Osaka 532, Japan

Ruth Schmid ● SINTEF, Applied Chemistry, N-7034 Trondheim, Norway

William H. Scouten ● Chemistry Department, Biotechnology Center, Utah State University, Logan, Utah 84321

Cheryl Spence ● Protein Biochemistry, Molecular Sciences Department, Hoffmann-La Roche Inc., Nutley, New Jersey 07110

Apryll M. Stalcup ● Department of Chemistry, University of Hawaii at Manoa, Honolulu, Hawaii 96822

Alexandra Štambergova ● Institute of Toxicology and Forensic Chemistry, Faculty of General Medicine, Charles University, Prague, Czech Republic

Earle Stellwagen ● Department of Biochemistry, University of Iowa, Iowa City, Iowa 52242

Tetsuya Tosa ● Research and Development, Tanabe Seiyaku Co., Ltd., Osaka 532, Japan

Tsutomu Tsuji ● Division of Chemical Toxicology and Immunochemistry, Faculty of Pharmaceutical Sciences, The University of Tokyo, Tokyo 113, Japan

John Ugelstad ● SINTEF, Applied Chemistry, N-7034 Trondheim, Norway

Mathias Uhlén ● Department of Biochemistry and Biotechnology, The Royal Institute of Technology, S-100 44 Stockholm, Sweden

M. A. Vijayalakshmi ● Département Génie Biologique, Laboratoire d'Intéractions Moleculaires et de Technologie des Séparations, Université de Technologie de Compiègne, 60206 Compiègne Cédex, France

Taizo Watanabe • Research Laboratory of Applied Biochemistry, Tanabe Seiyaku Co., Ltd., Osaka 532, Japan

Howard H. Weetall • Biotechnology Division, Chemical Sciences and Technology Laboratory, National Institute of Standards and Technology, Gaithersburg, Maryland 20899

Karen L. Williams • Department of Chemistry, University of Hawaii at Manoa, Honolulu, Hawaii 96822

Günter Wulff • Institute of Organic Chemistry and Macromolecular Chemistry, Heinrich-Heine-University Düsseldorf, 40225 Düsseldorf, Germany

Kazuo Yamamoto • Division of Chemical Toxicology and Immunochemistry, Faculty of Pharmaceutical Sciences, The University of Tokyo, Tokyo 113, Japan

Victor C. Yang • College of Pharmacy, The University of Michigan, Ann Arbor, Michigan 48109-1065

Tai-Tung Yip • Protein Structure Laboratory, USDA/ARS Children's Nutrition Research Center, Department of Pediatrics, Baylor College of Medicine, Houston, Texas 77030

David Zopf • Neose Pharmaceuticals, Inc., Horsham, Pennsylvania 19044

Preface

Rapid advances in molecular biology have accelerated the production of a great number of protein-based therapeutic agents. The major cost in producing these proteins appears to be associated with their purification from the complex mixture of the crude extract. A major challenge to the protein biochemist and the biochemical engineer is the development of rapid, efficient, and cost-effective purification systems.

This volume presents state-of-the-art reviews of current methods used in the purification of biological macromolecules that are based on molecular interactions. Thus, the major emphasis is placed on affinity-related techniques. Part I provides a general introduction to affinity chromatography and includes a chapter describing an interesting new technique called "slalom chromatography" for DNA fractionations. Affinity chromatography using molecules of biological origin as the affinity ligand is covered in Part II. Part III describes the use of a special class of biomolecules, antibodies, as affinity ligands. Affinity chromatography with biomimetic ligands is discussed in Part IV. Newer concepts and their applications in bioseparation are presented in Part V. Part VI covers affinity-related techniques such as affinity-based extracorporeal shunts, affinity electrophoresis, affinity precipitation, and affinity extraction.

I would like to express my sincere thanks to all the authors, who are recognized experts in their respective fields, for their cooperation and contributions. I thank the editorial staff of Plenum Press for their professionalism, and Mary Phillips Born, Senior Editor, for her encouragement. The support of my family (Ping and Peilin) made it possible to complete editing this book.

That T. Ngo

Irvine, California

Contents

Part III: Immunoaffinity Separation

Part VI: Affinity-Related Techniques

I

General Introduction

Affinity Chromatography
An Overview

Indu Parikh and Pedro Cuatrecasas

1. INTRODUCTION

Since its conception 25 years ago (Cuatrecasas *et al.*, 1968), the technique of affinity chromatography has seen tremendous growth in its applications and vast technical improvements. Besides the conceptual contributions of Cuatrecasas *et al.*, the availability of solid supports such as beaded agarose and the simple CNBr derivatization method greatly helped the initial development of affinity chromatography (Axen *et al.*, 1967; Cuatrecasas, 1970).

Affinity chromatography exploits the ubiquitous and highly specific binding interaction between a ligand and a macromolecule for the purpose of purification. The biological interactions and molecular specificity underlying these interactions usually involve noncovalent, reversible processes in nature. It is this noncovalent but specific interaction between two molecules upon which the original concept of affinity chromatography was founded (Cuatrecasas and Wilchek, 1968; Cuatrecasas, 1969; Jacoby and Wilchek, 1974). The successful purification of a macromolecule by affinity chromatography will obviously depend upon the specificity of the interaction, as well as the affinity between the interacting species. In most cases, these two factors dictate whether affinity chromatography provides a homogeneous product by a single-step procedure or whether the technique is one of several steps in the purification of a macromolecule.

During the past three decades, there have been a tremendous number of extensions and modifications of affinity chromatography, many of which were basically described in the original conceptual work. The original cyanogen bromide coupling technology, although still widely used, has been supplemented by coupling reagents and coupling strategies that provide a more defined and stable product (Sica *et al.*, 1973; Parikh *et al.*, 1974; Cuatrecasas and Parikh, 1972, 1974). The following list includes some of the well-recognized modifications and extensions of the classical affinity chromatography:

- Hydrophobic interaction chromatography
- Metal-chelate affinity chromatography

Indu Parikh • Research Triangle Pharmaceuticals Ltd., Durham, North Carolina 27713.   *Pedro Cuatrecasas* • Parke Davis Pharmaceutical Division, Warner Lambert Company, Ann Arbor, Michigan 48106.

Molecular Interactions in Bioseparations, edited by That T. Ngo. Plenum Press, New York, 1993.

- Immunoaffinity chromatography
- Covalent affinity chromatography
- Immobilized dye affinity chromatography
- Group-specific affinity chromatography
- Affinity partitioning
- Affinity electrophoresis
- Subunit exchange chromatography
- High-performance affinity chromatography
- Quantitative affinity chromatography
- Membrane-based affinity chromatography
- Displacement chromatography

Some of the above-listed modifications of affinity chromatography are discussed elsewhere in this volume, and others have been extensively reviewed in the recent literature (Chaiken *et al.*, 1983). In view of the extraordinary proliferation and extensiveness of publications that deal with affinity chromatography in its various forms, the citations and work referenced in this short chapter will be, by necessity, very brief and arbitrary.

2. APPLICATIONS IN RECOMBINANT DNA RESEARCH

Some of the more fascinating and ingenious recent applications of affinity chromatography are in the area of recombinant DNA technology. Although immobilized DNA and DNA fragments have been used in the past, it is only recently that the technology has been sufficiently defined to allow isolation of small quantities of specific DNA-binding proteins. A large variety of proteins interact with DNA within the cell in order to make possible such basic genetic processes as DNA repair, recombination, and transcription. In order to fully understand these processes at the molecular level, many of these "regulatory" proteins must be isolated and characterized and their mechanism of action delineated.

Nuclear proteins such as histones, nonhistones, and chromosomal proteins are closely associated with DNA of eucaryotic cells. Isolation and characterization of some of the nonhistone proteins have gained special importance because these proteins are found to be involved in the control of a variety of transcription processes. The nonhistone proteins are generally complex and are responsible for a variety of functions. Most of the nonhistone proteins are poorly characterized, and some are identified solely by their relative mobilities on polyacrylamide gels. Isolation and characterization of these proteins have remained a challenge to protein chemists since they were first discovered more than three decades ago. The difficulties in isolation of these proteins are mainly due to their extremely low abundance and poor solubility and lack of *in vivo* assays. One of these nonhistone proteins, the promoter specific transcription factor Sp1, was recently purified by Briggs *et al.* (1986) as well as by Reeves and Elton (1987). The Sp1 transcription factor, which binds to a promoter sequence in DNA, was isolated from the nuclear extracts of HeLa cells by affinity chromatography on an agarose matrix containing a covalently attached oligonucleotide with analogy to the promoter sequence.

Another group of nonhistone chromatin proteins is called the high-mobility group (HMG) because of its rapid migration during polyacrylamide electrophoresis. Although one subclass of these proteins, HMG1, binds specifically to certain A–T-rich DNA sequences, its exact *in vivo* function is still not quite clear. One member of this HMG1 subclass of proteins was isolated and purified to homogeneity from nuclear extracts of African green monkey cells by a combination of two separate chromatographic steps with affinity chromatography on a DNA-Sephacryl column (Strauss and Varshavsky, 1984).

DNA topoisomerases are enzymes that control and modify the topological states of DNA. The two known topoisomerases, Topo I and Topo II, which are present in both procaryotes and eucaryotes, are distinguished by the fact that Topo I induces modification and cleavage of single-stranded DNA while Topo II cleaves double-stranded DNA. Each enzyme has a multifunctional catalytic role and induces relaxation–supercoiling, knotting–unknotting, and catenation–decatenation of its respective DNA substrates. Immobilized single- or double-stranded DNA is the obvious choice for purification of these enzymes. Greater than 90% purity of both Topo I and Topo II from yeast homogenates was achieved (Goto *et al.*, 1984) by a series of chromatographic steps with a final affinity purification on a double-stranded DNA–cellulose column.

Recombinant DNA technology has allowed *in vitro* fusions of genes and gene fragments. The technology of gene fusion has been extensively used not only for expression but also for selective purification of expressed fused proteins. The hybrid gene allows selective expression of two fused proteins, one of which can be used as a convenient "affinity handle" for purification of the fused protein (Nilsson *et al.*, 1985). Such affinity handle-based fusion techniques have been used for isolation of a variety of proteins and peptides. Staphylococcal protein A is easily expressed in various bacteria and has many other characteristics which make it a popular affinity handle for a gene fusion expression system. Because of its high binding affinity (K_d 10^{-8} M), immobilized immunoglobulin G (IgG) is used for the purification of Staph A fused protein. As Staph A itself is proteolytically stable, the proteolytic stability of the fused protein is often significantly improved, resulting in increased overall recovery.

Germino and Bastia (1984) demonstrated that β-galactosidase can also be used as a fused affinity handle to provide proteolytic stability to a highly labile replication initiation factor called π protein. The fusion protein can be isolated on a column with an immobilized inhibitor of β-galactosidase (Steers *et al.*, 1971). If the fusion of the π protein with β-galactosidase is effected via a collagen linker, the two moieties can be separated from each other by controlled cleavage of the linker with collagenase. The technique should be generally applicable provided the protein to be isolated does not contain collagenlike amino acid sequences.

A newer concept recently described by Hammarberg *et al.* (1989) involves a dual affinity fusion in which the gene of interest is fused between two heterologous domains, the expression products of which have specific affinity for two different ligands. The full-length protein can be selectively recovered by two sequential affinity purification steps. While the approach is promising for proteins that are highly susceptible to proteolysis, it may not be quite applicable where exact N- and/or C-terminal sequences of a protein are important for biological activity.

Transcription of DNA is often regulated by specific DNA-binding proteins. The recent surge of interest in the search for specific DNA-binding proteins for possible therapeutic intervention by gene regulation has prompted the use of affinity chromatography for their isolation and characterization. One of the earliest examples of the purification of a specific DNA-binding protein was provided by Herrick (1980), who elegantly purified the *lac* repressor protein by taking advantage of the fact that it bound more tightly to immobilized

DNA containing the *lac* operator gene sequence than to a DNA lacking such a sequence. The DNA- binding proteins also frequently bind to nonspecific DNA, although with lower affinity. Nonspecific competitor DNA is often added to the crude proteins during affinity chromatography to prevent losses of specific DNA-binding protein. One modification involves the incorporation of multiple copies of the recognition sequence of the specific DNA in a plasmid. After amplification and purification of the plasmid, it can be coupled to a solid support for affinity purification of the specific DNA-binding protein (O'Neill and Kelly, 1988). A nonspecific DNA affinity column is often used to remove proteins that nonspecifically bind to DNA prior to specific affinity purification. In order to increase the selectivity of the procedure and the purity of the protein, sometimes a biotinylated DNA is used. In this case, the protein–biotinylated DNA complex may be purified on a streptavidin-agarose column (Chodosh *et al.*, 1986). The specific protein is eluted by a salt gradient, thus leaving the biotinylated DNA on the avidin column.

3. ENZYMES AND OTHER PROTEINS

The vast majority of applications of affinity chromatography continue to be in the area of protein purification. Some complex and unusual enzymes have recently been isolated by affinity chromatography. These might be useful in enzyme replacement therapy and in the understanding of the pathophysiology of diseases they affect.

DT-diaphorase, which catalyzes the oxidation of NADH and NADPH at equal rates, is perhaps the most important member of a group of enzymes generally called diaphorases. This enzyme has been extensively studied in view of its potential protective effect against cytotoxicity and mutagenicity of quinone-derived oxygen radicals in addition to its involvement in vitamin K-dependent protein carboxylation. Liver cytoplasm is the richest source of this enzyme. DT-diaphorase was recently isolated with agarose-immobilized dicumarol, which is a potent inhibitor of the enzyme (Lind *et al.*, 1990). Warfarin and various other benzoquinones may also be used as affinity ligands.

Phospholipid hydroperoxide glutathione peroxidase, one of the only two seleno-enzymes known in mammals, reduces the hydroperoxy derivatives of phospholipids. The specificity of the enzyme for the peroxide substrate is rather broad. The enzyme, which is present in many tissues, was recently purified to about 50–70% purity on agarose-immobilized bromosulfophthalein-glutathione adsorbent (Maiorino *et al.*, 1990).

One of the most widely studied "detoxifying" enzymes is the extracellular superoxide dismutase (EC-SOD). The three known isoenzymes of EC-SOD are all tetrameric, Cu–Zn-containing glycoproteins of approximately 120 kDa. All three isoenzymes can be separated from each other and purified by heparin-agarose affinity chromatography (Karlsson and Marklund, 1988). All EC-SOD isoenzymes, being glycoproteins, are easily distinguishable from each other by affinity chromatography on concanavalin A (Con A)- or other lectin-immobilized agarose adsorbents, unlike the cytoplasmic and mitochondrial isoenzymes.

In a recent elegant application of an established affinity adsorbent, Gerard *et al.* (1989) separated and purified individual members of a complex mixture of human bronchial mucus proteins. Thiol-agarose covalent affinity chromatography following deglycosylation was successfully utilized to analyze the protein core of the complex bronchial epithelial mucus glycoproteins. Several of the mucin-associated proteins were resolved on the thiol-agarose column without being actually bound to the adsorbent, while others were covalently retarded and differentially eluted with a gradient of a reducing agent.

It has been known for some time that application of pressures between 1000 and 5000 atm to proteins in solution can be used to dissociate the individual subunits of a multimeric protein without denaturation while pressures of 6000 atm or more may irreversibly denature a protein. In a very dramatic example of how apparently purely physical changes can be effective in elution of a protein from an affinity column, Olson *et al.* (1989) used high pressures ranging up to 2000 atm to elute bovine serum albumin from an immunoaffinity column. Over 75% of the protein was eluted following a single 15-minute incubation at 2000 atm, and over 90% recoveries were obtained by repeated pressurizations. These treatments did not have any detrimental effect on the efficacy of the immunoadsorbent, whereas routinely used acidic elution methods usually reduce the capacity and often the specificity of an immunoadsorbent (Olson *et al.*, 1989).

Lipoxygenases are a group of closely related enzymes that are widely distributed in both plant and animal tissue. They catalyze oxygenation of polyunsaturated fatty acids such as arachidonic acid. One of the physiologically important lipoxygenases is 5-lipoxygenase, which catalyzes the first step in the leukotriene biosynthesis. Because the leukotrienes are potent mediators of inflammation, a detailed understanding of lipoxygenases (and cyclooxygenases) would provide novel and safe nonsteroidal anti-inflammatory drugs. The 5-lipoxygenase has been purified from a variety of sources over the past ten years. More recently, purification of 5-lipoxygenase from porcine leukocytes has been achieved by immunoaffinity chromatography (Ueda *et al.*, 1986). Studies of the highly purified enzyme showed that both 5-lipoxygenase and leukotriene synthetase activities reside in the same protein.

Many of the peptide and nonpeptide hormones have been shown to stimulate tyrosine phosphorylation of intracellular substrates. Isolation and characterization of specific substrates that are tyrosine-phosphorylated in response to hormonal stimulation are important in understanding the mechanism of signal transduction. Some hemopoietic hormones such as interleukins (IL) and granulocyte–macrophage colony-stimulating factor (GM-CSF) are known to stimulate tyrosine phosphorylation of a series of intracellular substrates. Recently, Duronio *et al.* (1992) have used antiphosphotyrosine antibodies to affinity-purify two of the proteins that are tyrosine-phosphorylated as a result of IL-3 and GM-CSF stimulation.

After decades of fruitful investigation of the proteins isolated from various snake venoms, more recently a series of interesting and therapeutically promising proteins have been isolated from the saliva or salivary glands of the exotic and deadly caterpillar *Lanomia achelous* (Amarant *et al.*, 1991) and certain flies. The blood-sucking flies have been found to produce highly potent inhibitors of human factor Xa to disrupt its host's normal hemostatic mechanism. Recently, Jacobs *et al.* (1990) have isolated a novel 18-kDa protein inhibitor of factor Xa by affinity chromatography on an Xa-agarose column from the salivary gland extracts of the black fly, *Simulium vittatum*. The interaction of the inhibitor protein with human factor Xa was stoichiometric while no inhibition of thrombin was detected.

The term group-specific affinity chromatography was coined by the late Nathan Kaplan in the mid-1970s. This ingenious but obvious extension of affinity chromatography illustrates that a single immobilized ligand may be used to purify sequentially a variety of enzymes (Kaplan *et al.*, 1974). Thus, for example, an affinity matrix containing immobilized adenosine monophosphate (AMP) is likely to bind any enzyme that uses AMP, ADP, ATP, NAD^+, NADP, or coenzyme A as a substrate or cofactor. In spite of an often low degree of affinity and specificity, a large number of enzymes can be conveniently separated and purified on a single AMP-agarose column. Often a salt, a substrate, or a cofactor gradient is used to accomplish the separation of enzymes from a mixture. Other ligands that are used for such group-specific affinity purification include lectins, heparin, Cibacron blue or other dyes, and various other nucleosides and nucleotides.

4. PURIFICATION OF RECEPTORS

Some of the most rewarding applications of affinity chromatography have been in the area of hormone- and drug-receptor purification. As soon as affinity chromatography was discovered, it was put to test in the purification of hormone receptors which are present in very minute quantities in the respective target tissues. In the case of estradiol receptor, the first cytoplasmic hormone receptor to be purified (Sica *et al.*, 1973) from calf uterus, more than 300,000-fold purification was achieved on an estradiol affinity column after a preliminary enrichment on heparin-agarose adsorbent.

The membrane-bound receptors, such as insulin receptors, being insoluble in physiological buffers, posed a different problem of solubilization prior to affinity chromatography. Each membrane-bound receptor possesses a unique set of physical and chemical characteristics and requires a continuous presence of detergents to keep it in solution (for a recent review on purification of membrane-bound proteins, see Thomas and McNamee, 1990). Most of the groundwork for solubilizing, coupling chemistry, and elution strategies were accomplished with the purification of insulin receptors (Cuatrecasas, 1970). Through a systematic study of various ionic and nonionic detergents and protease inhibitors, Cuatrecasas established means to solubilize insulin receptor without altering its hormone-binding characteristics. Protease inhibitors and metal chelators are routinely included in the solubilizing buffers, to prevent proteolysis during isolation of hormone receptors. A large variety of both cytoplasmic and membrane-bound receptors have been purified so far, and this application of affinity chromatography has been extensively reviewed in the recent literature (Jacobs and Cuatrecasas, 1981). Affinity chromatography continues to be the only method available for purification of hormone receptors as attested to by the numerous successful examples in the current literature.

The existence of multiple types of opioid receptors is well supported by a variety of biochemical and pharmacological studies (Chang, 1984). The molecular basis of the opioid receptor multiplicity is still unknown, mainly because of the lack of biochemical characterization of any of these receptors. In spite of continuous efforts by a number of laboratories to isolate and clone these receptors, the amino acid composition and sequence of opioid receptors remain elusive. Recently, Li *et al.* (1992) isolated the detergent-solubilized μ opioid receptor by affinity chromatography under very ingenious experimental conditions. Sodium ions are known to decrease the binding of agonists and increase the binding of antagonists to opioid receptors. The G-protein-associated μ opioid receptor was solubilized with a mixture of nonionic detergents such as CHAPS and digitonin in the presence of protease inhibitors and $1M$ NaCl. The 62-kDa receptor was initially enriched by lectin affinity chromatography followed by further purification on an opioid antagonist-immobilized affinity column in the presence of $1M$ NaCl. The authors elegantly exploited an old observation that sodium ions increase the binding of the opioid receptor to opiate antagonists with retention of its affinity to G-protein, a property which is known to endow the receptor with its biological properties.

5. PEPTIDES

Specific peptides have been isolated by affinity chromatography in the past by Wilchek, Givol, and co-workers. The classical examples include the purification of dinitrophenyl (DNP)-labeled peptides from affinity-labeled anti-DNP antibodies (Weinstein *et al.*, 1969) and from DNP-binding mouse myeloma protein (MOPC-315) on an anti-DNP-agarose column (Givol *et al.*, 1971). There is a renewed interest in the affinity chromatography of peptides with the recent advances in multiple-peptide synthesis and recombinant technology. Parmley and Smith (1988) recently developed a bacteriophage expression vector that can synthesize specific peptide epitopes on its surface. Their vector system allows one to construct large collections of bacteriophages that could include virtually all possible sequences of a short or medium-sized peptide. Such an epitope library may contain tens to hundreds of millions of short and different amino acid sequences that are displayed on the surface of a bacteriophage. The phages displaying foreign epitopes on their surface can then be purified by "biopanning" with immobilized specific antibody (Smith, 1991). Identification of highly selective peptide ligands for specific antibodies is of great importance in vaccine design, epitope mapping, identification of genes, drug design, identification of diagnostic markers, and many other applications (Birnbaum and Mosbach, 1992). Rapid identification of a short-chain peptide ligand for a hormone receptor, for example, would greatly benefit drug discovery and drug design research. Thus, peptide ligands could be readily identified from epitope libraries and be used as lead compounds for further development of pharmacological agents (Scott, 1992).

6. SMALL-MOLECULAR-WEIGHT LIGANDS

Affinity chromatography has been generally classified as a method for purification of proteins with the use of immobilized small-molecular-weight protein-specific ligands. However, there have been many recent reports in which a small-molecular-weight ligand has been purified with the use of an immobilized ligand-specific protein. The often-encountered complex nature of and need for purification of small amounts of a small-molecular-weight compound is exemplified by arachidonic acid metabolites.

Separation and quantitative analysis of arachidonic acid and its structurally related metabolites usually involve a combination of thin-layer chromatography, gas chromatography, and high-performance liquid chromatography (HPLC). Immunoaffinity purification is an attractive alternative technology. Polyclonal antibodies to various prostaglandins and leukotrienes have appreciable cross-reactivity with other prostaglandins and metabolites. This observation has been exploited by Vrbanac *et al.* (1988) and Krause *et al.* (1985) for initial purification of these metabolites by immunoaffinity chromatography. The partially purified metabolites can then be subjected to gas chromatography/mass spectrometry (GC/MS) for further analysis and characterization. Similarly, Chiabrando *et al.* (1987) isolated thromboxane B_2 and one of its urinary metabolites by using agarose-immobilized anti-thromboxane B_2 antibodies. In forensic chemistry, large numbers of urine samples, often in large quantities, must be analyzed for minute quantities of drugs of abuse. Initial isolation of such drugs or their metabolites by immunoaffinity chromatography prior to their characterization by GC/MS is illustrated by Lemm *et al.* (1985) with the example of tetrahydrocannabinol.

7. BIOTECHNOLOGY APPLICATIONS

There have been major advances during the past decade in recombinant DNA technology, and scores of companies are preparing proteins for therapeutic applications. Biotechnology companies producing proteins, by genetic engineering or from natural resources, in most cases use affinity chromatography as one of the purification steps. A typical downstream purification process for a pharmaceutical-grade protein involves a multiple series of chromatographic steps. Purification of a protein can often contribute more than 50% of the total cost of the manufactured protein. Each step in the processing, carefully controlled by the guidelines of Good Manufacturing Practice (GMP), must be validated for purity, accuracy, and reproducibility. The validation process is time-consuming and costly. Fewer chromatographic steps in a processing scheme require fewer steps to validate. Affinity chromatography, capable of providing a few thousand-fold purification in a single step, can thus substantially bring down the cost of a protein pharmaceutical. Moreover, the stringent requirements for purity from regulatory agencies for the use of recombinantly made proteins as human therapeutic agents has substantially increased the need for efficient processes such as large-scale affinity chromatography.

On a laboratory scale, agarose is routinely used for affinity chromatography. However, because of its intrinsic compressibility, agarose has severe limitations for devising efficient production-scale separation systems. Although the use of silica-based affinity supports permits circumvention of the compressibility issue, one still must face the question of nonspecific adsorption and possible requirement of excessive operating column pressure. The more recently introduced solid support Poros, although not yet sufficiently well tested in industrial-scale applications, might provide certain advantages. In order to at least partially resolve such limitations of agarose, silica, and numerous other affinity supports, the newer trend among the "affinity engineers" is to modify bed geometry and column configurations (Brandt et al., 1988; Clonis, 1987).

Because of the finite and continuous ligand leakage and relative lack of chemical and physical integrity of the supports during repeated usage, all affinity chromatography supports have a limited life span. Although multifunctional ligand coupling strategies (Parikh et al., 1974) substantially reduce ligand leakage in addition to providing superior extension arms, the chemistry of ligand coupling is continually improving. Documenting the chemical and physical stability of an affinity support during the production process is an extremely important and expensive undertaking. As required by regulatory agencies, these issues become part of the above-mentioned validation criteria. Some chromatography media manufacturers provide leakage information, which would reduce the costs of overall validation of the downstream purification process.

8. FUTURE PROSPECTS

It is expected that the future will see an even greater array of basic applications of affinity chromatography to help solve fundamental problems in biology. For example, microtechniques for the isolation of minute quantities of peptides and proteins of uncertain function will be coupled with advanced analytical techniques. The current extraordinary advances in the application of affinity approaches to establishing the peptide specificity of major histocompatibility complexes (MHCs) is such an example.

Molecular and recombinant biology are identifying an increasingly large number of proteins and receptors whose function is not known. Many of these are referred to as "orphan" receptors or enzymes, especially when they belong to a known class or super-family (e.g., steroid hormone receptors or tyrosine kinase receptors). As gene cloning and the accumulation of gene sequences (e.g., from the Human Genome Project) progress, the pressure to identify the corresponding function of the encoded proteins will increase. In these challenges, affinity techniques loom as among the few capable of providing sufficient simplicity, rapidity, and specificity.

Affinity methods will also be pivotal to further exploration and elucidation of the complex set of sequential interactions initiated by ligand-stimulated perturbations of cell functions. Cells are "touched" exquisitively and carefully by other cells, extracellular matrix proteins, or extracellular hormones (big and small). These interactions with membrane-bound receptors initiate a complex network of signal-transducing interactions that culminate in changes in metabolic or gene expression processes. All these phenomena are based on equivalent protein–protein or protein–nucleic acid interacting systems, where affinity procedures are optimal to dissect the precise molecular components.

Affinity chromatography will continue to be a tool on which drug discovery and development will depend critically. Already, many sophisticated systems have been devised for the extraordinarily rapid screening and detection of vast mixtures of ligands (often in the millions of recombinations, such as with peptides) for potentially therapeutically useful receptors or enzymes. Some new companies have been founded uniquely based on such principles (e.g., Affimax of Palo Alto, California, borrowing its name from affinity chromatography). The approaches are already also being used to screen the multitude of natural substances present in our oceans and plants. The cumbersome approach of using *in vivo* or complex pharmacological screens, as was done a few decades ago for finding lead compounds, had been long abandoned. However, the search in nature or in synthetic libraries for biological and chemical diversity by screening methods is beginning to see a rebirth, in large part because of the availability of new tools that make the process fast, cheap, precise, and meaningful.

The use of affinity methodologies in large-scale protein/nucleic acid/small-molecule (e.g., specific enantiomers) purification will continue to increase with improvements in engineering and physical supporting technologies.

These are but a few examples. The list could be expanded as our imagination turns to our vision of future advances in biology and chemistry. There are hardly any of these where affinity chromatography will not likely play a major role. This is a fundamental conceptual technology which has now been totally accepted and integrated into the way that biologists think about problems, as well as how they execute the technical solutions to those problems.

REFERENCES

Amarant, T., Burkhart, W., LeVine, H., Arocha-Pinango, C. L., and Parikh, I., 1991, Isolation and complete amino acid sequence of two fibrinolytic proteases from the toxic saturmid caterpillar, *Lanomia achelous*, *Biochim. Biophys. Acta* **1079**:214–221.

Axen, R., Porath, J., and Ernback, S., 1967, Chemical coupling of peptides and proteins to polysaccharides by means of cyanogen halides, *Nature (London)* **214**:1302–1304.

Birnbaum, S., and Mosbach, K., 1992, Peptide screening, *Curr. Opin. Biotechnol.* **3**:49–54.

Brandt, S., Goffe, R. A., Kessler, S. B., O'Connor, J. L., and Zale, S. E., 1988, Membrane-based affinity technology for commercial scale purifications, *Bio/Technology* **6:**779–782.

Briggs, M. R., Kadonaga, J. T., Bell, S. P., and Tijan, R., 1986, Purification and biochemical characterization of the promoter-specific transcription factor, Sp1, *Science* **234:**47–52.

Chaiken, I. M., Wilchek, M., and Parikh, I. (eds.), 1983, *Affinity Chromatography and Biological Recognition*, Academic Press, New York.

Chang, K. J., 1984, Opioid Receptors: Multiplicity and Sequelae of Ligand-Receptor Interactions, in: *The Receptors*, Vol. 1 (P. M. Conn, ed.) Academic Press, Orlando, Florida, pp. 1–81.

Chiabrando, C., Benigni, A., Piccinelli, A., Carminati, C., Cozzi, E., Remuzzi, G., and Fanelli, R., 1987, Antibody mediated extraction/negative ion chemical ionization mass spectrometric measurement of thromboxane B2 and 2,3-dinor-thromboxane B2 in human and rat urine, *Anal. Biochem.* **163:**255–262.

Chodosh, L. A., Carthew, R. W., and Sharp, P. A., 1986, A single polypeptide possesses the binding and transcription activities of the adenovirus major late transcription factor, *Mol. Cell. Biol.* **6:**4723–4733.

Clonis, Y. D., 1987, Large-scale affinity chromatography, *Bio/Technology* **5:**1290–1293.

Cuatrecasas, P., 1969, Insulin-Sepharose: Immunoreactivity and use in the purification of antibody, *Biochem. Biophys. Res. Commun.* **35:**531–537.

Cuatrecasas, P., 1970, Protein purification by affinity chromatography: Derivatizations of agarose and polyacrylamide beads, *J. Biol. Chem.* **245:**3059–3065.

Cuatrecasas, P., 1972, Affinity chromatography and purification of the insulin receptor of liver cell membranes, *Proc. Natl. Acad. Sci. U.S.A.* **69:**1277–1281.

Cuatrecasas, P., and Parikh, I., 1972, Adsorbents for affinity chromatography: Use of N-hydroxysuccinimide esters of agarose, *Biochemistry* **11:**2291–2299.

Cuatrecasas, P., and Parikh, I., 1974, Insulin receptors, *Methods Enzymol.* **34:**653–670.

Cuatrecasas, P., Parikh, I., and Hollenberg, M. D., 1973, Affinity chromatography and structural analysis of *Vibrio cholerae* enterotoxin-ganglioside agarosa and the biological effects of ganglioside-containing soluble polymers, *Biochemistry* **12:**4253–4264.

Cuatrecasas, P., and Wilchek, M., 1968, Single-step purification of avidine from egg white by affinity chromatography on biocytin-Sepharose columns, *Biochem. Biophys. Res. Commun.* **33:**235–239.

Cuatrecasas, P., Wilchek, M., and Anfinsen, C. B., 1968, Selective enzyme purification by affinity chromatography, *Proc. Natl. Acad. Sci. U.S.A.* **61:**636–643.

Duronio, V., Clark-Lewis, I., Federsppiel, B., Wieler, J. S., and Schrader, J. W., 1992, Tyrosine phosphorylation of receptor beta subunits and common substrates in response to interleukin-3 and granulocyte-macrophage colony-stimulating factor, *J. Biol. Chem.* **267:**21856–2863.

Gerard, C., Bhaskar, K. R., Gerard, N. P., Drazen, J. M., and Reid, L., 1989, Studies on the peptide core of human bronchial mucus glycoprotein, *Symp. Soc. Exp. Biol.* **43:**221–230.

Germino, J., and Bastia, D., 1984, Rapid purification of a cloned gene product by genetic fusion and site-specific proteolysis, *Proc. Natl. Acad. Sci. U.S.A.* **81:**4692–4696.

Givol, D., Strausbach, P. H., Hurwitz, E., Wilchek, M., Haimovich, J., and Eisen, H. N., 1971, Affinity labeling and cross-linking of the heavy and light chains of a myeloma protein with anti-DNP activity, *Biochemistry* **10:**3461–3466.

Glencross, R. G., Abeywardene, S. A., Corney, S. J., and Morris, H. S., 1981, The use of oestradiol-17-beta antiserum covalently coupled to Sepharose to extract oestradiol-17-beta from biological fluids, *J. Chromatogr.* **223:**193–197.

Goto, T., Laipis, P., and Wang, J. C., 1984, The purification and characterization of DNA topoisomerases I and II of yeast *S. cerevisiae*, *J. Biol. Chem.* **259:**10422–10429.

Hammarberg, B., Nygren, P. A., Holmgren, E., Elmblad, A., Tally, M., Hellman, U., Moks, T., and Uhlen, M., 1989, Dual affinity fusion approach and its use to express recombinant human insulin-like growth factor II, *Proc. Natl. Acad. Sci. U.S.A.* **86:**4367–4371.

Herrick, G., 1980, Site-specific DNA affinity chromatography of the lac repressor, *Nucleic Acids Res.* **8:**3721–3728.

Jacobs, J. W., Cupp, E. W., Sardana, M., and Friedman, P. A., 1990, Isolation and characterization of a coagulation factor Xa inhibitor from black fly salivary glands, *Thromb. Haemost.* **64:**235–238.

Jacobs, S., and Cuatrecasas, P., 1981, Affinity chromatography for membrane receptors purification, in: *Receptors and Recognition*, Vol. 11, Series B (S. Jacobs and P. Cuatrecasas, eds.), Chapman and Hall, London, pp. 61–86.

Jacobs, S., Shechter, Y., Bissell, K., and Cuatrecasas, P., 1977, Purification and properties of insulin receptors from rat liver membranes, *Biochem. Biophys. Res. Commun.* **77:**981–988.

Jakoby, W. B., and Wilchek, M. (eds.), 1974, Affinity Techniques, *Methods Enzymol.* **34**.

Kaplan, N. O., Everse, J., Dixon, J. E., Stolzenbach, F. E., Lee, C. Y., Lee, C. L. T., Taylor, S. S., and Mosbach, K., 1974, Purification and separation of pyridine nucleotide-linked dehydrogenases by affinity chromatography techniques, *Proc. Natl. Acad. Sci. U.S.A,*. **71:**3450–3454.

Karlsson, K., and Marklund, S. L., 1988, Extracellular superoxide dismutase in the vascular system of mammals, *Biochem. J.* **255:**223–228.

Krause, W., Jakobs, U., Schulze, P. E., Nieuweboer, B., and Hümpel, M., 1985, Development of antibody mediated extraction followed by GC/MS and its application to iloprost determination in plasma, *Prostaglandins Leukotrienes Med.* **17:**167–182.

Lemm, U., Tenczer, J., Baudisch, H., and Krause, W., 1985, Antibody mediated extraction of the main tetrahydrocannabinol metabolite, 11-nor-delta-9-tetrahydrocannabinol-9-carboxylic acid, from human urine and its identification by GC-MS in the sub-nanogram range, *J. Chromatogr.* **342:**393–398.

Li, L.-Y., Zhang, Z.-M., Su, Y.-F., Watkins, W. D., and Chang, K.-J., 1992, Purification of opioid receptors in the presence of sodium ions, *Life Sci.* **51:**1177–1185.

Lind, C., Cadenas, E., Hochstein, P., and Ernster, L., 1990, DT-diaphorase: Purification, properties and function, *Methods Enzymol.* **186:**287–301.

Maiorino, M., Gregolin, C., and Ursini, F., 1990, Phospholipid hydroperoxide glutathione peroxidase, *Methods Enzymol.* **186:**448–457.

Nilsson, B., Abrahmsen, L., and Uhlen, M., 1985, Immobilization and purification of enzymes with staphylococcal protein A gene fusion vectors, *EMBO J.* **4:**1075–1080.

Olson, W. C., Leung, S. K., and Yarmush, M. L., 1989, Recovery of antigens from immunoadsorbents using high pressure, *Biotechnology* **7:**369–373.

O'Neill, E. A., and Kelly, T. J., 1988, Purification and characterization of nuclear factor III (origin recognition protein C), a sequence specific DNA binding protein required for efficient initiation of adenovirus DNA replication, *J. Biol. Chem.* **263:**931–937.

Parikh, I., March, S., and Cuatrecasas, P., 1974, Topics in the methodology of substitution reactions with agarose, *Methods Enzymol.* **34:**77–102.

Parmley, S. F., and Smith, G. P., 1988, Antibody selectable filamentous fd phage vectors: Affinity purification of target genes, *Gene* **73:**305–318.

Reeves, R., and Elton, T. S., 1987, Non-histone chromatin proteins that recognize specific sequences of DNA, *J. Chromatogr.* **418:**73–95.

Scott, J. K., 1992, Discovering peptide ligands using epitope libraries, *Trends Biochem. Sci.* **17:**241–245.

Sica, V., Parikh, I., Nola, E., Puca, G. A., and Cuatrecasas, P., 1973, Affinity chromatography and the purification of estrogen receptor, *J. Biol. Chem.* **248:**6543–6558.

Smith, G. P., 1991, Surface presentation of protein epitopes using bacteriophage expression systems, *Curr. Opin. Biotechnol.* **2:**668–673.

Steers, E., Cuatrecasas, P., and Pollard, H. B., 1971, The purification of beta-galactosidase from *E. coli* by affinity chromatography, *J. Biol. Chem.* **246:**196–200.

Strauss, F., and Varshavsky, A., 1984, A protein binds to a satellite DNA repeat at three specific sites that would be brought into mutual proximity by DNA folding in the nucleosome, *Cell* **37:**889–901.

Thomas, T. C., and McNamee, M. G., 1990, Purification of membrane proteins, *Methods Enzymol.* **182:**499–520.

Ueda, N., Kaneko, S., Yoshimoto, T., and Yamamoto, S., 1986, Purification of arachidonate 5-lipoxygenase from porcine leukocytes and its reactivity with hydroperoxyeicosatetranoic acids, *J. Biol. Chem.* **261:**7982–7988.

Vrbanac, J. J., Eller, T. D., and Knapp, D. R., 1988, Quantitative analysis of 6-keto-prostaglandin F1α using immunoaffinity purification and gas chromatography–mass spectrometry, *J. Chromatogr.* **425:**1–9.

Weinstein, Y., Wilchek, M., and Givol, D., 1969, Affinity labeling of anti-dinitrophenyl antibodies with bromoacetyl derivatives of homologous haptenes, *Biochem. Biophys. Res. Commun.* **34:**694–701.

2

Weak-Affinity Chromatography

Sten Ohlson and David Zopf

1. INTRODUCTION

Classical adsorption/desorption affinity chromatography (Turkova, 1978) relies upon high-affinity interactions between an immobilized ligand and a soluble ligate to form stable (bio)molecular complexes that will withstand washing to remove soluble impurities. Disruption of such complexes often requires harsh buffer conditions that lead to loss in (bio)activity of the purified ligate. For this reason, and because of its high cost, low throughput, limited reproducibility, and lack of resolving power, affinity chromatography has remained primarily a tool of the research laboratory. As technologies have emerged for economical production of relatively large quantities of homogeneous, bioactive proteins via hybridoma and recombinant molecular genetic methods, it has become feasible to incorporate into affinity matrices high concentrations of protein ligands that recognize small-molecular-weight ligates with weak affinity ($K_a = 10^2 - 10^4\ M^{-1}$) (Ohlson *et al.*, 1988; Zopf and Ohlson, 1990). The rapid dynamics of weak-affinity binding provide the possibility for true, high-resolution chromatography of multiple analytes on a crude matrix in nearly physiological buffers under isocratic conditions.

A major advantage of weak-affinity chromatography (WAC) is that ligates elute as retarded peaks whose shape and retention times can be understood in terms of basic chromatographic theory. This chapter will present a brief description of selected theoretical and practical aspects of weak-affinity chromatography as well as some speculations as to the future of this emerging technology.

2. WHY STUDY WEAK AFFINITY?

In natural biological systems, weak-affinity interactions between molecular species are common. Biological activity is often triggered by a multitude of weak interactions, which mediate cooperative binding between ordered assemblies of biomolecules. There are numerous examples from nature where weak binding plays an important role: during

Sten Ohlson • HyClone Laboratories, Inc., Logan, Utah 84321. *Present address*: Department of Natural Sciences, University of Kalmar, S-39129 Kalmar, Sweden.   *David Zopf* • Neose Pharmaceuticals, Inc., Horsham, Pennsylvania 19044.

Molecular Interactions in Bioseparations, edited by That T. Ngo. Plenum Press, New York, 1993.

initiation of complement activation, hexameric C1q molecules bind to aggregated immuno-globulins present on foreign microorganisms or to immune complexes; multivalent weak affinity immunoglobulin M (IgM) antibodies make effective agglutination agents; and microbial pathogens specifically bind to target cells and cells communicate with each other and with the extracellular matrix via a whole spectrum of weak-affinity interactions. Furthermore, to secure a dynamic control of cellular metabolism, for example, in traffick-ing of metabolites, molecular interactions frequently are weak and governed to a large extent by concentration gradients. An example illustrative of this is the weak binding of substrates and inhibitors to various enzymes. The introduction of weak-affinity chroma-tography has provided analytical and preparative tools to facilitate studies of these and related phenomena.

3. CHROMATOGRAPHIC THEORY

Basic chromatographic theory developed over the past several decades (Kubin, 1965; Kucera, 1965) accounts for separations based upon a wide range of molecular interactions.

Weak binding between complex biomolecules may include effects of solvent displace-ment as well as stereospecific contributions from hydrogen bonds, van der Waals inter-actions, ion exchange, and hydrophobic interactions. Making the assumption that the sum of these interactions can be expressed as an empirically measured equilibrium binding constant vastly simplifies analysis of the kinetic parameters driving chromatographic separation. This approach provides not only an experimental tool for investigating the physical and chemical nature of "recognition" between biomolecules, but also a rational basis for mathematical modeling to design chromatography columns that meet specific performance objectives.

In most cases the interaction between a ligate (Lt) dissolved in the bulk phase and an immobilized ligand (Ld) can be assumed to result in equilibrium binding kinetics typical of a second-order, reversible reaction:

$$\text{Lt} + \text{Ld} \overset{K_a}{\leftrightarrow} \text{LtLd} \tag{1}$$

where K_a is the association constant (M^{-1}).

Retention of the ligate is usually described by the capacity factor, k', defined as n_s/n_m, where n_s is the total number of moles of ligate bound to the stationary phase, and n_m is the total number of moles of ligate in the mobile phase. The quantity n_s can be expressed as q^*M_s, where q^* is the concentration of adsorbed ligate at equilibrium (moles per kilogram of solid phase), and M_s is the weight of the stationary phase (kg); likewise, n_m can be expressed as c^*V_m, where c^* is the concentration of free ligate at equilibrium, and V_m is the volume of the mobile phase.

Thus, retention can be expressed simply in terms of the ratio of adsorbed to free ligate concentrations at equilibrium, i.e., q^*/c^*, and a constant C ($C = M_s/V_m$) that accounts for mechanical characteristics of the column such as void fraction, porosity, and density of the stationary phase:

$$k' = Cq^*/c^* \tag{2}$$

C commonly has a numerical value in the range of 0.5–1.0.

From the expression for the Langmuir adsorption isotherm,

$$q^* = \frac{Q_{max}K_a c^*}{1 + K_a c^*}$$ (3)

where Q_{max} is the maximum concentration of accessible ligand sites (moles per kilogram of solid phase), it is seen that when $K_a c^* \ll 1$, $q^* = Q_{max}K_a c^*$. Thus, under conditions that usually pertain when performing weak-affinity chromatography (the adsorption isotherm is in the linear range),

$$k' = CQ_{max}K_a$$ (4)

Equation (4) provides a clear statement of the basic premise of weak-affinity chromatography: significant retention can be achieved for weak-affinity systems by increasing the ligand load. In practical terms, for an analytical column where $C = 1$, a retention value of $k' = 1$ can be achieved, for example, for a monovalent soluble antigen bound by an immobilized monoclonal antibody at $K_a = 10^3 \, M^{-1}$ by loading the column with an antibody at a concentration of approximately 75 mg/ml and injecting the sample ligates at about $<10^{-4}M$. Interestingly, process scaleup for preparative recovery of purified ligate is favored at weaker affinities; that is, despite lower retention, weaker affinity systems accommodate higher concentrations of injected ligate without loss of performance associated with sample overload. Not contained in these calculations is the notion that weaker affinities ($K_a < \sim 10^3 \, M^{-1}$) also minimize kinetic contributions to band spreading, permitting chromatographic performance commonly associated with standard high-performance liquid chromatography (HPLC) methods such as reversed-phase and ion-exchange chromatography.

4. SIMULATION OF WAC

As weak-affinity chromatography is a rather new technique and its applicability has been demonstrated in only a few cases, we developed a simulation model to more thoroughly understand the basics of this technology (Wikstrom and Ohlson, 1992). This model is based on a mathematical model for liquid chromatography originally developed by Kubin (1965) and Kucera (1965). The model takes into account the kinetic sorption equations as described above, the longitudinal dispersion in the mobile phase, the radial diffusion inside the porous particles, and the rate of mass transfer through the boundary surrounding each particle. We have applied the model on the antigen–antibody system with which most experience has been gained on WAC: the weak-affinity retention of small carbohydrate antigens on solid phases with bound monoclonal IgG antibodies.

4.1. Computer Simulation of WAC: The Influence of Kinetics

Kinetic characteristics are of great importance in determining the overall efficiency of a weak-affinity chromatography system. Generally, when performing traditional moderate- to high-affinity chromatography, efficiency in terms of "narrow peaks" is lost, mainly because of the slow dissociation of ligate from the ligand–ligate complex. To illustrate the influence of kinetics, Fig. 1 shows how performance deteriorates with decreasing association and dissociation rate constants.

This model is in good agreement with experimental results from weak-affinity

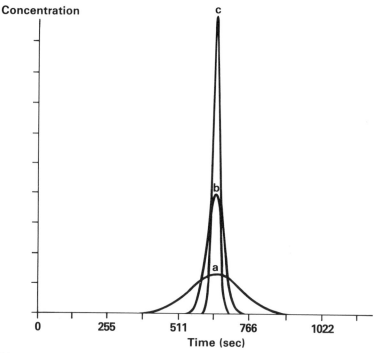

FIGURE 1. Computer simulation of WAC: The Influence of kinetics. Ten microliters of a carbohydrate antigen ($M_r \approx 700$, concentration = $1.4 \times 10^{-6}M$) was chromatographed simulated on a 25×0.5 cm column (particle size, 10 μm; porosity, 0.64; void volume, 0.42; density, 2.42 g/cm³). Isocratic elution was simulated at 1 ml/min. Q_{max} was 10^{-3} mol/kg gel. Values of k_a were 10^2 (a), 10^3 (b), and 10^4 (c) M^{-1} s^{-1}. [See Wikstrom and Ohlson (1992) for details.]

separation of carbohydrate antigens on immobilized monoclonal IgG antibodies (K_a = 0.8×10^3–5.2×10^3 M^{-1}), where we found (Wikstrom and Ohlson, 1992) that the association rate constants (k_a) were in the range of 0.4×10^3–2.7×10^3 M^{-1} s^{-1} and the corresponding dissociation rate constants (k_d) were in the range of 0.5–0.9 s^{-1}. The kinetics of interaction between ligate and ligand contribute significantly to the performance of WAC and can exert a major influence upon the extent of peak broadening.

4.2. Simulation of WAC: Peak Capacity

One of the important benefits of WAC is the capacity to resolve similar ligates present in a crude extract; however, resolution is limited by a number of practical considerations including the size of the column, the amount of immobilized ligand, parameters for elution, and design of the solid phase.

Another matter that is especially of concern in WAC is the negative impact of nonspecific adsorption, which should be generally defined as the amount of nondesired interactions, for example, those of ligates and other solutes in the sample with the support surface or with spacer molecules. It is generally seen in liquid chromatography that many nondesired interactions are of weak affinity. This does not mean that weak-affinity interactions are by nature nonspecific or nondesirable. The specific retention seen in WAC is the result of cumulative action of many weak binding events which are likely to differ from nonspecific adsorption effects both in terms of the number of sites and in their response to changes in running conditions.

Figure 2 depicts a computer simulation experiment, where we studied the requirements for specificity in chromatographic separation of ligates bound with weak to very weak affinities. A suitable separation "window" was defined from $k' = 1$ at approximately 8 min to $k' = 22$ (approximately 90 min). Retention of $k' < 1$ was considered to be nonspecific and so weak that it has no significance.

As can be seen from Fig. 2, the peak capacity is about ten, which means that ten similar ligates in the affinity range $K_a = 6 \times 10^2$–$1 \times 10^4 \, M^{-1}$ can be separated isocratically to more than 90%. In other words, WAC can take advantage of subtle but specific differences in binding behavior for a series of similar biomolecules to effect their chromatographic resolution without interference from nonretarded solutes or from nondesired interactions with the solid support.

5. APPLICATIONS OF WEAK-AFFINITY CHROMATOGRAPHY

5.1. Weak-Affinity Separation of Enzymatic Inhibitors

Trafficking of metabolites by the concerted action of a multitude of enzymes is often under a dynamic weak-affinity control. As the concentrations of substrates and inhibitors

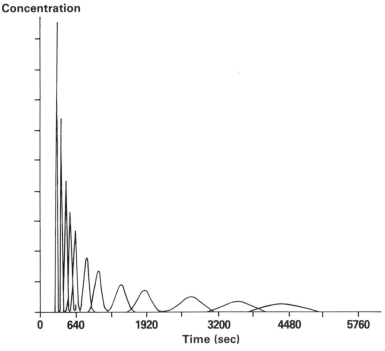

FIGURE 2. Computer simulation of WAC: peak capacity. The same parameters were used as in Fig. 1. K_a was varied between 2×10^2 and $1 \times 10^4 \, M^{-1}$ [See Wikstrom and Ohlson (1992) for details.]

are typically in the millimolar range, there is enough driving force for the weak-affinity enzymatic reactions to take place at an appreciable rate.

In our studies we selected the proteolytic enzyme α-chymotrypsin as a model for a weak-affinity interaction since many of its inhibitors and substrates are known to bind with weak affinities ($K_a = 10^1 - 10^4 \, M^{-1}$) (Barman, 1969). We also studied this system because the inhibitors were easily detected by ultraviolet (UV) absorption and, most importantly, we were able to bind significant amounts of α-chymotrypsin (44.5 mg per milliliter of column volume) at high enzymatic activity (approximately 50% of total bound).

Figure 3 shows an example of separation of the inhibitors N-acetyl-L-tryptophan and indole in a crude extract of human serum. The components of serum did not bind to the column and appeared in the void volume. The separation was performed on a small (5 cm) HPLC column under isocratic conditions. A reference column with no bound enzyme did not bind the inhibitors. The amount of active immobilized enzyme and binding constants for the enzyme inhibitors were estimated by frontal analysis (Kasai et al., 1986). The amount of immobilized α-chymotrypsin capable of interacting with the inhibitor was found to be 38.5 mg/ml of column volume (87% of total). The N-acetyl-L-tryptophan showed slight retardation with a binding constant of $K_a = 1.5 \times 10^2 \, M^{-1}$ [literature value, $K_a = 0.57 \times 10^2 \, M^{-1}$ (Boyer, 1970)], whereas the indole was retarded significantly with a binding constant of $K_a = 1.1 \times 10^3 \, M^{-1}$ [literature value, $K_a = 1.4 \times 10^3 \, M^{-1}$ (Boyer,

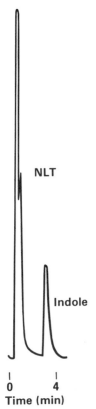

NLT

Indole

0	4

Time (min)

FIGURE 3. Separation of enzyme inhibitors with WAC. N-Acetyl-L-tryptophan (NLT) and indole (1.7 μg/ml of each in fetal bovine serum diluted 1:120 with mobile phase) on a 5 × 0.5 cm microparticulate silica column (particle size, 10 μm; porosity, 300 Å) with immobilized α-chymotrypsin. Isocratic chromatography was performed at 20°C using Tris (0.08M, pH 7.8) with CaCl₂ (0.1M) as the mobile phase (1 ml/min). Injection volume was 100 μl, and emerging peaks were monitored at 274 nm.

1970)]. The binding constants estimated by WAC showed good correlation with the binding constants determined for interactions in free solution. This preliminary study showed that inhibitors in a crude environment can be separated by biospecific weak-affinity chromatography with their corresponding immobilized enzymes.

5.2. Separation of Antigens with Weak-Affinity Monoclonal Antibodies

Monoclonal antibodies are routinely produced against a variety of substances (Kennett *et al.*, 1980). Even when a deliberate attempt is made to generate high-affinity antibodies, screening of hybridomas often gives a substantial number of clones that secrete weaker binding immunoglobulins. These weak-affinity antibodies may be useful reagents for WAC when they exhibit suitable specificity for antigen. For exploratory studies, it is often convenient and economical to immobilize monoclonal antibodies noncovalently via an attachment protein bonded to the solid phase. For example, most IgM, IgG, or IgA antibodies can be immobilized at physiological pH and salt concentrations on macroporous HPLC columns containing covalently linked concanavalin A, protein G, or jacalin, respectively. Both concanavalin A and jacalin are lectins that can themselves interact with certain oligosaccharides (Merkle and Cummings, 1987; Hortin, 1990), so that it is essential to evaluate oligosaccharide ligate binding to a reference column containing the lectin prior to loading the column with the antibody. Noncovalently bound antibodies can exhibit greater immunoreactivity than antibodies that undergo chemical coupling to activated solid supports, and they can be removed from the column under mild conditions to permit reuse of both antibody protein and affinity support.

Alternatively, monoclonal antibodies can be covalently coupled at high density (50–100 mg/ml of column bed) to activated silica supports, often with preservation of more than 50% of ligate binding activity. For example, Ohlson *et al.* (1988) employed an immobilized IgG monoclonal antibody that binds a tetrasaccharide found in human urine (Hallgren *et al.*, 1974). Ligand–ligate binding in this system exhibited a significant dependence on temperature, permitting straightforward study of the effects of binding affinity on both retention and chromatographic performance (see Fig. 4). The improvement in chromatographic performance demonstrated at weaker affinities (Ohlson *et al.*, 1988) offers the possibility of optimizing chromatography by selecting the affinity at which a ligate will be analyzed. Many monoclonal antibodies that bind oligosaccharides show enhanced binding at lower temperatures (Ohlson *et al.*, 1988; Dakour *et al.*, 1987, 1988), approximately doubling their affinity for the ligate for each downward shift of 8°C.

Such temperature effects are useful for expanding the range of ligates that can be analyzed by WAC: raising the column temperature can often cause earlier elution and improved peak shape for ligates that are strongly bound at lower temperatures (Dakour *et al.*, 1988).

The opportunity to manipulate retention by changing the temperature also makes it simple to find conditions for separation of ligate from a very large void volume peak when measuring a minor component in a complex biological fluid. For example, Wang *et al.* (1989) analyzed $(Glc)_4$ in human urine and serum by WAC, optimizing separation by adjusting column temperature. As shown in Fig. 5, retarded peaks containing nanogram amounts of oligosaccharide (analyzed by a pulsed amperometric detector) were well separated from the very large unbound fractions containing hundreds of noninteracting compounds. Interestingly, urine contains minor amounts of the disaccharide, β-maltose,

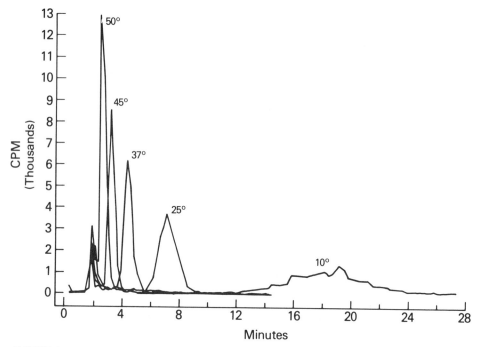

FIGURE 4. Weak-affinity chromatography of (Glc)$_4$ on a column containing immobilized monoclonal antibody. A monoclonal antibody (86 mg), prepared against oligosaccharide (Glc)$_4$ linked to a carrier protein, was coupled to a tresyl-activated macroporous silica column (10 × 0.5 cm), and the oligosaccharide was run as its tritiated alditol. The system was eluted isocratically at 1.0 ml/min with 0.2M NaCl plus 0.02M sodium phosphate buffer, pH 7.5. Samples (10 μl containing 1.5 ng) were run at the temperatures shown, and fractions were collected and counted by liquid scintillation. [See Ohlson *et al.* (1988) for details.]

which interacts more weakly with the antibody than (Glc)$_4$ and emerges from the WAC column as a sharp peak prior to (Glc)$_4$ (Fig. 5A). For analysis of serum (Fig. 5B), which contains smaller amounts of both oligosaccharides, the temperature of the system was raised to improve sensitivity for (Glc)$_4$, the analyte of interest.

WAC also provides a rapid method to characterize the fine specificity of monoclonal antibodies. Affinities for soluble antigens with related chemical structures are measured according to chromatographic retention. Figure 6 illustrates how differences in K_a, calculated from k' values, for a series of oligosaccharides can provide information regarding antibody binding specificity to various core structures that occur among complex sugar chains linked to mucin molecules.

The immobilized antibody originally had been selected for binding to the mucin core trisaccharide, Galβ1-3(GlcNAcβ1–6)GalNAc. From the retention times for the compounds indicated in Fig. 6, it is apparent that substitution at the 6-position of galactose has a much smaller effect on antibody binding than substitution at the 3-position and that replacement of the galactosyl residue with GlcNAc produces a marked reduction in antibody binding affinity. These results illustrate that WAC can very quickly yield detailed information regarding binding specificity for a family of ligates. WAC is much simpler than standard, multitube radioimmunoassay, enzyme-linked immunosorbent assay (ELISA), or quantita-

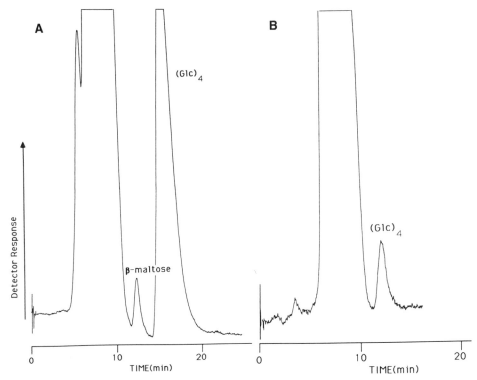

FIGURE 5. Weak-affinity chromatography analysis for (Glc)$_4$ in urine and serum. Using the same column as described in the legend to Fig. 4, a 25-μl sample of ultrafiltered and deionized urine from a patient with acute pancreatitis was run isocratically at 0.2 ml/min in 0.1M Na$_2$SO$_4$ plus 0.02M phosphate, pH 7.5, at a column temperature of 30°C (A). Serum from the same patient was ultrafiltered, deionized, and extracted on a C$_{18}$ column prior to analysis (25 μl) under the same conditions (B). The column effluent was alkalinized by postcolumn addition of NaOH prior to analysis with a pulsed amperometric detector. Integration of the peaks labeled "(Glc)$_4$" gave 1.01 ng and 0.35 ng in urine and serum, respectively. [See Wang *et al.* (1989) for details.]

tive precipitin inhibition experiments, where compounds typically must be tested in triplicate over a range of concentrations, resulting in voluminous data that must be reduced by extensive calculations to obtain binding constants.

6. FUTURE TRENDS

The metamorphosis of affinity chromatography into a high-performance technique began during the 1980s and is likely to accelerate during the years ahead. New opportunities will be recognized for replacing the conventional "lock and key" model for affinity chromatography with "dynamic" procedures that give true chromatographic separations. Tools to develop efficient weak-affinity chromatography are already in place, including HPLC supports of different kinds, mild immobilization methods that can give high concentrations of active ligand (e.g., the use of attachment proteins discussed above),

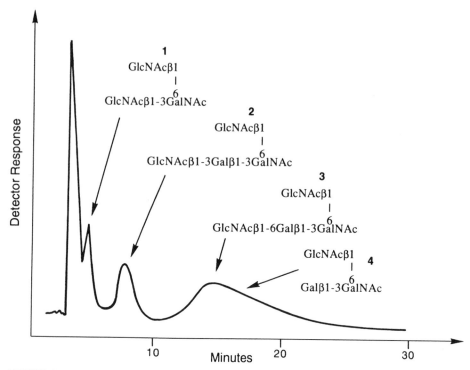

FIGURE 6. Weak-affinity chromatography of a mixture of four oligosaccharides on a column containing immobilized monoclonal antibody. A murine IgG_1 antibody (60 mg) prepared against a synthetic oligosaccharide, $Gal\beta1-3(GlcNAc\beta1-6)GalNAc\alpha1-R$, linked to a carrier protein, was coupled to a tresyl-activated macroporous silica column (10 × 0.5 cm), and the oligosaccharides (2 nmol each) shown were run as their α-benzyl glycosides. The system was eluted isocratically at 0.5 ml/min with $0.2M$ Na_2SO_4 plus $0.02M$ phosphate, pH 7.5, at 10°C. Arrows indicate the apex of the peak obtained when each oligosaccharide was run separately. Values for the association constants (K_a) calculated from the retention (k') from runs with each oligosaccharide are as follows: compound 1, 2.5×10^3 M^{-1}; compound 2, 7.7×10^3 M^{-1}; compound 3, 1.9×10^4 M^{-1}; compound 4, 2.2×10^4 M^{-1}.

sophisticated elution procedures (e.g., temperature elution), and, perhaps most importantly, new technologies to produce weak-affinity ligands.

Continued progress in ligand design will be critical to the success of future developments of weak-affinity chromatography. Promising approaches include (1) advances in genetic engineering of active sites, including V_H immunoglobulin fragments (Sastry *et al.*, 1989), and perhaps engineering of high-affinity sites to give weaker affinity ligands with similar or improved specificity for ligate binding; (2) new procedures to define small organic molecules that mimic binding of bioactive proteins (Saragovi *et al.*, 1991); (3) production of small peptide "paraloges" to mimic the binding sites of immunoligands (Kauvar *et al.*, 1990); and (4) a more futuristic endeavor, molecular imprinting to provide a range of weak-affinity polymeric ligands specific to small molecules (Ekberg and Mosbach, 1989).

It seems likely that WAC will find many diverse applications, both in basic research to identify and characterize the components of natural weak-affinity recognition systems and in industry for purification schemes whereby products with high bioactivity can be obtained by chromatography under "mild" elution conditions. However, perhaps the greatest potential for WAC will be as a new tool for immunoassays where similar antigens can be separated and analyzed in crude samples. Traditional immunoassays sometimes fail due to high cross-reactivity or inability to distinguish between target molecules and similar nontarget substances bound with similar affinities. WAC will here show great promise as an analytical method to recognize subtle molecular differences and to discriminate among several similar analytes recognized with similar but nonidentical affinities.

REFERENCES

Barman, T. E., 1969, in: *The Enzyme Handbook*, Vol. 2 (T. E. Barman, ed.), Springer-Verlag, New York.

Boyer, P. D., 1970, in: *The Enzymes*, Vol. 3 (P. D. Boyer, ed.), Academic Press, New York.

Dakour, J., Lundblad, A., and Zopf, D., 1987, Separation of blood group A-active oligosaccharides by high-pressure liquid affinity chromatography using a monoclonal antibody bound to concanavalin A silica, *Anal. Biochem.* **161:**140–143.

Dakour, J., Lundblad, A., and Zopf, D., 1988, Detection and isolation of oligosaccharides with Le[a] and Le[b] blood group activities by affinity chromatography using monoclonal antibodies, *Arch. Biochem. Biophys.* **264:** 203–213.

Ekberg, B., and Mosbach, K., 1989, Molecular imprinting: A technique for producing specific separation materials, *Trends Biotechnol.* **7:**92–96.

Hallgren, P., Hansson, G., Henriksson, K. G., Hager, A., Lundblad, A., and Svensson, S., 1974, Increased excretion of a glucose-containing tetrasaccharide in the urine of a patient with glycogen storage disease type II (Pompe's disease), *Eur. J. Clin.* **4:**429–436.

Hortin, G. L., 1990, Isolation of glycopeptides containing O-linked oligosaccharides by lectin affinity chromatography on jacalin-agarose, *Anal. Biochem.* **191:**262–267.

Kasai, K. I., Oda, Y., Nishikata, M., and Ishii, S. I., 1986, Frontal affinity chromatography: Theories for its application to studies on specific interactions in biomolecules, *J. Chromatogr.* **376:**33–47.

Kauvar, L. M., Cheung, P. Y. K., Gomer, R. H., and Fleischer, A. A., 1990, Paralog chromatography, *Biochromatography* **5:**22–26.

Kennett, R. H., McKearn, T. J., and Bechtol, K. B. (eds.), 1980, *Monoclonal Antibodies*, Plenum Press, New York.

Kubin, M., 1965, Beitrag zur theorie der chromatographie, *Collect. Czech. Chem. Commun.* **30:**1104–1118.

Kucera, E., 1965, Contribution to the theory of chromatography linear nonequilibrium elution chromatography, *J. Chromatogr.* **19:**237–248.

Merkle, R. K., and Cummings, R. D., 1987, Lectin affinity chromatography of glycopeptides, *Methods Enzymol.* **138:**232–259.

Ohlson, S., Lundblad, A., and Zopf, D., 1988, Novel approach to affinity chromatography using "weak" monoclonal antibodies, *Anal. Biochem.* **169:**204–208.

Saragovi, H. U., Fitzpatrick, D., Raktabutr, A., Nakanishi, H., Kahn, M., and Greene, M. I., 1991, Design and synthesis of a mimetic from an antibody complementarity-determining region, *Science* **253:**792–795.

Sastry, L., Alting-Mebs, M., Huse, W. D., Short, J. M., Sorge, J. A., Hay, B. N., Janda, K. D., Benkovic, S. J., and Lerner, R. A., 1989, Cloning of the immunological repertoire in *Escherichia coli* for generation of monoclonal catalytic antibodies: Construction of a heavy chain variable region-specific cDNA library, *Proc. Natl. Acad. Sci. U.S.A.* **86:**5728–5832.

Turkova, J., 1978, *Affinity Chromatography*, Elsevier, Amsterdam.

Wang, W.-T., Kumlien, J., Ohlson, S., Lundblad, A., and Zopf, D., 1989, Analysis of a glucose-containing tetrasaccharide by high performance liquid affinity chromatography, *Anal. Biochem.* **182:**45–53.

Wikstrom, M., and Ohlson, S., 1992, Computer simulation of weak affinity chromatography, *J. Chromatogr.* **597:**83–92.

Zopf, D., and Ohlson, S., 1990, Weak-affinity chromatography, *Nature* **346:**87–88.

3

Affinity Chromatography on Inorganic Support Materials

Howard H. Weetall

1. INTRODUCTION

Inorganic support materials have been used for chromatography for over 75 years. However, it has only been within the last 25 years that these materials have been applied to affinity separations in which the affinity ligand is covalently bonded to the support. This chapter will cover the application of inorganic silicas and metal oxides for coupling to ligands of various types for the isolation and purification of antigens, antibodies, enzymes, nucleic acids, and related species. Ion chromatography or the use of hydrocarbon-bonded phases for protein isolations will not be covered here, nor will the application of ion exchange as a separation and purification method on inorganic supports.

Before choosing a material to be used as an affinity support, one must ask several important questions:

1. Should the support material be porous or nonporous?
2. If the support material is porous, does the pore morphology permit the entry of the component to be purified?
3. Is the support material durable in the solvents utilized (acids, bases, high salt, organic solvents)?
4. Can the support material be conveniently handled?
5. Does the support material have high compression strength?
6. Can one attach enough ligand to give adequate loading per unit volume?
7. Does the carrier tolerate maximum pressure drop?
8. How are all the above factors affected by particle size, flow rates, and particle shape?
9. Can the derivatized support material be easily stored?

Each of the above questions must be answered in choosing an affinity material. In many cases, inorganic materials may fit the criteria more closely than other materials. Inorganics have an advantage in that they come as particles of all shapes and sizes as well

Howard H. Weetall • Biotechnology Division, Chemical Sciences and Technology Laboratory, National Institute of Standards and Technology, Gaithersburg, Maryland 20899.

Molecular Interactions in Bioseparations, edited by That T. Ngo. Plenum Press, New York, 1993.

as in the shape of monolithic porous supports or filter materials. They can withstand extremely high pressure drops with no compression. They are not susceptible to microbial attack, and they do not shrink or swell in organic solvents or with pH change. In addition and most importantly, they have loading capacities high enough for a variety of applications.

2. CHARACTERISTICS OF INORGANIC AFFINITY SUPPORT MATERIALS

2.1. Pore Size

Most inorganic support materials consist of either porous or nonporous glasses, silicas, or metal oxides. Examples of some porous inorganic support materials are shown in Table 1. These support materials are all controlled-pore ceramics or glasses. They can be prepared in a variety of pore morphologies and particle sizes (Messing, 1975). The controlled-pore glasses, unlike the ceramics, can be prepared in pore sizes ranging from 30 to 2000 Å with narrow distribution ranges (± 5%), rather than the much broader pore ranges seen for the ceramic carriers (Weetall and Filbert, 1974).

2.2. Chemical Durability

Chemical durability is of importance for materials that are subjected to prolonged or repeated exposure to solvents. Table 2 shows the durability of several inorganic support materials under different conditions of acid and base treatment (Eaton, 1974; Weetall, 1976). These data show that glass is not the most durable material under basic conditions whereas the metal oxides are extremely durable under these conditions.

2.3. Capacity of Inorganic Supports

When using inorganic support materials as affinity carriers, one must consider capacity for ligand binding. Nonporous materials in the 40–80 U.S. mesh size (177–420 μ) range have an available surface area of only 1–5 m^2 per gram. Extremely small nonporous

TABLE 1
Physical Characteristics of Several Ceramic Support Materials

Composition	Size (U.S. mesh)	Pore diameter (Å) Range	Pore diameter (Å) Average	Pore volume (ml/g)
TiO_2 98%, MgO 2%	30/45	205–500	410	0.53
SiO_2 75%, Al_2O_3 25%	30/60	205–275	435	0.89
SiO_2 89.3%, ZrO_2 10.7%	30/60	110–575	235	1.30
SiO_2 100%	30/60	185–700	435	0.76
SiO_2 100%[a]	30/60	310–655	550	2.2
SiO_2 90%, ZrO_2 10%	30/60	185–700	435	0.76
SiO_2 75%, TiO_2 25%	30/45	875–205	465	0.76
Controlled-pore glass	30/60	450–600	550	0.50

[a]Modified binder used.

TABLE 2
Durability Test Results for Supports

Material description	Static test (mg/m² per 16 h)		Dynamic test (mg/m² per day)		
	1% NaOH	5% HCl	pH 4.5	pH 7.0	pH 8.2
TiO_2	0.2	0.8	0.05	0.05	ND[a]
ZrO_2	0.2	1.1	0.004	0.004	0.01
Al_2O_3	0.6–0.8	2.0	0.056–0.086	0.01	0.01
Al_2O_3–SiO_2	1.85	3.65	0.08–0.1	0.02–0.05	0.06
CPG^b–SiO_2	1.3–2.0	0.2	0.03	0.7	0.7–0.09
CPG^b	3.06	0.08	0.02	0.5	0.3

[a]ND: Not done.
[b]CPG, Controlled-pore glass.

particles in the 5–10-μ size range have surface areas from one to two orders of magnitude greater than those of the large particles. However, because of the packing densities observed with these fine materials, they are generally reserved for high-performance liquid chromatography (HPLC) types of applications. For standard low-pressure affinity chromatography applications, high capacity can be achieved with larger particles by using porous materials. Table 3 shows the average pore diameter and available surface areas for several porous glasses. The surface area is inversely proportional to the pore diameter at constant pore volume. However, in choosing a pore diameter, one must also consider the size of the molecules one wishes to couple and purify. Molecules to be coupled must have access to the surface, and the molecules to be purified must have access to the ligand within the pores of the support. Figure 1 shows the binding of a protein to a porous support as a function of surface area and pore diameter. The data show that there is a pore diameter at which the protein is excluded from the pores of the support and couples only to the external surfaces. The figure also shows that there is an optimal pore diameter for the protein at which maximal coverage is achieved.

TABLE 3
Comparison of Surface Area Calculated for Two Different
Pore Volumes for a Controlled-Pore Inorganic Support

Pore diameter (Å)	Surface area (m²/g) calculated for a pore volume of:	
	0.70 ml/g	1.0 ml/g
75	249	356
125	149	214
175	107	153
240	78	111
370	50	72
700	27	38
1250	15	21
2000	9	13

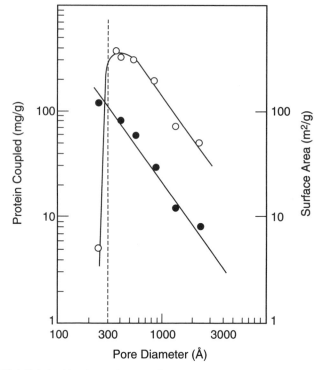

FIGURE 1. Relationship of pore size to surface area (●) and quantity of protein coupled (○).

3. PREPARATION OF INORGANIC AFFINITY SUPPORTS

The most often used method of preparing inorganic support materials for affinity chromatography is to either adsorb an organic polymer to the support or silanize the support before coupling the ligand or protein to be used for purification. Silanization methods are of two types: aqueous and nonaqueous. Aqueous silanization offers the advantage of more even coverage and an apparently thinner silane layer on the support material. Organic silanization, whether in a water-miscible or -immiscible solvent, produces a thicker, uneven, more loosely bound but higher capacity coating.

3.1. Silane Coupling

3.1.1. Pretreatment

Before any inorganic material is silanized, it should be cleaned to remove any adsorbed organics. This can be accomplished by either heating the material to a temperature that will burn off the adsorbed organics or boiling it in 5% nitric acid solution for 45 min followed by exhaustive washing in distilled water to remove any residual acid. Chlorosilanes should always be used in organic solvents while the methoxy- and ethoxysilanes can be coupled by either organic or aqueous methods.

3.1.2. Available Silanes

Available silanes include epoxysilanes, aminosilanes, cyanosilanes, sulfhydrylsilanes, phenylic silanes, and glycidoxysilanes, in methoxy, ethoxy, and chloro forms. In many cases, after silanization with one of the commercially available silanes, chemical modifications are performed to change the functional group for later coupling. Some of these methods will be discussed later in this chapter.

3.1.3. Organic Silanization

To 1 g of clean inorganic support material is added 75 ml of 10% silane dissolved in toluene or similar solvent. After reflux for 12–24 h, the material is washed with toluene followed by acetone. The product can be air-dried. Stability can be improved by heating the dried product to 110°C for 4 to 8 h.

3.1.4. Silane Evaporative Coupling

The support material, previously cleaned as described and dried, is added to enough silane dissolved in acetone or other volatile solvent to cover the particles thoroughly. Generally, a 1% solution of silane is sufficient for coupling. The solution is evaporated to dryness, leaving behind the silane adsorbed to and reacted with the support material. The support must be washed with the same solvent to remove unreacted silane and dried.

3.1.5. Aqueous Silanization

If the product has been fired, it must be thoroughly wetted before silanization. This is accomplished by boiling in distilled water for 30 min to 1 h before carrying out the silanization steps. The washed support is suspended in 75 ml of a 10% solution of silane dissolved in distilled water. The pH of the solution is lowered to pH 4.0 with HCl (usually 6*N*), and the pH-adjusted solution is heated in a water bath at 75°C for 3–5 h. The product is then filtered or decanted followed by washing with two volumes of distilled water. The product should be dried overnight at 110°C to increase stability. With micron-size particles, drying should be avoided. Drying these samples causes irreversible clumping.

3.1.6. Curing

When proteins are coupled to silanized supports, it may be necessary to remove loosely coupled or bound material from the support material before it can be used. In some cases, this also applies to low-molecular-weight ligands. This "curing" process is accomplished by placing the coupled support material in a water bath at 56°C for several hours to several days or until no protein or ligand can be detected in the concentrated supernatant fluid.

3.1.7. General Comments on Silane Coupling

It is well known that the siloxane bond between the silanol residue and the silane coupling agent is labile. Data in our and other laboratories have shown that monoethoxy-, monomethoxy-, or monochlorosilanes have poor stability on inorganic support materials and are lost rapidly over time. Trifunctional silanes, on the other hand, produce products which have greater stability. These silanes not only couple to the silanol residue on the support material's surface, but they bind to each other, forming a polymer across the surface of the carrier. The heating step after silanization aids in the polymerization by driving off water. The reaction between a silica support and a triethoxysilane is depicted in Fig. 2.

3.2. Carrier Activation (Direct, Nonsilane Methods)

The silanol residues on the surface of glass and silica as well as metal oxide groups on ceramic surfaces appear capable of reaction with several organic activating agents. The activated carriers react with other organic functional groups and ligands, forming permanent linkages.

3.2.1. Cyanogen Bromide

The coupling of cyanogen bromide (CNBr) to a support material was described by Weetall and Detar (1975). To 1 g of support material suspended in 5–10 ml of distilled water, cooled to 4°C, is added 250 mg of solid CNBr. The solid is added slowly with the pH maintained constant between pH 10 and 11 with NaOH solution. After the CNBr has been added, the reaction mixture is allowed to stand in the ice bath for an additional 30 min. The product is washed until the pH is near neutrality. The procedure should be carried out in a chemical fume hood since CNBr is highly toxic. The washed and activated support material is ready for coupling to a ligand or protein. It may be advisable to predissolve the CNBr in a water-miscible solvent prior to addition. The solvent should not be reactive with CNBr. The mechanism of action is not understood. However, it may be similar to that observed with conventional hydroxyl-containing compounds that give reactive imidocarbonates with CNBr. In this case, however, the hydroxyl groups are in the form of silanol residues. A possible mechanism for the activation of CNBr is presented in Fig. 3.

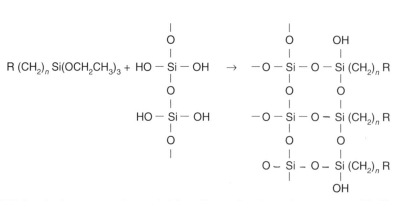

FIGURE 2. Reaction between γ-aminopropyltriethoxysilane and an inorganic support material. The ethoxy group reacts with the available silanol residues, forming a siloxane linkage. In addition, the silane appears to polymerize over the support surface. A product of this reaction is ethanol.

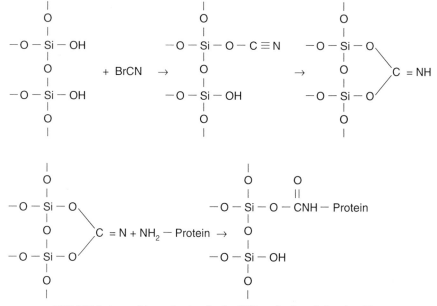

FIGURE 3. A possible mechanism for the CNBr activation of silanol residues.

3.2.2. Bis-Diazotized Silica

Bis-diazotized silica may be prepared according to a method described by Messing and Stinson (1974). A solution of 0.01% 4,4′-bis(2-methoxybenzene)diazonium chloride (commercially available) is prepared in distilled water. To 500 mg of inorganic support material is added 2 ml of the 0.01% coupling reagent. The reaction is allowed to proceed at room temperature for 20 min, after which the liquid is decanted off the support material. The product is washed with distilled water. At this point it is ready for coupling at pH 8–9 via the azo linkage to a ligand or protein. The mechanism, like the CNBr mechanism, is obscure. However, one possible mechanism is presented in Fig. 4.

3.2.3. Aldehyde Derivative

A method for preparation of an active aldehyde derivative was described by Bursecz (1983). The product is prepared by mixing 6 ml of a 2% aqueous solution of epiamine per gram of support material for 1 h. The epiamine is prepared by heating ethylenediamine to 100–110°C and slowly adding, with stirring, epichlorohydrin homopolymer at a rate of 1 mole of chloromethyl group to 6 moles of ethylenediamine. These reactants are heated for an additional 3 h with stirring. The mixture is neutralized with NaOH and filtered, and the excess ethylenediamine is removed by vacuum distillation. After reaction of the support material with the epiamine, it is filtered and can be dried. This product can be reacted with glutaraldehyde for coupling to protein or other ligand.

A simpler approach is to prepare a solution of polylysine at a ratio of 100 mg/g of support material (Royer and Green, 1975). This is allowed to adsorb at room temperature in

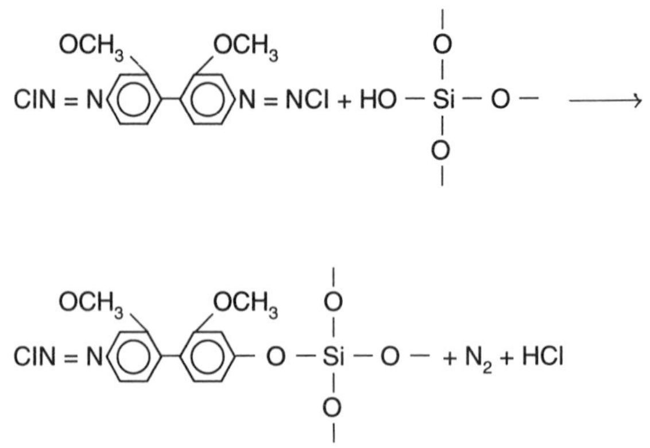

$$\text{CIN} = \text{N}\underset{\text{OCH}_3}{\bigcirc}\underset{\text{OCH}_3}{\bigcirc}\text{N} = \text{NCl} + \text{HO} - \overset{\text{O}}{\underset{\text{O}}{\text{Si}}} - \text{O} - \longrightarrow$$

$$\text{CIN} = \text{N}\underset{\text{OCH}_3}{\bigcirc}\underset{\text{OCH}_3}{\bigcirc} - \text{O} - \overset{\text{O}}{\underset{\text{O}}{\text{Si}}} - \text{O} - + \text{N}_2 + \text{HCl}$$

FIGURE 4. A possible coupling mechanism of the bis-diazotized reagent coupling to a silanol residue.

10mM phosphate buffer at near-neutral pH overnight. The product is washed with distilled water. It can be dried and stored for later activation with glutaraldehyde.

3.2.4. Metal Bridge Activation

Proteins as well as low-molecular-weight ligands can be covalently coupled to inorganic support materials through a metal bridge (Emery *et al.*, 1975; Baker *et al.*, 1971). The activated carrier is prepared by steeping the support material in a solution of the metal salt for a short time. It can generally be dried and stored for later coupling at neutral pH. The most successfully used metal salts are $TiCl_4$, $TiCl_3$, $SnCl_4$, $SnCl_2$, $ZnCl_4$, VCl_3, $FeCl_2$, and $FeCl_3$. The best results observed in our laboratory with an inorganic support material have been with $SnCl_2$ (Messing, 1974). To 500 mg of inorganic support material is added 20 ml of a 1% solution of stannous chloride dihydrate solution at 37°C. This is allowed to react for 45 min, decanted, and washed with three 20-ml aliquots of water. The product can be directly coupled at neutral pH. Like the other direct coupling mechanisms, this one is also unknown. A proposed mechanism is presented in Fig. 5.

4. ACTIVATION OF AFFINITY SUPPORTS

The previous sections of this chapter have dealt with characteristics of inorganic supports and silanization or direct activation of these inorganic support materials. In this section we will deal with activation and will assume that the molecules to be covalently coupled are proteins or ligands containing similar functional groups. Since proteins contain typical reactive groups, it is reasonable that methods used for protein coupling will be useful for the covalent attachment of other types of molecules including carbohydrates, lipids, nucleic acids, and low-molecular-weight hapten or inhibitor molecules.

Proteins contain several types of reactive groups that can be used to form covalent bonds to solid supports. The most frequently used are the amino groups, followed by sulfhydryl, carboxyl, and aromatic groups. The functional group on the support material can be (among others) an aliphatic amine, sulfhydryl, aromatic amine, or epoxy group. Consequently, a considerable variety of methods are available for coupling proteins to

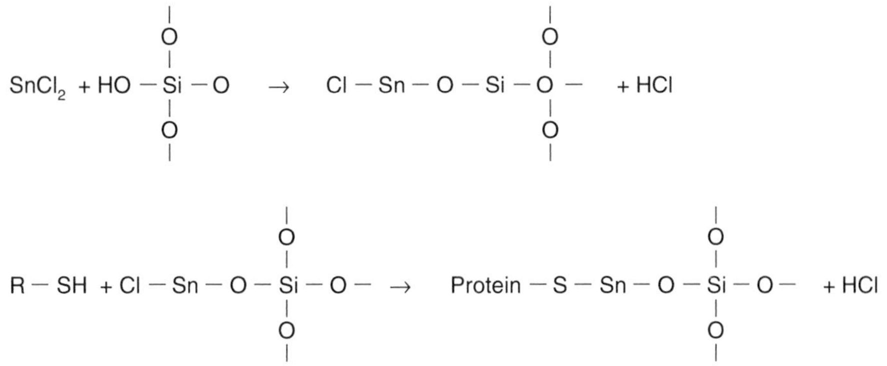

$$\text{SnCl}_2 + \text{HO} - \overset{\overset{\displaystyle |}{O}}{\underset{\underset{\displaystyle |}{O}}{\underset{|}{Si}}} - O \quad \rightarrow \quad \text{Cl} - \text{Sn} - \text{O} - \overset{\overset{\displaystyle |}{O}}{\underset{\underset{\displaystyle |}{O}}{\underset{|}{Si}}} - O - \quad + \text{HCl}$$

$$\text{R} - \text{SH} + \text{Cl} - \text{Sn} - \text{O} - \overset{\overset{\displaystyle |}{O}}{\underset{\underset{\displaystyle |}{O}}{\underset{|}{Si}}} - O - \quad \rightarrow \quad \text{Protein} - \text{S} - \text{Sn} - \text{O} - \overset{\overset{\displaystyle |}{O}}{\underset{\underset{\displaystyle |}{O}}{\underset{|}{Si}}} - O - \quad + \text{HCl}$$

FIGURE 5. A proposed mechanism for the stannous bridge coupling to a silanol residue.

inorganic supports. The methods vary in complexity from simple one-step processes to sequences involving three or more steps.

One-step methods, primarily with derivatization of the support with a group capable of direct coupling to the protein, include use of oxiranes (epoxides) and imidocarbonates.

Two-step processes, which are the most commonly used, either employ an initial activation of the functional group on the solid support, so as to react with the protein, or they involve the preliminary attachment of a bifunctional "linker" substance to the support, followed by coupling to the linker. The linker may have the same reactive function on each end (homobifunctional), or the two ends may have different functional groups (heterobifunctional). In either case, the free group is capable of reacting with the protein or ligand. In some cases the linker is first coupled to the protein before attachment to the solid support. In this case it must be of a heterobifunctional type, since the free end must not be capable of reacting with a second protein or ligand, but only with the support.

In the three-step methods, the functional group on the solid support is first coupled to a linker molecule that is not capable of further reaction. Subsequently, the linker attached to the solid support or, alternatively, the protein or ligand must be modified so that the coupling reaction will take place.

The survey of coupling methods below, although in no way complete, provides useful examples of methods for the coupling of both large and small molecules to porous and nonporous inorganic support materials.

4.1. Coupling of Alkylamine-Containing Inorganic Support to Amino Group on the Ligand

Two-step methods using homobifunctional linkers are presented first. Later in this chapter, a three-step procedure that utilizes an active ester which can react with amino groups will be presented.

4.1.1. Glutaraldehyde

To 500 mg of inorganic support material previously silanized with an alkylamine silane or aminated in some other fashion is added 50 ml of 2.5% glutaraldehyde in 0.1M phosphate buffer, pH 7.0. The reaction is allowed to continue for 60 min followed by extensive washing with phosphate buffer. The product is ready for immediate coupling at pH 6–8 (Fig. 6).

4.1.2. Disuccinimidyl Suberate

Five hundred milligrams of alkylamine inorganic solid support is suspended in 50 ml of methanol containing 50 mg of disuccinimidyl suberate. The mixture is shaken for 60 min and washed twice with 50 ml of methanol, and the product is immediately coupled to an amine-containing ligand or protein (Fig. 7).

4.1.3. Succinylation: Formation of an Active Ester

One gram of alkylamine inorganic support material is suspended in 100 ml of 0.05M sodium phosphate buffer, pH 6.0. To this is added 0.3 g of succinic anhydride. The mixture is allowed to react for 15 h at room temperature and is then washed three times with 100 ml of water, twice with 100 ml of methanol, and once with 100 ml of dioxane. The particles are then suspended in 1.2 g of N-hydroxysuccinimide (NHS) and 1.9 g of 1-ethyl-3-(3-dimethylaminopropyl) carbodiimide hydrochloride (EDC) in 100 ml of dioxane. This suspension is shaken for 90 min at room temperature, washed twice with 100 ml of dioxane and twice with 100 ml of methanol, dried, and stored in a moisture-free sealed container. The particles are ready for coupling to amino-containing compounds, generally in aqueous suspension at pH 8 (Fig. 8).

4.2. Coupling of Alkylamine Inorganic Support to Sulfhydryl Groups on the Ligand

The methods described below for the coupling of an alkylamine inorganic support to sulfhydryl groups on the ligand use heterobifunctional linkers.

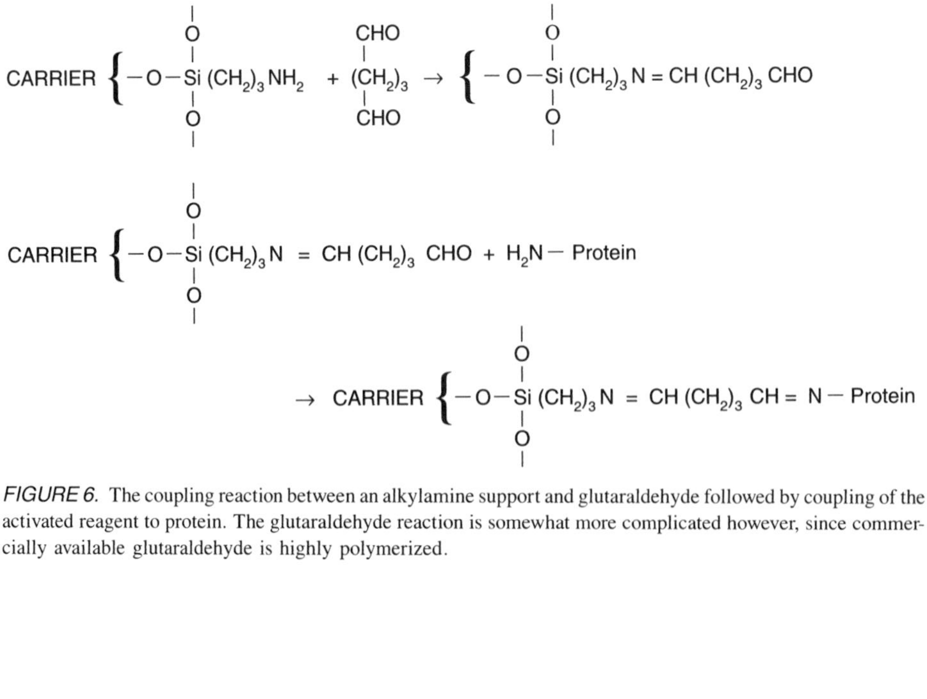

FIGURE 6. The coupling reaction between an alkylamine support and glutaraldehyde followed by coupling of the activated reagent to protein. The glutaraldehyde reaction is somewhat more complicated however, since commercially available glutaraldehyde is highly polymerized.

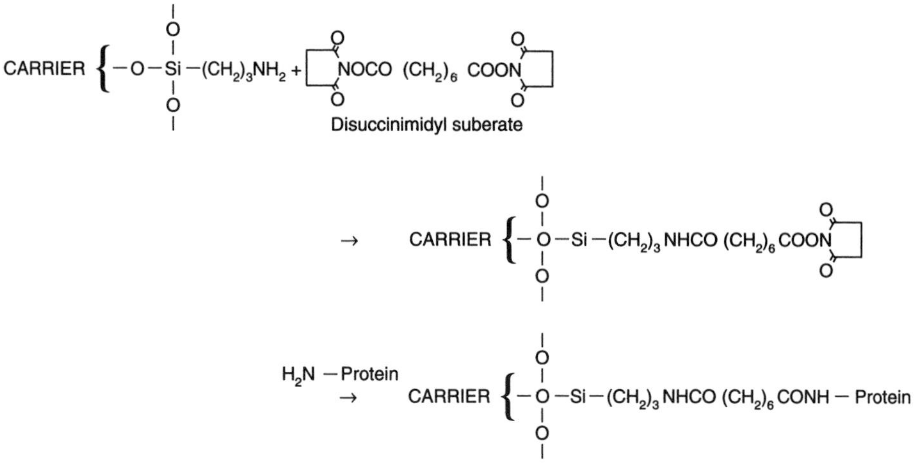

FIGURE 7. The coupling of disuccinimidyl suberate to an alkylamine inorganic support and subsequent coupling to a protein.

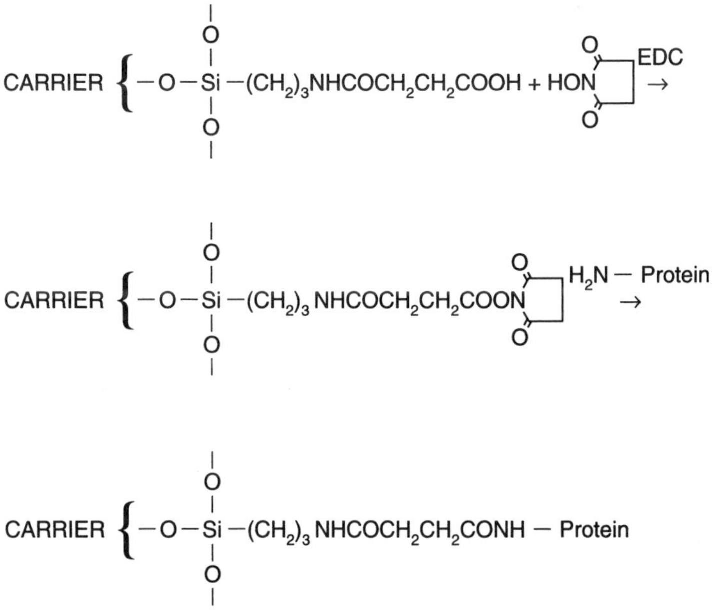

FIGURE 8. The coupling of a carboxy-derivatized carrier to *N*-hydroxysuccinimide using a carbodiimide coupling reagent [1-ethyl-3-(3-dimethylaminopropyl)carbodiimide hydrochloride (EDC)] followed by coupling to a protein via the active ester.

4.2.1. Succinimidyl 3-(-2-Pyridyldithio)propionate (SPDP)

To 500 mg of inorganic alkylamine support material suspended in 50 ml of 0.1M phosphate buffer, pH 8.0, is added, dropwise, 50 mg of SPDP dissolved in 5 ml of methanol. The suspension is agitated at room temperature for 60 min and then washed three times with 50 ml of 0.1M phosphate, pH 7.4, containing 1mM EDTA. The product is ready for coupling to available sulfhydryl groups at pH 7.4 (Fig. 9).

4.2.2. Succinimidyl 4-(N-Maleimidomethyl)cyclohexane 1-Carboxylate (SMCC)

To 500 mg of alkylamine inorganic support material suspended in 50 ml of 0.1M phosphate buffer, pH 7.0, at 30°C is added 50 mg of SMCC dissolved in 5 ml of dioxane. The reactants are agitated for 2 h followed by three washes with 50 ml of phosphate buffer, pH 7.4, containing 1mM EDTA. The product is ready immediately for coupling to sulfhydryl-containing ligands or protein at neutral pH (Fig. 10).

4.3. Arylamine Inorganic Supports Coupled to Different Functional Groups (Diazotization)

The diazotization method utilizes principally tyrosine residues or other aromatic compounds. However, histidine, lysine, cysteine, tryptophan, and arginine will also react to a lesser extent. The inorganic solid support can be converted to an arylamine derivative (Weetall, 1976), or the ligand, if it contains a diazotizable amine, can be coupled to a tyrosine or similar group contained on the support material.

Diazotization is carried out as follows. To 500 mg of arylamine support suspended

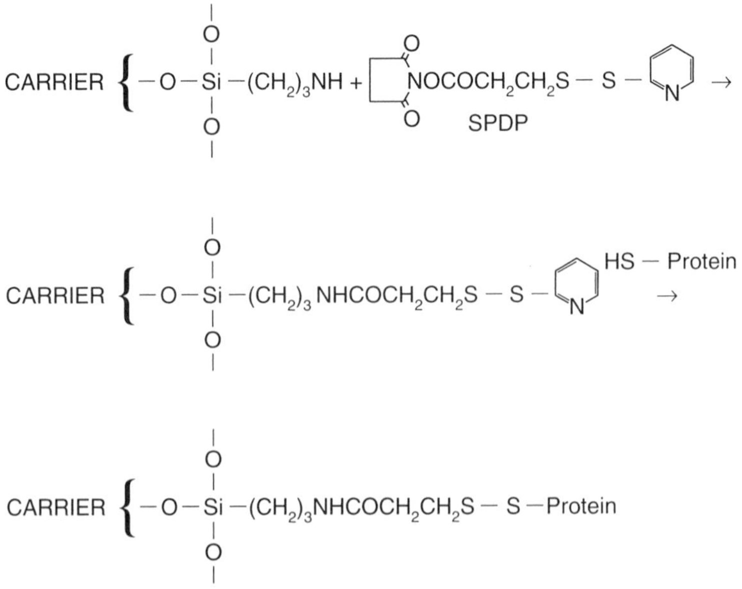

FIGURE 9. The coupling of a protein to an alkylamine support via activation with SPDP.

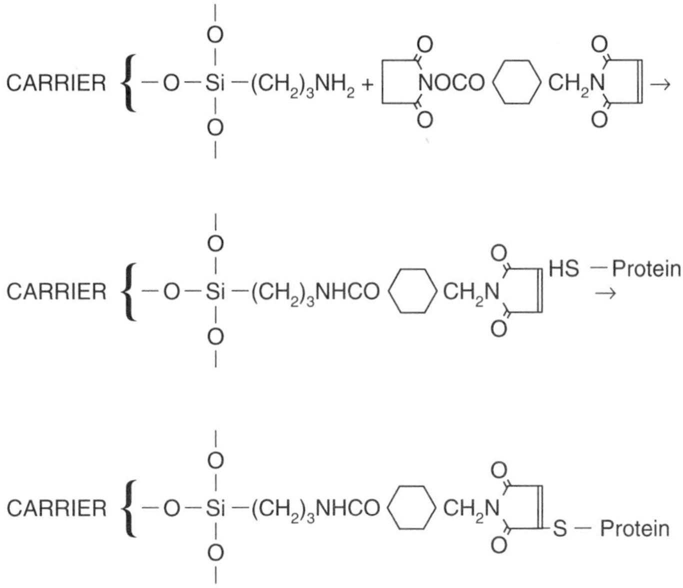

FIGURE 10. The coupling of an alkylamine support by activation with SMCC followed by coupling to a protein.

in 50 ml of 2N HCl in an ice bath is added 5 ml of a 1% solution of sodium nitrite. After 30 min, the excess nitrous acid is eliminated by the addition of solid urea or sulfamic acid until starch–iodide paper no longer shows a purple color when a drop of the reaction mixture is placed on the paper. The product is washed with ice-cold water. It is ready for coupling to a protein or aromatic compound at pH 8–10 (Fig. 11).

4.4. Coupling of Sulfhydryl Inorganic Support to Sulfhydryl Ligand or Protein

The method described below uses a homobifunctional linker. It should be remembered that sulfhydryl compounds are very susceptible to oxidation on storage and need to be reduced prior to carrying out the reactions described below. The carrier material is reduced by suspending it in 0.03M dithiothreitol in 0.1M Tris-HCl, pH 8.0, containing 1mM EDTA for 30 min, followed by extensive washing.

4.4.1. 1,6-Bismaleimidohexane (BMH)

To 500 mg of sulfhydryl inorganic support is added 50 ml of 0.1M phosphate buffer, pH 7.0, containing 1mM EDTA. To this is then added 50 mg of BMH dissolved in 5 ml of acetone. The reactants are agitated for 60 min followed by three 50 ml washes with the phosphate–EDTA buffer. The product is ready for coupling at pH 7.0 (Fig. 12).

4.4.2. 2,2'-Dipyridyl Disulfide (DPDS)

Fully reduced sulfhydryl inorganic support material is suspended in 50 ml of 0.1M Tris-HCl buffer, pH 8.0, containing 1mM EDTA and 1.5mM DPDS. The reactants are

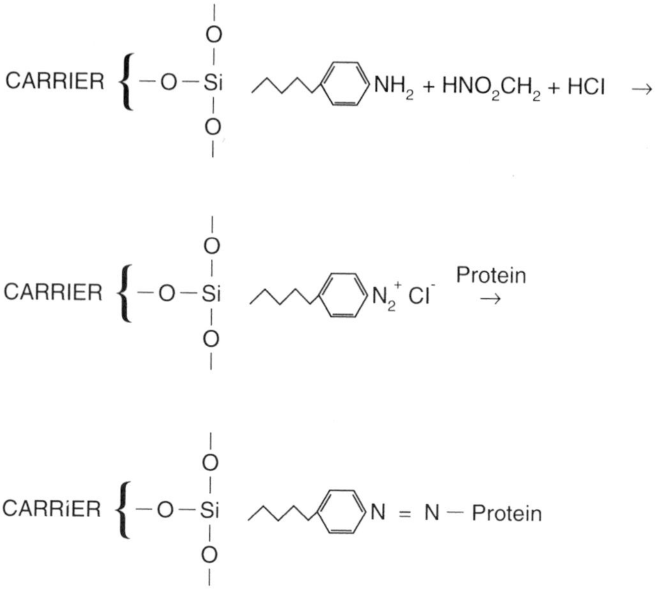

FIGURE 11. The diazotization of an arylamine support followed by coupling to a protein.

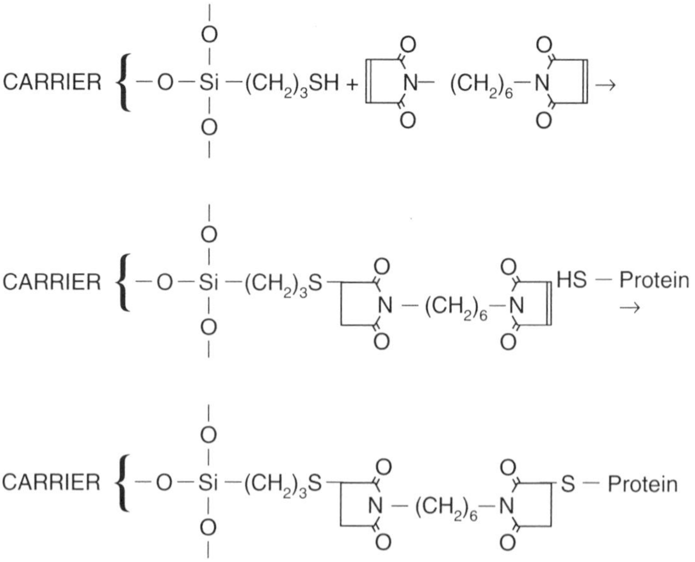

FIGURE 12. The coupling of a fully reduced sulfhydryl support to protein via activation with BMH.

agitated for 90 min and then washed repeatedly with the Tris buffer containing 1mM EDTA. The product is ready for coupling at pH 8.0 (Fig. 13).

4.5. Coupling of Sulfhydryl Inorganic Support to Amino-Containing Ligand: Sulfosuccinimidyl 4-(N-Maleimidomethyl)cyclohexane 1-Carboxylate (Sulfo-SMCC)

To 1.5 ml of a 1% solution of the ligand to be coupled dissolved in 0.1M phosphate buffer, pH 7.0, at 30°C is added, dropwise, 10 mg of SMCC in 1.5 ml of dioxane, with stirring. After 2 h, the solution is added to the sulfhydryl support and agitated for 16 h, washed, and stored. The molar concentration of the SMCC should be less than that of the amino-containing ligand so that uncoupled linker will not be available for coupling to the support material. If a protein is the agent to be coupled, then one can add excess SMCC and separate the unreacted portion on a Sephadex G-25 column before further reaction (Fig. 14).

4.6. Coupling of Oxirane Inorganic Support to Amino Group on Ligand or Protein

The oxirane group is coupled to the inorganic support by direct attachment of the epoxysilane as previously described. Thus, the oxirane group itself can be reacted directly with amino groups in a one-step procedure. To 500 mg of an oxirane support is added a solution containing a large excess of the amino-containing ligand or protein in 0.2M carbonate–bicarbonate buffer, pH 9.0. The mixture is agitated for 48 h at room temperature and washed with the same buffer. The product is ready for use in an affinity procedure (Fig. 15).

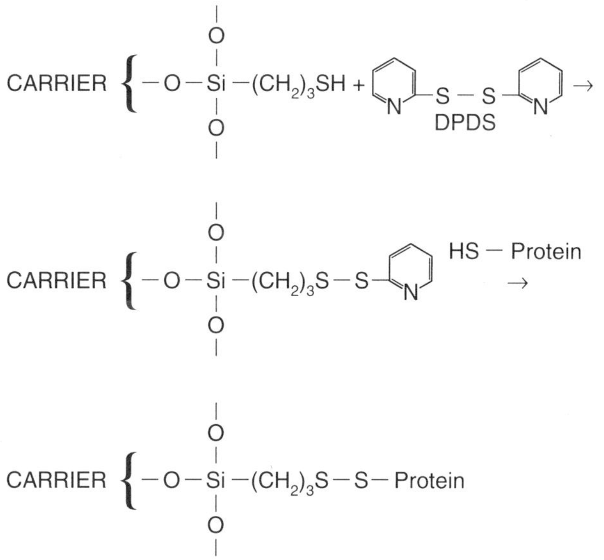

FIGURE 13. The coupling of a fully reduced sulfhydryl support to protein via activation with DPDS.

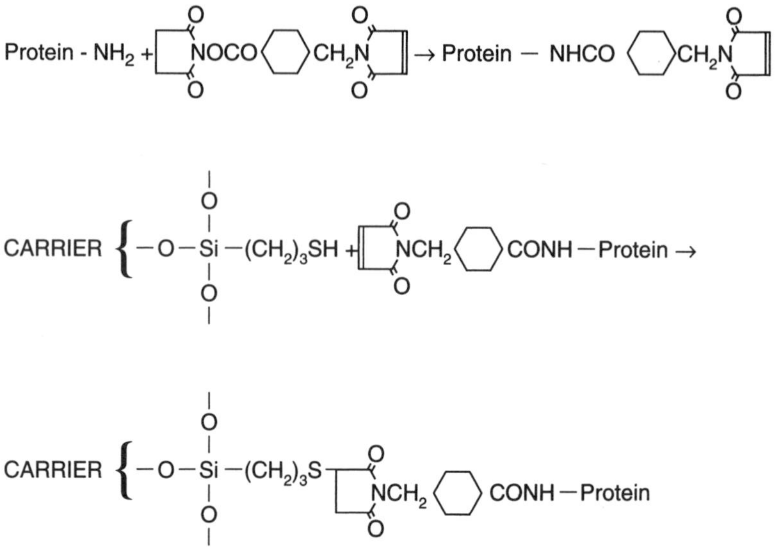

FIGURE 14. The coupling of a protein to a fully reduced sulfhydryl carrier via activation with SMCC. In this example the protein is activated before the carrier.

5. AFFINITY APPLICATIONS OF INORGANIC SUPPORT MATERIALS

5.1. Purity of the Isolated Product

As is the case with the organic polymeric support materials, inorganic supports have been used for the isolation and purification of a variety of compounds. In most cases the purity of the isolated products ranges from 70% to the high 90% range. The variation in range is determined by the nonspecific binding and release of nonspecific (NS) bound

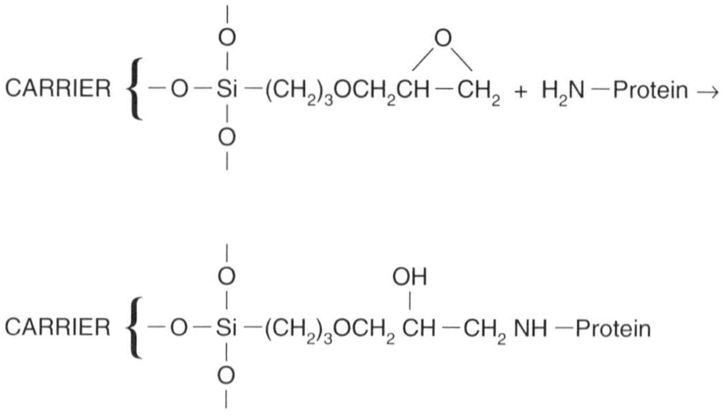

FIGURE 15. The coupling of a protein to an epoxide (oxirane) activated inorganic support.

compounds. Generally, the quantity of NS components bound is a percentage of the total quantity of material offered during the affinity separation. On a well-prepared support, where all sites have been filled or blocked, the nonspecific binding (NSB) ranges from 0.01% to 0.001% of the total quantity of material offered. This means that if a 1-g affinity support column is offered 1000 mg of protein, approximately 0.1 to 1.0 mg of NS compounds will bind. Assuming that the compound of interest is at a concentration of 1.0 mg/g of starting material and that it is totally bound by the support, then upon elution the best that one could expect would be 1.0 mg of specific component and 0.1 to 1.0 mg of NS protein. The purity, therefore, could be as high as 90% or as low as 10%. Improvement in purity of the isolated component can be achieved by decreasing the NSB or by increasing the concentration of the component to be purified as a percentage of the offered protein. Similar results are observed in cases where low-molecular-weight components are the products of the desired isolation. Improvements in purity are sometimes achieved by gradient elution such that either the NS compounds elute either over the entire range of the gradient while the product elutes as a sharp peak or the product elutes before the NS compounds. Generally, the former approach seems to work best.

5.2. Enzyme Purification

A wide variety of ligands have been used for the isolation of enzymes by affinity chromatography on inorganic supports. The most common ligands used are the triazine dyes. These dyes have been successfully applied to the isolation of polynucleotide kinase from phage T_4 (Marcisauskas et al., 1988) as well as several dehydrogenases (Clonis, 1987; Anspach et al., 1988; Kroviarski et al., 1989; Livingston and Chase, 1989). Additionally, dyes have been used successfully for the purification of lysozyme (Livingston and Chase, 1989), human fibroblast membrane protein (Kinkel et al., 1984), and erythropoietin (Krystal et al., 1984) as well as other enzymes (Small et al., 1981). The application of triazine dyes on inorganic supports has been previously reviewed (Atkinson et al., 1982). Enzymes have also been isolated on antibiotic supports (Stepanov et al., 1981; Rudenskaya et al., 1983; Ignatchenko et al., 1987; Gaida et al., 1987, 1989; Van den Burg et al., 1989). In the case of plant and fungal peroxidases, vanillin was used as the affinity ligand (Lobarzewski, 1981; Rudenskaya et al., 1983). For exonucleases, nucleotide phosphates (Varlamov et al., 1986), nucleotides (Bannikova et al., 1982), and cytidine (Bannikova et al., 1983) have been used successfully. Trypsin inhibitors, both soybean (Walters, 1983) and pancreatic (Ernst-Cabrera and Wilchek, 1983), have been used for trypsin purification. The purification of carbonhydrases has been accomplished on a variety of ligands, including α-keratins (Lobarzewski et al., 1985; Ulezlo et al., 1990), sugar alcohols (Ulezlo et al., 1990), and propionic acid (Baum, 1975).

5.3. Glycoproteins and Nucleic Acids

Glycoproteins of all types have been isolated using concanavalin A (Borchert et al., 1982; Muller and Carr, 1984; Renauer et al., 1985; El Rassi et al., 1988; Bergold and Carr, 1989). The isolation of DNA, DNA fragments, and mRNA has also been successfully achieved on inorganic support materials (Akashi et al., 1988; Chow et al., 1988; Goss et al., 1990; Zarytova and Shishkina, 1990).

5.4. Antigens and Antibodies

Immobilized antibodies were first covalently coupled to an inorganic support by this author (Weetall, 1970) for the isolation of antibodies and immunoglobulins. Since that successful attempt, antibodies on inorganic supports have been used for the isolation and purification of haptens (Blanch *et al.*, 1984), bovine serum albumin and immunoglobulins (Weetall, 1970, 1972; Ernst-Cabrera and Wilchek, 1986), thromboxane B_2 (Hubbard *et al.*, 1987), blood group A-active oligosaccharide (Dakour *et al.*, 1987), arginosuccinase (Massom and Jarrett, 1989), lysozyme (Liapis *et al.*, 1989), human epidermal growth factor (Hayashi *et al.*, 1989), receptors (Phillips *et al.*, 1988), viruses (Kiselev *et al.*, 1983), antigen B27 (Babsshak and Phillips, 1988) and cerebrosides and glycolipids (Lingwood, 1984).

5.5. Summary of Affinity Chromatographic Applications

A tabulated summary of affinity applications on inorganic supports may be found in Table 4. From the variety of applications presented, it is obvious that inorganic supports are logical candidates for the isolation and purification of most materials amenable to this form of separation. General problems associated with affinity chromatography include: conformational changes on attachment to surfaces, pH modifications at the support surface, and nonspecific binding (NSB). Elution kinetics are for inorganic supports, similar in nature to their organic relatives. However, for support materials with good capacity, good compression strength, high pressure drop resistance, and solvent compatibility, one cannot do better than choosing an inorganic support material.

TABLE 4
Applications of Inorganic Supports for Affinity Chromatography[a]

Coupled ligand	Purified product	Reference(s)
Aminophosphomanno-pyranoside	Concanavalin A	Ulezio *et al.*, 1990
Antibiotics		
Gramicidin S	Bacterial protease	Ignatchenko *et al.*, 1987; Rudenskaya *et al.*, 1983; Stepanov *et al.*, 1981
Bacitracin	Thrombin, bacterial neutral protease	Gaida *et al.*, 1987, 1989; Van den Burg *et al.*, 1989
Neomycin	Phospholipids	Bannikova *et al.*, 1983
Antibodies	Thromboxane B_2	Stepanov *et al.*, 1981
	Blood group A-active oligosaccharide	Small *et al.*, 1981
	IgG, bovine serum albumin	Weetall, 1970; Ernst-Cabrera and Wilchek, 1986; Weetall, 1972
	Arginosuccinase	Massom and Jarrett, 1989
	Lysozyme	Liapis *et al.*, 1989
	Epidermal growth factor	Hayashi *et al.*, 1989
	Lymphocyte receptor antigen	Phillips *et al.*, 1988
	B27 antigen	Babsshak and Phillips, 1988
	Aromatic haptens	Blanch *et al.*, 1984
	Cerebrosides, glycolipids	Lingwood, 1984

TABLE 4 (Continued)

Coupled ligand	Purified product	Reference(s)
	Influenza virus	Kiselev *et al.*, 1983
Boronic acid	Nucleotides, alcohols, carbohydrates, tRNA	Okayama, 1980; Hagemeier *et al.*, 1983; Payne and Ames, 1981; Glad *et al.*, 1980
Concanavalin A	Glycoproteins	Kinkel *et al.*, 1984; Ernst-Cabrera and Wilchek, 1983; Borrebaeck *et al.*, 1984; El Rassi *et al.*, 1988; Renauer *et al.*, 1985
DNA, oligonucleotides		
Nucleotide phosphates	Exonucleases	Varlamov *et al.*, 1986; Bannikova *et al.*, 1982
Cytidine	Exonuclease A5	Bannikova *et al.*, 1983
Oligonucleotides	DNA, DNA fragments	Goss *et al.*, 1990; Chow *et al.*, 1988; Blanch *et al.*, 1984
Poly(dt)	Poly(A) mRNA	Akashi *et al.*, 1988
Fibrin	Thrombin	Allary *et al.*, 1990
Ganglioside GM	Antibodies	Evstratova *et al.*, 1980
Heparin	Antibodies	Zhou *et al.*, 1989
Isoproterenol	β-Adrenergic receptor adenylate cyclase	Ventner and Kaplan, 1974
Keratins	Pectinases, α-galactosidase	Ulezio *et al.*, 1990; Lobarzewski *et al.*, 1985
Metillin	Calmodulin	Bannikova *et al.*, 1982
Neurophysin	Vasopressin	Swaisgood and Chaiken, 1986
Oxidized cigitonin	Cholesterol	Berezin *et al.*, 1980
Phenothiazine	Calmodulin	Evstratova *et al.*, 1980
Phospholipids	Phospholipase	Evstratova *et al.*, 1980
Propyllipoamide	Lipoamide dehydrogenase	Scouten *et al.*, 1973
Protein A	Immune complexes	Krystal *et al.*, 1984; Ohlson and Niss, 1988
	IgG	Anspach *et al.*, 1988; Hage *et al.*, 1986; Crowley and Walters, 1980
Sugar alcohols	Glucose isomerase	Kroviarski *et al.*, 1989
Triazine dyes	Polynucleotide kinase from T4	Marcisauskas *et al.*, 1988
	Lactate dehydrogenase	Marcisauskas *et al.*, 1988; Anspach *et al.*, 1988; Clonis, 1987; Livingston and Chase, 1989
	Lysozyme	Livingston and Chase, 1989
	Phosphogluconate dehydrogenase	Kroviarski *et al.*, 1989
	Malate dehydrogenase	Anspach *et al.*, 1988
	Human fibroblast membrane proteins	Bergold and Carr, 1989
	Erythropoietin	Krystal *et al.*, 1984
Tomatine	Sterols	Caiky and Hansson, 1986
Toxin	Antitoxin (IgG)	Gaida *et al.*, 1989
Trypsin inhibitor	Trypsin	Walters, 1983; Borchert *et al.*, 1982
Vanillin	Fungal and plant peroxidases	Rudenskaya *et al.*, 1983; Lobarzewski, 1981

[a]Supports used in these studies were porous and nonporous glasses or silica gel. All but a few of the studies listed used some form of silane coupling for attachment of the ligand to the support.

REFERENCES

Akashi, M., Yamaguchi, M., Miyata, H., Hayashi, M., Yashima, E., and Miyauchi, N., 1988, Affinity chromatographic separation. II. Novel stationary phases for affinity chromatography. Nucleobase-selective recognition of nucleosides and nucleotides on poly(9-vinyladenine)-supported silica gel, *Chem. Lett.* **7:**1093.

Allary, M., Boschetti, E., and Lorne, J. L., 1990, Isolation by affinity chromatography with a silica support, of human thrombin for use in the preparations of fibrin glue, *Ann. Pharm.* **48:**129.

Anspach, B., Unger, K. K., Davies, J., and Hearn, M. T. W., 1988, Affinity chromatography with triazine dyes immobilized onto activated non--porous monodisperse silicas, *J. Chromatogr.* **457:**195.

Atkinson, A., McArdell, J. E., Scawen, M. D., Sherwood, R. F., Small, D. A. P., Lowe, C. R., and Burton, C. J., 1982, The potential of organic dyes as affinity ligands in protein studies, *Anal. Chem. Symp. Ser.* **9:**399.

Babsshak, J. V., and Phillips, T. M., 1988, Use of avidin-coated glass beads as a support for high-performance immunoaffinity chromatography, *J. Chromatogr.* **444:**21.

Baker, S. A., Emery, A. N., and Novais, J. M., 1971, Development of insolubilized enzymes, *Process Biochem.* **5:**11.

Bannikova, G. E., Varlamov, V. P., Samsonova, O. L., and Rogozhin, S. V., 1982, Biospecific nuclease chromatography 2. Purification of exonuclease A5 from actinomycetes on organosilicic sorbents with immobilized 5-hydroxyuridine 2'(3'),5'-diphosphate, *Bioorg. Khim.* **8:**212.

Bannikova, G. E., Varlamov, V. P., and Rogozhin, S. V., 1983, Biospecific chromatography of nucleases. Synthesis and use of organosilica sorbents with immobilized derivatives of cytidine for the purification of exonuclease A5 from actinomycetes, *Bioorg. Khim.* **9:**1515.

Baum, G., 1975, Affinity chromatography of beta-galactosidase on controlled-pore glass derivatives, *J. Chromatogr.* **104:**105.

Berezin, I. V., Lopukhin, Yu. M., Andrianova, I. P., Lapuk, Ya. I., Alekseeva, L. B., Sergienko, V. I., Borodin, E. A., Khalilov, E. M., and Archakov, A. I., 1980, Affinity chromatography in elimination of cholesterol, *Vopr. Med. Khim.* **26:**843.

Bergold, A. F., and Carr, P. W., 1989, Improved resolution of glycoproteins by chromatography with concanavalin A immobilized on microparticulate silica via temperature-programmed elution, *Anal. Chem.* **61:**1117.

Blanch, H. W., Arnold, F. H., and Wilke, C. R., 1984, Mass transfer effects in affinity chromatography, *Proceedings of the European Congress on Biotechnology*, Vol. 1, p. 613.

Borchert, A. L., Larsson, P. O., and Mosbach, K., 1982, High-performance liquid affinity chromatography on silica-bound concanavalin A, *J. Chromatogr.* **244:**49.

Borrebaeck, C. A. K., Soares, J., and Mattiasson, B., 1984, Fractionation of glycoproteins according to lectin affinity and molecular size using a high-performance liquid chromatography system with sequentially coupled columns, *J. Chromatogr.* **284:**187.

Bursecz, C. F., 1983, Epiamine coupling to inorganic supports, U.S. Patent 4,415,664.

Chow, T., Juby, C., and Yuen, L., 1988, A high capacity, reusable oligodeoxythymidine affinity column, *Anal. Biochem.* **175:**63.

Clonis, Y. D., 1987, Matrix evaluation for preparative high-performance affinity chromatography, *J. Chromatogr.* **407:**179.

Crowley, S. C., and Walters, R. R., 1980, Determination of immunoglobulins in blood serum by high-performance affinity chromatography, *J. Chromatogr.* **266:**157.

Csiky, I., and Hansson, L., 1986, High-performance liquid affinity chromatography (HPLAC) of sterols with tomatine chemically bonded to microparticulate silica, *J. Liq. Chromatogr.* **9:**875.

Dakour, J., Lundblad, A., and Dopf, D., 1987, Separation of blood group A-active oligosaccharides by high-pressure liquid affinity chromatography using a monoclonal antibody bound to concanavalin A silica, *Anal. Biochem.* **161:**140.

Eaton, D. A., 1974, Immobilized enzymes, in: *Immobilized Biochemicals and Affinity Chromatography* (R. B. Dunlap, ed.), Plenum Press, New York, pp. 241–258.

El Rassi, Z. T., Yunghuoy, M., Fen, Y., and Horvath, C., 1988, High-performance liquid chromatography with concanavalin A immobilized by metal interactions on the stationary phase, *Anal. Biochem.* **169:**172.

Emery, A. N., Hough, J. S., Novais, J. M., and Lyons, T. P., 1975, Immobilization of a glucose isomerase, *Chem. Eng. (London)* **1975:**71.

Ernst-Cabrera, K., and Wilchek, M., 1983, Coupling of ligands to primary hydroxy-containing silica for high-performance affinity chromatography. Optimization of conditions, *J. Chromatogr.* **397:**187.

Ernst-Cabrera, K., and Wilchek, M., 1986, Silica containing primary hydroxyl groups for high-performance affinity chromatography, *Anal. Biochem.* **159:**267.

Evstratova, N. G., Vasilenko, I. A., Kondrat'eva, N. Yu., Serebrennikova, G. A., Evstigneeva, R. P., Varlamov, V. P., Semenova, N. N., and Rogozhin, S. V., 1980, Synthesis of biospecific sorbents for the isolation of proteins, *Bioorg. Khim.* **6:**1355.

Gaida, A. V., Magerovskii, Yu. V., and Monastyrskii, V. A., 1987, Bacitracin and gramicidin S as biospecific ligands for affinity chromatography of human thrombin, *Biokhimiya* **52:**69.

Gaida, A. V., Monastyrskii, V. A., Magerovskii, Yu. V., Staroverov, S. M., and Lisichkin, G. V., 1989, Chromatography of thrombin on modified silica: The role of matrix spacer and pH value, *Biotekhnologiya* **5:**449.

Glad, M., Ohlson, S., Hansson, L., Mansson, M.-O., and Mosbach, K., 1980, High-performance liquid affinity chromatography of nucleosides, nucleotides and carbohydrates with boronic acid-substituted microparticulate silica, *J. Chromatogr.* **200:**254.

Goss, T. A., Bard, M. J., and Harry, W., 1990, High-performance affinity chromatography of DNA, *J. Chromatogr.* **508:**279.

Hage, D. S., Walters, R. R., Rodney, R., and Hethcote, H. W., 1986, Split-peak affinity chromatographic studies of the immobilization-dependent adsorption kinetics of protein A, *Anal. Chem.* **58:**274.

Hagemeier, E., Boos, K. S., Schlimme, E., Lechtenboerger, K., and Kettrup, A., 1983, Synthesis and application of a boronic acid-substituted silica for high-performance liquid affinity chromatography, *J. Chromatogr.* **268:**291.

Hayashi, T., Sakamoto, S., Shikanabe, M., Wada, I., and Yoshida, H., 1989, HPLC analysis of human epidermal growth factor using immunoaffinity precolumn. I. Optimization of immunoaffinity column, *Chromatographia* **27:**11.

Hubbard, H. L., Eller, T. D., Mais, D. E., Halushka, P. V., Baker, R. H., Blair, I. A., Vrbanac, J. J., and Daniel, R., 1987, Extraction of thromboxane B$_2$ from urine using an immobilized antibody column for subsequent analysis by gas chromatography–mass spectrometry, *Prostaglandins* **33:**149.

Ignatchenko, A. P., Bogomaz, V. I., Tugai, V. A., and Chuiko, A. A., 1987, Isolation and purification of proteolytic enzymes on organo-silica supports with immobilized gramicidin S, *Ukr. Biokhim. Zh.* **59:**28.

Kinkel, J. N., Anspach, B., Unger, K. K., Wieser, R., and Bruner, G., 1984, Separation of plasma membrane proteins of cultured human fibroblasts by affinity chromatography on bonded microparticulate silicas, *J. Chromatogr.* **297:**167.

Kiselev, A. V., Kolikov, V. M., Mchedlishvili, B. V., Nikitin, Yu. S., and Khokholva, T. D., 1983, Chromatography of viruses on chemically modified macroporous silicas, *Dokl. Akad. Nauk. SSSR* **272:**1158.

Kroviarski, O., Santarelli, Y., Chochet, X., Muller, S., Arnaud, D., Boivin, T., and Bertrand, P., 1989, Evaluation of chromatography supports prepared by grafting reactive dyes onto dextran coated silica beads, *Proceedings of the International Conference on Protein–Dye Interactions*, p. 115.

Krystal, G. E., Eaves, C. J., and Allen, C., 1984, CM Affi-Gel Blue chromatography of human urine: A Simple one-step procedure for obtaining erythropoietin suitable for *in vitro* erythropoietic progenitor assays, *Br. J. Haematol.* **58:**533.

Kundu, S. K., and Roy, S. K., 1979, Aminopropyl silica gel as a solid support for preparation of glycolipid immunoadsorbent and purification of antibodies, *J. Lipid Res.* **20:**825.

Liapis, A. I., Anspach, B., Findley, M. E., Davies, J., Hearn, M. T. W., and Unger, K. K., 1989, Biospecific adsorption of lysozyme onto monoclonal antibody ligand immobilized on nonporous silica particles, *Biotechnol. Bioeng.* **34:**467.

Lingwood, C. A., 1984, Production of glycolipid affinity matrices by use of heterobifunctional crosslinking agents, *J. Lipid Res.* **25:**1010.

Livingston, A. G., and Chase, H. A., 1989, Preparation and characterization of adsorbents for use in high-performance liquid affinity chromatography, *J. Chromatogr.* **481:**159.

Lobarzewski, J., 1981, Affinity chromatography method for the separation of vanillic acid induced forms of fungal peroxidase from *Trametes versicolor* on porous beads activated with vanillin, *J. Biol. Macromol.* **3:**77.

Lobarzewski, J., Fiedurek, J., Ginalska, and Wolski, T., 1985, New matrices for the purification of pectinases by affinity chromatography, *Biochem. Biophys. Res. Commun.* **131:**666.

Marcisauskas, I., Karalite, R., Barilkaite, D., Sudziuviene, I., and Pesliakas, O., 1988, Application of immobilized triazine dyes for purifying polynucleotide kinase of phage T4, *Biotekhnologiya* **4:**97.

Massom, L. R., and Jarrett, H. W., 1989, Purification of argininosuccinase by high-pressure immunoaffinity chromatography on monoclonal anti-argininosuccinase-silica, *J. Chromatogr.* **482:**252.

Messing, R. A., 1974, Enzymes immobilized on porous ceramics, *Biotechnol. Bioeng.* **16:**1419.

Messing, R. A., 1975, Porous inorganic carrier materials, U.S. Patent 3,892,580.

Messing, R. A., and Stinson, H., 1974, *ortho*-Dianisidine as a coupling agent for porous glass, *Mol. Cell. Biochem.* **4:**217.

Muller, A. J., and Carr, P. W., 1984, Chromatographic study of the thermodynamic and kinetic characteristics of silica-bound concanavalin A, *J. Chromatogr.* **284**:33.

Ohlson, S., and Niss, U., 1988, Profiling of circulating immune complexes by high performance liquid affinity chromatography (HPLAC) on a protein-A silica matrix, *J. Immunol. Methods* **106**:225.

Ohlson, S., and Weislander, J., 1987, High-performance liquid affinity chromatographic separation of mouse monoclonal antibodies with protein A silica, *J. Chromatogr.* **397**:207.

Okayama, H., 1980, Application of boronic acid derivatives to affinity chromatography, *Tanpakushitsu Kakusan Koso, Bessatsu* **22**:72.

Payne, S. M., and Ames, B. N., 1981, A procedure for rapid extraction and high-pressure liquid chromatographic separation of the nucleotides and other small molecules from bacterial cells, *Anal. Biochem.* **123**:151.

Phillips, T. M., Faantz, S. C., and Chmielinska, J. J., 1988, Isolation of bioactive lymphocyte receptors by high performance immunoaffinity chromatography, *BioChromatography* **3**:149.

Renauer, D., Oesch, F., Kinkel, J., Unger, K. K., and Wieser, R. J., 1985, Fractionation of membrane proteins on immobilized lectins by high-performance liquid affinity chromatography, *Anal. Biochem.* **151**:424.

Royer, G. P., and Green, G. M., 1975, Immobilized pronase, *Biochem. Biophys. Res. Commun.* **44**:426.

Rudenskaya, G. N., Gaida, A. V., Osterman, A. L., and Stepanov, V. N., 1983, Affinity chromatography of proteinases on sorbents prepared by the addition of peptide antibiotics to silica derivatives, *Biokhim. Biofiz. Mikroorganizmov. Gorkii* **11**:36.

Scouten, W. H., Torok, F., and Gitomer, W., 1973, Purification of lipoamide dehydrogenase by affinity chromatography on propyllipoamide-glass columns, *Biochim. Biophys. Acta* **309**:521.

Small, D. A. P., Atkinson, T., and Lowe, C. R., 1981, High-performance liquid affinity chromatography of enzymes on silica-immobilized triazine dyes, *J. Chromatogr.* **216**:175.

Stepanov, V. M., Rudenskaya, G. N., Gaida, A. V., and Osterman, A. L., 1981, Affinity chromatography of proteolytic enzymes on silica-based biospecific sorbents, *J. Biochem. Biophys. Methods* **5**:177.

Swaisgood, H. E., and Chaiken, I. M., 1986, Analytical high-performance affinity chromatography: Evaluation by studies of neurophysin self-association and neurophysin–peptide hormone interaction using glass matrixes, *Biochemistry* **25**:4148.

Ulezlo, I. V., Ananichev, A. V., and Bezborodov, A. M., 1985, Isolation and purification of glucose isomerase by affinity chromatography, *Prikl. Biokhim. Mikrobiol.* **21**:445.

Ulezlo, I. V., Ananichev, A. V., Zaprometova, O. M., and Bezborodov, A. M., 1990, Affinity chromatography of *Cephalosporium acremonium* alpha-galactosidase, *Prikl. Biokhim. Mikrobiol.* **26**:616.

Van den Burg, B. E., Vincent, G. H., Stulp, B. K., and Venema, G., 1989, One-step affinity purification of bacillus neutral proteases using bacitracin-silica, *J. Biochem. Biophys. Methods* **18**:209.

Varlamov, V. P., Bannikova, G. E., Lopatin, S. A., and Rogozhin, S. V., 1986, Specific sorbents for chromatography of nucleolytic enzymes, *Symp. Biol. Hung.* **31**:127.

Ventner, J. C., and Kaplan, N. O., 1974, Partial purification of the beta-adrenergic receptor adenylate cyclase complex by affinity chromatography to glass bead-immobilized isoproterenol, *Methods Enzymol.* **38**:187.

Walters, R. R., 1983, Rapid quantitative analysis using high-performance affinity chromatography minicolumns, in: *Proceedings of the Fifth International Symposium on Affinity Chromatography and Biological Recognition* (Jerusalem) (I. Chaiken, M. Wilchek, and I. Parikh, eds.), pp. 261–264.

Weetall, H. H., 1970, Preparation and characterization of antigen and antibody adsorbents covalently coupled to an inorganic carrier, *Biochem. J.* **117**:257.

Weetall, H. H., 1972, Insolubilized antigens and antibodies, in: *Chemistry of Biosurfaces*, Vol. 2 (M. L. Hair, ed.), Marcel Dekker, New York, pp. 597–631.

Weetall, H. H., 1976, Covalent coupling methods for inorganic supports, *Methods Enzymol.* **44**:134.

Weetall, H. H., and Detar, C. C., 1975, Covalent attachment of proteins to inorganic supports directly by activation with cyanogen bromide, *Biotechnol. Bioeng.* **17**:295.

Weetall, H. H., and Filbert, A. M., 1974, Porous glass for affinity chromatography applications, *Methods Enzymol.* **34**:59.

Zarytova, V. F., and Shishkina, I. G., 1990, Affinity chromatography of DNA fragments and P-modified oligonucleotides, *Anal. Biochem.* **188**:214.

Zhou, F. L., Muller, D., Santarelli, X., and Jozefonvicz, J., 1989, Coated silica supports for high-performance affinity chromatography of proteins, *J. Chromatogr.* **476**:195.

4

2-Fluoro-1-methylpyridinium (FMP) Salt-Activated Gels

Properties and Uses in Affinity Chromatography and Enzyme Immobilization for Analytical Applications

Dyer Narinesingh and That T. Ngo

1. INTRODUCTION

There are many techniques currently available for linking bioligands to water-insoluble matrices to produce either catalytically active bioreactors or matrices for use in covalent and affinity chromatography. The basic strategies that can be utilized to covalently link bioligands to supports include the use of activated groups on the side chains of the polymeric supports, partial modification of the polymer backbone to produce activatable groups, and the use of activated residues on the bioligand to couple to the supports.

1.1. Types of Supports

A wide variety of natural and synthetic polymers have been employed as supports. These include the naturally occurring polymers such as agarose, dextran, and cellulose and their derivatives Sephadex, Sepharose, and Sephacel. Examples of the synthetic polymers include the Eupergit gels, which are copolymers of methacrylamide, allylglycidyl ether, and *N*-methylene-bisacrylamide (Hannibal-Friedrich *et al.*, 1980; Butler *et al.*, 1984; Bamberger *et al.*, 1986; Solomon *et al.*, 1986); Fractogel TSK, a semirigid gel made of a copolymer of oligo(ethylene glycol), glycidyl methacrylate, and pentaerythritol dimethacrylate (Kato *et al.*, 1981); the HEMA gels, which are macroporous hydrophilic

Dyer Narinesingh • Department of Chemistry, The University of the West Indies, St. Augustine, Trinidad, West Indies. *That T. Ngo* • BioProbe International, Inc., Tustin, California 92680. *All correspondence should be addressed to T. T. Ngo at*: Department of Developmental and Cell Biology, University of California, Irvine, California 92717.

Molecular Interactions in Bioseparations, edited by That T. Ngo. Plenum Press, New York, 1993.

copolymers of 2-hydroxyethyl methacrylate and ethylene dimethacrylate; Trisacryl GF-2000 gels, which are nonionic and are prepared by copolymerization of N-acryloyl-2-amino-2-hydroxymethyl–1,3-propanediol and a bifunctional cross-linker; the Ultrogels and Magnogels, which are mixtures of polyacrylamide and agarose gelled together; and the amide- and ester-containing matrices including polyacrylamide (Inman and Dintzis, 1969), gelatin (Coulet $et\ al.$, 1980), and nylon (Hornby and Goldstein, 1976). Among the inorganic supports which have been used are derivatized porous glass (Hatchikian and Monsan, 1980) and the transition-metal oxides of titanium and zirconium (Dale and White, 1979; Kennedy $et\ al.$, 1981).

1.2. Coupling Chemistries

The various coupling chemistries which have been developed have their genesis in the need to utilize only those residues on the bioligand that are not "involved" in binding/ catalysis for covalent linkage to the support and to employ coupling conditions which do not result in denaturation of the bioligand or inactivation of the support.

One of the earliest methods employed for immobilizing bioligands to polyhydroxy supports involved the use of cyanogen bromide (CNBr) as the activating agent. The method is relatively simple and rapid. However, it is inferior to many of the more recently developed activation methods. Not only does CNBr coupling produce substantial leakage of the bioligand due to the electrophilic nature of the isourea formed (Lasch and Koelsch, 1978), but also the reaction between the activated polymer and the ligand frequently results in the introduction of charged species that interfere with utilization of the reaction product in affinity adsorption. In addition, CNBr is a noxious and lachrymatory chemical which requires special care in its handling. The use of the water-soluble, nonvolatile CNBr analogues 1-cyano-4-dimethylaminopyridinium tetrafluoroborate (CDAP) (Kohn and Wilchek, 1983) and p-nitrophenyl cyanate (Kohn $et\ al.$, 1983) reduces toxicity and results in much better yields. However, leakage is still a problem.

Efforts to find activating agents other than CNBr have resulted in the use of a number of reagents including various sulfonyl chlorides, the chlorocarbonates, carbonyldiimidazole, triazine, and epoxy compounds. With the sulfonyl chlorides (tresyl and tosyl), both the coupling efficiency and the catalytic activity are generally high (Nilsson and Mosbach, 1981). Unlike the CNBr-activated supports, tresyl-activated supports can be stored in dilute HCl (1mM) at room temperature without any appreciable hydrolysis. In the freeze-dried form, it can be stored desiccated for extended periods without any appreciable loss of coupling efficiency. The use of chromophoric sulfonyl chlorides (Scouten and van der Tweel, 1984) such as dabsyl, dipsyl, and diabsyl chlorides, lissamine rhodamine β sulfonyl chloride, and pentafluorobenzene sulfonyl chloride allows rapid qualitative and quantitative monitoring of the coupling reaction.

The chlorocarbonates which have found widespread use include p-nitrophenyl chlorocarbonate and N-hydroxysuccinimide chlorocarbonate (Wilchek and Miron, 1982). Such chlorocarbonates react with polyalcohols to produce activated carbonates which readily react with amino ligands to produce stable, uncharged carbamates. As with the sulfonyl chlorides, activation with the chlorocarbonates must be performed in dry, nonaqueous solvents to avoid hydrolysis of the chlorocarbonate as well as the activated support. The latter is highly stable when stored at 4°C in solvents such as isopropanol. The residual reactive groups that do not react with the bioligand during the coupling process can be readily regenerated to the original hydroxyl group on the carrier.

The use of carbonyldiimidazole (CDI), 1,1'-carbonyldi-1,2,3-benzotriazole (CDB), and carbonyldi-1,2–4-triazole to activate hydroxy groups results in neutral, activated supports which, when stored in the anhydrous form, are very stable (Hearn, 1986; Bethell *et al.*, 1981). These activated supports, especially CDI and CDB, react very rapidly with nucleophiles.

The use of epoxy compounds was described by Porath and Axen (1976). Epichlorohydrin or 1,4-bis(2,3-epoxypropoxy)butane reacts with hydroxyl groups of supports to form epoxide gels. These activated gels can then be reacted with sodium thiosulfate followed by reduction with dithiothreitol to give thiol gels. These thiol gels can be used in covalent chromatography for the purification of molecules containing the thiol group.

In addition to the above methods for producing activated supports for subsequent covalent linkage to bioligands, there are a number of less frequently used methods. These include the introduction on the support of the following activated groups: oxirane (Sundberg and Porath, 1974); aziridine (Porath and Axen, 1976); pyrimidines (Gribnau, 1978); benzoquinones (Brandt *et al.*, 1975); hydrazides (Turkova, 1976); activated double bonds (Porath *et al.*, 1975), and activated halogens (Jagendorf *et al.*, 1963). These activated groups react primarily with amino- and thiol-containing bioligands.

1.3. 2-Fluoro-1-methylpyridinium (FMP) Activation

In 1975, Mukaiyama and co-workers reported on the use of 2-halopyridinium salts in a wide range of reactions such as in the synthesis of esters, amides, and thioesters (Mukaiyama *et al.*, 1975; Bald *et al.*, 1975). In 1977, Hojo *et al.* demonstrated the conversion of various alcohols to the corresponding thioalcohols by reaction with 2-fluoro-1-methylpyridinium (FMP) salts and sodium N,N'-dimethyldithiocarbamate followed by reductive cleavage. The alcohols used in these studies were low-molecular-weight carbohydrates, steroids, and monomeric alcohols.

In 1986, Ngo (1986a,b) reported on a facile method using FMP that can be used for activating hydroxyl groups in naturally occurring polymers such as dextran, agarose, and cellulose or in synthetic polymers such as poly(vinyl alcohol) or poly(hydroxyalkyl methacrylate). Such activated polymers are useful for immobilization of enzymes, antibodies, and other biomolecules and for affinity matrix development. The activated hydroxyls react readily with amino or thiol ligands in the pH range 6–10 to form stable covalent bonds between the support and the ligand (Fig. 1). The activated supports can be stored desiccated at $-10°C$ or in inert anhydrous organic solvents or in dilute HCl (1mM) at 4°C. Under these various storage conditions, the activated polymer retains its activity for several months. Also, because of the relatively long "half-life" of the FMP-activated polymers in non-nucleophilic buffers, it is possible to prepare bioreactors and affinity matrices with good ligand coupling and greater reproducibility.

The use of FMP appears to offer several additional advantages over other activating agents. It is obtainable in a pure, dry crystalline form; it is not very corrosive, moisture-sensitive, or toxic; the process of the coupling reaction in some cases can be monitored spectrophotometrically; and FMP conjugates can be used to develop techniques for affinity-directed immobilization of ligands.

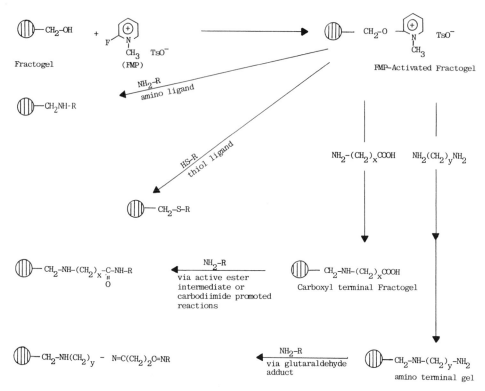

FIGURE 1. FMP activation of polyhydroxy groups and their reactions with amino- and thiol-containing ligands.

2. EXPERIMENTAL

2.1. General Procedure for the Activation of Polyhydroxyl Groups with FMP

The gel (200 ml) is successively washed with 200 ml each of distilled water, 25%, 50%, and 75% acetone/water (v/v) mixtures, acetone, and finally N,N'-dimethylformamide (DMF). To the washed gel is added DMF (5 ml) containing 4-dimethylaminopyridine (6 mmol) followed by DMF (15 ml) containing FMP (5 mmol). The gel suspension is tumbled end over end at room temperature for 30–60 min. The activated gel is finally washed twice with DMF (50 ml) and then with acetone (100 ml). The activated gels can be stored either dry and desiccated at 4°C (as for the Fractogel gels) or in boric acid (5mM) or in phosphoric acid solutions (10mM) at 4°C (as for the agarose gels).

2.2. General Procedure for Coupling Ligands to FMP-Activated Gel

The activated support (2.5 ml) is washed with 10 ml each of HCl (1mM), distilled, deionized water, and a non-nucleophilic coupling buffer (0.05M) in the pH range 5–10 but usually between pH 7 and 9. A solution of the bioligand in the coupling buffer (5 ml) is added to the washed gel, and the suspensions tumbled end over end at 4°C or at room temperature for 6–20 h.

After three washes with the coupling buffer (25 ml), the gel is resuspended in 25 ml of Tris-HCl ($0.05M$, pH 8.0) containing ethanolamine ($0.1M$) and tumbled at room temperature for an additional 4–5 h for deactivating remaining activated groups. Finally, the gel is washed with coupling buffer, sodium acetate buffer ($0.1M$) containing NaCl ($0.5M$), and then the storage buffer containing NaN_3 (0.1%). In cases where removal of all traces of the bioligand is essential, the gel can be further washed twice with 25 ml each of the following distilled, deionized water; phosphate buffer ($0.1M$, pH 7.5) containing NaCl ($1M$); KSCN ($1.5M$); urea ($8M$); sodium dodecyl sulfate (10%); distilled water; and finally storage buffer.

3. RESULTS AND DISCUSSION

3.1. Activation Density

The density of FMP-activated hydroxyls (micromoles of FMP per gram of dry activated gel) can be readily quantified by reacting the activated gel with suitable nucleophiles and monitoring the absorbance of the liberated α-pyridone at 297 nm ($\epsilon = 5900$ mol^{-1} liter cm^{-1}) (Ngo, 1986a).

Routinely, 0.1 g of the dry activated gel is treated with NaOH ($0.2M$) or Tris buffer ($0.1M$, pH 9.0), and the suspension tumbled at room temperature for about 24 h. The suspension is then centrifuged, the supernatant suitably diluted, and the absorbance read at 297 nm using the solvent as the blank. The activation densities of a number of FMP-activated supports are shown in Table 1. These results show that FMP-activated polymers possess relatively high activation densities.

3.2. Ligand-Binding Capacities of FMP-Activated Gels

Also included in Table 1 are data on the ligand binding capacities of various proteins onto FMP-activated gels. These results indicate that FMP-activated gels possess relatively high binding capacities. In view of the similarities in experimental procedure between FMP and tresyl (TC) activation methods and the identical linkages in the resulting immobilized

TABLE 1
Density of Activation and Ligand Binding Capacity of Various FMP-Activated Gels

Matrix	FMP content (μmol/g dry gel)	Bovine serum albumin coupled (mg/g dry gel)	Bovine α-casein coupled (mg/g dry gel)	Heparin coupled (mg/ml gel)
Fractogel TSK HW-75F	155	235	75	10
Sepharose CL-4B	260	58	75	12
Trisacryl GF 2000	55	108	—	—
Trisacryl GF 2000 LS	45	25	—	—
Glyceryl controlled pore glass 3000-200	3	109	—	—
Cellufine GH 2000	20	63	—	—
Filter paper	—	1.2[a]	—	—

[a]Milligrams of BSA bound to 1 cm^2 of filter paper.

ligand prepared from either FMP- or TC-activated gels, it was found useful to determine if there is any difference between gels activated by these two methods.

Figure 2 shows that FMP-activated Fractogel was able to bind more than twice as much bovine serum albumin (BSA) as TC-activated Fractogel. Activation of Fractogel with TC was carried out under identical conditions to those for activation with FMP except that FMP was replaced by TC.

3.3. Storage Stability of FMP-Activated Gels

FMP-activated gels can be conveniently stored desiccated under anhydrous conditions at $-10°C$ as well as in aqueous acid media at $4°C$ for extended periods without significant loss of activity. When FMP Fractogel is stored in $5mM$ boric acid solution at $4°C$, it shows a $t_{1/2}$ of > 70 days (Fig. 3). Longer $t_{1/2}$ is observed when it is stored in $10mM$ phosphoric acid.

3.4. Stability of Matrix–Ligand Linkage

The stability of the linkage between the ligand and the matrix is of primary importance in developing an efficient affinity matrix. The linkages generated from reacting nucleophilic affinity ligands with FMP-activated gels are stable covalent $C—NR_2$ or C-S-R bonds. This stability is borne out by experiments on the stability of $[^{125}I]$-BSA bound to FMP-

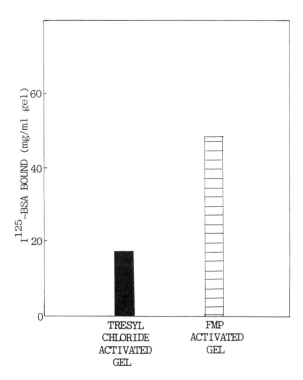

FIGURE 2. Comparison of the $[^{125}I]$-BSA binding capacity of tresyl chloride (TC)-activated Fractogel with that of FMP-activated Fractogel. Fractogel TSK HW75 was activated by TC or FMP under identical experimental conditions and procedures on the same day. Experimental details are described in the text.

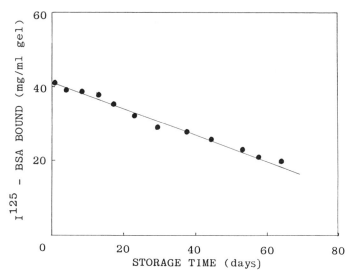

FIGURE 3. Stability of FMP-activated Fractogel in 5mM boric acid at 4°C. At the indicated intervals, FMP-activated Fractogel was tested for [^{125}I]-BSA binding capacity.

activated Fractogel and Trisacryl GF 2000, where no detectable leakage was observed over a period of at least 100 days (Fig. 4).

3.5. Influence of pH, Ionic Strength, and Temperature on the Degree of Hydrolysis, Amminolysis, and Thiolysis of FMP-Activated Fractogel

In these sets of experiments, FMP-activated Fractogel [16.1 mg (150 μmol) of FMP/g dry gel] was suspended in 25 ml of pure or amine-containing or thiol-containing phosphate/borate buffer (0.01M in each buffer component) in the pH range 6.0–10.0. The thiol nucleophiles investigated were ethanethiol, 2-mercaptoethanol, and 3-mercaptopropanoic acid whereas the amine-containing nucleophiles used were ethanolamine and glycine. The agitated reaction mixtures were thermostated in a water bath at 25°C. At various time intervals, the reaction mixtures were vortexed, 0.5-ml aliquots were pipetted out and centrifuged at 14,000 rpm for exactly 1 min, and 0.2-ml aliquots of the supernatants were diluted to 1.0 ml with the buffer of appropriate pH. Absorbances were then read at λ_{297nm}. The spectrophotometric blank consisted of all the components of the reaction mixture except the activated gel. Reactions were also carried out in phosphate/borate buffers of varying concentrations (0.01–0.10M) as well as over the temperature range 25–60°C.

The results (Table 2) show that the rate of hydrolysis increases with increasing pH, being very slow at the lower pHs but much faster at higher pHs. In fact, hydrolysis occurs about 3800 times faster at pH 10.0 than at pH 6.0. Both amminolysis (ethanolamine as nucleophile) and thiolysis (β-mercaptoethanol as nucleophile) proceed by a first-order rate law and are faster than hydrolysis at all pHs, the differences being more pronounced at pHs above 7.0. Thus, it is possible to carry out effective coupling with thiol and amino ligands in the pH range 7–9.

Compared to amminolysis, thiolysis proceeds at a faster rate. Thus, it appears that, at

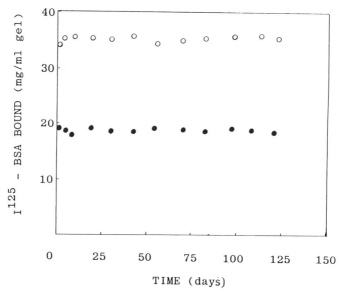

FIGURE 4. Stability of [^{125}I]-BSA covalently bound to FMP-activated Fractogel TSK HW75 (○) and FMP-activated Trisacryl GF 2000 (●).

least for small nucleophiles, electronic considerations appear to be the dominant factor in determining the rate at which the N-methylpyridoxyl moiety is displaced from the activated support. This is further borne out by the observation that for the thiol-containing ligands investigated the order of reactivity is ethanethiol > 2-mercaptoethanol > 3-mercapto-propanoic acid and for the amino ligands the order is ethanolamine > glycine.

TABLE 2

Rate Constants for the Hydrolysis, Amminolysis, and Thiolysis of FMP-Activated Fractogel at Various pHs[a]

pH	$10^4 k_{hydrolysis}$ (s^{-1})	$10^4 k_{amminolysis}$ (s^{-1})	$10^4 k_{thiolysis}$ (s^{-1})
6.0	0.01	0.015	0.02
6.5	0.03	0.05	0.07
7.0	0.08	0.32	0.68
7.5	0.32	5.30	9.50
8.0	0.60	19.53	21.68
8.5	3.20	30.06	35.48
9.0	7.50	69.73	80.29
9.5	22.00	246.18	308.96
10.0	38.10	600.72	895.09

[a]Reaction conditions: $T = 25°C$; buffer composition = $0.01M$ $NaH_2PO_4 + 0.01M$ borax; amine = ethanolamine; thiol = β-mercaptoethanol; amount of FMP-activated Fractogel = 16.1 mg (150 μmol per gram of dry gel).

It was also found that increasing the ionic strength of the coupling medium causes a marked decrease in the rate of displacement of the *N*-methylpyridoxyl moiety from the FMP-activated gel. For example, addition of 0.1*M* NaCl to phosphate/borate buffer (0.01*M*) caused a 4-fold decrease in the rate of amminolysis whereas changing the buffer concentration from 0.01*M* to 0.1*M* caused a 2.5-fold decrease in the rate of hydrolysis (Table 3). Thus, it is generally recommended that coupling reactions with FMP-activated supports be carried out in buffers of relatively low ionic strengths (<0.05*M*).

3.6. Solvent Effects

The reactions of amines and thiols with FMP-activated gels can also be conveniently carried out in the same organic solvents used in the preparation of the activated support. Comparison of the results in Table 4 shows that coupling reactions are faster in DMF than in aqueous medium. However, the order of reactivity is the same in both types of media.

3.7. Temperature Effects

The rate constants for hydrolysis, amminolysis, and thiolysis all increase with increasing temperatures, as expected. The rate constants and activation energies for the coupling of FMP-activated Fractogel with β-mercaptoethanol and ethanolamine are given in Table 5.

3.8. Coupling of Enzymes to FMP-Activated Supports

A wide variety of enzymes have been successfully immobilized to various FMP-activated supports such as Trisacryl, Fractogel, poly(vinyl alcohol), nylon, glyceryl glass, collagen, and gelatin. These enzymes include papain (Narinesingh and Ngo, 1987), urease (Narinesingh *et al.*, 1990a, 1991a), β-galactosidase (Narinesingh *et al.*, 1991b), coimmobilized β-galactosidase/mutarotase (Narinesingh *et al.*, 1991b), coimmobilized glucose oxidase/peroxidase (Narinesingh *et al.*, 1991c), alkaline phosphatase (Narinesingh *et al.*,

TABLE 3
Rate Constants for the Hydrolysis and Amminolysis of FMP-Activated Fractogel in the Presence of Added Salt and Varying Buffer Concentrations at pH 9.0[a]

Buffer composition	Added NaCl (*M*)	$10^4 k_{hydrolysis}$ (s^{-1})	$10^4 k_{amminolysis}$ (s^{-1})
0.01*M* NaH$_2$PO$_4$ + 0.01*M* borax	—	7.50	69.73
	0.02	—	45.62
	0.05	—	20.60
	0.10	—	17.80
	0.20	—	13.50
0.05*M* NaH$_2$PO$_4$ + 0.05*M* borax	—	4.20	25.30
0.10*M* NaH$_2$PO$_4$ + 0.10*M* borax	—	3.10	14.70

[a]Other conditions: *T* = 25°C; amount of FMP-activated Fractogel = 16.1 mg; amine = ethanolamine; reaction volume = 25 ml.

TABLE 4

Effect of Solvent on the Rates of Amminolysis
and Thiolysis of FMP-Activated Fractogel

Nucleophile	10^4k (s^{-1}) for coupling in:	
	DMF[a]	Phosphate/borate buffer[b]
CH_3CH_2SH	132.5	91.7
$HOCH_2CH_2SH$	120.3	80.3
$HOCH_2CH_2NH_2$	90.0	69.7
$HSCH_2CH_2COOH$	25.7	9.4

[a]Conditions: [Nucleophile] = 0.05M; [Triethylamine] = 0.055M;
16.1 mg of FMP-activated Fractogel in 25 ml of solvent; T = 25°C.
[b]Phosphate (0.01M)/borate (0.01M) buffer; conditions: [Nucleo-
phile] = 0.05M; T = 25°C; 16.1 mg of FMP-activated Fractogel
in 25 ml of buffer.

1990), linamarase (Narinesingh *et al.*, 1988), trypsin (Narinesingh *et al.*, 1989), pepsin (Ngo, 1991), arginase (Mungal, 1991), galactose oxidase (Ramtahal, 1992), xanthine oxidase (Balladin, 1993), *Aspergillus ficuum* phytase (Ullah and Cummins, 1987; Ullah and Phillippy, 1988), and *Aspergillus ficuum* optimum acid phosphatase (Ullah and Cummins, 1988).

Using β-galactosidase (*Escherichia coli*) as a representative example, it was found that FMP-activated Trisacryl GF 2000 was able to bind up to 7 mg of enzyme per milliliter of gel (Fig. 5). The time course of the binding showed that the coupling reaction proceeds rapidly, with most of the reaction occurring in the first 30 min (Fig. 6). The effect of coupling pH on the amount of bound protein was studied over the pH range 6.0–10.0. The amount of enzyme coupled increases with increasing pH and reaches a maximum above pH 8.0. Table 6 and Fig. 7 show the relationship between the activity of bound β-galactosidase and the amount of protein offered for coupling to FMP-activated Fractogel (Davis, 1991). Similar relationships have been observed for papain (Narinesingh and Ngo, 1987), trypsin (Narinesingh *et al.*, 1989), and urease (Narinesingh, 1990a) bound to this support.

The solid-phase immobilization of pepsin is of considerable interest because of the

TABLE 5

The Effect of Temperature on the Rates of Hydrolysis,
Amminolysis, and Thiolysis of FMP-Activated Fractogel[a]

Temperature (°C)	$10^3k_{hydrolysis}$ (s^{-1})	$10^3k_{amminolysis}$ (s^{-1})	$10^3k_{thiolysis}$ (s^{-1})
25	0.75	6.97	8.01
40	3.10	19.15	24.15
50	8.90	42.86	51.03
60	20.71	85.32	93.00
Activation energy (kJ mol^{-1})	78.38	59.95	57.80

[a]Conditions: [Nucleophile] = 0.05M; phosphate (0.01M)/borate (0.01M) buffer; pH 9.0;
16.1 mg FMP-activated Fractogel in 25 ml of buffer.

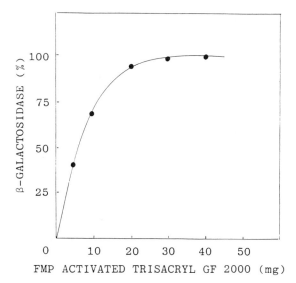

FIGURE 5. Coupling of β-galactosidase to FMP-activated Trisacryl GF 2000. In these experiments a fixed amount of the enzyme was added to varying amounts of the support, and the percent protein coupled determined in each case.

utility of immobilized pepsin in the production of F(ab')$_2$ fragments from intact immuno-globulin G (IgG) as well as the challenge of immobilizing a protein that is stable only in acidic pH. A systematic study on the immobilization of pepsin on FMP-activated Trisacryl GF 2000 showed that the coupling efficiency using 3 mg of pepsin and varying amounts of activated gels (5–50 mg of dry gel) ranges from 20% with 5 mg of gel to 99% with 50 mg of gel. The resulting specific activity of pepsin immobilized under such conditions is found to be a high 2358 units per milliliter of gel. This contrasts with a specific activity of only 2.2

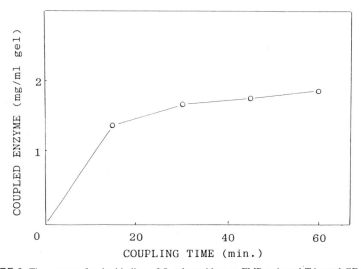

FIGURE 6. Time course for the binding of β-galactosidase to FMP-activated Trisacryl GF 2000.

TABLE 6

Coupling Efficiency of β-Galactosidase Immobilized
on an FMP-Activated Fractogel Support as a Function of Protein Offered

Protein offered for coupling (mg)	Specific activity of enzyme offered for coupling[a] (units)	Protein bound (μg/ml gel)	Percentage bound protein	Specific activity of immobilized enzyme[a] (units/ml gel)	Coupling efficiency[b] (%)
0.026	29.8	3.9	14.9	4.3	97
0.052	59.3	15.9	30.5	10.1	56
0.078	89.0	31.9	40.7	14.8	41
0.104	118.7	57.4	55.0	21.8	33
0.157	178.0	121.0	77.3	34.4	25
0.313	356.1	217.0	69.3	51.2	21

[a]Defined as the number of micromoles of O-nitrophenol liberated per minute at 35°C and pH 7.3

[b]Coupling efficiency $= \dfrac{\text{Specific activity of the immobilized enzyme}}{\text{Specific activity of the enzyme offered for coupling}} \times 100.$

units per milliliter of gel for this same enzyme immobilized to this same support but preactivated with carbonyldiimidazole (Hearn, 1986).

3.9. Applications of Enzymes/Proteins Coupled to FMP-Activated Fractogel in Bioanalysis

Fractogel as a support has many desirable properties. It is nonionic and hydrophilic, synthetic and nonbiodegradable, and chemically, thermally, and mechanically stable. It also possesses good porosity and yields high flow rates. These properties, together with the high coupling capacity with amino- and thiol-containing nucleophiles under relatively mild conditions when its hydroxyl groups are activated with FMP, make it ideal for use in

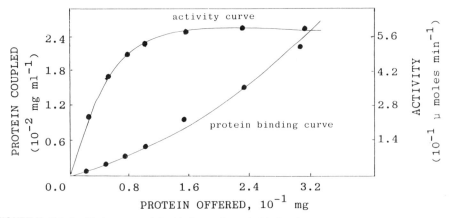

FIGURE 7. Relationship between activity binding and protein binding for β-galactosidase immobilized to an FMP-activated Fractogel support.

bioreactors in flow analysis. Table 7 lists a number of enzymes and other proteins which have been successfully immobilized to FMP-activated Fractogel for the quantitation of a variety of biomolecules.

3.10. Affinity Supports

A variety of commercially available affinity matrices have been prepared using FMP-activated supports. The affinity coupling capacity of these supports is dependent upon a number of factors including the biochemical nature of the affinity ligand, ligand concentration, pH, and ionic strength. Typical ligand coupling densities range from 1 to 40 mg of ligand per milliliter of swollen gel. Ideal ligand concentrations should always be determined experimentally. In the preparation of these affinity matrices, usually 5–10 mg of the ligand is initially reacted per milliliter of the FMP-activated gel. Following the determination of the gel's overall ligand density and biological affinity activity, the ligand density is then optimized by increasing/decreasing the amount used during the coupling. Ligand density can also be controlled by varying the amount of salt used in the coupling buffer. Salt at concentrations greater than $0.1M$ decreases the overall ligand density.

Since coupling of the affinity ligands to FMP-activated supports occurs under mild conditions (pH 7–9), there is minimum ligand denaturation, thus maximizing the affinity gel's overall biological capacity. Also, since the resulting ligand–support bonds are highly stable alkylamine or thioether bonds, minimal ligand leaching and a long gel lifetime are ensured.

TABLE 7
Applications of Enzymes/Proteins Immobilized to FMP-Activated Fractogel in Bioanalysis

Bioligand	Application[a]	Reference
Urease	FIA analysis of urea in serum	Narinesingh et al., 1990a
β-Galactosidase/mutarotase and glucose oxidase/peroxidase	FIA analysis of lactose in milk	Narinesingh et al., 1991b
Glucose oxidase/peroxidase	FIA analysis of glucose in serum	Narinesingh et al., 1991c
Alkaline phosphatase-labeled collagen	Single and multi-enzyme-amplified assay of clostridiopeptidase A (collagenase)	Narinesingh and Ngo, 1990
Urease-labeled collagen	Enzyme-amplified potentiometric determination for collagenase	Narinesingh and Ngo, 1989
Linamarase	FIA determination of bound cyanide in cassava	Narinesingh et al., 1988
Galactose oxidase	FIA determination of trace levels of Cu^{2+}	Ramtahal, 1992
Fluorescein-labeled bovine serum albumin	FIA analysis of proteolytic activity of various enzymes	Narinesingh et al., 1990b
Protein A	Low-pressure liquid affinity chromatographic quantitation of IgG using FIA	Narinesingh and Ngo, 1991
Cibacron blue	Low-pressure liquid affinity chromatographic quantitation of bovine serum albumin using FIA	Narinesingh et al., 1992
Xanthine oxidase	FIA analysis of the freshness of fish	Balladin, 1993
Urease	Urease electrode for the potentiometric determination of urea in serum	Narinesingh et al., 1991a

[a]FIA, Flow injection analysis.

A variety of affinity supports prepared from FMP-activated gels are available. These include *N*-ε-2,4-dinitrophenyl (DNP)-L-lysine-coupled Sepharose CL-4B, which can be used as the affinity sorbent for purifying rabbit anti-DNP serum (Fig. 8) (Ngo, 1986a). Eluting the antibodies with 10% tetrahydrofuron (THF) in $0.1M$ glycine-HCl buffer (pH 2.5) does not cause leakage of DNP from the matrix nor does storage of DNP-Sepharose CL-4B in PBS at 4°C.

Direct coupling of tobramycin to FMP-activated Sepharose CL-4B results in an affinity matrix which efficiently purifies antitobramycin antibodies from rabbit antisera. The elution patterns are shown in Fig. 9 (Ngo, 1986a). Since no antibody was eluted by the high-salt-concentration buffer, the possibility of antibody retention on the gel by mechanisms involving ionic interactions can be ruled out. The selective recognition of the antibody is thus an affinity biospecific process.

The coupling of human IgG to FMP-activated Sepharose CL-4B resulted in an affinity gel having 2.9 mg of human IgG covalently bound to the gel. This IgG gel was shown to be a good affinity sorbent for purifying anti-human IgG. Using an affinity-purified IgG to test the efficiency of the affinity gel, the recovery of purified IgG was 94% (Ngo, 1986a).

α-Casein-Sepharose and α-casein-Fractogel are effective adsorbents for immuno-

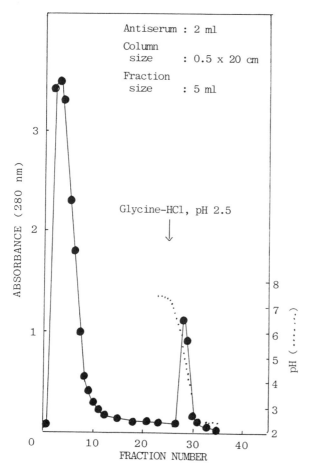

FIGURE 8. Affinity purification of rabbit antiserum to 2,4-dinitrophenyl (DNP)-bovine serum albumin using 2,4,-DNP-conjugated Sepharose CL-4B as the affinity matrix.

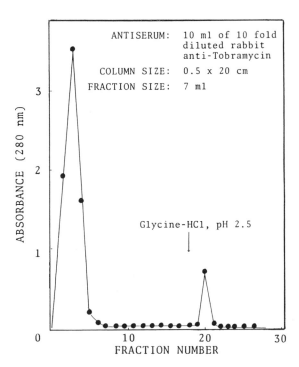

ANTISERUM: 10 ml of 10 fold
             diluted rabbit
             anti-Tobramycin

COLUMN SIZE: 0.5 x 20 cm

FRACTION SIZE: 7 ml

Glycine-HCl, pH 2.5

FIGURE 9. Affinity purification of rabbit antitobramycin in serum using tobramycin directly conjugated to Sepharose CL-4B as the affinity matrix.

affinity purification of rabbit anti-bovine α-casein antibody. Each milliliter of α-casein antibody bound to Sepharose CL-2B was able to bind 0.6 mg of specific antibody (Fig. 10).

Protein A bound to FMP-activated Fractogel can be used to isolate monoclonal and polyclonal antibodies. Protein A is a 42,000-Da protein that binds the Fc region of immunoglobulins, especially the IgGs of most mammalian species. One milliliter of protein A gel can bind approximately 16–18 mg of human IgG, 8–9 mg of murine IgG, 8–12 mg of rabbit IgG, 2–4 mg of goat IgG, and 8–12 mg of bovine IgG.

Katari *et al.* (1990) as well as Sheer *et al.* (1988) have reported on the isolation of the tumor-associated antigen TAG-72 directly secreted from effusions of ovarian, colorectal, pancreatic, and endometrial carcinoma patients. The affinity adsorbent used was mono-clonal antibody CC 49 covalently linked to FMP-activated Fractogel and resulted in a 300–800-fold purification of TAG-72. Yoneda *et al.* (1988) have reported on the purification of antibodies specific to the synthetic peptides DDDED and EEEDE on an affinity matrix consisting of these peptides covalently linked to FMP-activated Fractogel. Monosized superparamagnetic polymer particles containing secondary hydroxyls at the surface have been activated with FMP and used to chemically couple to IgG for use in immunomagnetic cell separation processes (Ugelstad and Nustad, 1989).

3.11. Second-Generation Immobilization Techniques: Affinity-Directed Immobilization of Ligands

Current techniques for immobilizing macromolecules to a solid support do not provide a means for binding a particular site of the macromolecule to the solid support. Instead, the ligands are randomly immobilized. In such cases, the immobilized molecules may have

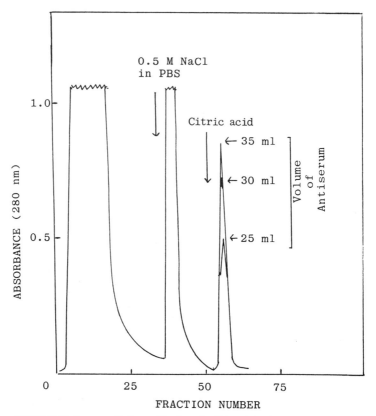

FIGURE 10. Immunoaffinity purification of rabbit anti-bovine α-casein antibody.

their active/binding sites facing the support or even bound to the activated support, thus rendering the immobilized systems ineffective as biocatalysts or affinity matrices.

One approach currently under investigation to minimize this type of problem is the use of the affinity-directed immobilization technique. In the case of FMP-activated supports, the method requires the synthesis of a unique FMP-guide conjugate which can serve a dual function. Firstly, it should serve as an analog of FMP capable of activating OH groups on the solid support. Secondly, it should serve as a guiding ligand (G) capable of specific binding interactions with the receptor ligand (L). The overall concept is summarized in Fig. 11. It involves the activation of the OH group of the solid support with the FMP-guide conjugate (reaction a) followed by addition of the receptor ligand (L) to the activated gel (reaction b), which allows the binding of the receptor ligand to the guiding molecule portion of the conjugate. Simultaneously, an amino or any other nucleophilic group of the receptor ligand reacts with the activated hydroxyl group of the support and results in the displacement of the FMP moiety of the conjugate from the support (reaction c). Meanwhile, the conjugate is still firmly bound to the receptor ligand and can subsequently be dissociated from the receptor by using an eluent (reaction d).

Rines *et al.* (1989) have reported on the use of DNP-CMP [5-amino-2-chloro-*N*-(2,4-dinitrophenyl)-1-methylpyridinium tosylate]-activated Sepharose for the purification of

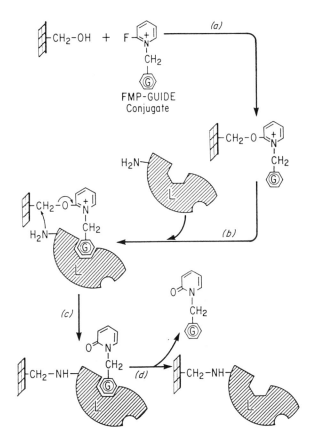

FIGURE 11. Second-generation immobilization techniques: affinity-directed
immobilization of ligands.

anti-DNP antibodies. The avidin/α-lipoic acid pair is currently being used in our laboratories as a model system to demonstrate this concept of an affinity-directed specific
immobilization. Here, FMP will be covalently linked together with a derivative of α-lipoic
acid to form the FMP-guide conjugate.

4. CONCLUSIONS

FMP is an efficient and versatile agent for activating polymeric hydroxyl groups for
affinity chromatography and enzyme immobilization. It leads to high densities of activation
and coupling capacities. The coupling reactions can be carried out under relatively mild
conditions at 4°C or at room temperature at a pH in the range 6–10; activation with FMP
can be carried out in anhydrous organic solvents; linkages between the matrix and the
coupled ligands are stable bonds (secondary amine or thioether) without the introduction of
charges; FMP-guide conjugates can be prepared and used in affinity-directed ligand
immobilization; and FMP is less hydroscopic, less expensive, less corrosive, and easier to
handle than the widely used tresyl chloride. The latter advantage is of special importance in
relation to industrial, large-scale preparation of preactivated supports.

These advantages associated with the use of FMP make it a very viable alternative to the widely used cyanogen bromide, which is highly toxic and produces unstable linkages between ligand and support. Also, recent studies by Miron and Wilchek (1987) have convincingly demonstrated severe limitations (instability and high rate of leakage of immobilized ligands) of several commercially available gels that use *N*-hydroxysuccinamide esters in the activation process which severely limit their use in affinity chromatography and protein immobilization.

REFERENCES

Bald, E., Saigo, K., and Mukaiyama, T., 1975, A facile synthesis of carboxamides by using 1-methyl-2-halopyridinium iodides as coupling reagents, *Chem. Lett.* **1975**:1163–1166.

Balladin, D., 1993, The kinetics and applications of immobilized xanthine oxidase, Ph.D. Thesis, University of the West Indies, St. Augustine, Trinidad and Tobago.

Bamberger, U., Scheuber, P. H., Sailer-Kramer, B., Bartsch, K., Hartman, A., Beck, G., and Hammer, D. K., 1986, Anti-idiotypic antibodies that inhibit immediate type skin reactions in unsensitized monkeys on challenge with staphylococcal enterotoxin, *Proc. Natl. Acad. Sci. U.S.A.* **83**:7054–7058.

Bethell, G. S., Ayers, J. S., Hearn, T. W., and Hancock, W. S., 1981, Investigation of the activation of cross-linked agarose with carbonylating reagents and the preparation of matrices for affinity chromatography purifications, *J. Chromatogr.* **219**:353–359.

Brandt, J., Anderson, L.-O., and Porath, J., 1975, Covalent attachment of proteins to polysaccharide carriers by means of benzoquinone, *Biochim. Biophys. Acta* **386**:196–202.

Butler, P. E., Fairhurst, D., and Beynon, R. J., 1984, Purification of glycogen phosphorylase from small quantities of mouse skeletal muscles, *Anal. Biochem.* **141**:494–498.

Coulet, P. R., Sternberg, R., and Trevenot, D. R., 1980, Electrochemical study of reactions at interfaces of glucose oxidase collagen membranes, *Biochim. Biophys. Acta* **612**:317–327.

Dale, B. E., and While, D. H., 1979, Degradation of ribonucleic acid by immobilized ribonuclease, *Biotechnol. Bioeng.* **21**:1639–1648.

Davis, G., 1991, The kinetics and applications of immobilized β-galactosidase, Ph.D. Thesis, University of the West Indies, St. Augustine, Trinidad and Tobago.

Gribnau, T. C. J., 1978, Alternatives to the cyanogen bromide activation for ligand immobilization on agarose, in: *Chromatography of Synthetic and Biological Polymers*, Vol. 2 (R. Epton and E. Horwood, eds.), The Chemical Society, London, pp. 258–264.

Hannibal-Friedrich, O., Chun, M., and Sernetz, M., 1980, Immobilization of β-galactosidase, albumin and γ-globulin on epoxy activated acrylic beads, *Biotechnol. Bioeng.* **22**:157–175.

Hatchikian, E. C., and Monsan, P., 1980, Highly active immobilized hydrogenase from *Desulfovibrio gigas*, *Biochem. Biophys. Res. Commun.* **92**:1091–1096.

Hearn, T. W., 1986, Application of 1;1′-carbonyldiimidazole activated matrices for purification of proteins, *J. Chromatogr.* **376**:245–257.

Hojo, K., Yoshino, H., and Mukaiyama, T., 1977, New synthetic reactions based on 1-methyl-2-fluoropyridinium salts: Stereospecific preparation of thioalcohols from alcohols, *Chem. Lett.* **1977**:437–440.

Hornby, W. E., and Goldstein, L., 1976, Immobilization of enzymes on nylon, *Methods Enzymol.* **44**:118–134.

Inman, J. K., and Dintzis, H. M., 1969, Derivatization of cross-linked polyacrylamide beads. Controlled introduction of functional groups for the preparation of special-purpose biochemical absorbents, *Biochemistry* **8**:4074–4082.

Jagendorf, A. T., Pathornik, A., and Sela, M., 1963, Use of antibody bound to modified cellulose as an immuno-specific absorbent of antigens, *Biochim. Biophys. Acta* **78**:516–528.

Katari, R. S., Fernsten, P. D., and Schlom, J., 1990, Characterization of the shed form of human tumor-associated glycoproteins (TAG-72) from serous effusions of patients with different types of carcinomas, *Cancer Res.* **50**:4885–4890.

Kato, Y., Komiya, K., Iwaeda, T., Sasaki, H., and Hashimoto, T., 1981, Packing of Toyopearl column for gel filtration, *J. Chromatogr.* **205**:185–188.

Kennedy, J. F., Humphreys, J. D., and Baker, S. A., 1981, Zirconium hydroxide-mediated alkaline hydrolysis of phosphate esters, *Enzyme Microb. Technol.* **3**:129–136.

Kohn, J., and Wilchek, M., 1983, New approaches for the use of cyanogen bromide and related cyanylating agents for the preparation of activated polysaccharide resins, in: *Affinity Chromatography and Biological Recognition* (I. Chaiken, M. Wilchek, and I. Parikh, eds.), Academic Press, New York, pp. 197–207.

Kohn, J., Lengder, R., and Wilchek, M., 1983, *p*-Nitrophenylcyanate—an efficient, convenient, and nonhazardous substitute for cyanogen bromide as an activating agent for Sepharose, *Appl. Biochem. Biotechnol.* **8:**227–235.

Lasch, J., and Koelsch, R., 1978, Enzyme leakage and multipoint attachment of agarose bound enzyme preparation, *Eur. J. Biochem.* **82:**181–186.

Miron, T., and Wilchek, M., 1987, Immobilization of proteins and ligand using chlorocarbonates, *Methods Enzymol.* **135:**84–90.

Mukaiyama, T., Usui, M., Shimada, E., and Saigo, K., 1975, A convenient method for the synthesis of carboxylic esters, *Chem. Lett.* **1975:**1045–1048.

Mungal, R., 1991, The kinetics and applications of free and immobilized urease, M. Phi. Thesis, University of the West Indies, St. Augustine, Trinidad and Tobago.

Narinesingh, D., and Ngo, T. T., 1987, Papain covalently immobilized on Fractogel derivative: Preparation, bioreactor flow kinetics and stability, *Biotechnol. Appl. Biochem.* **9:**450–461.

Narinesingh, D., and Ngo, T. T., 1989, Enzyme amplified potentiometric assay for collagenase using ureaslabelled, solid supported substrate, *Anal. Lett.* **22:**323–338.

Narinesingh, D., and Ngo, T. T., 1990, Solid-supported single- and multi-enzyme amplified assays for clostridiopeptidase A, *Anal. Chim. Acta* **232:**273–280.

Narinesingh, D., and Ngo, T. T., 1991, Combining low pressure liquid affinity chromatography and flow injection analysis for the quantitation of immunoglobulins, *Anal. Lett.* **24:**2147–2155.

Narinesingh, D., Jaipersad, D., and Chang-Yen, I., 1988, Immobilization of linamarase and its use in the determination of bound cyanide in cassava using flow injection analysis, *Anal. Biochem.* **172:**89–95.

Narinesingh, D., Stoute, V. A., Shaama, F., and Ngo, T. T., 1989, The bioreactor flow kinetics of covalently immobilized trypsin on Fractogel support, *J. Mol. Catal.* **54:**147–164.

Narinesingh, D., Mungal, R., and Ngo, T. T., 1990a, Flow injection analysis of serum urea using urease covalently immobilized on 2-fluoro-1-methylpyridinium salt-activated Fractogel and fluorescence detection, *Anal. Biochem.* **188:**325–329.

Narinesingh, D., Pope, A., and Ngo, T. T., 1990b, Rapid, sensitive fluorimetric assay for proteases using fluorescein-labelled albumin coupled to Fractogel activated with 2-fluoro-1-methylpyridinium salt in flow injection analysis, *Anal. Chim. Acta* **230:**131–136.

Narinesingh, D., Mungel, R., and Ngo, T. T., 1991a, Urease coupled to poly(vinylalcohol) activated by 2-fluoro-1-methylpyridinium salt; preparation of a urease potentiometric electrode and application to the determination of urea in serum, *Anal. Chim. Acta* **249:**387–393.

Narinesingh, D., Stoute, V. A., Davis, G., and Ngo, T. T., 1991b, Flow injection analysis of lactose using covalently immobilized β-galactosidase, mutarotase and glucose oxidase/peroxidase on 2-fluoro-1-methylpyridinium salt-activated Fractogel support, *Anal. Biochem.* **194:**16–24.

Narinesingh, D., Stoute, V. A., Davis, G., Shaama, F., and Ngo, T. T., 1991c, Combining flow injection analysis and immobilized enzymes for rapid and accurate determination of serum glucose, *Anal. Lett.* **24:** 727–747.

Narinesingh, D., Pope, A., and Ngo, T. T., 1992, Flow injection analysis of human serum albumin using a stabilized immobilized Cibacron Blue support, *Anal. Lett.* **25:**1721–1739.

Ngo, T. T., 1986a, Facile activation of Sepharose hydroxyl groups by 2-fluoro-1-methylpyridinium toluene-4-sulphonate: Preparation of affinity and covalent chromatographic matrices, *Bio/Technology* **4:**134–137.

Ngo, T. T., 1986b, U.S. Patent 4,582,875 issued to Bioprobe International Inc., Tustin, California.

Ngo, T. T., 1991, Facile activation of Trisacryl gels with 2-fluoro-1-methylpyridinium salt (FMP): Applications in affinity chromatography and enzyme immobilization, in: *Cosmetic and Pharmaceutical Applications of Polymers* (C. G. Gebelein, T. C. Cheng, and V. C. Yang, eds.), Plenum Press, New York, pp. 385–398.

Nilsson, K., and Mosbach, K., 1981, Immobilization of enzymes and affinity ligands to various hydroxy group carrying supports using highly reactive sulphonyl chlorides, *Biochem. Biophys. Res. Commun.* **102:**449–457.

Porath, J., and Axen, R., 1976, Immobilization of enzymes to agar, agarose and Sephadex supports, *Methods Enzymol.* **44:**19–45.

Porath, J., Laas, T., and Janson, J.-C., 1975, Rigid agarose gels cross-linked with divinyl sulphone, *J. Chromatogr.* **103:**49–62.

Ramtahal, E., 1992, Amperomeric biosensing and quantitation of Cu(II) ions using immobilized galactose oxidase and flow injection analysis, Final year analytical chemistry project, University of the West Indies, St. Augustine, Trinidad and Tobago.

Rines, R., Mulder, A. H. L., Scouten, W. H., and Loy, R., 1989, The use of dyes in affinity ligand and protein immobilization, in: *Protein–Dye Interactions: Development and Applications* (M.A. Vijayalakshmi and O. Bernard, eds.), Elsevier Applied Science, Amsterdam, pp. 137–146.

Scouten, W. H., and van der Tweel, W., 1984, Coloured sulphonyl chlorides as activating agents for hydroxylic matrices, *Ann. N.Y. Acad. Sci.* **434:**249–254.

Sheer, D. G., Schlom, J., and Cooper, H. L., 1988, Purification and composition of human tumor-associated glycoprotein (TAG-72) defined by monoclonal antibodies CC 49 and B 72.3, *Cancer Res.* **48:**6811–6818.

Solomon, B., Koppel, R., Pines, G., and Katchalski-Katzir, E., 1986, Enzyme immobilization via monoclonal antibodies: Preparation of a highly active immobilized carboxypeptidase, *Biotechnol. Bioeng.* **28:**1213–1221.

Sundberg, L., and Porath, J., 1974, Attachment of group containing ligands to insoluble polymers by means of bifunctional oxiranes, *J. Chromatogr.* **90:**87–98.

Turkova, J., 1976, Immobilization of enzymes on hydroxyalkyl methacrylate gels, *Methods Enzymol.* **44:**82–83.

Ugelstad, J., and Nustad, K., 1989, New developments in preparation and application of monosized polymer particles, in: *Protein Recognition of Immobilized Ligands* (T. W. Hutchens, ed.), Alan R. Liss, New York, pp. 267–275.

Ullah, A. H. J., and Cummings, B. J., 1987, Immobilization of *Aspergillus ficuum* extracellular phytase on Fractogel, *Biotechnol. Appl. Biochem.* **9:**380–388.

Ullah, A. H. J., and Cummins, B. J., 1988, Extracellular pH 2.5 optimum acid phosphatase from *Aspergillus ficuum.* Immobilization on modified Fractogel, *Prep. Biochem.* **18:**473–481.

Ullah, A. H. J., and Phillippy, B. Q., 1988, Immobilization of *Aspergillus ficuum* phytase: Product characterization of bioreactor, *Prep. Biochem.* **18:**483–489.

Wilchek, M., and Miron, T., 1982, Immobilization of enzymes and affinity ligands on to agarose via stable and uncharged carbamate linkages, *Biochem. Int.* **4:**629–635.

Yoneda, Y., Imamoto-Sonobe, N., Matsuoka, Y., Iwamoto, R., Kiho, Y., and Uchida, T., 1988, Antibodies to Asp-Asp-Glu-Asp can inhibit transport of nuclear proteins into the nucleus, *Science* **242:**275–278.

5

Slalom Chromatography

A Size-Dependent Separation Method for DNA Molecules Based on a Hydrodynamic Principle

Jun Hirabayashi and Ken-ichi Kasai

1. INTRODUCTION

High-performance liquid chromatography has rarely been applied to size-dependent separation of relatively large DNA molecules though it has greatly contributed to research on other biomolecules. This is mainly because of the low efficiency of gel permeation chromatography in comparison with gel electrophoresis, which has been the most popular and efficient procedure for the separation of DNA fragments according to their sizes. One of the limitations of gel permeation chromatography is that separation must take place within a small volume, that is, the volume of the liquid retained in the stationary phase, or, in other words, the total volume of the pores of the packing particles. Moreover, it is essential that macromolecules diffuse into the pores of the packing particles and that equilibrium is maintained between the moving phase and the stationary phase. This has made it difficult to increase the number of theoretical plates under relatively fast flow conditions. Nevertheless, gel permeation chromatography is indispensable for size separation of native proteins because of the lack of any appropriate alternative.

In addition to the poor efficiency in comparison with electrophoresis, there are other reasons why application of high-performance gel permeation chromatography to size-dependent separation of DNA has not been preferred. The possible risk of cleavage of long DNA molecules due to shear force generated by laminar flow has also prevented the use of chromatography. DNA molecules have also been believed to be too large for conventional gel permeation media. Even small DNA fragments have Stokes radii considerably larger than those of proteins of similar molecular weight. Is there any separation mode other than gel permeation?

Recently, we discovered by chance that double-stranded DNA molecules could be

Jun Hirabayashi and Ken-ichi Kasai • Department of Biological Chemistry, Faculty of Pharmaceutical Sciences, Teikyo University, Sagamiko, Kanagawa 199-01, Japan.

Molecular Interactions in Bioseparations, edited by That T. Ngo. Plenum Press, New York, 1993.

separated according to their sizes by using a system for high-performance gel permeation chromatography, though the order of elution was opposite to that expected for gel permeation chromatography; that is, larger fragments were eluted later than smaller ones (Hirabayashi and Kasai, 1988, 1989; Hirabayashi et al., 1990). A similar observation was independently made by Boyes et al. (1988). The separation significantly depended on the flow rate and the particle size of the packings but did not depend on the pore size or chemical nature of the packings. This phenomenon could not be explained in terms of any separation mode previously known. Therefore, we devised the model shown in Fig. 1.

The column, which is closely packed with hard and spherical beads, has narrow and tortuous open spaces. When DNA molecules are applied to the column, they are unfolded and extended due to laminar flow generated by the solvent passing through the narrow channels. These extremely extended molecules must flex quickly under a fast flow of moving phase to pass through the openings. If we use a 30-cm column packed with particles of 10-μm diameter, a DNA molecule should turn as many as 36,000 times because the

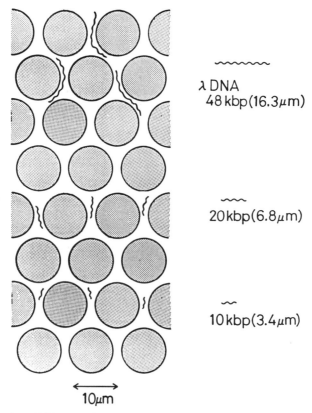

λ DNA
48 kbp (16.3 μm)

20 kbp (6.8 μm)

10 kbp (3.4 μm)

10μm

FIGURE 1. Illustration of DNA separation in slalom chromatography. DNA fragments are extended due to the occurrence of laminar flow. They turn frequently in passing through the narrow and tortuous openings between closely packed spherical particles (indicated by shaded particles). The larger the fragments, the more difficult it is for them to turn around the particles. If the DNA molecules are extended to the maximum extent, their length will be comparable to the diameter of the packing particles. The distance between particles is exaggerated. [Reprinted with permission from Hirabayashi et al. (1990).]

number of layers of particles reaches 1200 per centimeter of column length. It is quite possible that the longer the DNA molecule, the more difficulty it encounters in passing through the openings. Therefore, size-dependent separation in the opposite order to that in gel permeation chromatography should occur. We named this new mode of separation "slalom chromatography," because the proposed model reminds us of a skier who goes down a slope turning quickly around flags. The results of all our experiments showed that this explanation is satisfactory as a first approximation for the mechanism of separation.

2. DISCOVERY OF A NOVEL MODE OF CHROMATOGRAPHY

In the course of purification by high-performance liquid chromatography of a recombinant λ phage DNA, which was cloned from λgt11 cDNA library of human placenta, we found a curious phenomenon (Hirabayashi and Kasai, 1988, 1989). The DNA extracted from the recombinant phage was contaminated with a considerable amount of RNA. Since RNA often inhibits restriction enzymes, we attempted to remove it by high-performance gel permeation chromatography on a column packed with polymer-based packing (Asahipak GS-510; 9-μm diameter; exclusion limit for proteins, 3×10^5 Da; column size, 7.6 × 500 mm; Asahi Chemical Industry Co. Ltd). The resulting chromatogram is shown in Fig. 2A.

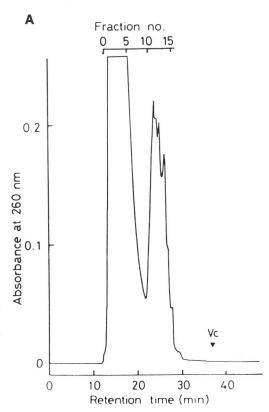

FIGURE 2. (A) Chromatography of crude λ phage DNA on an Asahipak GS-510 column (7.6 × 500 mm) equilibrated with 0.1M Tris-HCl, pH 7.5, 0.2M NaCl, and eluted with the same buffer at a flow rate of 0.6 ml/min. (B) Agarose gel electrophoresis of fractionated DNA and RNA. M, Size marker (λ/HindIII); λ, λDNA (wild type, 48.5 kbp). [Reprinted with permission from Hirabayashi and Kasai (1989).] (Continued)

B

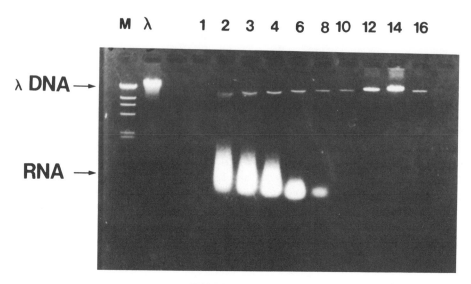

FIGURE 2. (*Continued*)

We expected that the recombinant λ DNA would be eluted in the first peak (retention time 12–20 min) and be separated from the smaller RNA eluted later (after 22 min). However, agarose gel electrophoresis of the separated fraction revealed that, contrary to our expectation, the contaminating RNA was eluted in the first peak (Fig. 2B, fractions 2 to 8). On the other hand, the recombinant λ DNA was eluted much later (fraction 14 in Fig. 2B). The size of the RNA was no more than 2 kilobase (kb) (6×10^5 Da) while that of the recombinant λ DNA was approximately 45 kbp (2.9×10^7 Da). Thus, this result could not be explained in terms of the known separation mechanism of gel permeation chromatography.

Figure 2B also shows that each fraction contained a small amount of DNA fragments slightly smaller than the recombinant DNA, which might be degradation products of the intact λ DNA. Their size increased with increasing retention time. On the other hand, the contaminating RNA was separated on the basis of normal gel permeation; that is, the smaller the size, the greater the retardation. This finding called for more extensive studies on the DNA separation. Therefore, mixtures of DNA fragments that could be used for calibration were prepared and applied to various columns. Figure 3A shows the elution profile of the *Hin*dIII digest of λ DNA, which contains eight fragments (0.13, 0.56, 2.03, 2.32, 4.36, 6.56, 9.42, and 23.13 kbp). Three peaks appeared when chromatography was carried out in the absence of EDTA.

Size analysis by gel electrophoresis (Fig. 3B) showed that the first peak (peak a) contained all seven fragments smaller than 9.5 kbp. The second peak (peak b) contained the 23-kbp fragment. An additional peak (peak c) did not contain any fragment longer than 23 kbp but contained two fragments, 23 kbp and 4.4 kbp. This was quite reasonable because

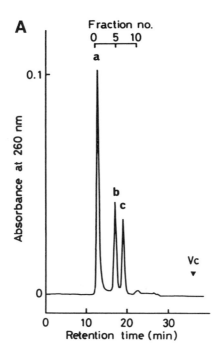

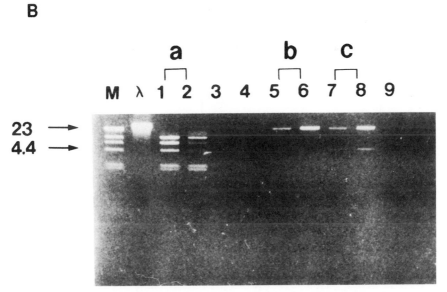

FIGURE 3. (A) Chromatography of *Hin*dIII digest of λDNA on an Asahipak GS-510 column. (B) Agarose gel electrophoresis of fractionated λ/*Hin*dIII digest. M, Size marker (λ/*Hin*dIII); λ, λDNA (wild type, 48.5 kbp). [Reprinted with permission from Hirabayashi and Kasai (1989).]

these two fragments were derived from both ends of the λ DNA and have cohesive ends (*cos* site), which are single-stranded stretches of 12 nucleotides complementary to each other (Sanger *et al.*, 1982). Thus, they must have formed a complex and behaved as a longer fragment having a size of 27 kbp. In the presence of EDTA, this peak disappeared. It is well known that formation of the double strand between the cohesive ends is unfavorable in the absence of Mg^{2+} ions. Thus, the column proved to be able to fractionate DNA fragments larger than 10 kbp in the opposite order to that in gel permeation chromatography.

3. EXAMINATION OF EFFECTS OF VARIOUS CHROMATOGRAPHIC FACTORS ON DNA SEPARATION

3.1. Effect of the Particle Size of the Column Packing

The unexpected nature of the separation described above suggested that it is based on a new principle that would be affected by the physical properties of column packings. Therefore, the effect of the particle size of the packing was first investigated (Hirabayashi and Kasai, 1988, 1989). Figure 4 shows the separation patterns of a mixture of eight DNA fragments, ranging from 10 to 38 kbp, on two different columns. One was packed with 5-μm particles (Fig. 4A, Asahipak GFT-510), and the other with 9-μm particles (Fig. 4B, Asahipak GS-510). They are both made of a synthetic polymer and have the same fractionation range as high-performance gel permeation chromatography media (exclusion

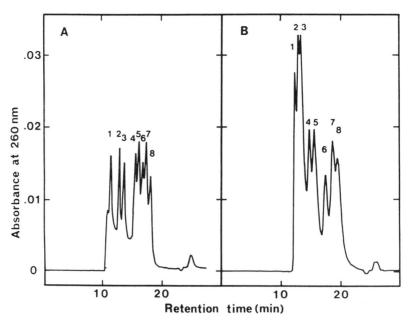

FIGURE 4. Chromatography of restriction fragments of λDNA on Asahipak GFT-510 (5 μm; A) and GS-510 (9 μm; B). Column sizes were the same (7.6 × 250 mm). A set of DNA fragments was applied. Flow rate was 0.3 ml/min. Fragment size: 1, 10.09 kbp; 2, 15.00 kbp; 3, 17.05 kbp; 4, 22.63 kbp; 5, 24.77 kbp; 6, 29.95 kbp; 7, 33.50 kbp; 8, 38.42 kbp. [Reprinted with permission from Hirabayashi and Kasai (1989).]

limit for proteins, 3×10^5 Da). A buffer of low ionic strength containing EDTA was used for most experiments to prevent undesirable interaction between fragments carrying complementary cohesive ends.

Both columns separated eight fragments sharply according to their sizes. The order of elution was again opposite to that expected for normal gel permeation chromatography. Thus, this phenomenon was again confirmed to be based on a novel mode of separation.

Although the pore sizes are the same for the two columns, the column packed with 5-μm particles proved to be superior for the separation of smaller fragments (<20 kbp), while larger fragments (>20 kbp) were better separated by the column packed with 10-μm particles. This result suggested that the fractionation range depends on the particle size of the column packing, or, more properly, on the size of the openings or gaps between particles in the column.

Next, more systematic experiments were carried out (Hirabayashi *et al.*, 1990). Four columns packed with packings of different sizes (average particle diameter, 5.0, 9.0, 13.1, and 19.1 μm; Asahipak GS-310 series) were used. Chromatograms of λ/*Hind*III digests obtained for columns packed with 5-μm and 9-μm packings are shown as examples in Fig. 5. It is evident that when the 5-μm packing was used and a flow rate of 1.2 ml/min applied, resolution of the fragments of less than 10 kbp was considerably improved.

To compare the data obtained on different columns, a relative retention time (R_{RT}) was calculated for each DNA fragment as follows:

$$R_{RT} = R/R_0$$

where R is the retention time of a fragment, and R_0 is that corresponding to the void fraction for each column. R_0 was experimentally determined by using *Hind*III fragments of λ DNA

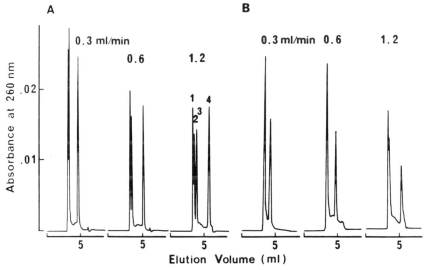

FIGURE 5. Chromatography of λ/*Hind*III digests on Asahipak GS-310 columns packed with 5-μm (A) and 9-μm packings (B). DNA fragments were eluted at a flow rate of 0.3, 0.6, or 1.2 ml/min. Fragments contained in peak 1 are 0.13, 2.03, 2.32, and 4.36 kbp. Peak 2, 3, and 4 correspond to 6.56, 9.42, and 23.13 kbp, respectively. [Reprinted with permission from Hirabayashi *et al.* (1990).]

smaller than 4.4 kbp, which are always eluted in the void volume. The data are summarized in Fig. 6.

In Fig. 6, the relative retention times are plotted versus DNA length. Size-dependent separation in an order opposite to that expected for gel permeation occurred in all columns examined. Figure 6 also shows that the four columns have different ranges of resolution; for example, at a flow rate of 0.6 ml/min (middle curves in Fig. 6A–D), the 5-, 9-, 13-, and 19-μm columns resolved DNA fragments larger than 7, 9, 13, and 17 kbp, respectively, from the fragments that appeared in the void volume ($R_{RT} = 1.0$). In addition, smaller packings showed better resolution for smaller DNA fragments, while larger ones were better for larger fragments. Thus, high-resolution zones can be assigned as 9–17 kbp, 15–30 kbp, 23–40 kbp, and 35–50 kbp for the 5-, 9-, 13-, and 19-μm packings, respectively, in the case of a flow rate of 0.6 ml/min. Each curve can apparently be divided into three zones according to the magnitude of the slope, that is, a poor-resolution zone, a high-resolution zone, and a moderate-resolution zone from left to right in Fig. 6. The high-resolution zone (middle zone) shows the steepest slope. In the third or final zone, the relationship between relative retention time and DNA length seems to be almost linear. Thus, each column

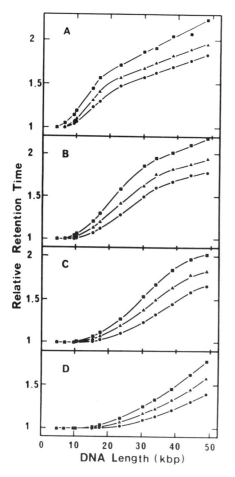

FIGURE 6. Dependence of relative retention times of DNA fragments on their lengths. Packings of different particle size were used, and different flow rates were applied. Relative retention time (R_{RT}) is plotted as a function of DNA length (kbp). Asahipak GS-310 columns were used. Particle sizes are 5 (A), 9 (B), 13 (C), and 19 μm (D). Flow rates are 0.3 (●), 0.6 (▲), and 1.2 ml/min (■). [Reprinted with permission from Hirabayashi et al. (1990).]

packed with a particular size of packing particles has an optimum size range for DNA resolution.

3.2. Effects of Flow Rate

A lower flow rate is usually preferable to obtain a higher number of theoretical plates in chromatography because equilibrium between the stationary phase and the moving phase should be maintained. In gel permeation chromatography, flow rate does not affect the range of fractionation size. However, in the present novel mode of chromatography, the size range of DNA fragments that could be well separated depended on the flow rate. Therefore, flow rate dependency was examined in greater detail by using columns of different particle sizes (Fig. 7) (Hirabayashi et al., 1990).

Application of a higher flow rate resulted in a significant increase in the relative retention time of DNA for all columns (Fig. 7). This enabled us to obtain better resolution of a fragment of particular interest from the void fraction, though it is improbable that a higher flow rate increases the number of theoretical plates. On the other hand, at the lowest flow rate applied (0.03 ml/min, corresponding to a linear flow rate of 0.067 cm/min), all the DNA fragments were eluted almost in the void fraction (Fig. 7B). These results strongly suggest that when a higher flow rate is applied, DNA molecules are more extended and consequently more retarded. It is notable that there are two zones in the curves; that is, up to a flow rate of 0.3 ml/min, relative retention time increased steeply with the increase of flow rate, while after that, the increase was less steep and almost linear. The existence of such highly flow rate-dependent zones seems to correspond to drastic conformational changes of DNA molecules occurring in a certain range of flow rates. This point will be discussed later. Change in flow rate does not seem to shift the high-resolution zone characteristic of each column.

3.3. Effect of Pore Size

Next, the effect of pore size of the packing was examined (Hirabayashi and Kasai, 1989; Hirabayashi et al., 1990). Columns packed with packings of the same particle size

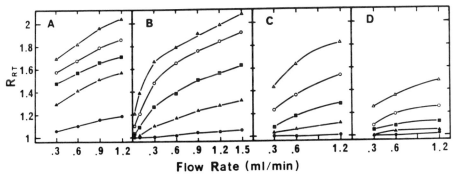

FIGURE 7. Relationship between retardation of the fragments and flow rate. Relative retention time (R_{RT}) is plotted as a function of flow rate. Asahipak GS-310 columns were used. Particle sizes are 5 (A), 9 (B), 13 (C), and 19 μm (D). The fragments used were 10.09 (●), 17.05 (▲), 23.13 (■), 29.95 (○), and 38.42 kbp (△). [Reprinted with permission from Hirabayashi et al. (1990).]

(9 μm) but differing in pore size (Asahipak GS-220, -310, -320, -510, and -710) were used. GS-310 and GS-320 are supposed to have the same pore size but differ subtly in chemical nature. No substantial difference in relative retention times for five DNA fragments was found among the columns except for GS-710 (Table 1). This indicated that pore size does not affect the separation, suggesting that DNA fragments do not permeate into the pores. On the other hand, the relative retention times obtained for the GS-710 column were somewhat short. This may be explained by partial permeation of some fragments into pores of the packing, because GS-710 has the widest pores among the packings compared (exclusion size determined for pullulan, 1×10^7 Da). The use of packings having extremely large pores seems to be undesirable for slalom chromatography. In such a case, dual modes of separation (i.e., gel permeation mode and slalom mode) would occur at the same time, resulting in a complex separation pattern.

3.4. Effect of the Chemical Nature of the Packings

The results described above were obtained by using only columns packed with polymer particles. It is extremely important to verify that slalom mode separation also occurs in columns containing packing other than polymer particles. Therefore, columns of TSK G2000SW and TSK G3000SW (Tosoh Co. Ltd.), both packed with silica gel particles of 10-μm diameter, were used (Hirabayashi et al., 1990). Although the two columns have different exclusion sizes (G2000SW, 1×10^5 Da for dextran; G3000SW, 5×10^5 Da), both the separation patterns and the flow rate dependency obtained with these columns were almost the same (Fig. 8B), and also very similar to those obtained for the column packed with 9-μm polymer particles (Fig. 8A, broken line). Thus, it was confirmed that the separation pattern depends on the particle size and flow rate, but not on the pore size or chemical nature of the packing.

3.5. Effect of Salt and Organic Solvent

There remained the possibility that larger DNA fragments interact more strongly with the packing matrix, for example, by electrostatic or hydrophobic interaction. Therefore, the effect of addition of salt or organic solvent was examined (Hirabayashi and Kasai, 1989;

TABLE 1

Comparison of Relative Retention Times (R_{RT}) of Several DNA Fragments Eluted from Columns (7.6 × 250 mm) Packed with 9-μm Particles of Different Pore Size at a Flow Rate of 0.6 ml/min

Fragment (size in kbp)	R_{RT} for column packing:				
	GS-220	GS-320	GS-310	GS-510	GS-710
10.09	1.01	1.02	1.03	1.01	1.00
17.09	1.17	1.16	1.19	1.14	1.06
23.13	1.39	1.36	1.42	1.34	1.19
29.95	1.62	1.59	1.66	1.57	1.32
38.42	1.74	1.75	1.82	1.71	1.49

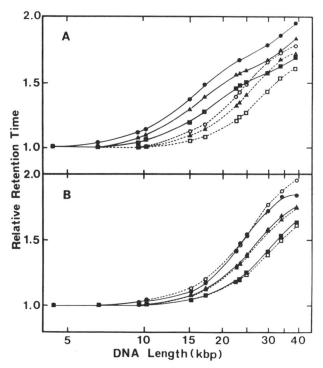

FIGURE 8. Comparison of polymer-based columns and silica gel-based columns. Relative retention time (R_{RT}) is plotted against DNA length (kbp). (A) Elution curves obtained with porous polymer columns Asahipak GFT-510 and Asahipak GS-510 (7.9 × 250 mm) are shown by solid lines and broken lines, respectively. Flow rates are 0.3 (■, □), 0.6 (▲, △), and 0.9 ml/min (●, ○). (B) Elution curves obtained with silica gel-based columns TSK G2000SW and TSK G3000 SW (7.8 × 600 mm) are shown by solid lines and broken lines, respectively. Flow rates are 0.3 (■, □), 0.6 (▲, △), and 1.2 ml/min (●, ○). [Reprinted with permission from Hirabayashi and Kasai (1989).]

Hirabayashi *et al.*, 1990). The retention times of DNA fragments were not significantly affected by the addition of up to 0.5M NaCl to the elution buffer. Thus, size-dependent separation of DNA is not attributable to electrostatic interaction. Next, the effect of the addition of an organic solvent to the elution buffer was examined. The addition of up to 20% (v/v) acetonitrile had virtually no effect on the retention times of DNA fragments. These results again confirmed that there is no interaction between the chromatographic media and DNA fragments, and DNA fragments are separated only by a hydrodynamic phenomenon.

3.6. Chromatography on Porous and Nonporous Cation-Exchange Columns

The above results raised the possibility that slalom chromatography could also be attained by using columns other than those for gel permeation chromatography. Since cation-exchange columns are of interest because nucleic acids are not likely to bind, but rather are repelled, two strongly acidic cation-exchange columns bearing sulfopropyl groups, TSK-SP-5PW (7.5 × 75 mm) and TSK-SP-NPR (4.6 × 35 mm) (Tosoh Co. Ltd), were examined (Hirabayashi *et al.*, 1990). The latter is of special interest because it is a column packed with a nonporous polymer having a particle size as small as 2.5 μm (Kato *et al.*, 1987).

Although the columns were very small, they were able to fractionate DNA fragments as well as the longer Asahipak columns (Fig. 9). Curves of elution volume versus DNA size obtained with the SP-5PW column are essentially similar to those obtained with the Asahipak GS-310 column packed with 9-μm packing, reflecting the similar particle sizes. Both of them resolved a DNA fragment of approximately 10 kbp from the void fraction and showed a high-resolution zone in the size range of 15–30 kbp. On the other hand, the curves obtained for the SP-NPR column (Fig. 9B) seem to resemble to some degree those for the Asahipak GS-310 column packed with 5-μm packing. Thus, we were able to demonstrate that slalom chromatography is achievable even with cation-exchange and nonporous packings. This result again confirmed that there is no interaction between DNA molecules and the chromatographic media.

3.7. Effect of Temperature

Temperature had a significant effect on the slalom mode separation. Elution of the λ/*Hind*III digest at various temperatures was analyzed. Fragments smaller than 10 kbp (peak 1 in Fig. 10) were all eluted in the void fraction (approximately at 6 min) at every temperature examined. Peak 4 (small compounds such as salt) was also not very sensitive to temperature. On the other hand, the elution of a 23-kbp fragment (peak 2 in Fig. 10) was strongly temperature-dependent. Its relative retention time significantly increased when the temperature was lowered. Under the conditions employed, some of the 23-kbp and 4.4-kbp fragments formed a complex via *cos* sites, and elution of this complex (27.5 kbp; peak 3

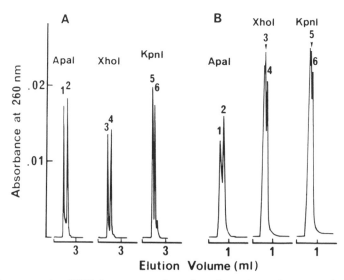

FIGURE 9. Chromatography of DNA fragments on columns for cation-exchange chromatography. The columns used were TSK-SP-5PW (7.5 × 75 mm, 10 μm, porous) (A) and TSK-SP-NPR (4.6 × 35 mm, 2.5 μm, nonporous) (B). Fragment size: 1, 10.09 kbp; 2, 38.42 kbp; 3, 15.00 kbp; 4, 33.50 kbp; 5, 17.05 kbp; 6, 29.95 kbp. Flow rate was 0.29 ml/min (A) or 0.11 ml/min (B). [Reprinted with permission from Hirabayashi *et al.* (1990).]

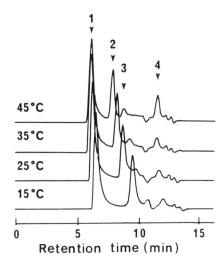

FIGURE 10. Chromatography of the λ/HindIII digest at various temperatures. λ/HindIII fragments were applied to a column of Asahipak GS-310 packed with 9-μm packing. Elution was performed in 0.1M Tris-HCl, pH 7.5, 0.2M NaCl at a flow rate of 0.6 ml/min. Peak 1, fragments smaller than 10 kbp; peak 2, 23.13 kbp; peak 3, 23.13 kbp + 4.36 kbp; peak 4, small compounds such as salt. [Reprinted with permission from Hirabayashi et al. (1990).]

in Fig. 10) was also found to be temperature-dependent. Thus, at lower temperature, retardation of DNA fragments became more significant. Temperature should affect significantly various physicochemical factors such as solvent viscosity, flexibility of DNA, and size of packing (packing ratio) and thus result in altered retardation of DNA molecules.

3.8. Application to Studies on Physicochemical Properties of DNA

The utility of this new procedure is not limited to preparative purposes. This method will undoubtedly provide us valuable information on DNA molecules almost equivalent to the information that gel filtration has provided on the properties of proteins. Not only the sizes of DNA molecules but also physicochemical properties, such as conformation, topology, and rigidity, can be analyzed, because this method is based on a hydrodynamic phenomenon.

Though the experimental results described above were obtained only for linear DNA molecules, the behavior of circular DNA molecules is also interesting. When the circular replicative form of M13 phage DNA (7 kbp) and its linearized form were compared, the former was eluted faster than the latter. This showed that even DNA molecules of the same size can be distinguished, if their conformations are different. Longer circular DNAs were also analyzed. Even at the highest flow rate examined, 42-kbp circular DNA was eluted in a similar elution volume as 20-kbp linear DNA. This result was quite reasonable because the length of circular DNAs in a maximally stretched state will be equal to that of linear DNAs of half size.

Even within circular DNAs, different flow rate dependency was observed for supercoiled, relaxed, and single-stranded DNAs. Flow rate dependency was not marked for both supercoiled and single-stranded DNAs. These results might be attributed to their rigidity or special three-dimensional structure formed by an intrachain double helix. Thus, it will be possible to separate and analyze DNA molecules according to their three-dimensional structures. If the efficiency of slalom chromatography becomes much higher, isolation of topoisomers of supercoiled DNAs according to the extent of coiling might become possible.

4. MECHANISM OF SLALOM CHROMATOGRAPHY

4.1. Characteristics of Slalom Chromatography

From the studies described above, the characteristics of slalom chromatography can be summarized as follows:

1. DNA fragments do not interact with the packing matrix, because polymer-based and silica-based packings gave essentially similar separation patterns, and addition of a salt or organic solvent had little effect. Even anion-exchange columns, which should repel DNA molecules, showed the slalom-mode separation.
2. The pore size of the packing particles is not important.
3. Only the particle size of the packing is important. The size range of DNA fragments that are efficiently separated was found to depend largely on the particle size. Smaller particles could resolve smaller DNA fragments while larger particles could resolve larger DNA fragments.
4. Flow rate has a significant effect on the separation range. At a faster flow rate, relative retention time increased and better resolution was attained.
5. Temperature has a significant effect on the extent of retardation.

4.2. Possible Mechanism of Slalom Chromatography

The characteristics listed above cannot be explained by any separation mode previously known. Therefore, we proposed a new possible mechanism. When DNA molecules are applied to a column for high-performance liquid chromatography, they are stretched because of the laminar flow generated by the solvent passing through the narrow channels between packing particles. For example, the end-to-end distance of the maximally stretched 23-kbp λ/*Hin*dIII fragment is as long as 7.9 μm, which is comparable to the diameter of the packing particle used.

The DNA molecules are too large to permeate into the pores and can only pass through the narrow openings between closely packed spherical particles. Since these channels are extremely tortuous, the molecules have to turn very frequently. It will be more difficult for longer DNA strands to turn quickly, and thus this will result in retardation of longer DNA strands. A simple calculation shows that DNA fragments turn about 36,000 times if applied to a column of 300-mm length packed with 10-μm packing. Under these conditions, the 23-kbp fragment turns as frequently as 70 times per second, when a flow rate of 0.6 ml/min is applied. If a column of 5-μm particles is used at a flow rate of 1.2 ml/min, the fragment has to turn 72,000 times at a frequency of 280 times per second. On each turn, DNA strands should be exposed to a significant frictional force against the solvent, and this force would increase with the increase in DNA length so that larger DNA molecules are retarded (Fig. 1).

As to the effect of flow rate, application of a faster flow rate increases the average end-to-end distance of DNA, and, therefore, DNA retardation becomes greater (Fig. 7). When smaller packing materials are used, DNA strands should turn more frequently in the openings, which would be narrower and more tortuous. Thus, a stronger frictional force should be generated between DNA strands and solvent. The possible occurrence of a steeper flow-rate gradient in the laminar flow may also contribute by extending the DNA molecules more efficiently. These considerations may explain why the use of smaller packings and a higher flow rate is advantageous for the separation of smaller DNA molecules. Although the explanation presented here is undoubtedly oversimplified, it should be a useful model in elucidating the precise mechanism of slalom chromatography.

It was the combination of hard, spherical packings and the application of high flow rates that allowed this new mode of separation to be established. This phenomenon should be investigated in terms of hydrodynamic theory rather than equilibrium kinetics.

A mode of separation without packings, named capillary hydrodynamic chromatography, which is based on a similar mechanism, has also been reported (Noel *et al.*, 1978; Tijssen *et al.*, 1983). The new mode of chromatographic separation described in this chapter is, however, based on a different principle.

4.3. Dynamic Behavior of DNA Molecules

DNA molecules are known to show dynamic behavior in solution. This was first visualized by a method for observing individual DNA molecules under a fluorescence microscope by using a DNA-binding fluorescent probe, 4',6-diamino-2-phenylindole (Yanagida *et al.*, 1982). DNA molecules were observed to have a folded structure in the absence of external force. However, the conformation of DNA in solution is very flexible and is never fixed. Therefore, it fluctuates rapidly due to thermal motion. However, in the presence of laminar flow, the DNA molecule is extended. As the flow rate increases, its shape changes gradually from an ellipsoid to a solenoid, a thick filament, and finally a thin filament (Fig. 11). This observation explains well the flow rate dependency of slalom chromatography.

5. ASPECTS OF SLALOM CHROMATOGRAPHY

5.1. Advantages of Slalom Chromatography

Slalom chromatography can be used for both analytical and preparative purposes. It has advantages in comparison with other experimental procedures for nucleic acids research such as gel electrophoresis. The advantages of slalom chromatography, in terms of both basic and practical aspects, include the following:

1. As an analytical procedure, it provides data of extremely high quality on the size of molecules. Only gel permeation chromatography can provide similar information. Other chromatographic modes such as ion exchange or hydrophobic inter-

FIGURE 11. Dynamic structural change of a DNA molecule in a laminar flow. As the flow rate is increased (from left to right), a DNA molecule having a folded structure which fluctuates rapidly due to thermal motion is gradually extended and finally becomes a thin filament.

action give only limited information about the molecular nature of the target materials.

2. It is also useful for physicochemical studies on DNA, because separation depends not only on the molecular weight but also on the shape of the DNA.
3. It provides a new effective tool for hydrodynamic study of fibrous macromolecules.
4. There is no limitation on the volume in which separation must take place, which is the case in gel permeation chromatography.
5. The experimental procedure is very simple and rapid.
6. Reproducible, accurate, and quantitative data can be easily obtained.
7. Prediction of retention time is easy once the size of the DNA fragment, the particle size of the packing, and the flow rate are known.
8. The solvent system is simple. An isocratic elution system is sufficient, and there is no need for column washing or regeneration.
9. Isolation of DNA fragments can be completed in a shorter time than in the case of agarose gel electrophoresis.
10. Isolated DNA can be recovered immediately because processes such as extraction from agarose gel are not needed.
11. Recovered DNA does not contain any undesirable contaminating materials such as are often present in agarose gel and tend to inhibit restriction endonucleases.
12. DNA can be detected without the use of a harmful reagent such as ethidium bromide.
13. More accurate quantification than with electrophoresis is possible.
14. No special equipment is needed; the ordinary apparatus for high-performance liquid chromatography suffices.
15. No special packing material is needed; ordinary gel permeation media for high-performance liquid chromatography can be used.
16. Chromatography is more amenable to automation than electrophoresis.

5.2. Limitations of Slalom Chromatography

1. The range of sizes of DNA molecules that can be separated is limited. It is primarily determined by the size of the packing particles. At present, from 5- to 50-kbp fragments can be practically separated by using particles from 5 to 20 μm. Although resolution of fragments smaller than 5 kbp may be theoretically possible if extremely small packing particles (e.g., less than 1 μm) are available, application of high flow rates will be practically impossible due to increased back pressure. On the other hand, if large packing particles are used with the aim of separating fragments larger than 50 kbp, a very long column will be required in order to increase the frequency of turning. Therefore, it seems difficult to overcome these limits by using packings presently available. It will be necessary to find other ways in which the slalom effect can be achieved more efficiently. The use of packings of different shape such as coral-type hydroxyapatite or media for perfusion chromatography may be one possibility.

2. The resolution is inferior to that of gel electrophoresis.

3. A relatively high flow rate, which may cause physical degradation of DNA, is required. This presents a dilemma in that a higher flow rate is favorable for better resolution, but selection of a low flow rate is preferable in order to avoid fragmentation of DNA.

However, no significant cleavage of DNA fragments even as long as 50 kbp was observed in almost all experiments described in this chapter. Thus, it is improbable that DNA fragments of less than 50 kbp are extremely unstable under the conditions used (for example, flow rate of 1.2 ml/min).

    4. Damage by ultraviolet light is possible during passage through the flow cell of the UV detector.

5.3. Possible Fields of Application of Slalom Chromatography

Slalom chromatography can be used mainly for DNA studies, but its application is not limited only to DNA. It will provide an efficient tool in the following areas of basic and applied science:

1. Size-dependent separation of DNA.
2. Estimation of size of DNA.
3. Monitoring and analysis of changes in the size of DNA upon, for example, degradation by a restriction enzyme or a smaller chemical species such as bleomycin, polymerization by polymerase, or hybridization of cohesive ends.
4. Separation of DNA based on conformation or topology.
5. Analysis of interaction of DNA with other molecules such as DNA-binding proteins.
6. Distinction of types of circular DNAs, for example, supercoil, relaxed, and single strand.
7. Estimation of stability of intramolecular complementarity of single-stranded nucleic acids.
8. Evaluation of physicochemical properties of nucleic acids and other fibrous macromolecules, such as rigidity, elasticity, and bendability.
9. Hydrodynamic studies of nucleic acids and other fibrous macromolecules.

5.4. Comparison with Other Size Separation Methods for Macromolecules

Among size-fractionation methods for macromolecules, one called hydrodynamic chromatography is of particular interest in connection with slalom chromatography. It has been developed as a size-determination method for submicron to micron-sized colloid particles, such as latices (Small, 1974). It is generally performed by using a column packed with nonfunctionalized, nonporous particles of around 15–20 μm in diameter. Its mechanism was explained as follows (Small and Langhorst, 1982); large colloid particles are excluded from the interface, where the fluid velocity is lowest. The larger the particle, the higher is its mean velocity. Consequently, larger particles are eluted faster than smaller ones, as in gel permeation chromatography and opposite to slalom chromatography. There is no direct interaction of the colloid particles with the packing material. The packing seems to serve to produce a microscopic heterogeneity in the flow rate. A procedure named capillary hydrodynamic chromatography, which does not use packing, has been reported as a variation of this technique (Noel *et al.*, 1978; Tijssen *et al.*, 1983).

As a size separation procedure for DNA, pulsed-field gel electrophoresis is very efficient for extremely large DNA molecules (Schwarz and Cantor, 1984; Carle and Olson, 1984; Cantor *et al.*, 1988; Mathew *et al.*, 1988a,b). This procedure is based on the ability of

DNA molecules to change their direction of migration in response to frequently changing electric field. Larger DNA molecules have longer reorientation times. Separation depends on various physicochemical factors, namely, agarose concentration, temperature, pulse time, electric field strength, electric field shape, and DNA topology. Thus, there are some common features between pulsed-field gel electrophoresis and slalom chromatography; for example, both methods are based on the ability of DNA molecules to adapt to a frequently changing environment (direction of electric field or flow) that depends on their size. The size range of DNA molecules that can be separated by pulsed-field electrophoresis is much larger than in the case of slalom chromatography, but the latter has its own merits, as already pointed out. Therefore, these two methods are complementary to each other.

6. CONCLUSION

The principles of all chromatographic separation methods currently applied to bio-molecules are based on equilibrium phenomena between the moving phase and the stationary phase which are governed by, for example, electrostatic interaction, hydrophobic interaction, or differential solubility. A certain packing material that serves as a support on which one of these phenomena can take place is always needed. Slalom chromatography, however, is based on a completely different principle, that is, the hydrodynamic phenomenon which is caused by liquid currents passing through narrow spaces having peculiar shape. The column packing serves only for construction of these spaces. It is very likely that slalom chromatography is not the only mode that can be used in such a situation. The discovery of slalom chromatography suggests the possibility that a variety of unknown separation mechanisms exists and probably marks the beginning of an expansion of the horizon of chromatography.

REFERENCES

Boyes, B. E., Walker, D. G., and McGeer, P. L., 1988, Separation of large DNA restriction fragments on a size-exclusion column by a nonideal mechanism, *Anal. Biochem.* **170:**127–134.

Cantor, C. R., Gaal, A., and Smith, C. L., 1988, High-resolution separation and accurate size determination in pulsed-field gel electrophoresis. 3. Effect of electrical field shape, *Biochemistry* **27:**9216–9221.

Carle, G. F., and Olson, M. V. G., 1984, Separation of chromosomal DNA molecules from yeast by orthogonal-field-alteration gel electrophoresis, *Nucleic Acids Res.* **12:**5647–5664.

Hirabayashi, J., and Kasai, K., 1988, Slalom chromatography. A new size-dependent separation method for DNA, *Nucleic Acid Res. Symp. Ser.* **20:**67–68.

Hirabayashi, J., and Kasai, K., 1989, Size-dependent, chromatographic separation of double-stranded DNA which is not based on gel permeation mode, *Anal. Biochem.* **178:**336–341.

Hirabayashi, J., Ito, N., Noguchi, K., and Kasai, K., 1990, Slalom chromatography: Size-dependent separation of DNA molecules by a hydrodynamic phenomenon, *Biochemistry* **29:**9515–9521.

Kato, Y., Kitamura, T., Mitui, A., and Hashimoto, T., 1987, High-performance ion-exchange chromatography of protein on nonporous ion exchanger, *J. Chromatogr.* **398:**327–334.

Mathew, M. K., Smith, C. L., & Cantor, C. R., 1988a, High-resolution separation and accurate size determination in pulsed-field gel electrophoresis. 1. DNA size standards and the effect of agarose and temperature, *Biochemistry* **27:**9204–9210.

Mathew, M. K., Smith, C. L., and Cantor, C. R., 1988b, High-resolution separation and accurate size determination in pulsed-field gel electrophoresis. 2. Effect of pulse time, field strength and implications for models of the separation process, *Biochemistry* **27:**9210–9216.

Noel, R. J., Gooding, K. M., Regnier, F. E., Ball, D. M., Orr, C., and Mullins, M. E., 1978, Capillary hydrodynamic chromatography, *J. Chromatogr.* **166:**373–382.

Sanger, F., Coulson, A. R., Hong, G. F., Hill, O. F., and Petersen, G. B., 1982, Nucleotide sequence of bacteriophage λDNA, *J. Mol. Biol.* **162:**729–773.

Schwarz, D. C., and Cantor, C. R., 1984, Separation of yeast chromosome-sized DNA by pulse field gradient gel electrophoresis, *Cell* **37:**67–75.

Small, H., 1974, Hydrodynamic chromatography. Technique for size analysis of colloidal particles, *J. Colloid Interface Sci.* **48:**147–161.

Small, H., and Langhorst, M. A., 1982, Hydrodynamic chromatography, *Anal. Chem.* **54:**892A–898A.

Tijssen, R., Bleumer, J. P. A., and van Kreveld, M. E., 1983, Separation by flow (hydrodynamic chromatography) of macromolecules performed in open microcapillary, *J. Chromatogr.* **260:**297–304.

Yanagida, M., Hiraoka, Y., and Katsura, I., 1982, Dynamic behaviors of DNA molecules in solution studied by fluorescence microscopy, *Cold Spring Harbor Symp. Quant. Biol.* **47:**177–187.

Affinity Chromatography with Biological Ligands

Applications of Bacterial Immunoglobulin-Binding Proteins to the Purification of Immunoglobulins

Michael D. P. Boyle, Ervin L. Faulmann, and Dennis W. Metzger

1. INTRODUCTION

The specificity of the antibody molecule has proved extremely valuable for many applications in affinity chromatography. For these applications to be effective, it is necessary to be able to isolate antibodies of the desired specificity free from other serum components or nonantibody plasma proteins. The majority of the earlier studies utilized polyclonal antibodies generated in a variety of mammalian species, in particular, rabbits and goats. With the advent of monoclonal antibody technology and the ability to produce large quantities of antibodies of a single specificity, these reagents have found broad applications in affinity purification studies. Monoclonal antibodies have an advantage over polyclonal antibodies in that the majority of antibodies in the preparation are directed against the antigen of interest. By contrast, specific antibodies present in a polyclonal hyperimmune serum represent only 5–10% of the total immunoglobulin present. The ability to separate antibodies, monoclonal and polyclonal, from complex mixtures of other plasma proteins or tissue culture fluids is extremely important in obtaining pure sources of antibody for immobilization and use in affinity chromatography.

A number of bacterial surface proteins have been described that react with immunoglobulin molecules in a nonimmune manner (Langone, 1982a; Boyle, 1984; Boyle and Reis, 1987; Boyle *et al.*, 1990a). These bacterial immunoglobulin-binding proteins display selective, high-affinity binding for conserved regions in the Fc portion of the antibody molecule. To date, six types of immunoglobulin G (IgG)-binding proteins have been described on gram-positive bacteria (Myhre and Kronvall, 1981; Reis *et al.*, 1988). These different types have been classified by their ability to bind with the Fc region of different mammalian species and subclasses of IgG. The most extensively studied IgG-binding protein is the type I receptor associated with the majority of human *Staphylococcus aureus* isolates and more frequently designated as protein A (Langone, 1982a; Boyle, 1990). The

Michael D. P. Boyle, Ervin L. Faulmann, and Dennis W. Metzger • Department of Microbiology, Medical College of Ohio, Toledo, Ohio 43699.

Molecular Interactions in Bioseparations, edited by That T. Ngo. Plenum Press, New York, 1993.

type II receptor is found on the surface of the majority of clinical human isolates of group A streptococci and displays a large amount of variability in its immunoglobulin binding profile (Faulmann and Boyle, 1990; Raeder et al., 1991a). The type III binding protein is found on the majority of human groups C and G streptococcal isolates and displays the widest range of species and subclass reactivities of any of the immunoglobulin-binding proteins identified to date (Reis et al., 1984a,b, 1985, 1986; Björck and Kronvall, 1984). This protein, more frequently designated as streptococcal protein G, has been shown to also bind to serum albumin for a limited range of species (Åkerström and Björck, 1986; Faulmann et al., 1989; Raeder et al., 1991b). Five distinct protein G variants associated with human isolates of group C and G streptococci have been characterized (von Mering and Boyle, 1986; Sjöbring et al., 1988, 1989a, 1991; Otten and Boyle, 1991a). The type IV (Raeder et al., 1991c), type V (Yarnall and Widders, 1990), and type VI (Reis and Boyle, 1990a) immunoglobulin-binding proteins are associated with animal isolates of strepto-cocci and have not been as extensively studied.

IgG-binding proteins have also been described on the surface of other bacteria including Haemophilus somnus (Yarnall et al., 1988), Clostridium perfringens (Lindahl and Kronvall, 1988; Lindahl, 1990), Brucella abortus (Bricker et al., 1991), and Taylorella equigenitalis (Widders et al., 1985). It is not clear whether these represent a subset of one of the types I through VI IgG-biding proteins or are members of a larger family of bacterial IgG-binding molecules. In addition to bacterial binding proteins reported to react with the Fc region of IgG, a variety of other proteins have been reported that react with constant regions in IgD (Forsgren and Grubb, 1979; Ruan et al., 1990; Akkoyunlu et al., 1991) or IgA (Russell-Jones et al., 1984; Brady and Boyle, 1989; Lindahl, 1989; Lindahl and Åkerström, 1989; Hedén et al., 1991). Other proteins have been recognized that bind to regions within constant portions of the F(ab')$_2$ regions (Myhre, 1990; Lindahl, 1990) as well as molecules that show selective reactivity with immunoglobulin light chains (Myhre and Erntell, 1985; Björck, 1988; Åkerström and Björck, 1989). To date, a bacterial protein with absolute selectivity for the Fc region of IgM from any species has not been described. A Clostridium difficile toxin protein that displays preferential reactivity with murine IgM (Lyerly et al., 1989) and a Brucella abortus protein with reactivity for certain bovine IgM allotypes (Nielsen et al., 1981) have been reported.

2. EXPERIMENTAL CONSIDERATIONS

The purpose of this chapter is to provide an overview of the general application of bacterial IgG-binding proteins to the isolation of immunoglobulins from serum and complex solutions. The basic approaches utilized are similar in all cases, and consequently only representative procedures have been included. The general procedure for isolation of immunoglobulin molecules by bacterial binding proteins is very similar to approaches used in all affinity chromatography and is summarized schematically in Fig. 1.

In this scheme, a high-affinity IgG-binding protein is immobilized on an inert support, and then the solution containing the antibodies is passed over the column under condi-tions of pH and ionic strength that facilitate the selective interaction of the immobilized binding protein with the Fc region of antibody molecules. Proteins having no affinity for the immobilized ligand are removed by washing the column with the application buffer. The selectively bound material is then eluted by changing the conditions of pH and/or ionic

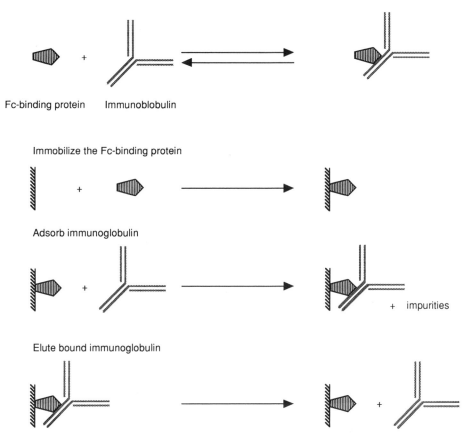

Fc-binding protein   Immunoblobulin

Immobilize the Fc-binding protein

Adsorb immunoglobulin

+ impurities

Elute bound immunoglobulin

FIGURE 1. Schematic representation of the use of immobilized bacterial immunoglobulin-binding proteins for the affinity purification of antibodies.

strength to reverse the binding of the immunoglobulin to the immobilized bacterial immunoglobulin-binding protein.

There are a number of considerations that facilitate the selection of appropriate reagents for purifying the immunoglobulin of interest. The first is the selection of the immunoglobulin-binding protein to be immobilized. This selection is based on identifying a binding protein that demonstrates reactivity with the immunoglobulin species or subclass that is to be purified. The specificities of the two most widely used bacterial IgG-binding proteins, protein A and protein G, are provided in Table 1. It should be noted that Table 1 shows the reactivity of these binding proteins with representative samples of the species and subclasses shown. This reactivity is not, however, absolute, and variations have been reported (Rousseaux *et al.*, 1981; Langone, 1982a; Nilsson *et al.*, 1983; Villemez *et al.*, 1984; Boyle and Reis, 1987). This is particularly relevant for mouse monoclonal antibodies, and it is not possible to predict with absolute certainty the reactivity of any class or subclass of monoclonal antibody with either protein A or protein G.

Variations in the reactivity of different mammalian species and subclasses of immunoglobulins with bacterial IgG-binding proteins have also been reported to depend on the

TABLE 1

Comparison of Immunoglobulin Species and Subclass Reactivities
of Protein A and Protein G[a,b]

Species	Subclass	Reactivity[c] with:		Species	Subclass	Reactivity[c] with:	
		Protein A	Protein G			Protein A	Protein G
Human	IgG1	+++	+++	Rabbit	IgG	+++	+++
	IgG2	+++	+++	Cow	IgG1	−	++
	IgG3	−[d]	+++		IgG2	+	+++
	IgG4	+++	+++	Sheep	IgG1	−	++
	IgA	+/−[e]	−		IgG2	+	+++
	IgM	+/−[e]	−	Goat	IgG1	+/−	++
Mouse	IgG1	+/−	++		IgG2	+	+++
	IgG2a	++	+++	Horse	IgG(ab)	+	+++
	IgG2b	++	+++		IgG(c)	+	+++
	IgG3	++	+++	Dog	IgG	+	NT[f]
Rat	IgG1	−	+/−		IgM	+	NT
	IgG2a	−	+		IgA	+	NT
	IgG2b	−	+				
	IgG2c	+/−	+				

[a]The majority of these studies have been carried out with myeloma proteins, and many instances have been noted in which the expected reactivity with a given mammalian subclass was not observed. This has been particularly evident in studies of mouse monoclonal antibodies. Consequently, this table should only be used as a guide for probable reactivities. Each antibody should be checked directly for reactivity with protein A or protein G prior to its use for any immunochemical procedure.

[b]Derived from studies by Langone (1982a), Reis *et al.* (1984b), Boyle *et al.* (1985), Wallner *et al.* (1987), Reis *et al.* (1986), Nilsson *et al.* (1982), and Nilsson *et al.* (1986).

[c]+++, Strong reactivity; ++, intermediate reactivity; +, weak reactivity; +/−, very low afffinity reactivity; −, no reactivity.

[d]Certain IgG3 allotypes have been reported to react with the type I Fc-binding protein (Haake *et al.*, 1982).

[e]Certain IgM and IgA samples have been reported to react with protein A (Harboe and Fölling, 1974; Inganäs, 1981; Sasso *et al.*, 1989).

[f]NT, Not tested.

pH of the buffer used to apply the sample. It is also important to be aware that certain animals may have naturally occurring antibodies in their serum which recognize bacterial antigens (including IgG-binding proteins) in an antigenically specific manner. For example, in one study, we analyzed the interaction of ten normal goat sera with protein A and found one animal with a very reactive immunoglobulin fraction (Boyle *et al.*, 1985). This activity was attributable to the presence of a natural antibody that bound to protein A through $(Fab')_2$ fragments rather than via a conserved constant-region domain. This presumably occurred because of a natural infection of this goat with a protein A-positive staphylococcal organism.

Other differences in reactivity of bacterial IgG-binding proteins owing to a small change in primary sequence of conserved regions of immunoglobulins have also been noted. This phenomenon is best illustrated by the interaction of human IgG3 antibodies with protein A. IgG3 antibodies of s^-t^- allotype do not react with protein A while those of the s^+t^+ allotype are reactive (Haake *et al.*, 1982). This difference can be attributed to a single amino acid substitution of a histidine (s^+t^+) for an arginine residue (s^-t^-) at position 435 of the heavy chain. Subtle allotypic differences or small changes in immunoglobulin domains may also account for variation among subclasses of murine monoclonal antibodies in their reactivities with either protein A or protein G.

In addition to variation in the reactivity of protein A and protein G with different species and subclasses of IgG, reactivity with other antibody isotypes and nonimmunoglobulin binding proteins has also been described. For example, protein A has been reported to react with certain subclasses of human IgA and IgM as well as displaying an alternative reactivity with $F(ab')_2$ of human IgG (Myhre, 1990). These non-IgG Fc reactivities are of lower affinity than the IgG Fc binding properties and can be eliminated or minimized by careful selection of application and elution buffers. Protein G has been shown to bind other non-IgG plasma proteins including certain species of albumin, kininogen, and α_2-macroglobulin (Sjöbring et al., 1989b). The only reactivity that is of equivalent affinity to the IgG Fc binding is protein G's reactivity with human and primate serum albumin. Protein G does not react with serum albumin from goat, bovine, horse, mouse, or rat, and consequently immobilized forms of native protein G can be used effectively for purification of immunoglobulins from these species (Raeder et al., 1991b). The albumin- and IgG-binding domains of protein G have been shown to occur in distinct regions of the molecule. A non-albumin-binding form of protein G has been prepared by deletion of the DNA sequence encoding the albumin-binding region of the molecule (Fahnestock, 1987). This form of recombinant protein G is commercially available and is the form most usually immobilized for antibody purification procedures. While the investigator should be aware of the potential for binding of plasma proteins other than IgG to immobilized protein A or protein G, the conditions used to apply and recover the desired antibody type from the bacterial binding protein columns have been optimized for most species to minimize any undesired contaminants.

A variety of technical considerations are involved in the use of bacterial immunoglobulin-binding proteins for antibody purification: (1) preparation of immobilized immunoglobulin-binding protein supports, (2) selection of conditions to apply and elute the antibody sample, and (3) monitoring purity of the resulting product. In the next sections, each of these considerations is reviewed.

2.1. Preparation of Immunoglobulin-Binding Protein Supports

Bacterial immunoglobulin-binding proteins are commercially available as both wild-type and recombinant proteins from a number of suppliers (for a listing, see *Linscott's Directory of Immunological and Biological Reagents*). For researchers interested in the extraction and isolation of bacterial immunoglobulin-binding proteins, methods describing appropriate procedures have been reported by Boyle et al. (1990b). In general, most of the available bacterial binding proteins can be covalently coupled to matrices with retention of functional activity. It has been noted that the efficiency of immunoglobulin binding to immobilized proteins is greater when a matrix with an 8–10-carbon spacer arm is used. Suitable activated supports that couple by amino or carboxyl groups on the immunoglobulin-binding proteins are available from a number of companies including Pharmacia (Piscataway, New Jersey), Bio-Rad Laboratories (Richmond, California) and the Pierce Chemical Company (Rockford, Illinois). Each of these companies offers a variety of high-capacity, activated gels suitable for immobilization of bacterial immunoglobulin-binding proteins. The choice of the most appropriate matrix will depend on the chromatographic system being used by the investigator. In addition, there are a number of commercial sources of already immobilized protein A and protein G (see Linscott, 1991).

2.2. Selection of Conditions for Application and Elution of Antibody Samples

The efficacy of bacterial immunoglobulin-binding proteins in affinity chromatography is restricted by the range of species and subclasses that will bind the immobilized ligand. As presented in Table 1, protein A and protein G have differences in their overall species reactivity profiles, which provide the investigator with a guide to the selection of appropriate immobilized binding protein. One should keep in mind that these reactivity profiles are only generally representative and that variations within and between immunoglobulins of a given species or subclass have been noted. In this regard, it has generally been found that murine monoclonal antibodies of the IgG1 subclass fail to bind well to protein A, while the majority of these antibodies show some affinity for protein G. A number of reports have indicated that the use of a selected application buffer system can increase the efficacy of binding of IgG1 monoclonal antibodies to protein A. These buffer systems have included a proprietary MAPS buffer system available from Bio-Rad Laboratories (Scott and Juarez-Salinas, 1990) as well as a high-salt, high-pH buffer [e.g., $1.5M$ glycine/NaOH, pH 8–9, $3M$ NaCl, (Pharmacia, 1986)], both of which demonstrate enhanced binding of murine IgG1 to protein A.

The affinity of protein A or protein G for IgG from most mammalian species, under physiological conditions, is sufficiently high that complex buffers to promote their binding to an affinity column are not required (see Tables 2 and 3). In general, we have found that the use of phosphate-buffered saline in the range pH 7.0–7.5 facilitates the binding of reactive mammalian immunoglobulins to immobilized protein A or protein G.

A number of methods have been used for eluting bound proteins (for a summary, see Tables 2 and 3). In our experience, $0.1M$ glycine-HCl, pH 2, applied as a single step, leads to a good quantitative recovery of functional immunoglobulins from bacterial IgG-binding protein columns. Other investigators have described the use of less harsh elution conditions including chaotropic agents, high salt, or selective dipeptides (Bywater et al., 1978, 1983; MacSween and Eastwood, 1977, 1981). In addition to single-step elution procedures, a number of investigators have demonstrated successful resolution of different reactive immunoglobulin subclasses by the use of pH gradients (for examples, see Tables 2 and 3). The selection of elution conditions will be influenced by the ultimate use of the isolated antibody. In most applications in which specific immunoglobulins are being isolated for use in affinity chromatography, the specificity of the antibody rather than the subclass distribution is more important and, therefore, a one-step elution buffer can be used. However, if a specific subclass is desired, gradient elution would be preferable.

2.3. Analysis of Immunoglobulin Recovery and Purity

In all purification procedures involving bacterial immunoglobulin-binding proteins, it is valuable to be able to monitor the recovery and purity of different column fractions. In particular, in the isolation of specific antibodies, it is often valuable to be able to monitor the recovery of functional antibody as well as the recovery of total immunoglobulin. A variety of different immunoassays [e.g., radioimmunoassay (Langone et al., 1977), enzyme-linked immunosorbent assay (ELISA; Reis et al., 1983), and less sensitive radial immunodiffusion

TABLE 2

Isolation of Immunoglobulin Classes and Subclasses Using Chromatography on Protein A-Sepharose[a]

Species (class or subclass)	Application conditions		Elution conditions		Comments	Reference
	pH	Buffer	pH	Buffer		
Human						
Whole IgG	7.0	0.1 M NaH$_2$PO$_4$	3.0	0.1M glycine-HCl		Hjelm et al., 1972
IgG2 IgG1	7.0	0.2M NaH$_2$PO$_4$ adjusted to pH 7 with 0.1M citric acid	4.7 4.3	0.2M phosphate/0.1M citric acid	90–95% pure; ≤5% contamination with other IgG isotypes	Duhamel et al., 1979
IgG3	7.0	0.1M phosphate	7.0	0.1M phosphate	IgG3 in effluent, further purified on Sephadex G-150	Hjelm, 1975
IgM	7.5	Phosphate-buffered saline	2.5	0.1M glycine-HCl	>80% of IgM in G-20 void volume peak was isolated	Balint et al., 1981
Rabbit						
IgG[b]	7.4	Phosphate-buffered saline	3.0	0.58% acetic acid containing 0.15M sodium chloride	Yield: 5–6 mg/ml serum; antibodies retain antigen-binding and antigenic properties	Goding, 1978
IgG	8.0	0.016M boric acid, 0.012M NaCl, 0.025M NaOH	3.0	0.1M glycine-HCl	Good yield of specific antibody	Miller and Stone, 1978
Goat						
IgG1 IgG2	9.1	0.1M Na$_2$HPO$_4$	9.1 5.9	0.1M phosphate	70% yield; 5–6% IgM and IgA plus trace of albumin removed on Sephadex G-200. Only 2–4% IgG1 contamination	Delacroix and Vaerman, 1979
IgG1 IgG2	7.0	0.2M Na$_2$HPO$_4$ adjusted to pH 7.0 with 0.1M citric acid	6.7 5.8	0.1M phosphate	Application of >2 mg IgG gave significant IgG in effluent	Duhamel et al., 1980

(Continued)

TABLE 2 (Continued)

Species (class or subclass)	Application conditions pH	Application conditions Buffer	Elution conditions pH	Elution conditions Buffer	Comments	Reference
Mouse						
IgG1	8.0	0.1M Na$_2$HPO$_4$	7–6	0.14M phosphate or 0.1M citrate/citric acid	IgM, IgA, and IgE in effluent. Purity ≥90%; yield 90–100% except IgG2b obtained in 60% yield	Ey et al., 1978
IgG2a			5–4.5			
IgG2b			4–3.5			
IgG1	8.0	0.1M Phosphate	6.0	0.1M phosphate or 0.1M citrate/citric acid (step gradient)		Seppälä et al., 1981
IgG2a						
Allotype b			4.5			
Allotype a,j			5.0			
IgG3			4.5			
IgG2b			3–5			
IgG1			6.6–5.8	(continuous gradient)		
IgG2a			4.8–4.5			
IgG2b			4.4–4.0			
IgG3			4.7–4.4			
IgG2a + IgG2b	7.4	Phosphate-buffered saline	—	1.5–2.0M NaSCN		Mackenzie et al., 1978
IgG1			—	0.5M NaSCN		
IgG1	7.3	Phosphate-buffered saline	—	0–3.0M linear NaSCN gradient	73% yield, 94% pure	Chalon et al., 1979
IgG2a + IgG2b + IgG3			—		Free of IgG1 and IgM	
Rat						
IgG1	8.0	Phosphate-buffered saline	7.0	0.14M phosphate or 0.1M citrate/citric acid or, in first step, column equilibrated with 0.01M borate containing 0.015M sodium chloride, pH 9.0		Rousseaux et al., 1981
IgG2c			4.0–3.0			

Isotype	Application buffer	Application pH	Eluant pH	Eluant	Comments	Reference
IgG1	10mM sodium phosphate, 0.15M sodium chloride, 0.1M glycine		—	0.5M NaSCN		Nilsson et al., 1982, 1983
IgG2c			—	1.0M NaSCN		
IgG2a			8.0	5mM phosphate	Isolated by DEAE cellulose chromatography of SpA-Sepharose effluent	
IgG2b			7.0	0.14M phosphate	From SpA-Sepharose after application of 15mM phosphate fraction from DEAE cellulose	
Guinea pig IgG1 + IgG2	0.1M Na_2HPO_4	9.0	First: 4.5 Second: 3.2	0.1M phosphate (step gradient)	Each IgG subclass showed similar heterogeneity: 60% elution at pH 4.5, 40% at pH 4.5, 40% at pH 3.2. Pure subclasses cannot be separated	Ricardo et al., 1981
Syrian hamster IgG2 IgG1	0.2M sodium acetate	8.0	5.9 5.3	0.2N acetate/acetic acid gradient from pH 7.5–3.0	Equilibrate at pH 8.0	Coe et al., 1981
Dog IgM	Phosphate-buffered saline	7.5	2.5	0.1M glycine-HCl	>80% yield of G-200	Balint et al., 1981
Rhesus monkey IgGI IgGII	0.2M Na_2HPO_4 adjusted to pH 7.3 with 0.1M acetic acid	7.3	5.0–4.72 5.1–4.65	0.1M citrate 0.2M phosphate	SpA-Sepharose combined with DEAE-cellulose chromatography. Subclass-enriched, but not highly purified fractions were obtained	Martin, 1982

[a] Reproduced from Langone (1982b) with permission.
[b] Distinct isotypes of rabbit IgG have not been identified.

TABLE 3
Use of Protein G for Affinity Chromatography

Species	Application buffer	Elution buffer	Comments	Reference(s)
Human IgG	0.05M sodium phosphate containing 0.25M NaCl pH 7.6	Glycine-HCl, pH 2–3	79% recovery	Mandaro et al., 1987
Human IgG	Optimize around pH 5	Variable	Detailed analysis of effect of pH on binding of human IgG to protein G-Sepharose	Åkerstrom et al., 1985; Åkerstrom and Björck, 1986
Human (monoclonal)	0.1M glycine-sodium hydroxide, pH 9.0	Gradient elution, pH 4.0 → 2.5 Single-step elution at pH 2.5	~50% recovery ~70% recovery	Jungbauer et al., 1989 Jungbauer et al., 1989
Bovine IgG	Optimize around pH 5	Glycine-HCl, pH 2–3	>90% recovery	Mandaro et al., 1987
Rabbit IgG	Optimize around pH 5 Phosphate-buffered saline, pH 7.4	Glycine-HCl, pH 2–3 Glycine-HCl, pH 2.0	>90% recovery when applied at pH 5.0	Mandaro et al., 1987 Mandaro et al., 1987
Goat IgG	0.1M sodium phosphate, pH 7.0	0.5M acetic acid adjusted to pH 3.0 with sodium hydroxide		Webb-Walker, 1990
	0.1M sodium acetate, 0.1M NaCl pH 5.0	0.5M acetic acid adjusted to pH 3.0 with sodium hydroxide	Higher binding capacity when applied at pH 5.0	Webb-Walker, 1990
Mouse				
IgG1	0.05M sodium phosphate, pH 6.5	0.05M glycine-HCl, pH 2.5	85% recovery	Ohlson et al., 1988
IgG2	0.05M sodium phosphate, pH 6.5	0.5M glycine-HCl, pH 2.5	80% recovery; nonreactive with protein A	Ohlson et al., 1988
Rat				
IgG1	0.5M sodium phosphate, pH 6.5	0.05M glycine-HCl pH 2.5	50% recovery; nonreactive with protein A	Ohlson et al., 1988
IgG2a	0.5M sodium phosphate, pH 6.5	0.05M glycine-HCl, pH 2.5	100% recovery; nonreactive with protein A	Ohlson et al., 1988
IgG2c	0.5M sodium phosphate, pH 6.5	0.05M glycine-HCl, pH 2.5	100% recovery	Ohlson et al., 1988
Rat IgG1 or IgG2a (monoclonal)	Phosphate-buffered saline, pH 7.4	0.1M glycine-HCl, pH 2.5	Nonreactive with protein A	Monestier et al., 1989

assays (Hudson, 1986)] are available for quantifying total immunoglobulin. In general, it has been found that >95% of the total IgG present can be removed from either sera, ascites fluids, or tissue culture media using immobilized bacterial immunoglobulin-binding proteins. IgG recoveries from these columns are routinely >80% of applied immunoglobulin.

The presence of nonimmunoglobulin contaminating proteins eluted from immunoglobulin-binding protein columns can be monitored by a variety of procedures, including immunoelectrophoresis (Lawman *et al.*, 1985) and sodium dodecyl sulfate-polyacrylamide gel electrophoresis (SDS-PAGE) under reducing or nonreducing conditions followed by protein staining (Reis *et al.*, 1988). Examples of this latter procedure are shown in Figs. 2 and 3. In general, there are few, if any, nonimmunoglobulin contaminating proteins in fractions eluted from these binding protein columns under optimized conditions.

3. APPLICATIONS OF BACTERIAL IMMUNOGLOBULIN-BINDING PROTEINS

3.1. Use of Bacterial Immunoglobulin-Binding Proteins to Isolate Polyclonal Mammalian Immunoglobulins

Immobilized protein A and protein G have been used to isolate a wide variety of mammalian immunoglobulins based on the reactivity pattern described in Table 1. Initially, because of its commercial availability since 1978, protein A-Sepharose was used to isolate different immunoglobulin classes and subclasses from many species (Goding, 1978; Langone, 1982b; Schwartz, 1990; Scott and Juarez-Selinas, 1990).

The use of protein A and protein G for purification of mammalian immunoglobulins has followed a very similar strategy to that outlined in Fig. 1. Initially, the bacterial immunoglobulin-binding protein columns should be washed with the eluting buffer to remove all proteins that are noncovalently attached. The column should then be equilibrated in the sample application buffer. For most applications, 5 ml of immobilized protein A or protein G is sufficient to purify 10–20 mg of IgG or 1–2 ml of hyperimmune serum. The sample should be applied to the column slowly at a rate of half a milliliter per minute or less. The optical density at 280 nm (OD_{280}) of the elution fractions should be monitored to determine when the absorbance returns to baseline. At that point, specifically bound proteins can be eluted from the column by treatment with a variety of agents. We have found $0.1M$ glycine-HCl, pH 2, used as a simple step leads to efficient recovery of bound immunoglobulins. The eluted fractions can be monitored for protein by following their absorbance at 280 nm, and a protein peak is usually noted in the first acidic samples eluting from the column. It is advisable to neutralize fractions as quickly as possible, and we have found collection of fractions into tubes containing a suitable neutralizing buffer to be effective.

A variety of application and elution conditions have been reported for the isolation of different mammalian IgG antibodies using immobilized protein A or protein G. A list of representative examples of purification procedures utilizing protein A was compiled by Langone (1982b) and has been included, with permission, in Table 2. With the more recent availability of protein G and various immobilized forms of this protein, a wider range of immunoglobulin species are amenable to this isolation procedure. A representative listing of purification procedures involving immobilized protein G is shown in Table 3. The

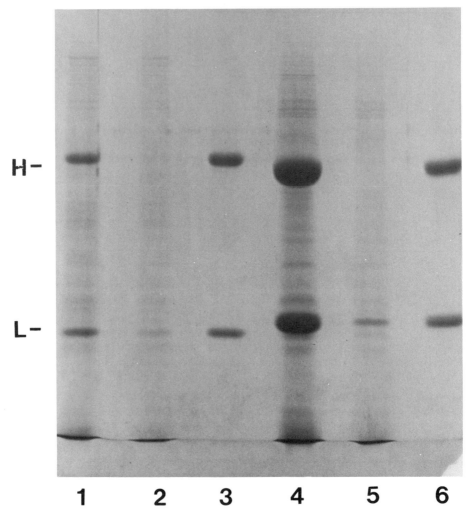

H–

L–

1   2   3   4   5   6

FIGURE 2. Purification of murine monoclonal antibodies by protein G conjugated to Sepharose 4B. The locations of the antibody H and L chains on the gel are indicated. Lanes 1 and 4 represent supernatants (concentrated 20× and 30×, respectively) from hybridomas grown in protein-free culture medium. Lanes 2 and 5 represent these same supernatants after passage over protein G columns. Lanes 3 and 6 represent protein eluted from the columns with 0.2M glycine-sulfate buffer, pH 2.3. Lanes 1, 2, 4, and 5 were loaded with 100 μl of supernatant fluid; lanes 3 and 6 were loaded with 10 μg of protein. Monoclonal antibodies used: lanes 1–3, BALB/c IgG1 from cell line AF6-78.25.4; lanes 4–6, CWB IgG2a from cell line Ig (5a) 7.2.

approaches summarized in Tables 2 and 3 are both rapid and efficient, and there is no need for extensive sample preparation. The serum samples can be applied directly to the column without dialysis or dilution. The affinity columns can be regenerated and usually have effective working lives of 15–25 runs before they lose capacity for IgG. Between runs, we normally store our affinity columns in the presence of 20% ethanol or in phosphate buffer containing 0.2% sodium azide to minimize bacterial contamination.

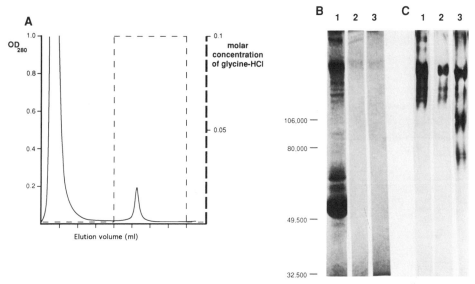

FIGURE 3. Purification of human IgA by affinity chromatography on protein B-Sepharose. Column profile of affinity purification of human IgA on protein B-Sepharose (A) and SDS-PAGE separation of human serum and purified IgA followed by Coomassie staining (B) and Western blotting (C). Lane 1, Human serum (precolumn material); lane 2, material bound to the protein B-Sepharose column and subsequently eluted with 0.1M glycine-HCl (pH 2.0); lane 3, commercially obtained purified human IgA.

In addition to the use of a single-step elution as noted above, a variety of investigators have used pH gradients to elute bound immunoglobulins (see Tables 2 and 3). This has allowed separation of reactive immunoglobulin into subclasses. Based on the subclass reactivities shown in Table 1, it is possible to combine the use of protein A and protein G columns to isolate different subclasses of mammalian immunoglobulins. For example, we have demonstrated that protein A is highly selective in its reactivity for bovine IgG2, while protein G demonstrates high affinity for both subclasses (Wallner *et al.*, 1987; Lawman *et al.*, 1985). Bovine serum samples passed over protein G columns can deplete all the IgG from serum; the IgG can then be recovered in a simple one-step elution procedure. The eluted proteins, containing both IgG1 and IgG2, can be applied following neutralization to a protein A-Sepharose column. Proteins passing directly through that column are exclusively of the IgG1 subclass while those eluting from the column are exclusively of the IgG2 subclass.

3.2. Use of Bacterial Immunoglobulin-Binding Proteins to Isolate Monoclonal Antibodies

With the description by Kohler and Milstein (1975) of a method to generate hybridoma cell lines producing a single antibody isotype and specificity, much interest was focused on the use of monoclonal antibodies. In the early stages of the development of monoclonal antibody technology, it became clear that efficient methods for screening of hybridoma supernatants for antibody activity and recovery of those selective antibodies from either ascites fluid, tissue culture medium, or serum would be a critical factor in the rate at which this new technology could advance. With the commercial availability of protein A at the same time, the advantage of a bacterial binding protein for screening and isolating monoclonal antibodies became obvious. Indeed, had a protein A equivalent been available for rat immunoglobulins, the monoclonal antibody technology might well have been developed using rat hybridomas rather than mouse because of the more stable characteristics and growth requirements of rat hybridomas (Milstein, 1986).

Immobilized protein A and protein G have been the reagents of choice for the purification of murine monoclonal antibodies. However, as shown in Table 1, protein A fails to react with the majority of murine IgG1 antibodies. This is particularly troublesome since many murine antiprotein responses primarily consist of IgG1 antibodies (Metzger *et al.*, 1984). With the discovery and availability of protein G, many investigators reported that this binding protein showed a broader range of reactivity with murine monoclonal antibodies, and many laboratories switched to using this form of the reagent for both screening and isolating their monoclonal antibodies.

The principle of isolating monoclonal antibodies using bacterial IgG-binding proteins is identical to that used for isolating polyclonal antibodies. The sample is applied to an immobilized column and then eluted selectively. Many investigators have compared different methods of applying monoclonal antibodies to IgG-binding columns, and a variety of different buffer systems have been described that facilitate binding of monoclonal antibodies to bacterial immunoglobulin-binding supports. In particular, the poor binding of IgG1 to protein A can be enhanced by a variety of buffer systems as described above, and a similar ability to enhance the reactivity of certain monoclonal antibodies with protein G has been reported (Webb-Walker, 1990).

An example of the results obtained with the purification of mouse monoclonal antibodies on columns of immobilized protein G are shown in Fig. 2. In this particular experiment, the antibodies were of the IgG1 and IgG2a isotypes, and the form of protein G used was a recombinant protein that contains all of the immunoglobulin-binding and albumin-binding domains that are present in the native molecule (Otten and Boyle, 1991b). [Most commercial forms of recombinant protein G are produced from plasmids from which the coding sequences for the albumin-binding domains have been deleted (Fahnestock, 1987). However, the presence of the albumin-binding domain does have advantages for certain experimental procedures (Sisson, 1990). Since the albumin binding activity is restricted and shows no reactivity with rodent albumins, we used the recombinant form of protein G that contains the albumin-binding domains in this study; however, this form of protein G does bind to primate albumin and would, therefore, be of limited value for purification of human or monkey immunoglobulin.]

The hybridoma cells were obtained from the American Type Culture Collection, grown in RPMI 1640 containing 15% fetal bovine serum, and, after reaching confluence, were washed and cultured for an additional 3–4 days in serum-free RPMI 1640. The monoclonal antibodies were then precipitated from the tissue culture supernatants by dialysis against 70% ammonium sulfate. The precipitates were dialyzed against $0.02M$ sodium phosphate buffer, pH 7.0, and applied to immobilized protein G columns that were equilibrated with the same buffer. The unbound proteins were washed through the column with the application buffer, and the bound proteins were then eluted with $0.2M$ glycine-sulfate buffer, pH 2.3. The fractions containing protein were pooled, dialyzed against physiological saline, and analyzed by SDS-PAGE under reducing conditions.

The results showed that the hybridoma cells grown in serum-free medium produced predominantly immunoglobulin heavy and light chains (Fig. 2, lanes 1 and 4). However, there was significant contamination with other proteins in the unseparated tissue culture supernatants; these contaminating proteins were probably released from dead cells. To isolate the immunoglobulins, the supernatants were applied to immobilized protein G. Analysis of the material that did not bind to the protein G showed that the contaminants were still present, but all of the immunoglobulin heavy chains and most of the light chains were removed (lanes 2 and 5). The band at approximately 25,000 daltons probably represents free light chains that were produced by the hybridomas in the absence of associated heavy chain. The material that was bound to the protein G and eluted at low pH contained only immunoglobulin chains and was totally free of any detectable contaminants (lanes 3 and 6).

Recently, plasmid constructs containing the coding sequences for the binding domains of protein A and protein G have been generated at the gene level (Eliasson *et al.*, 1988, 1989). The resulting protein A/protein G recombinant hybrid protein displays the combined range of reactivity of protein A and protein G with monoclonal antibodies (Eliasson *et al.*, 1988, 1989). Monoclonal antibodies that react with protein A but not with protein G are extremely rare; however, the use of a protein A/protein G hybrid would ensure that those antibodies could also be identified. In general, the methods used to screen monoclonal antibodies and the rapid methods available to determine whether any given monoclonal antibody reacts with protein A or protein G (Boyle and Faulmann, 1990) allow the selection of the appropriate immobilized bacterial binding protein to be made in an efficient manner.

3.3.   Isolation of Human Serum IgA on a Column of Immobilized Protein B

The applications described thus far have focused on the use of bacterial immunoglobulin-binding proteins for the isolation of IgG molecules. Bacterial surface molecules with selective reactivity with the Fc region of IgA (Russell-Jones *et al.*, 1984; Brady and Boyle, 1990; Lindahl *et al.*, 1991; Hedén *et al.*, 1991) and IgD (Forsgren and Grubb, 1979; Akkoyunlu *et al.*, 1991) have been described as well as a bacterial protein from *Peptococcus magnus*, protein L, which reacts selectively with human immunoglobulin light chains (Myhre and Erntell, 1985; Björck, 1988; Åkerström and Björck, 1989). As would be anticipated, these ligands, once immobilized, can be used for isolation of appropriate immunoglobulin isotypes or, in the case of protein L, antibodies and (Fab′)$_2$ fragments of any isotype (Björck, 1988). Recently, an IgA-Fc-binding protein, protein B, has become commercially available (Blake Laboratories, Cambridge, Massachusetts), and studies in our laboratory have demonstrated that protein B can be used as an effective tracer for human serum IgA (Faulmann *et al.*, 1991). Protein B reacts with both IgA subclasses (Russell-Jones *et al.*, 1984; Faulmann *et al.*, 1991), which make this ligand superior for purification of IgA to the IgA-binding lectin jacalin, which demonstrates preferential reactivity with IgA1 (for reviews, see Aucouturier *et al.*, 1987, 1988).

Protein B coupled to Sepharose was provided to us by Blake Laboratories, Cambridge, Massachusetts. We have analyzed this immobilized ligand for the purification of human serum IgA. In the experiments shown in Fig. 3, human serum containing IgA was applied to a column of immobilized protein B in phosphate-buffered saline at pH 7.4. Unbound proteins were removed by washing in this buffer. Bound proteins were subsequently eluted from the protein B column using 0.1M glycine-HCl, pH 2. The OD$_{280}$ profile of the affinity chromatography procedure is shown in Fig. 3A. The eluted proteins were analyzed for total proteins by staining with Coomassie brilliant blue (Fig. 3B) and for IgA following Western blotting, using a monospecific goat anti-human-α-chain antibody and radiolabeled protein G (see Fig. 3C). These results demonstrate that the protein B affinity column effectively removed IgA from serum, and it could be recovered from the column in a pure form. Our protein B column has been used to purify serum IgA on ten consecutive occasions with similar results. This example demonstrates that there are selective bacterial binding proteins for different immunoglobulin isotypes that can be used to isolate the corresponding protein from serum or other biological fluids.

3.4. Fc-Directed Immobilization of Specific Antibody for Use in Affinity Chromatography

The efficiency of immobilized antibodies for affinity purification procedures can be increased when the specific immunoglobulin is immobilized via the Fc region (Little *et al.*, 1988). This enables the antibody molecule to be oriented with respect to the support in such a manner that the antigen-combining F(ab')$_2$ regions are available to interact with the antigen. Recently, much interest has been focused on the use of affinity supports in which antibodies are coupled via a hydrazide linkage to the carbohydrate side chain localized in the Fc region of the immunoglobulin. Bacterial immunoglobulin-binding proteins can also be used for immobilizing antibody molecules in this orientation. A representative list of affinity columns using specific antibodies immobilized to bacterial IgG-binding protein columns is provided in Table 4. These examples include the use of specific antibodies bound to immobilized bacterial IgG-binding molecules to detect IgM and IgA in serum and other secretions as well as specific antibody methods for covalently coupling the Fc-binding protein–IgG complex for antigen purification.

3.5. Use of Bacterial Binding Proteins for Preparation of Clinically Useful Monoclonal Antibodies

Recent studies on the use of monoclonal antibodies to treat infectious diseases (Fant, 1990; Chmel, 1990) and cancer (Goldenberg, 1989) have shown therapeutic promise. In any

TABLE 4

Use of Bacterial IgG-Binding Proteins to Immobilize Specific Antibodies for Affinity Chromatography

Bacterial binding protein support	Specific antibody source	Antigen purified or detected	Reference
Protein A	Polyclonal rabbit	Human IgA	Russell *et al*, 1984
	Polyclonal rabbit	Human IgM	Russell *et al.*, 1984
	Mouse monoclonal	Common acute leukemia-associated antigen	Schneider *et al.*, 1982
	Mouse monoclonal	HLA-AB antigen	Schneider *et al.*, 1982
	Transferrin rabbit polyclonal antigen–antibody complex	Transferrin receptor	Schneider *et al.*, 1982
	Mouse monoclonal	Transferrin receptor	Schneider *et al.*, 1982
Protein G	Mouse monoclonal	Hen egg lysozyme	Webb-Walker, 1990

procedure involving injection of antibody products into people, a number of important factors must be considered. First, it is essential to ensure that during the purification of any of these therapeutic monoclonal antibodies, no bacterial product leaches from the column that could cause toxic or anaphylactic reactions in patients. Previous studies with extracorporeal treatment of plasma from cancer and autoimmune patients over protein A columns have identified problems with protein A leaching from these columns and causing toxic reactions in the patients (Ainsworth *et al.*, 1990). It is therefore important to guarantee that any therapeutic antibody be free of bacterial binding protein contaminant.

Detection of bacterial binding proteins in the presence of reactive immunoglobulins is technically extremely complex. Based on the functional activity of these particular molecules, it is difficult to distinguish antigen-specific reactions from reactions with the Fc regions of antibodies. For this reason, we have generated a panel of specific antibody reagents in chickens, whose nonimmune immunoglobulins fail to react with any of the bacterial binding proteins thus far studied (Boyle and Reis, 1990). Using these selective chicken antibodies, we have been able to distinguish different types of immunoglobulin-binding proteins serologically as well as to develop quantitative methods for the detection of bacterial binding proteins (Reis and Boyle, 1990b). These antibody methods can be adapted to measure the presence of bacterial binding proteins in solutions containing large quantities of reactive immunoglobulins. As the therapeutic application of monoclonal antibodies becomes more accepted clinically, the problems associated with injecting mouse antibodies will become more critical. Consequently, much emphasis is now being placed on generation of humanized monoclonal antibodies. As this trend continues, there will be an increased need for selective bacterial binding proteins that can identify antibodies of each human IgG subclass and that, once immobilized, can be used for their purification. Recent studies on type II immunoglobulin-binding proteins associated with group A streptococci have suggested that subclass-specific bacterial binding proteins for human IgG subclasses do exist (Yarnall and Boyle, 1986a,b; Faulmann and Boyle, 1990; Raeder *et al.*, 1991a) and that these proteins will be of value to detect and to isolate specific monoclonal antibodies of any human subclass. Evaluation of the potential clinical use of human monoclonal antibodies or human antibodies engineered in bacteria will await future clinical trials and additional research efforts. It is clear that antibodies expressing the desired specificity, generated by any of these methods, will require efficient methods of purification. Bacterial immunoglobulin-binding proteins remain viable candidates for these purification procedures provided the corresponding binding site remains in whatever form of specific humanized antibody is generated.

4. FUTURE DIRECTIONS FOR USE OF BACTERIAL Fc BINDING PROTEINS

The examples provided in this chapter demonstrate the usefulness of high-affinity bacterial binding proteins for the isolation of a variety of different isotypes, classes, and subclasses of immunoglobulins. The continual discovery of new bacterial immunoglobulin-binding proteins with different functional activities [for example, proteins with selective reactivity for human IgG3 (Yarnall and Boyle, 1986a,b) or rat immunoglobulins (Reis *et al.*, 1988; Reis and Boyle, 1990a)] suggests that more applications will be developed. Molecular biological techniques are available for combining different coding sequences for immunoglobulin-binding proteins to generate hybrids, for example, protein A/protein G hybrids (Eliasson *et al.*, 1988, 1989), and the potential for manipulating the affinity of specific proteins at the gene level (Fahnestock *et al.*, 1990) is becoming increasingly practical.

REFERENCES

Ainsworth, S. K., Chen, Z., and Pilia, P., 1990, Therapeutic extracorporeal immunoadsorption with type I Fc receptor in systemic lupus erythematosus, in: *Bacterial Immunoglobulin Binding Proteins*, Vol. 1 (M. D. P. Boyle, ed.) Academic Press, San Diego, pp. 335–346.

Åkerström, B., and Björck, L., 1986, A physicochemical study of protein G, a molecule with unique immuno-globulin G-binding properties, *J. Biol. Chem.* **261:**10240–10247.

Åkerström, B., and Björck, L., 1989, Protein L: An immunoglobulin light chain-binding bacterial protein, *J. Biol. Chem.* **264:**19740–19746.

Åkerström, B., Brodin, T., Reis, K., and Björck, L., 1985, Protein G: A powerful tool for binding and detection of monoclonal and polyclonal antibodies, *J. Immunol.* **135:**2589–2592.

Akkoyunlu, M., Ruan, M., and Forsgren, A., 1991, Distribution of protein D, an immunoglobulin D-binding, among *Haemophilus* strains, *Infect. Immun.* **59:**1231–1238.

Aucouturier, P., Mihaesco, E., Mihaesco, C., and Preud'homme, J. L., 1987, Characterization of jacalin, the human IgA and IgD binding lectin from jackfruit, *Mol. Immunol.* **24:**503–511.

Aucouturier, P., Pineau, N., and Preud'homme, J. L., 1988, A simple procedure for the isolation of human secretory IgA of IgA1 and IgA2 subclass by a jackfruit lectin, jacalin, affinity chromatography, *Mol. Immunol.* **25:**321–322.

Balint, J. P., Ideka, Y., Nagai, T., and Terman, D. S., 1981, Isolation of human and canine IgM utilizing protein A affinity chromatography, *Immunol. Commun.* **10:**533–540.

Björck, L., 1988, Protein L: A novel bacterial cell wall protein with affinity for IgG light chains, *J. Immunol.* **140:** 1194–1197.

Björck, L., and Kronvall, G., 1984, Purification and some properties of streptococcal protein G, a novel IgG binding reagent, *J. Immunol.* **133:**969–974.

Björck, L., Kastern, W., Lindahl, G., and Widebäck, K., 1987, Streptococcal protein G, expressed by streptococci or by *Escherichia coli*, has separate binding sites for human albumin and IgG, *Mol. Immunol.* **24:**1113–1122.

Boyle, M. D. P., 1984, Applications of bacterial Fc receptors in immunotechnology, *Biotechniques* **2:**334–340.

Boyle, M. D. P., 1990, The type I bacterial immunoglobulin-binding protein: Staphylococcal protein A, in: *Bacterial Immunoglobulin Binding Proteins*, Vol. I (M. D. P. Boyle, ed.), Academic Press, San Diego, pp. 17–28.

Boyle, M. D. P., and Faulmann, E. L., 1990, Application of bacteria expressing immunoglobulin-binding proteins to immunoprecipitation reactions, in: *Bacterial Immunoglobulin-Binding Proteins*, Vol. II (M. D. P. Boyle, ed.), Academic Press, San Diego, pp. 273–289.

Boyle, M. D. P., and Reis, K. J., 1987, Bacterial Fc receptors, *Bio/Technology* **5:**697–703.

Boyle, M. D. P., and Reis, K. J., 1990, Antigenic relationships among bacterial immunoglobulin-binding proteins, in: *Bacterial Immunoglobulin-Binding Proteins*, Vol. I (M. D. P. Boyle, ed.), Academic Press, San Diego, pp. 175–186.

Boyle, M. D. P., Wallner, W. A., von Mering, G. O., Reis, K. J., and Lawman, M. J. P., 1985, Interaction of bacterial Fc receptors with goat immunoglobulins, *Mol. Immunol.* **22:**1115–1121.

Boyle, M. D. P., Faulmann, E. L., Otten, R., and Heath, D. H., 1990a, Streptococcal immunoglobulin binding proteins, in: *Pathogenic Mechanisms and Host Responses* (E. M. Ayoub, G. H. Cassell, W. C. Branche, and T. J. Kenny, eds.), ASM Press, Washington, D.C., pp. 273–289.

Boyle, M. D. P., Reis, K. J., and Faulmann, E. L., 1990b, Isolation and functional characterization of bacterial immunoglobulin binding proteins, in: *Bacterial Immunoglobulin Binding Proteins*, Vol. II (M. D. P. Boyle, ed.), Academic Press, pp. 71–89.

Brady, L. J., and Boyle, M. D. P., 1989, Identification of non IgA Fc binding forms and low molecular weight secreted forms of the group B streptococcal b antigen, *Infect. Immun.* **57:**1573–1581.

Bricker, B., Tabatabai, L. B., and Mayfield, J. E., 1991, Immunoglobulin G binding activity of *Brucella abortus*, *Mol. Immunol.* **28:**35–39.

Bywater, R., Horwood, E., and Chichester, U. K., 1978, Elution of immunoglobulins from protein A-Sepharose CL-4B columns, in: *Chromatography of Synthetic and Biological Polymers* (R. Epton, ed.), Halsted Press, New York, pp. 337–340.

Bywater, R., Eriksson, G. B., and Ottosson, T., 1983, Desorption of immunoglobulins from protein A-sepharose CL-4B under mild conditions, *J. Immunol. Methods* **64:**1–6.

Chalon, M. P., Milne, R. W., and Vaerman, J. P., 1979, Interactions between mouse immunoglobulins and staphylococcal protein A, *Scand. J. Immunol.* **9:**359–364.

Chmel, H., 1990, Role of monoclonal antibody therapy in the treatment of infectious disease, *Am. J. Hosp. Pharm.* **3:**11–15.

Coe, J. E., Coe, P. R., and Ross, M. J., 1981, Staphylococcal protein A purification of rodent IgG$_1$ and IgG$_2$ with particular emphasis on syrian hamsters, *Mol. Immunol.* **18:**1007–1012.

Delacroix, D., and Vaerman, J. P., 1979, Simple purification of goat IgG1 and IgG2 subclasses by chromatography on protein A-Sepharose at various pH, *Mol. Immunol.* **16:**837–840.

Duhamel, R. C., Schur, P. H., Brendel, K., and Meezan, E., 1979, pH gradient elution of human IgG$_1$-, IgG$_2$ and IgG$_4$ from protein A-Sepharose, *J. Immunol. Methods* **31:**211–217.

Duhamel, R. C., Meezan, E., and Brendell, K., 1980, The pH dependent binding of gout IgG$_1$ and IgG$_2$ to protein A-Sepharose, *Mol. Immunol.* **17:**29–36.

Eliasson, M., Olsson, A., Palmcrantz, E., Wiberg, K., Inganäs, M., Guss, B., Lindberg, M., and Unlén, M., 1988, Chimeric IgG-binding receptors engineered from staphylococcal protein A and streptococcal protein G, *J. Biol. Chem.* **263:**4323–4327.

Eliasson, M., Andersson, R., Olsson, A., Wigzell, H., and Uhlén, M., 1989, Differential IgG-binding characteristics of staphylococcal protein A, streptococcal protein G, and a chimeric protein AG, *J. Immunol.* **142:**575–581.

Ey, P. L., Prowse, S. J., and Jenkin, C. R., 1978, Isolation of pure IgG$_1$, IgG$_{2a}$ and IgG$_{2b}$ immunoglobulins from mouse serum using protein A–Sepharose, *Immunochemistry* **15:**429–436.

Fahnestock, S. R., 1987, Cloned streptococcal protein G genes, *Trends Biotechnol.* **5:**79–84.

Fahnestock, S. R., Alexander, P., Filpula, D., and Nagle, J., 1990, Structure and evolution of the streptococcal genes encoding protein G, in: *Bacterial Immunoglobulin-Binding Proteins*, Vol. I (M. D. P. Boyle, ed.), Academic Press, San Diego, California, pp. 133–148.

Fant, W. K., 1990, Selecting patients for monoclonal antibody therapy, *Am. J. Hosp. Pharm.* **47:**16–19.

Faulmann, E. L., and Boyle, M. D. P., 1990, Type IIa and type IIb immunoglobulin binding proteins associated with group A streptococci, in: *Bacterial Immunoglobulin Binding Proteins*, Vol. I (M. D. P. Boyle, ed.), Academic Press, pp. 69–83.

Faulmann, E. L., Otten, R. A., Barrett, D. J., and Boyle, M. D. P., 1989, Immunological applications of type III Fc binding proteins. Comparison of different sources of protein G. *J. Immunol. Methods* **123:**269–281.

Faulmann, E. L., Duvall, J. L., and Boyle, M. D. P., 1991, Protein B: A versatile bacterial Fc-binding protein selective for human IgA, *Bio-Techniques* **10:**748–755.

Forsgren, A., and Grubb, A., 1979, Many bacterial species bind human IgD, *J. Immunol.* **122:**1468–1472.

Goding, J. W., 1978, Use of staphylococcal protein A as an immunological reagent, *J. Immunol. Methods* **20:**241–253.

Goldenberg, D. M., 1989, Targeted cancer treatment, *Immunol. Today* **10:**286–288.

Haake, D. A., Franklin, E. C., and Frangione, B., 1982, The modification of human immunoglobulin binding to staphylococcal protein A using diethylpyrocarbonate, *J. Immunol.* **129:**190–192.

Harboe, M., and Fölling, I., 1974, Recognition of two distinct groups of human IgM and IgA based on different binding to staphylococci, *Scand. J. Immunol.* **3:**471–482.

Hedén, L.-O., Frithz, E., and Lindahl, G., 1991, Molecular characterization of an IgA receptor from group B streptococci: Sequence of the gene, identification of a proline-rich region with unique structure and isolation of N-terminal fragments with IgA-binding capacity, *Eur. J. Immunol.* **21:**1481–1490.

Hjelm, H., 1975, Isolation of IgG$_3$ from normal human sera and from a patient with multiple myeloma by using protein A-sepharose 4B, *Scand. J. Immunol.* **4:**633–640.

Hjelm, H., Hjelm, K., and Sjöquist, J., 1972, Protein A from *Staphylococcus aureus*. Its isolation by affinity chromatography and its use as an immunosorbent for isolation of immunoglobulins, *FEBS Lett.* **28:**73.

Hudson, G. A., 1986, Characterization of antisera, in: *Manual of Clinical Laboratory Immunology*, 3rd ed. (N. R. Rose, H. Friedman, and J. L. Fahey, eds.), ASM Publications, Washington, D.C., pp. 9–13.

Inganäs, M., 1981, Comparison of mechanisms of interaction between protein A from *Staphylococcus aureus* and human monoclonal IgG, IgA and IgM in relation to the classical Fc and the alternative F(ab')$_2$ protein A interactions, *Scand. J. Immunol.* **13:**343–352.

Jungbauer, A., Tauer, C., Reiter, M., Purtscher, M., Wenisch, E., Steindl, F., Buchacher, A., and Katinger, H., 1989, Comparison of protein A, protein G and copolymerized hydroxyapatite for the purification of human monoclonal antibodies, *J. Chromatogr.* **476:**257–268.

Kohler, G., and Milstein, C., 1975, Continuous cultures of fused cells secreting antibody of predefined specificity, *Nature (London)* **256:**495–497.

Langone, J., 1982a, Protein A of *Staphylococcus aureus* and related immunoglobulin receptors produced by streptococci and pneumococci, *Adv. Immunol.* **32:**157–252.

Langone, J. J., 1982b, Applications of immobilized protein in immunochemical techniques, *J. Immunol. Methods* **55:**277–296.

Langone, J. J., Boyle, M. D. P., and Borsos, T., 1977, ^{125}I protein A: Applications to the quantitative determination of fluid phase and cell-bound IgG, *J. Immunol. Methods* **18:**281–293.

Lawman, M. D. P., Joiner, S., Gauntlett, D. R., and Boyle, M. D. P., 1985, A rapid purification for bovine IgG$_2$. Evidence for IgG$_{2a}$ and IgG$_{2b}$ sub-classes, *Comp. Immunol. Microbiol. Infect. Dis.* **8:**1–8.

Lindahl, G., 1989, Cell surface proteins of a group A streptococcus type M4: The IgA receptor and a receptor related to M proteins are coded for by closely linked genes, *Mol. Gen. Genet.* **216:**372–379.

Lindahl, G., 1990, Receptor for immunoglobulins in *Clostridium perfringens*: Binding in the F(ab')$_2$ region, in: *Bacterial Immunoglobulin-Binding Proteins*, Vol. I (M. D. P. Boyle, ed.), Academic Press, San Diego, pp. 257–265.

Lindahl, G., and Åkerström, B., Receptor for IgA in group A streptococci: Cloning of the gene and characterization of the protein expressed in *Escherichia coli*, *Mol. Microbiol.* **3:**239–247.

Lindahl, G., Åkerström, B., Vaerman, J.–P., and Stenberg, L., 1990, Characterization of an IgA receptor from group B streptococci: specificity for serum IgA, *Eur. J. Immunol.* **20:**2241–2247.

Lindahl, G., and Kronvall, G., 1988, Nonimmune binding of Ig to *Clostridium perfringens*. Preferential binding of IgM and aggregated IgG, *J. Immunol.* **140:**1223–1227.

Linscott, W. D., 1991, *Linscott's Directory of Immunological and Biological Reagents*, Linscott, Mill Valley, California.

Little, M., Siebert, C., and Matson, R., 1988, Enhanced antigen binding to IgG molecules immobilized to a chromatographic support via the Fc domains, *Biochromatography* **3:**156–160.

Lyerly, D. M., Carrig, P. E., and Wilkins, T. D., 1989, Nonspecific binding of mouse monoclonal antibodies to *Clostridium difficile* toxins A and B, *Curr. Microbiol.* **19:**303–306.

MacKenzie, M. R., Warner, N. L., and Mitchell, G. F., 1978, The binding of murine immunoglobulins to staphylococcal protein A, *J. Immunol.* **120:**1493–1496.

MacSween, J. M., and Eastwood, S. L., 1977, Recovery of immunologically active antigen from staphylococcal protein A-antibody adsorbent, *J. Immunol. Methods* **23:**259–267.

MacSween, J. M., and Eastwood, S. L., 1981, Recovery of antigen from staphylococcal protein A-antibody adsorbents, *Methods Enzymol.* **73:**459–471.

Mandaro, R. M., Sujata, R., and Hou, K. C., 1987, Filtration supports for affinity separation, *Bio/Technology* **5:**928–932.

Martin, L. N., 1982, Chromatographic fractionation of rhesus monkey (*Macaca mulatta*) IgG subclasses using DEAE cellulose and protein A-Sepharose, *J. Immunol. Methods* **50:**319–329.

Metzger, D. W., Ch'ng, L.-K., Miller, A., and Sercarz, E. E., 1984, The expressed lysozyme-specific B-cell repertoire. I. Heterogeneity in the monoclonal anti-HEL specificity repertoire, and its difference from the *in situ* repertoire, *Eur. J. Immunol.* **14:**87–93.

Miller, T. J., and Stone, H. O., 1978, The rapid isolation of ribonuclease-free immunoglobulin G by protein A-Sepharose affinity chromatography, *J. Immunol. Methods* **24:**111–125.

Milstein, C., 1986, Overview: Monoclonal antibodies: in: *Handbook of Experimental Immunology*, Vol. 4 (D. M. Wier, ed.), Blackwell Scientific Publications, Oxford, 1–12.

Monestier, M., Debbas, M. E., and Zhou, S., 1989, Rat monoclonal antibodies to murine IgM determinants: Application to antibody purification, *Hybridoma* **8:**631–637.

Myhre, E. B., 1990, Interaction of bacterial immunoglobulin receptors with sites in the Fab region, in: *Bacterial Immunoglobulin Binding Proteins*, Vol. I (M. D. P. Boyle, ed.), Academic Press, San Diego, pp. 243–256.

Myhre, E. B., and Erntell, M., 1985, A non-immune interaction between the light chain of human immunoglobulin and a surface component of a *Peptococcus magnus* strain, *Mol. Immunol.* **22:**879–885.

Myhre, E. B., and Kronvall, G., 1981, Immunoglobulin specificities of defined types of streptococcal Ig receptors, in: *Basic Concepts of Streptococci and Streptococcal Diseases* (S. E. Holm and P. Christensen, eds.), Redbook, Ltd., Chertsey, Surrey, England, pp. 209–210.

Nielsen, K., Stilwell, K., Stemshorn, B., and Duncan, R., 1981, Ethylenediaminetetraacetic acid (disodium salt) labile bovine immunoglobulin M Fc binding to *Brucella abortus*: A cause of nonspecific agglutination, *J. Clin. Microbiol.* **14**:32–38.

Nilson, B., Björck, L., and Åkerström, B., 1986, Detection and purification of rat and goat immunoglobulin G antibodies using protein G-based solid-phase radioimmunoassays, *J. Immunol. Methods* **91**:275–281.

Nilsson, R., Myhre, E., Kronvall, G., and Sjögren, H. O., 1982, Fractionation of rat IgG subclasses and screening for IgG Fc-binding to bacteria, *Mol. Immunol.* **19**:119–126.

Nilsson, R., Myhre, E., Kronvall, G., and Sjögren, H. O., 1983, Different protein A immunosorbents may have different binding specificity for rat immunoglobulins, *J. Immunol. Methods* **62**:241–245.

Ohlson, S., Nilsson, R., Niss, U., Kjellbert, B.-M., and Freiburghaus, C., 1988, A novel approach to monoclonal antibody separation using high performance liquid affinity chromatography (HPLAC) with SelectiSpher-10 protein G, *J. Immunol. Methods* **114**:175–180.

Otten, R. A., and Boyle, M. D. P., 1991a, Characterization of protein G expressed by human group C and G streptococci, *J. Microbiol. Methods* **13**:185–200.

Otten, R. A., and Boyle, M. D. P., 1991b, The mitogenic activity of type III bacterial Ig binding proteins (protein G) for human peripheral blood lymphocytes is not related to their ability to react with human serum albumin or IgG, *J. Immunol.* **146**:2588–2595.

Pharmacia, 1986, Separation News 13.5, Pharmacia Laboratory Separation Division, Piscataway, New Jersey.

Raeder, R., Faulmann, E., and Boyle, M. D. P., 1991a, Evidence for functional heterogeneity in IgG Fc-binding proteins associated with group A streptococci, *J. Immunol.* **146**:1247–1253.

Raeder, R., Otten, R., and Boyle, M. D. P., 1991b, Comparison of albumin receptors expressed on bovine and human group C streptococci, *Infect. Immun.* **59**:609–616.

Raeder, R. A., Otten, R. A., and Boyle, M. D. P., 1991c, Isolation and partial characterization of a type IV bacterial immunoglobulin binding protein, *Mol. Immunol.* **28**:661–671.

Reis, K. J., and Boyle, M. D. P., 1990a, Isolation and characterization of a type VI bacterial immunoglobulin binding protein, in: *Bacterial Immunoglobulin Binding Proteins*, Vol. I (M. D. P. Boyle, ed.), Academic Press, San Diego, pp. 165–173.

Reis, K. J., and Boyle, M. D. P., 1990b, Production of polyclonal antibodies to immunoglobulin binding proteins, in: *Bacterial Immunoglobulin Binding Proteins*, Vol. II (M. D. P. Boyle, ed.), Academic Press, San Diego, pp. 105–124.

Reis, K. J., Ayoub, E. M., and Boyle, M. D. P., 1983, Detection of receptors for the Fc region of IgG on streptococci, *J. Immunol. Methods* **59**:83–94.

Reis, K. J., Ayoub, E. M., and Boyle, M. D. P., 1984a, Streptococcal Fc receptors. I. Isolation and partial characterization of the receptors from a group C streptococcus, *J. Immunol.* **132**:3091–3097.

Reis, K. J., Ayoub, E. M., and Boyle, M. D. P., 1984b, Streptococcal Fc receptors. II. Comparison of the reactivity of a receptor from a group C streptococcus with staphylococcal protein A, *J. Immunol.* **132**:3098–3102.

Reis, K. J., Ayoub, E. M., and Boyle, M. D. P., 1985, A rapid method for the isolation and characterization of a homogeneous population of streptococcal Fc receptors, *J. Microbiol. Methods* **4**:45–58.

Reis, K. J., Hansen, H. F., and Björck, L., 1986, Extraction and characterization of streptococcal IgG Fc receptors from group C and group G streptococci, *Mol. Immunol.* **23**:425–431.

Reis, K. J., Siden, E. J., and Boyle, M. D. P., 1988a, Selective colony blotting to expand bacterial surface receptor: Applications to receptors for rat immunoglobulins, *Bio/Techniques* **6**:130–136.

Ricardo, M. J., Trauy, R. L., and Grimm, D. T., 1981, Effect of pH on the binding between guinea pig 7sG isotypes and protein A: Evidence for intra-isotypic binding heterogeneity, *J. Immunol.* **127**:946–951.

Rousseaux, J., Picque, M. T., Bazin, H., and Biserte, G., 1981, Rat IgG subclasses: Differences in affinity to protein A-Sepharose, *Mol. Immunol.* **18**:639–645.

Ruan, M., Akkoyunlu, M., Grubb, A., and Forsgren, A., 1990, Protein D of *H. Influenza*; a novel bacterial surface protein with affinity for human IgD, *J. Immunol.* **145**:3379–3384.

Russell, S. H., Carter, R., and Firkin, F. C., 1984, Determination of immunoglobulin class by pattern of absorption to protein A-Sepharose coupled with immunoglobulin class specific antibodies, *J. Immunol. Methods* **66**:323–326.

Russell-Jones, G. J., Gotschlich, E. C., and Blake, M. S., 1984, A surface receptor specific for human IgA on Group B streptococci possessing the Ibc protein antigen, *J. Exp. Med.* **160**:1467–1475.

Sasso, E. H., Silverman, G. J., and Mannik, M. 1989, Human IgM molecules that bind staphylococcal protein A contain V_HIIIH. chains, *J. Immunol.* **142**:2778–2783.

Schneider, C., Newman, R. A., Sutherland, D. R., Asser, U., and Greavews, M. F., 1982, A one-step purification of membrane proteins using a high efficiency immunomatrix, *J. Biol. Chem.* **257**:10766–10769.

Schwartz, L., 1990, Use of immobilized protein A, to purify immunoglobulins, in: *Bacterial Immunoglobulin-Binding Proteins*, Vol. II (M. D. P. Boyle, ed.), Academic Press, San Diego, pp. 309–339.

Scott, S., and Juarez-Salinas, H., 1990, Purification and quantitation of monoclonal antibodies by affinity chromatography with immobilized protein A, in: *Bacterial Immunoglobulin-Binding Proteins*, Vol. II (M. D. P. Boyle, ed.), Academic Press, San Diego, pp. 341–354.

Seppälä, I., Sarvas, H., Péterfy, F., and Mäkelä, O., 1981, The four subclasses of IgG can be isolated from mouse serum by using protein A-Sepharose, *Scand J. Immunol.* **14**:335–342.

Sisson, S. N., 1990, Use of fluorescent-conjugated bacterial immunoglobulin-binding proteins, in: *Bacterial Immunoglobulin-Binding Proteins*, Vol. II (M. D. P. Boyle, ed.), Academic Press, San Diego, pp. 197–203.

Sjöbring, U., Falkenberg, C., Nielsen, E., Åkerström, B., and Björck, L., 1988, Isolation and characterization of a 14-kD albumin-binding fragment of streptococcal protein G., *J. Immunol.* **140**:1595–1599.

Sjöbring, U., Björck, L., and Kastern, W., 1989a, Protein G genes: Structure and distribution of IgG-binding and albumin-binding domains, *Mol. Microbiol.* **3**:319–328.

Sjöbring, U., Trojnar, J., Grubb, A., Åkerström, B., and Björck, L., 1989b, Ig-binding bacterial proteins also bind proteinase inhibitors, *J. Immunol.* **143**:2948–2954.

Sjöbring, U., Björck, L., and Kastern, W., 1991, Streptococcal protein G: Gene structure and protein binding properties, *J. Biol. Chem.* **266**:399–405.

Villemez, C. L., Russel, M. A., and Carlo, P. L., 1984, Mouse IgG$_1$ heterogeneity: Variable binding of monoclonal IgG$_1$ antibodies to protein A-Sepharose, *Mol. Immunol.* **21**:993–998.

von Mering, G. O., and Boyle, M. D. P., 1986, Comparison of type III Fc receptors associated with group C or group G streptococci, *Mol. Immunol.* **23**:811–821.

Wallner, W. A., Lawman, M. J. P., and Boyle, M. D. P., 1987, Reactivity of bacterial Fc receptors with bovine immunoglobulins: Implications and limitations for quantitative immunochemistry, *Appl. Microbiol. Biotechnol.* **27**:168–173.

Webb-Walker, B., 1990, Use of immobilized protein G to isolate IgG, in: *Bacterial Immunoglobulin-Binding Proteins*, Vol. II (M. D. P. Boyle, ed.), Academic Press, San Diego, pp. 355–367.

Widders, P. R., Stoke, C. R., Newby, T. J., and Bourne, F. J., 1985, Nonimmune binding of equine immunoglobulin by the causative organism of contagious equine metritis, *Taylorella equigenitalis*, *Infect. Immun.* **48**:417–421.

Yarnall, M., and Boyle, M. D. P., 1986a, Isolation and characterization of Type IIa and Type IIb Fc receptors from a group A streptococcus, *Scand. J. Immunol.* **24**:549–557.

Yarnall, M., and Boyle, M. D. P., 1986b, Identification of a unique receptor on a group A streptococcus for the Fc region of human IgG$_3$, *J. Immunol.* **136**:2670–2673.

Yarnall, M., and Widders, P. R., 1990, Type V Fc receptor from *Streptococcus zooepidemicus*, in: *Bacterial Immunoglobulin-Binding Proteins*, Vol. I (M. D. P. Boyle, ed.), Academic Press, San Diego, pp. 155–164.

Yarnall, M., Widders, P. R., and Corbeil, L. B., 1988, Isolation and characterization of Fc receptors from *Haemophilus somnus*, *Scand. J. Immunol.* **28**:129–137.

7

Affinity Chromatography of Oligosaccharides and Glycopeptides with Immobilized Lectins

Tsutomu Tsuji, Kazuo Yamamoto, and Toshiaki Osawa

1. INTRODUCTION

Lectins are proteins (glycoproteins) that specifically bind to a particular carbohydrate structure and have been shown to be widely distributed among plants, animals, and bacteria. Although the functional role of lectins in living organisms has not been well understood, over 100 kinds of lectins have been isolated and their binding specificities characterized (Goldstein and Hayes, 1978; Liener *et al.*, 1986; Lis and Sharon, 1986; Sharon and Lis, 1989). These lectins have been extensively used for studies on glycoconjugates, especially (1) histochemical detection of cell-surface carbohydrate chains, (2) staining and structural estimation of glycoproteins on Western-blotted membranes or of glycolipids on thin-layer chromatography (TLC) plates, (3) separation of cells with different cell-surface carbohydrate chains, and (4) isolation and fractionation of glycoproteins or oligosaccharides by using affinity chromatography. For the fractionation of glycoproteins and oligosaccharides, a variety of chromatographic techniques have been employed. These molecules can be separated on the basis of differences in ionic charge, degree of polymerization, or monosaccharide sequences. Affinity chromatography with immobilized lectins seems to be a quite effective technique, because it can achieve not only fractionation of glycoproteins or oligosaccharides but also their structural assessment on the basis of the elution profile from an immobilized lectin column. If affinity columns of several lectins with different binding specificities are serially combined, this technique would be a more powerful tool for the separation of glycoproteins and oligosaccharides. In this chapter, we focus on the separation of glycopeptides and oligosaccharides derived from glycoproteins by affinity chromatography on immobilized lectin columns. The lectins mentioned in this chapter are listed in Table 1.

Tsutomu Tsuji, Kazuo Yamamoto, and Toshiaki Osawa • Division of Chemical Toxicology and Immunochemistry, Faculty of Pharmaceutical Sciences, The University of Tokyo, Tokyo 113, Japan.

Molecular Interactions in Bioseparations, edited by That T. Ngo. Plenum Press, New York, 1993.

TABLE 1

Classification of Lectins Based on Carbohydrate Binding Specificity

Monosaccharide specificity	Lectin	Commonly used abbreviation	High-afinity interaction with
Sialic acid	*Allomyrina dichotoma* (beetle)	Allo-A	α2-3 or α2-6Galβ1-4GlcNAc
	Limax flavus (slug)	LFA	Neu5Ac
	Limulus polyphemus (horseshoe crab)	Limulin	
	Maackia amurensis	MAL or MAM	Neu5Acα2-3Galβ1-4GlcNAc
	Sambucus nigra (elderberry)	SNA	SAα2-6Gal/GalNAc
Fucose	*Aleuria aurantia*	AAL	
Mannose	Concanavalin A	Con A	Oligomannose type, biantennary complex type
	lens culinaris (lentil)	LCA	Biantennary complex type with a Fuc resudue in the core
	Pisum sativum (pea)	PSA	Similar to lentil
	Vicia faba (fava)	VFA	Similar to lentil
Galactose or GalNAc	*Agaricus bisporus* (mushroom)	ABA	Ser/Thr-linked sugar chain
	Arachis hypogaea (peanut)	PNA	Ser/Thr-linked sugar chain
	Bauhinia purpurea	BPA	Ser/Thr-linked sugar chain
	Glycine max (soybean)	SBA	Ser/Thr-linked sugar chain GalNAc > Gal
	Ricinus communis (castor bean)	RCA-I or RCA$_{120}$	Primarily complex type
	Vicia villosa	VVA	Ser/Thr-linked sugar chain GalNAc > Gal
GlcNAc	*Datura stramonium*	DSA	Poly-*N*-acetyllactosamine complex type with 2,6-di-O-substituted α-Man residue
	Erythrina variegata	EVA	Poly-*N*-acetyllactosamine
	Lycopersicon esculentum	LEA	Poly-*N*-acetyllactosamine
	Pokewood mitogen	PWM	Poly-*N*-acetyllactosamine
	Solanum tuberosum (potato)	STA	Poly-*N*-acetyllactosamine
	Wheat germ agglutinin	WGA	Poly-*N*-acetyllactosamine, sialic acid-containing sugar chain, hybrid type with a bisecting GlcNAc

2. CARBOHYDRATE CHAINS FOUND IN GLYCOPROTEINS

A number of recent studies have suggested that glycoproteins on the cell surface constitute a great number of surface markers of cells in various tissues at various differentiation stages and with various functions. Furthermore, it has become apparent that these cell surface markers control and determine cell–ligand and cell–cell interactions in many important cellular phenomena.

We can classify the structure of the carbohydrate chains of cell-surface glycoproteins into two groups, termed asparagine (Asn)-linked (*N*-linked) carbohydrate chains and serine (Ser)/threonine (Thr)-linked (*O*-linked) carbohydrate chains. The former group is further subdivided into three types: oligomannose type, complex type, and hybrid type. Typical carbohydrate structures of these types are shown in Fig. 1. An oligomannose-type chain is composed of mannose (Man) and N-acetylglucosamine (GlcNAc) residues and is a precursor to complex-type and hybrid-type carbohydrate chains. Actually, three types of carbohydrate chains share a common "core" pentasaccharide structure with a sequence of Manα1-6(Manα1-3)Manβ1-4GlcNAcβ1-4GlcNAcβ-Asn. An oligomannose-type carbohydrate chain has five to nine mannose residues; the one located adjacent to the GlcNAc-GlcNAc (N,N'-diacetylchitobiose) sequence is β-linked, and the other residues are α-linked. A complex-type carbohydrate chain is characterized by the presence of two to five outer chains with the sequence Galβ1-4GlcNAcβ1-(N-acetyllactosamine unit) attached to α-mannose residues of the core pentasaccharide structure to give either a branched structure or a linear structure with tandem repeats of N-acetyllactosamine units (Fig. 2). The galactose residues of a complex-type chain are often substituted by sialic acid residues via α2-3 or α2-6 linkages. Further structural variations are caused by the attachment of a fucose residue to the C-6 position of the innermost GlcNAc residue or to the C-3 position of a GlcNAc residue in an outer chain. Moreover, the β-linked mannose residue is sometimes substituted at the C-4 position by a GlcNAc residue. This GlcNAc residue is called a "bisecting" GlcNAc residue. A hybrid-type carbohydrate chain has characteristics of both oligomannose-type and complex-type chains. Some carbohydrate chains of this type have a bisecting GlcNAc residue linked to the C-4 position of the β-mannose residue.

Ser/Thr-linked carbohydrate chains are commonly found in mucins. Therefore, they are sometimes called mucin-type carbohydrate chains. The simplest structure of this type of carbohydrate chains is GalNAcα-Ser/Thr. In addition to this structure, galactosylation and/ or sialylation take place in most Ser/Thr-linked carbohydrate chains (Fig. 1). These carbohydrate chains are usually less polymerized than Asn-linked carbohydrate chains. However, some Ser/Thr-linked carbohydrate chains contain a long sugar sequence such as repeats of N-acetyllactosamine units as in the case of complex-type carbohydrate chains.

3. FRACTIONATION OF CARBOHYDRATE CHAINS BY LECTIN AFFINITY CHROMATOGRAPHY

Most of the well-characterized lectins and their immobilized derivatives on agarose are now commercially available. However, we can couple a lectin to agarose or other insoluble matrices for chromatography by the usual methods employed for the immobilization of various proteins. It is recommended that the coupling reaction be carried out in the presence of a mono- or a disaccharide specific for the lectin, which is often called a "haptenic sugar," to prevent the lectin from denaturation. Usually, the final product bears 1–10 mg of lectin per milliliter of gel. A longer column of immobilized lectin sometimes gives better chromatographic resolution for oligosaccharides or glycopeptides with rather low affinity for the lectin. A flow rate of 5–30 cm/h has been used in most cases. The carbohydrates tightly bound to a lectin column can be eluted with a buffer containing a haptenic sugar or its methylglycoside.

3.1. Separation of Sialic Acid-Containing Oligosaccharides or Glycopeptides

Most sialic acid (N-acetyl- or N-glycolylneuraminic acid) (SA) residues are present at the terminals of carbohydrate chains, in which they link to other sugar residues in a

I. Asn-linked Carbohydrate Chain

A. Oligomannose-type

$$
\begin{array}{l}
(\text{Man}\alpha1\text{-}2)\text{Man}\alpha1 \searrow_6 \\
\phantom{(\text{Man}\alpha1\text{-}2)}\text{Man}\alpha1 \nearrow^3 \text{Man}\alpha1 \searrow \\
(\text{Man}\alpha1\text{-}2)\text{Man}\alpha1 \phantom{\text{Man}\alpha1} {}_6 \\
\phantom{(\text{Man}\alpha1\text{-}2)\text{Man}\alpha1} {}_3\text{Man}\beta1\text{-}4\text{GlcNAc}\beta1\text{-}4\text{GlcNAc-Asn} \\
(\text{Man}\alpha1\text{-}2\ \text{Man}\alpha1\text{-}2)\ \text{Man}\alpha1 \nearrow
\end{array}
$$

B. Complex-type

$$
\begin{array}{l}
(\text{Fuc}\alpha1) \\
| \\
(\text{SA}\alpha2\text{-}3\ \text{or}\ 6)\text{Gal}\beta1\text{-}4\text{GlcNAc}\beta1\text{-}2\text{Man}\alpha1 \searrow 6 \\
{}_6 \\
(\text{GlcNAc}\beta1\text{-}4)\ \text{Man}\beta1\text{-}4\text{GlcNAc}\beta1\text{-}4\text{GlcNAc-Asn} \\
{}_3 \\
(\text{SA}\alpha2\text{-}3\ \text{or}\ 6)\text{Gal}\beta1\text{-}4\text{GlcNAc}\beta1\text{-}2\text{Man}\alpha1 \nearrow
\end{array}
$$

C. Hybrid-type

$$
\begin{array}{l}
\text{Man}\alpha1 \searrow_6 \\
\phantom{\text{Man}\alpha1 \searrow}{}_6\text{Man}\alpha1 \\
\text{Man}\alpha1 \nearrow^3 \phantom{\text{Man}\alpha1} \searrow \\
{}_6 \\
(\text{GlcNAc}\beta1\text{-}4)\ \text{Man}\beta1\text{-}4\text{GlcNAc}\beta1\text{-}4\text{GlcNAc-Asn} \\
{}_3 \\
\text{Gal}\beta1\text{-}4\text{GlcNAc}\beta1\text{-}2\text{Man}\alpha1 \nearrow
\end{array}
$$

II. Ser/Thr-linked Carbohydrate Chain

$$
\begin{array}{c}
\text{SA}\alpha2 \\
| \\
6 \\
\text{SA}\alpha2\text{-}3\text{Gal}\beta1\text{-}3\text{GalNAc}\alpha\text{-Ser/Thr}
\end{array}
$$

FIGURE 1. Typical carbohydrate chains found in cell-surface glycoproteins.

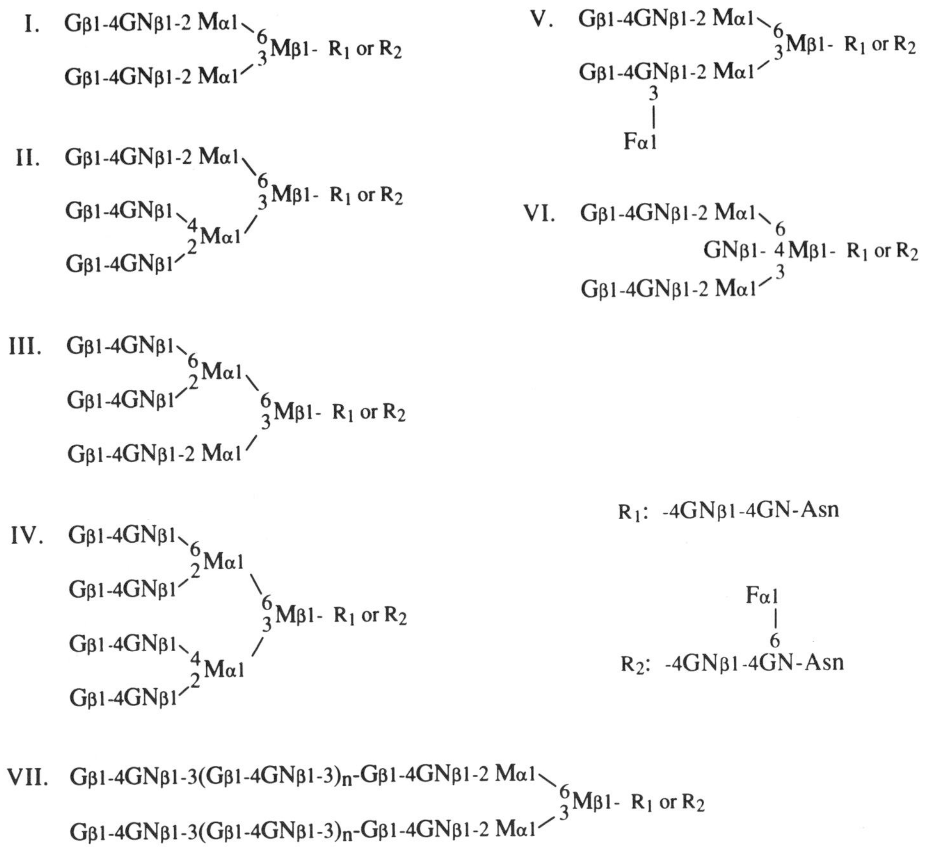

I. Gβ1-4GNβ1-2 Mα1
$\underset{3}{\overset{6}{\diagdown}}$Mβ1- R₁ or R₂
Gβ1-4GNβ1-2 Mα1

V. Gβ1-4GNβ1-2 Mα1
$\underset{3}{\overset{6}{\diagdown}}$Mβ1- R₁ or R₂
Gβ1-4GNβ1-2 Mα1
3
|
Fα1

II. Gβ1-4GNβ1-2 Mα1
$\underset{3}{\overset{6}{\diagdown}}$Mβ1- R₁ or R₂
Gβ1-4GNβ1
$\underset{2}{\overset{4}{\diagdown}}$Mα1
Gβ1-4GNβ1

VI. Gβ1-4GNβ1-2 Mα1
$\overset{6}{\diagdown}$
GNβ1- 4 Mβ1- R₁ or R₂
Gβ1-4GNβ1-2 Mα1
3

III. Gβ1-4GNβ1
$\underset{2}{\overset{6}{\diagdown}}$Mα1
Gβ1-4GNβ1
$\underset{3}{\overset{6}{\diagdown}}$Mβ1- R₁ or R₂
Gβ1-4GNβ1-2 Mα1

R₁: -4GNβ1-4GN-Asn

IV. Gβ1-4GNβ1
$\underset{2}{\overset{6}{\diagdown}}$Mα1
Gβ1-4GNβ1
$\underset{3}{\overset{6}{\diagdown}}$Mβ1- R₁ or R₂
Gβ1-4GNβ1
$\underset{2}{\overset{4}{\diagdown}}$Mα1
Gβ1-4GNβ1

Fα1
|
6
R₂: -4GNβ1-4GN-Asn

VII. Gβ1-4GNβ1-3(Gβ1-4GNβ1-3)ₙ-Gβ1-4GNβ1-2 Mα1
$\underset{3}{\overset{6}{\diagdown}}$Mβ1- R₁ or R₂
Gβ1-4GNβ1-3(Gβ1-4GNβ1-3)ₙ-Gβ1-4GNβ1-2 Mα1

G: Galactose   M: Mannose   GN: *N*-acetylglucosamine   F: Fucose

FIGURE 2. Variations of Asn-linked complex-type carbohydrate chains.

variety of linkages. The sialylated sequences most commonly found in glycoproteins are SAα2-3Galβ1-4GlcNAc- and SAα2-6Galβ1-4GlcNAc- in Asn-linked carbohydrate chains or SAα2-3Galβ1-3GalNAc- and SAα2-6GalNAc- in Ser/Thr-linked carbohydrate chains.

Since sialylated glycopeptides or oligosaccharides are negatively charged, anion-exchange chromatography or paper electrophoresis can be used for their separation. On the other hand, several lectins have been reported to interact with sialic acid-containing carbohydrate chains, e.g., wheat germ agglutinin (WGA; Bhavanandan and Katlic, 1979; Peters *et al.*, 1979), *Limax flavus* (slug) agglutinin (LFA; Miller *et al.*, 1982), *Sambucus nigra* (elderberry) agglutinin (SNA; Shibuya *et al.*, 1987a,b), *Allomyrina dichotoma* (beetle) lectin (Allo-A; Sueyoshi *et al.*, 1988a; Yamashita *et al.*, 1988), *Maackia amurensis* leukoagglutinin (mitogen) (MAL or MAM; Kawaguchi *et al.*, 1974; Wang and Cummings, 1988; Knibbs *et al.*, 1991), and *Limulus polyphemus* (horseshoe crab) lectin (limulin; Roche and Monsigny, 1974).

Among these sialic acid-specific lectins, several were shown to have somewhat distinct binding specificities (Table 1). LFA has a rather broad specificity; i.e., it can bind *N*-acetylneuraminic acid (Neu5Ac) in any linkage and in any position in a glycoprotein.

Allo-A firmly binds a carbohydrate chain having a sialylated Galβ1-4GlcNAc-sequence irrespective of whether the sialic acid residue is α2-3- or α2-6-linked. Although this lectin, like LFA, does not discriminate between α2-3 and α2-6 linkages, it does discriminate Asn-linked complex-type carbohydrate chains from Ser/Thr-linked mucin-type carbohydrate chains; namely, it does not bind a sialylated Galβ1-3GalNAc- sequence, which is a typical core structure of mucin-type carbohydrate chains, but does bind a sialylated Galβ1-4GlcNAc- sequence, which is commonly found in complex-type carbohydrate chains.

MAL has also been shown to react strongly with a complex-type carbohydrate chain with an SAα2-3Galβ1-4GlcNAc-sequence. However, it fails to bind to an SAα2-6Galβ1-4GlcNAc- sequence, which is also found in complex-type oligosaccharides, indicating that this lectin is useful for the separation of complex-type carbohydrate chains with a sialic acid residue α2-3-linked to the penultimate galactose residue from those with an α2-6-linked sialic acid residue. It has also been shown that this lectin column does not react with Ser/Thr-linked carbohydrate chains, even though they contain a sialic acid residue linked via α2-3 to a galactose residue.

In contrast to MAL, an immobilized SNA column binds sialylated carbohydrate chains containing either an SAα2-6Gal- or an SAα2-6GalNAc- sequence, whereas those containing an α2-3-linked sialic acid residue cannot be bound by the column. This lectin column binds Ser/Thr-linked carbohydrate chains, as well as Asn-linked complex-type carbohydrate chains if they have an SAα2-6Gal or SAα2-6GalNAc- sequence.

An immobilized lectin from *Agaricus bisporus* (mushroom) (Sueyoshi *et al.*, 1985), which has high affinity for a mucin-type glycopeptide with the sequence Galβ1-3GalNAcα1-Ser/Thr, can also bind its sialylated derivative, but has no apparent interaction with sialylated complex-type glycopeptides. Therefore, this lectin can be used for the separation of sialylated glycopeptides of the mucin type from those of the complex type. The specificity of this lectin will be discussed in detail in Section 3.3.

The combination of affinity chromatography with these immobilized sialic acid-specific lectins facilitates (1) the separation of sialic acid-containing oligosaccharides or glycopeptides from those lacking sialic acids, (2) the fractionation of oligosaccharides or glycopeptides which are heterogeneous in the linkage of sialic acid residues, and (3) the distinction between sialic acid-containing Ser/Thr-linked and Asn-linked carbohydrate chains.

3.2. *Separation of Asn-Linked Carbohydrate Chains*

As mentioned in the previous section, there are a number of structural variations of Asn-linked carbohydrate chains (Fig. 2). In this section, the possible scheme for separating these various carbohydrate chains by the use of immobilized lectins is discussed with reference to the detailed specificity of each lectin.

3.2.1. Separation of Complex-Type Chains from Oligomannose-Type Chains

To separate complex-type carbohydrate chains from oligomannose-type chains, galactose-binding lectins have been employed, since complex-type chains contain at least one Galβ1-4GlcNAc- sequence in their peripheral portions. For this purpose, *Ricinus communis* agglutinin (RCA-I or RCA$_{120}$) has been most commonly used (Baenziger and Fiete, 1979b). The complex-type chains can be bound by an RCA-I-agarose column and eluted with a buffer containing lactose, whereas oligomannose-type chains, which lack a galactose residue, cannot be retained on the column. It should be noted that most complex-type chains have sialic acid residues in their termini, which decrease their affinity for an immobilized RCA-I column. Therefore, the treatment of the samples with sialidase prior to application to the column should be considered if the oligosaccharides or the glycopeptides contain sialic acid residues. Conversely, sialic acid-specific lectins can also be used for the separation of complex-type chains as described in Section 3.1. Some hybrid-type carbohydrate chains carrying a Galβ1-4GlcNAc- sequence are also bound by an RCA-I-agarose column.

Concanavalin A (Con A) is a mannose-specific lectin and has much higher affinity for oligomannose-type chains than complex-type chains. By utilizing the difference in their affinities for Con A-agarose, we can separate oligomannose-type chains from complex-type chains. The binding specificity of immobilized Con A is discussed further in the following section.

3.2.2. Separation of Biantennary Complex-Type Chains from Tri- and Tetraantennary Chains

One of the variations of complex-type carbohydrate chains is caused by the difference in the number of outer chains consisting of a Galβ1-4GlcNAcβ1- sequence (*N*-acetyllactosamine unit) that are attached to α-linked mannose residues in the core pentasaccharide. The number of outer chains of a complex-type carbohydrate chain varies from two to four (some glycoproteins occasionally have five outer chains), and those having two, three, and four outer chains are called bi-, tri-, and tetraantennary complex-type carbohydrate chains, respectively (structures I–IV in Fig. 2).

The biantennary complex-type chains have been shown to be separated from tri- or tetraantennary chains on a Con A-agarose column. Con A has a high affinity for an α-linked mannose residue with unsubstituted hydroxyl groups at C-3, C-4, and C-6. It has been indicated that at least two residues of unsubstituted or 2-*O*-monosubstituted α-linked mannose are required for oligosaccharides or glycopeptides to be retained on a Con A-agarose column (Ogata *et al.*, 1975; Krusius *et al.*, 1976; Baenziger and Fiete, 1979a; Ohyama *et al.*, 1985; Narasimhan *et al.*, 1986). The biantennary complex-type chain has two 2-*O*-monosubstituted α-linked mannose residues, and it is bound by a Con A-agarose column. On the contrary, neither the triantennary structure, with only one 2-*O*-monosubstituted α-mannose residue, nor the tetraantennary structure, without a 2-*O*-monosubstituted α-mannose residue, can be retained on a Con A-agarose column.

Since the binding of a biantennary complex-type chain to Con A-agarose is weaker than that of an oligomannose-type chain, the biantennary chain bound to the column is easily eluted with a buffer containing a low concentration of a haptenic sugar (e.g., 5–

15mM methyl-α-mannoside). In contrast, for the elution of most of the oligomannose-type and some of the hybrid-type chains, a higher concentration of a haptenic sugar (e.g., 100–200mM methyl-α-mannoside) is required. It has also been shown that the attachment of a bisecting N-acetylglucosamine residue to a biantennary complex-type oligosaccharide (structure VI in Fig. 2) decreases its affinity for a Con A-agarose column, so that it is no longer bound by the column.

3.2.3. Separation of Two Isomers of Triantennary Complex-Type Chains

Two isomers of triantennary complex-type carbohydrate chains are found in many glycoproteins (structures II and III in Fig. 2); one contains an α-mannose residue substituted at the C-2 and C-4 positions by N-acetyllactosamine units, and the other contains one substituted at the C-2 and C-6 positions.

Phaseolus vulgaris leukoagglutinin (L-PHA; Cummings and Kornfeld, 1982a; Hammarström *et al.*, 1982) and *Datura stramonium* agglutinin (DSA; Cummings and Kornfeld, 1984; Yamashita *et al.*, 1987) can distinguish these isomers from each other and have been employed in lectin affinity chromatography. The triantennary structure with branches arising from the C-2 and C-6 positions of an α-mannose residue (structure III in Fig. 2) was found to have higher affinity for both lectin columns than that with branches arising from the C-2 and C-4 positions of an α-mannose residue (structure II in Fig. 2). The former carbohydrate chain shows high-affinity binding to or significant retardation on each lectin column. The tetraantennary chain with an α-mannose residue substituted at C-2 and C-6 by N-acetyllactosamine units (structure IV in Fig. 2) also has a high affinity for each lectin column as in the case of the triantennary chains with a 2,6-di-O-substituted α-mannose residue. However, it is quite difficult to separate a tetraantennary complex-type chain from a C-2,6-branched triantennary chain by lectin affinity chromatography, because no lectin has been reported to distinguish these carbohydrate chains. For this purpose, other chromatographic techniques should be used.

DSA, but not L-PHA, was reported to have a strong interaction with so-called poly-N-acetyllactosamine structures. Details are discussed in Section 3.2.6.

3.2.4. Separation of Glycopeptides with a Fucose Residue

A fucose residue is often attached to the C-6 position of the innermost N-acetylglucosamine residue of a complex-type glycopeptide (structures with the core hexasaccharide designated R_2 in Fig. 2). Three mannose-specific lectins from *Lens culinaris* (lentil), *Pisum sativum* (pea), and *Vicia faba* (fava) are suitable for the separation of a fucosylated biantennary glycopeptide (structure I with R_2 in Fig. 2) from one without a fucose residue (structure I with R_1 in Fig. 2) (Kornfeld *et al.*, 1981; Yamamoto *et al.*, 1982; Katagiri *et al.*, 1984). These three lectins have similar binding specificities, and their immobilized derivatives bind only the fucosylated glycopeptide but not the unfucosylated glycopeptide. They have also been shown to bind a triantennary complex-type glycopeptide with outer chains branching from the C-2 and C-6 positions of an α-mannose residue. However, neither a tetraantennary glycopeptide nor a triantennary glycopeptide with branches from C-2 and C-4 is retained on these columns, even though these glycopeptides have a fucose residue linked to the innermost N-acetylglucosamine residue. It should be noted that an oligosaccharide-alditol released from the glycopeptide (followed by reduction with NaBH$_4$, or NaB^3H_4 for radioisotopic labeling) cannot interact with these lectin columns, although the original glycopeptide shows high-affinity binding to the lectin columns.

A fucose-specific lectin from *Aleuria aurantia* (AAL; Kochibe and Furukawa, 1980; Yamashita *et al.*, 1985) has rather broad specificity, which makes it a very useful tool for the separation of fucosylated carbohydrate chains. An AAL column binds complex-type glycopeptides or oligosaccharides with an α-fucosyl residue in the core structure irrespective of the number of outer chains. Furthermore, it can interact with other fucosyl linkages such as a Fucα1-2Galβ1-4GlcNAc- sequence [blood group H(O) determinant] and a Galβ1-4(Fucα1-3)GlcNAc- sequence (Le^x determinant). Some complex-type carbohydrate chains occasionally have an Le^x determinant in an outer chain moiety (structure V in Fig. 2), and this type of carbohydrate chain is also bound by the immobilized AAL column.

3.2.5. Separation of Carbohydrate Chains with a Bisecting GlcNAc Residue

A bisecting GlcNAc residue, which links to the β-mannose residue in the core structure of an Asn-linked carbohydrate chain, is an important determinant for high-affinity binding to immobilized *Phaseolus vulgaris* erythroagglutinin (E-PHA) (Irimura *et al.*, 1981; Cummings and Kornfeld, 1982a). A complex-type glycopeptide with a bisecting GlcNAc residue (structure VI in Fig. 2) can be significantly retarded on an E-PHA-agarose column, whereas one without a bisecting GlcNAc (structure I in Fig. 2) is eluted in the void volume without retardation. In addition to the presence of a bisecting GlcNAc residue, it is essential for the high-affinity interaction that a Galβ1-4GlcNAcβ1-2Manα1-6 sequence, one of the outer branches, should not be substituted or modified by other carbohydrates such as a sialic acid residue, a fucose residue, or another outer chain originating from the α-mannose residue in the above sequence (Yamashita *et al.*, 1983). A tetraantennary complex-type carbohydrate chain, which has an additional outer chain linked to C-6 of the α-mannose residue in the essential sequence mentioned before, fails to interact with an E-PHA-agarose column, even though it has a bisecting GlcNAc residue.

Wheat germ agglutinin (WGA), a GlcNAc-binding lectin, was found to be useful for the separation of glycopeptides containing a bisecting GlcNAc residue. Yamamoto *et al.* (1981) reported that most hybrid-type glycopeptides having a bisecting GlcNAc residue have high affinity for a WGA-agarose column. However, if the glycopeptide has an α-fucose residue linked to the innermost GlcNAc residue (as in structure VI with R_2 in Fig. 2), it can interact with this lectin column only after the removal of the α-fucose residue. WGA-agarose binds not only glycopeptides containing a bisecting GlcNAc residue but also poly-*N*-acetyllactosamine-type carbohydrate chains, as mentioned in the following section.

3.2.6. Separation of Poly-*N*-acetyllactosamine-Type Carbohydrate Chains

A poly-*N*-acetyllactosamine-type carbohydrate chain is characterized by the presence of *N*-acetyllactosamine (Galβ1-4GlcNAc) repeating units (structure VII in Fig. 2). Oligosaccharides of this type are classified into two groups based on the branching pattern at galactose residues in the disaccharide repeats as shown in Fig. 3; one contains unsubstituted *N*-acetyllactosamine repeating units (blood group *i* antigenic structure), and the other contains branched *N*-acetyllactosamine units where another *N*-acetyllactosamine unit is linked to the C-6 position of galactose residues in the repeats (blood group *I* antigenic structure).

Galβ1-4GlcNAcβ1-3Galβ1-4GlcNAcβ1-   **i** antigenic structure

Galβ1-4GlcNAcβ1
              |
              6
Galβ1-4GlcNAcβ1-3Galβ1-4GlcNAcβ1-   **I** antigenic structure

FIGURE 3. Branched and unbranched poly-*N*-acetyllactosamine-type oligosaccharides.

Several poly-*N*-acetyllactosamine-binding lectins have been reported: WGA (Gallagher *et al.*, 1985; Renkonen *et al.*, 1988), pokeweed mitogen (PWM; Irimura and Nicolson, 1983), *Datura stramonium* agglutinin (DSA; Cummings and Kornfeld, 1984; Yamashita *et al.*, 1987), *Lycopersicon esculentum* (tomato) agglutinin (LEA; Merkle and Cummings, 1987), *Solanum tuberosum* (potato) agglutinin (STA; Allen and Neuberger, 1973), and *Erythrina variegata* (EVA; Li *et al.*, 1990). Kawashima *et al.* (1990) characterized the binding specificities of these lectins toward poly-*N*-acetyllactosamine-type carbohydrate chains by affinity chromatography on the immobilized derivative of each lectin. A DSA-agarose column binds oligosaccharides containing the i-antigenic structure with linear *N*-acetyllactosamine repeats, whereas for immobilized WGA, PWM, EVA, LEA, and STA columns, the *N*-acetyllactosamine repeating sequence is not sufficient for the interaction, but the presence of a GlcNAcβ1-6 linkage, which constitutes a part of the I-antigenic structure, is required. These lectins are therefore efficient tools for the separation and the structural characterization of poly-*N*-acetyllactosamine-type carbohydrate chains.

3.3. Separation of Ser/Thr-Linked Carbohydrate Chains

Affinity columns prepared from lectins that preferentially bind to Ser/Thr-linked carbohydrate chains (mucin-type carbohydrate chains) have been less precisely analyzed for their carbohydrate binding specificities than those specific for Asn-linked carbohydrate chains. Sueyoshi *et al.* (1988b) quantitatively determined carbohydrate binding specificities of several lectins specific for mucin-type carbohydrate chains, such as those listed in Table 1, by affinity column chromatography. Figure 4 shows the association constants of five lectins for a glycopeptide prepared from human erythrocyte glycophorin A and its degraded derivatives. On the basis of the results, these lectins can be classified into two groups with respect to reactivity with typical mucin-type glycopeptides, Galβ1-3GalNAcα1-Ser/Thr (glycopeptide **2** in Fig. 4) and GalNAcα1-Ser/Thr (glycopeptide **3** in Fig. 4). One group, which consists of *Agaricus bisporus* (mushroom) agglutinin (ABA-I), *Arachis hypogaea* (peanut) agglutinin (PNA), and *Bauhinia purpurea* agglutinin (BPA), preferentially binds Galβ1-3GalNAcα1-Ser/Thr, and the other group, which consists of *Glycine max* (soybean) agglutinin (SBA) and *Vicia villosa* agglutinin (VVA-B$_4$), shows higher affinity for GalNAcα1-Ser/Thr than for its galactosylated derivative. Thus, the immobilized derivatives of these five lectins can be applied to the separation of glycopeptides with sequences Galβ1-3GalNAcα1-Ser/Thr and GalNAcα1-Ser/Thr. For instance, if the mixture of glyco-

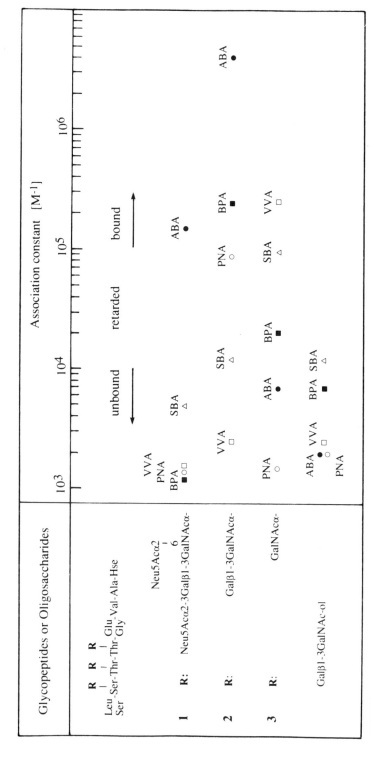

FIGURE 4. Association constants of five lectins specific for mucin-type carbohydrate chains for glycopeptides and a disaccharide obtained from human erythrocyte glycophorin A. ●, ABA; ■, BPA; ○, PNA; △, SBA; □, VVA.

peptides **1**, **2**, and **3** shown in Fig. 4 is subjected to a column of BPA-agarose, it is expected that (1) glycopeptide **1** (with sialic acid residues at the termini) is recovered in the void volume without retardation, (2) glycopeptide **2** (with galactose residues at the end) is bound to the column and eluted with a lactose-containing buffer, and (3) glycopeptide **3** (with only GalNAc residues linked to the peptide) is eluted with significant retardation from the column, between glycopeptides **1** and **2**.

A disaccharide-alditol with a Galβ1-3GalNAc-ol (structure **4** in Fig. 4) has much lower affinity for the lectins BPA, PNA, and ABA-1, which belong to one group according to the classification described above, when compared to the glycopeptide with the same sugar sequence. This result indicates the α-pyranoside structure of the GalNAc residue and/ or the linkage region between the GalNAc residue and Ser/Thr is important for the interaction between these lectins and the glycopeptide.

Among the immobilized lectins tested, only an ABA-I-agarose column retained a sialylated glycopeptide containing three tetrasaccharide chains with the sequence Neu5Acα2-3Galβ1-3(Neu5Acα2-6)GalNAcα1- (glycopeptide **1** in Fig. 4), with an association constant of $1.5 \times 10^5 \; M^{-1}$. An ABA-I-agarose column is therefore suitable for the separation of sialylated mucin-type glycopeptides from sialylated complex-type glycopeptides, since the latter glycopeptides have no apparent interaction with immobilized ABA-I.

4. CONCLUDING REMARKS

As described in the previous sections, affinity chromatography on lectin columns is a quite effective technique for the separation and structural assessment of oligosaccharides and glycopeptides. The use of a series of various lectin columns with different binding specificities enables us to fractionate a small amount of oligosaccharides or glycopeptides (usually labeled with radioisotopes or fluorescence tags) and also gives us valuable information about their structures, which facilitates subsequent structural analyses. The fractionation of a mixture of Asn-linked carbohydrate chains by serial lectin-agarose affinity chromatography, as originally proposed by Cummings and Kornfeld (1982b), is now employed in a wide range of glycoprotein studies.

For more detailed reviews on the carbohydrate-binding specificities of immobilized lectins, the reader is referred to two of our recent publications (Osawa and Tsuji, 1987; Osawa, 1989).

REFERENCES

Allen, A. K., and Neuberger, A., 1973, The purification and properties of the lectin from potato tubers, a hydroxyproline-containing glycoprotein, *Biochem. J.* **135**:307–314.

Baenziger, J. U., and Fiete, D., 1979a, Structural determinants of concanavalin A specificity for oligosaccharides, *J. Biol. Chem.* **254**:2400–2407.

Baenziger, J. U., and Fiete, D., 1979b, Structural determinants of *Ricinus communis* agglutinin and toxin specificity for oligosaccharides, *J. Biol. Chem.* **254**:9795–9799.

Bhavanandan, V. P., and Katlic, A. W., 1979, The interaction of wheat germ agglutinin with sialoglycoproteins: The role of sialic acid, *J. Biol. Chem.* **254**:4000–4008.

Cummings, R. D., and Kornfeld, S., 1982a, Characterization of the structural determinants required for the high affinity interaction of asparagine-linked oligosaccharides with immobilized *Phaseolus vulgaris* leukoagglutinating and erythroagglutinating lectins, *J. Biol. Chem.* **257**:11230–11234.

Cummings, R. D., and Kornfeld, S., 1982b, Fractionation of asparagine-linked oligosaccharides by serial lectin-agarose affinity chromatography: A rapid, sensitive and specific technique, *J. Biol. Chem.* **257**:11235–11240.

Cummings, R. D., and Kornfeld, S., 1984, The distribution of repeating [Galβ1,4GlcNAcβ1,3] sequences in asparagine-linked oligosaccharides of the mouse lymphoma cell line BW5147 and PHA[R]2.1: Binding of oligosaccharides containing these sequences to immobilized *Datura stramonium* agglutinin, *J. Biol. Chem.* **259**:6253–6260.

Gallagher, J. T., Morris, A., and Dexter, T. M., 1985, Identification of two binding sites for wheat-germ agglutinin on polylactosamine-type oligosaccharides, *Biochem. J.* **231**:115–122.

Goldstein, I. J., and Hayes, C. E., 1978, The lectins: Carbohydrate binding proteins of plants and animals, *Adv. Carbohydr. Chem. Biochem.* **35**:127–340.

Hammarström, S., Hammarström, M. L., Sundblad, G., Arnarp, J., and Lönngren, J., 1982, Mitogenic leukoagglutinin from *Phaseolus vulgaris* binds to a pentasaccharide unit in *N*-acetyllactosamine-type glycoprotein glycans, *Proc. Natl. Acad. Sci. U.S.A.* **79**:1611–1615.

Irimura, T., Tsuji, T., Tagami, S., Yamamoto, K., and Osawa, T., 1981, Structure of a complex-type sugar chain of human erythrocyte glycophorin A, *Biochemistry* **20**:560–566.

Irimura, T., and Nicolson, G. L., 1983, Interaction of pokeweed mitogen with poly(*N*-acetyllactosamine)-type carbohydrate chains, *Carbohydr. Res.* **120**:187–195.

Katagiri, Y., Yamamoto, K., Tsuji, T., and Osawa, T., 1984, Structural requirements for the binding of glycopeptides to immobilized *Vicia faba* (fava) lectin, *Carbohydr. Res.* **129**:257–265.

Kawaguchi, T., Matsumoto, I., and Osawa, T., 1974, Studies on hemagglutinins from *Maackia amurensis* seeds, *J. Biol. Chem.* **249**:2786–2792.

Kawashima, H., Sueyoshi, S., Li, H., Yamamoto, K., and Osawa, T., 1990, Carbohydrate binding specificities of several poly-*N*-acetyllactosamine-binding lectins, *Glycoconjugate J.* **7**:323–334.

Knibbs, R. N., Goldstein, I. J., Ratcliffe, R. M., and Shibuya, N., 1991, Characterization of the carbohydrate binding specificity of the leukoagglutinating lectin from *Maackia amurensis*, *J. Biol. Chem.* **266**:83–88.

Kochibe, N., and Furukawa, K., 1980, Purification and properties of a novel fucose-specific hemagglutinin of *Aleuria aurantia*, *Biochemistry* **19**:2841–2846.

Kornfeld, K., Reitman, M. L., and Kornfeld, R., 1981, The carbohydrate-binding specificity of pea and lentil lectins: Fucose is an important determinant, *J. Biol. Chem.* **256**:6633–6640.

Krusius, T., Finne, J., and Rauvala, H., 1976, The structural basis of the different affinities of two types of acidic N-glycosidic glycopeptides for concanavalin A-Sepharose, *FEBS Lett.* **71**:117–120.

Li, H., Yamamoto, K., Kawashima, H., and Osawa, T., 1990, Structural requirements for the binding of oligosaccharides to immobilized lectin of *Erythrina variegata* (Linn) var. *Orientalis*, *Glycoconjugate J.* **7**: 311–322.

Liener, I. E., Sharon, N., and Goldstein, I. J., 1986, *The Lectins: Properties, Functions and Applications in Biology and Medicine*, Academic Press, Orlando, Florida.

Lis, H., and Sharon, N., 1986, Lectins as molecules and as tools, *Annu. Rev. Biochem.* **56**:35–67.

Merkle, R. K., and Cummings, R. D., 1987, Relationship of the terminal sequences to the length of poly-*N*-acetyllactosamine chains in asparagine-linked oligosaccharides from the mouse lymphoma cell line BW5147: Immobilized tomato lectin interacts with high affinity with glycopeptides containing long poly-*N*-acetyllactosamine chains, *J. Biol. Chem.* **262**:8179–8189.

Miller, R. L., Collawn, J. F., and Fish, W. W., 1982, Purification and macromolecular properties of a sialic acid-specific lectin from the slug *Limax flavus*, *J. Biol. Chem.* **257**:7574–7580.

Narasimhan, S., Freed, J. C., and Schachter, H., 1986, The effect of a "bisecting" *N*-acetylglucosaminyl group in the binding of biantennary, complex oligosaccharides to concanavalin A, *Phaseolus vulgaris* erythroagglutinin (E-PHA), and *Ricinus communis* agglutinin (RCA-120) immobilized on agarose, *Carbohydr. Res.* **149**: 65–83.

Ogata, S., Muramatsu, T., and Kobata, A., 1975, Fractionation of glycopeptides by affinity column chromatography on concanavalin A-Sepharose, *J. Biochem.* **78**:687–696.

Ohyama, Y., Kasai, K., Nomoto, H., and Inoue, Y., 1985, Frontal affinity chromatography of ovalbumin glycoasparagines on a concanavalin A-Sepharose column: A quantitative study of the binding specificity of the lectin, *J. Biol. Chem.* **260**:6882–6887.

Osawa, T., 1989, Recent progress in the application of plant lectins to glycoprotein chemistry, *Pure Appl. Chem.* **61**:1283–1292.

Osawa, T., and Tsuji, T., 1987, Fractionation and structural assessment of oligosaccharides and glycopeptides by use of immobilized lectins, *Annu. Rev. Biochem.* **56**:21–42.

Peters, B. P., Ebisu, S., Goldstein, I. J., and Flashner, M., 1979, Interaction of wheat germ agglutinin with sialic acid, *Biochemistry* **18:**5505–5511.

Renkonen, O., Mäkinen, P., Hård, K., Helin, J., and Penttila, L., 1988, Immobilized wheat germ agglutinin separates small oligosaccharides derived from poly-*N*-acetyllactosaminoglycans of embryonal carcinoma cells, *Biochem. Cell Biol.* **66:**449–453.

Roche, A.-C., and Monsigny, M., 1974, Purification and properties of limulin: A lectin (agglutinin) from hemolymph of *Limulus polyphemus*, *Biochim. Biophys. Acta* **371:**242–254.

Sharon, N., and Lis, H., 1989, *Lectins*, Chapman & Hall, London.

Shibuya, N., Goldstein, I. J., Broekaert, W. F., Nsimba-Lubaki, M., Peeters, B., and Peumans, W. J., 1987a, The elderberry (*Sambucus nigra* L.) bark lectin recognizes the Neu5Ac(α2-6)Gal/GalNAc sequence, *J. Biol. Chem.* **262:**1596–1601.

Shibuya, N., Goldstein, I. J., Broekaert, W. F., Nsimba-Lubaki, M, Peeters, B., and Peumans, W. J., 1987b, Fractionation of sialylated oligosaccharides, glycopeptides and glycoproteins on immobilized elderberry (*Sambucus nigra* L.) bark lectin, *Arch. Biochem. Biophys.* **254:**1–8.

Sueyoshi, S., Tsuji, T., and Osawa, T., 1985, Purification and characterization of four isolectins of mushroom (*Agaricus bisporus*), *Biol. Chem. Hoppe-Seyler* **366:**213–221.

Sueyoshi, S., Yamamoto, K., and Osawa, T., 1988a, Carbohydrate binding specificity of a beetle (*Allomyrina dichotoma*) lectin, *J. Biochem.* **103:**894–899.

Sueyoshi, S., Tsuji, T., and Osawa, T., 1988b, Carbohydrate-binding specificities of five lectins that bind to *O*-glycosyl-linked carbohydrate chains: Quantitative analysis by frontal-affinity chromatography, *Carbohydr. Res.* **178:**213–224.

Wang, W.-C., and Cummings, R. D., 1988, The immobilized leukoagglutinin from the seeds of *Maackia amurensis* binds with high affinity to complex-type Asn-linked oligosaccharides containing terminal sialic acid-linked α-2,3 to penultimate galactose residues, *J. Biol. Chem.* **263:**4576–4585.

Yamamoto, K., Tsuji, T., Matsumoto, I., and Osawa, T., 1981, Structural requirements for the binding of liposaccharides and glycopeptides to immobilized wheat germ agglutinin, *Biochemistry* **20:**5894–5899.

Yamamoto, K., Tsuji, T., and Osawa, T., 1982, Requirement of the core structure of a complex-type glycopeptide for the binding to immobilized lentil and pea lectins, *Carbohydr. Res.* **110:**283–289.

Yamashita, K., Hitoi, A., and Kobata, A., 1983, Structural determinants of *Phaseolus vulgaris* erythroagglutinating lectin for oligosaccharides, *J. Biol. Chem.* **258:**14753–14755.

Yamashita, K., Kochibe, N., Ohkura, T., Ueda, I., and Kobata, A., 1985, Fractionation of L-fucose-containing oligosaccharides on immobilized *Aleuria aurantia* lectin, *J. Biol. Chem.* **260:**4688–4693.

Yamashita, K., Totani, T., Ohkura, T., Takasaki, S., Goldstein, I. J., and Kobata, A., 1987, Carbohydrate binding properties of complex-type oligosaccharides on immobilized *Datura stramonium* lectin, *J. Biol. Chem.* **262:**1602–1607.

Yamashita, K., Umetsu, K., Suzuki, T., Iwaki, Y., Endo, T., and Kobata, A., 1988, Carbohydrate binding specificity of immobilized *Allomyrina dichotoma* lectin II, *J. Biol. Chem.* **263:**17482–17489.

8

Selective Isolation of C-Terminal Peptides by Affinity Chromatography

Shin-ichi Ishii and Takashi Kumazaki

1. INTRODUCTION

Every protein expresses its function through its capability of specific molecular recognition, which depends on its three-dimensional structure. In order to understand the fundamental relation between function and structure, one must also know its amino acid sequence. With the recent advent of cloning techniques for protein-encoding genes, the amino acid sequence of a protein is now easily deducible from the base sequence of its cDNA. This is, however, an indirect approach. If the protein of interest had been subjected to any posttranslational proteolysis, which is a process frequently observable before maturation, the deduced sequence would be that of a precursor, and not that of the functional protein. Furthermore, mistakes may be made in determining the reading frame of a long base sequence. These problems may be overcome by the direct determination of amino acid sequences near the N- and C-termini of the mature protein. While the analysis of N-terminal sequences can be easily accomplished by the traditional Edman method, that of C-terminal sequences is very difficult to do. The thiocyanate degradation method (Hawke *et al.*, 1987) was recently improved for C-terminal sequence analysis, but its applicability to large proteins is not yet clear. The carboxypeptidase digestion method is frequently used, but it allows the reliable identification of only the first two or three amino acid residues from the C-terminus. For the determination of a C-terminal sequence of sufficient length, the best way at present would be to isolate the C-terminal peptide after fragmentation of the protein and then subject it to the Edman method. A variety of methods (Fong and Hargrave, 1977; Gibó *et al.*, 1986; Horn, 1975; Isobe *et al.*, 1986) have been developed to allow specific isolation of the C-terminal peptide from a peptide mixture generated by fragmentation of a protein with proteases or cyanogen bromide. These methods require a step involving either chemical modification of the amino groups before the fragmentation or carboxypeptidase treatment after the fragmentation, and their application to complex peptide mixtures requires particular skills.

Shin-ichi Ishii and Takashi Kumazaki • Department of Biochemistry, Faculty of Pharmaceutical Sciences, Hokkaido University, Sapporo, Hokkaido 060, Japan.

Molecular Interactions in Bioseparations, edited by That T. Ngo. Plenum Press, New York, 1993.

Several methods for isolation of the C-terminal peptide that require neither a chemical modification step nor carboxypeptidase treatment have been recently reported. Among these, the method described by Kawasaki *et al.* (1987) utilizing cation-exchange chromatography has the advantage of simultaneous isolation of C-terminal and blocked N-terminal peptides, but its applicability is still limited. On the other hand, isolation of the C-terminal peptide by affinity chromatography on immobilized anhydrotrypsin (Kumazaki *et al.*, 1986, 1987) and on immobilized anhydrochymotrysin (Kumazaki *et al.*, 1986) has already given successful results in its application to C-terminal sequence determination of several unknown proteins. The method using *p*-phenylenediisothiocyanate-controlled pore glass (DITC-CPG) (Ouchi *et al.*, 1988) as a specific adsorbent for the C-terminal peptides is also promising, though only a few instances of its application have been reported. This method is based on a kind of covalent chromatography.

In the following, we will explain the principle and practice of the latter two methods for specific isolation of C-terminal peptides from protease digests of proteins.

2. SELECTIVE ISOLATION OF C-TERMINAL PEPTIDES FROM PROTEOLYTIC DIGESTS OF PROTEINS BY THE USE OF IMMOBILIZED ANHYDROTRYPSIN

2.1. Anhydrotrypsin—How It Works

Anhydrotrypsin is a catalytically inert derivative of trypsin in which the active-site residue Ser-195 has been chemically converted to a dehydroalanine residue. This derivative shows a strong specific affinity for peptides containing Arg or Lys [or S-aminoethylated Cys (SAECys)] residues at their C-termini (the peptides corresponding to the products of tryptic hydrolysis) under slightly acidic conditions (Yokosawa and Ishii, 1977). The affinity is much higher than that shown by intact trypsin. Anhydrotrypsin immobilized on agarose gel can behave as an efficient adsorbent specific for these product-type peptides, and various biologically active peptides with Arg at their C-termini, such as tuftosin and Met-Lys-bradykinin, have been effectively isolated by affinity chromatography on immobilized anhydrotrypsin (Yokosawa and Ishii, 1976; Yokosawa and Ishii, 1979). Even oxy-hemoglobin (MW 65,000) was found to show strong interaction with this adsorbent via a C-terminal Arg residue of its α-subunit (Ishii *et al.*, 1988). The affinity is so specific that all other peptides, including substrate-type peptides (and proteins) which contain Arg or Lys residues at internal positions, do not show substantial affinity for this adsorbent. An exception is Kunitz soybean trypsin inhibitor (STI), which is classified as a substrate-type ligand but can bind to this adsorbent.

Why trypsin acquires enhanced binding ability toward product-type peptides after removal of a hydroxyl group at the active-site Ser residue is an intriguing problem. The answer is not yet known, but a key to this problem may be the possible dual functions of the active-site hydroxyl group. This group may have an important role not only in producing a tetrahedral intermediate by reacting with an acyl moiety of the substrate but also in driving the product out of the active site to ensure a high turnover rate of the enzymatic reaction. Regardless of the mechanism involved, we postulated that anhydrotrypsin, with this unique binding ability, would offer a good means for selective isolation of the C-terminal peptide fragment from any protein.

2.2. Strategy for the Isolation of the C-Terminal Peptide from a Protein Digest by Means of Immobilized Anhydrotrypsin

If the original protein is expected to have as its C-terminal residue any amino acid other than Arg and Lys (and SAECys), it may be digested with trypsin. When the tryptic digest is applied to a column of immobilized anhydrotrypsin, the C-terminal peptide will pass through the column. The rest of the peptides in the digest will be fixed because all of them should have Arg or Lys (or SAECys) at their C-termini. Lysylendopeptidase or arginylendopeptidase may be usable instead of trypsin.

On the other hand, in the special case in which the original protein has Arg or Lys at the C-terminus, it must be digested with chymotrypsin (or other endopeptidases that differ from trypsin in the substrate specificity). In this case, the C-terminal peptide fragment from the original protein can be selectively adsorbed on a column of immobilized anhydrotrypsin and recovered by elution under acidic conditions.

The C-terminal peptides obtained in these ways (Fig. 1) are further purified, when necessary, by reversed-phase high-performance liquid chromatography (HPLC), and their sequences are then determined by the Edman method.

2.3. Immobilized Anhydrotrypsin

Anhydrotrypsin was prepared from bovine trypsin through phenylmethylsulfonylation at the hydroxyl group of Ser-195 and β-elimination of the sulfonyl moiety and immobilized on Sepharose 4B by the cyanogen bromide method as described previously (Kumazaki *et al.*, 1986). The 1-ml volume of immobilized anhydrotrypsin used in this study showed the ability to adsorb 84 nmol of Kunitz soybean trypsin inhibitor (STI) at pH 8.2. The gel was treated with 2.7 m*M* L-1-chloro-3-tosylamide-7-amino-2-heptanone (TLCK) at pH 7.0 and 25°C for 4 h in order to diminish the hydrolytic activity due to a trace of contaminating

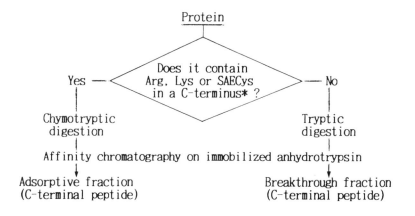

* When you do not have any information about the C-terminal residue of your protein sample, select "No".

FIGURE 1. Strategy for the isolation of C-terminal peptides from proteolytic digests of proteins by affinity chromatography on immobilized anhydrotrypsin.

trypsin to the lowest possible level. The remaining tryptic activity still detectable in 1 ml of the gel, with *N*-benzoyl-DL-arginine *p*-nitroanilide as substrate at pH 8.2, corresponded to that of 0.6 μg of native trypsin. Immobilized anhydrotrypsin of similar quality is now available from Takara Shuzo Co. Ltd. (Kyoto, Japan) as "anhydrotrypsin-agarose" and from Pierce Chemical Company (Rockford, Illinois) as "immobilized anhydrotrypsin."

2.4. Isolation of C-Terminal Peptides from a Tryptic Digest of SCM-SSI by Affinity Chromatography on Immobilized Anhydrotrypsin

Streptomyces subtilisin inhibitor (SSI) is a dimeric protein. The subunit consists of 113 amino acid residues and has Phe at the C-terminus (Ikenaka *et al.*, 1974). The reduced and S-carboxymethylated derivative of SSI (SCM-SSI) (58 nmol) was digested with L-1-chloro-3-tosylamide-4-phenyl-2-butanone (TPCK)-treated β-trypsin [enzyme:substrate = 1:100 (w/w)] in 0.1*M* sodium bicarbonate, pH 8.1, at 37°C for 10 h. The reaction was terminated with 1m*M* diisopropylphosphorofluoridate (DFP). The pH of the reaction mixture was adjusted to 5.0–5.5 by adding diluted acetic acid, and the resulting solution was applied to a column (4.7 ml) of immobilized anhydrotrypsin. The chromatogram is shown in Fig. 2. The adsorbed peptides were eluted with a decreasing pH gradient. The peptides recovered in the breakthrough fraction, indicated by the bar in Fig. 2, were pooled and subjected to reversed-phase HPLC. While the whole digest gave nine major peptide peaks (Fig. 3a), the breakthrough fraction gave only two, I and II (Fig. 3b). It was further confirmed that the adsorbed fraction afforded an HPLC pattern in which I and II had been exclusively deleted from the nine peaks of the whole digest (data not shown). The peptides I and II showed almost the same amino acid composition except that I contained less

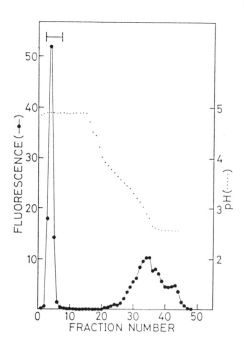

FIGURE 2. Affinity chromatography of a tryptic digest of SCM-SSI (58 nmol) on a column (0.86 × 8.1 cm) of immobilized anhydrotrypsin. The digest (0.27 ml, adjusted to pH 5) was applied to the column equilibrated with 0.05*M* sodium acetate, pH 5.0, containing 0.02*M* CaCl₂, and the column was washed with 20 ml of the same buffer. The adsorbed peptides were eluted with a decreasing pH gradient produced by mixing 20 ml of 0.1*M* sodium formate, pH 4.5, with 20 ml of 0.1*M* formic acid, pH 2.5. Two-milliliter fractions were collected at a flow rate of 10 ml/h. The breakthrough fractions indicated by the bar were pooled and subjected to reversed-phase HPLC. [From Kumazaki *et al.* (1987).]

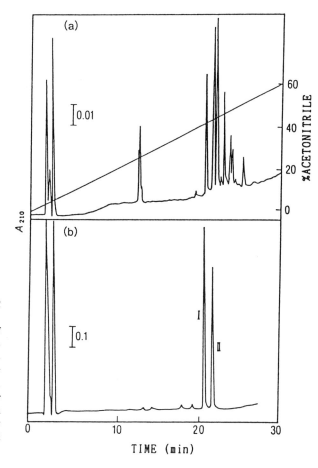

FIGURE 3. Reversed-phase HPLC on a Nucleosil $5C_{18}$ (column (0.4 × 15 cm) of the whole tryptic digest of SCM-SSI (0.4 nmol) (a) and the tryptic peptides recovered in the breakthrough fraction (0.5 ml) from affinity chromatography, shown in Fig. 2(b). Elution was carried out with a linear gradient of acetonitrile concentration from 1 to 65% in 0.1% TFA at a flow rate of 1 ml/min [From Kumazaki et al. (1987).]

methionine than II. Analytical results on the compositions and the sequences from the N-termini suggest that II is H-Val-Phe-Ser-Asn-Glu-SCMCys-Glu-Met-Asn-Ala-His-Gly-Ser-Ser-Val-Phe-Ala-Phe-OH, originating from the C-terminal portion (nos. 96–113) of SCM-SSI, and I is a derivative of II whose Met-103 has been accidentally S-carboxymethylated during the preparation of SCM-SSI from SSI. The recoveries of I and II were 43 and 29%, respectively.

In this series of experiments, we started with 58 nmol of SCM-SSI. Since this protein has six peptide bonds susceptible to trypsin, its tryptic digest should contain 406 nmol of peptides. Care must be taken not to apply amounts of peptides that exceed the capacity of the adsorbent. Peptides in the chromatographic effluents were precisely monitored by the fluorescamine reaction for amino groups as shown in Fig. 2. If a much smaller quantity of sample (<1 nmol) is applied to the column, however, the breakthrough fraction will have to be pooled without any monitoring. In such a case, careful consideration should be given to the possibility that some peptides, especially those of small size, may show weaker affinity for immobilized anhydrotrypsin even though they have Lys or Arg at their C-termini and may tend to leak from the column after appearance of the breakthrough fraction.

This method has been successfully applied to the isolation of the C-terminal peptides from tryptic digests of the following proteins: bacteriophage-T4 tail-sheath protein (MW 71,000) and tube protein (MW 19,000) (Kumazaki *et al.*, 1986) and aqualysine 1, a thermophilic alkaline serine protease from *Thermus aquaticus* (MW 28,000) (Kwon *et al.*, 1988).

Immobilized anhydrochymotrypsin showing high affinity for product-type peptides with C-terminal aromatic amino acids was also successfully used to isolate the C-terminal peptide from the tryptic digest of SCM-SSI (Kumazaki *et al.*, 1988). In this case, the C-terminal peptide, with a C-terminal Phe residue, was selectively recovered in the adsorptive fraction.

In the experiment shown in Fig. 2, a pH gradient was applied for the elution of adsorbed peptides. In order to improve the recoveries of such peptides, a step elution with 0.1M formic acid is recommended instead of the pH gradient elution. This is certainly true when the C-terminal peptide is recovered in the adsorptive fraction following immobilized anhydrochymotrypsin chromatography from a tryptic digest, as in the case mentioned above, or following immobilized anhydrotrypsin chromatography from a chymotryptic digest as described in the next section.

2.5. Isolation of C-Terminal Peptides from a Chymotryptic Digest of α_1-Antitrypsin

The whole amino acid sequence of human α_1-antitrypsin (MW 53,000) has been deduced from the cDNA base sequence (Long *et al.*, 1984). Because the deduced sequence has Lys at the C-terminus, all the tryptic peptides including that originating from the C-terminal portion must be adsorbed on a column of immobilized anhydrotrypsin. In order to isolate the C-terminal peptide selectively, we chromatographed the chymotryptic digest of this protein (obtained by incubating 31 nmol of the protein after heat denaturation with a 1:100 molar amount of TLCK-treated α-chymotrypsin at pH 8.1 and 37°C for 16 h, and then with 1mM DFP) on a column (4.7 ml) of the adsorbent at pH 5.0, with the expectation that the C-terminal peptide would be recovered in the adsorptive fraction. Most of the peptides in the digest passed through the column, while 2.7% of the peptides were recovered in the adsorptive fraction. This fraction was subjected to reversed-phase HPLC. Only one major peak (I) and a few minor ones (II–VI) appeared in the elution curves as shown in Fig. 4b, in contrast with the very complex chromatogram for the whole digest (Fig. 4a). The amino acid composition and the N-terminal sequence of peptide I corresponded to those of a C-terminal peptide H-Phe-Met-Gly-Lys-Val-Val-Asn-Pro-Thr-Gln-Lys-OH (nos. 384–394), produced by chymotryptic cleavage at the bond between Leu-383 and Phe-384. On the other hand, the analytical results suggested that II is another C-terminal peptide (nos. 385–394). Recoveries of I and II were 36 and 10%, respectively. Even when a smaller amount of sample (3 nmol) was applied, peptides I and II were obtained in almost the same yields. Minor components III, IV, V, and VI were also assigned to fragments of α_1-antitrypsin (nos. 384–387, nos. 261–274, nos. 156–168, and nos. 144–168, respectively) by amino acid composition analyses. The recoveries of these peptides were 5% or less. All of them contained Lys residues at their C-termini. The occurrence of these peptides in addition to the C-terminal ones in the adsorptive fraction would be due to either nonspecific cleavage during fragmentation of this protein by chymotrypsin or additional cleavage of the chymotryptic digest on the column by a trace of intact trypsin remaining in the immobilized

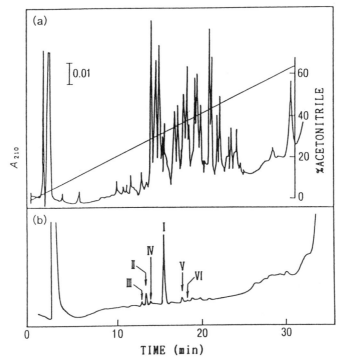

FIGURE 4. Reversed-phase HPLC on a Nucleosil 5C$_{18}$ column (0.4 × 15 cm) of the whole chymotryptic digest of α$_1$-antitrypsin (a) and the chymotryptic peptides recovered in the adsorptive fraction of affinity chromatography on immobilized anhydrotrypsin (b). Elution was carried out as described in the caption to Fig. 3. [From Kumazaki *et al.* (1987).]

anhydrotrypsin. In any case, these were only minor components easily removable by reversed-phase HPLC and, therefore, do not cause much trouble in the identification of C-terminal peptides.

2.6. Effects of Detergents on the Affinity and Binding Capacity of Immobilized Anhydrotrypsin

The first step in the isolation of the C-terminal peptide is enzymatic digestion of the protein. When the protein is hardly soluble in the usual buffer media, urea-containing media are frequently used for the digestion. We examined to what extent immobilized anhydrotrypsin can act as a specific affinity adsorbent when a urea-containing sample solution is applied to it (Kumazaki *et al.*, 1987). Bradykinin, a nonapeptide with a C-terminal Arg, in a buffer at pH 5.0 containing $4M$ urea was applied to a column of immobilized anhydrotrypsin previously washed with the same buffer without urea. It was found that, if the volume of sample solution did not exceed the column volume, the peptide could be adsorbed almost as tightly as in the absence of urea. Hirabayashi and Kasai (1990) have demonstrated that efficient purification of a recombinant protein (MW 14,000) artificially tagged with C-terminal Arg by affinity chromatography on immobilized anhydrotrypsin was only achieved when the protein sample was applied to the adsorbent column in a small volume of 4*M* urea-containing buffer. In this case, the effect of the addition of urea was probably to allow the C-terminal Arg to be exposed on the protein molecular surface, thus becoming accessible to the adsorbent.

Although most of the peptides adsorbed on the immobilized anhydrotrypsin column can be easily desorbed with 0.1*M* formic acid, some of them are bound to the column so tightly that complete desorption is not accomplished by the usual washing procedure. Accumulation of the peptide on the adsorbent after multiple operations causes deterioration of the binding capacity. The accumulated peptides would be removed with a urea-containing buffer. Thus, the stability of the adsorbent in urea solution was examined. The adsorbent column was washed with 20 times the column volume of a buffer, pH 5, containing 6*M* urea at 4°C, and then with the same buffer without urea. After the treatment, the binding capacities of the column for STI and for benzoyl-Gly-Arg-OH were determined in the absence of urea. The cycle of urea treatment and binding capacity determination was repeated five times. While the original binding capacity for STI was completely retained even after the fifth treatment, a gradual decrease (about 4% per treatment) was observed in the binding capacity for benzoyl-Gly-Arg-OH. We have found that STI, one of the strongest affinity ligands for immobilized anhydrotrypsin, can be effectively desorbed from the adsorbent column with 20 times the column volume of 3*M* urea-containing buffer at pH 5. Regeneration of deteriorated columns may, therefore, be done by washing with 3*M* urea-containing buffer first. The same concentration of guanidine hydrochloride can be used for the same purpose.

On the other hand, immobilized anhydrotrypsin is less stable against sodium dodecyl sulfate (SDS). Bradykinin dissolved in a buffer containing 0.1% SDS was adsorbed almost as tightly as in the absence of the detergent if the sample volume did not exceed the column volume. However, washing with a large volume of 0.1% SDS-containing buffer may cause a marked decrease in binding capacity of the adsorbent.

3. SELECTIVE ISOLATION OF C-TERMINAL PEPTIDES FROM LYSYLENDOPEPTIDASE DIGESTS OF PROTEINS BY THE USE OF DITC-CPG

3.1. Principle of the Isolation Method

p-Phenylenediisothiocyanate-controlled pore glass (DITC-CPG) is a solid support useful for the solid-phase Edman degradation method. According to the scheme shown in Fig. 5, the C-terminal peptide can be selectively separated from a lysylendopeptidase digest of a protein. All the peptides in the digest are allowed to react with the isothiocyanate groups of DITC-CPG under alkaline conditions. If the C-terminus of the original protein was not Lys, the peptide derived from the C-terminal portion of the protein would be covalently bound to the solid support only by its α-amino group. This situation must be quite distinct from that for all the peptides originating from the other parts of the protein. The latter peptides will be coupled to the support via both α- and ε-amino groups, because all of them contain Lys at their C-termini. The support to which the peptides are attached is then treated with trifluoroacetic acid (TFA) under dehydrated conditions. The N-terminal

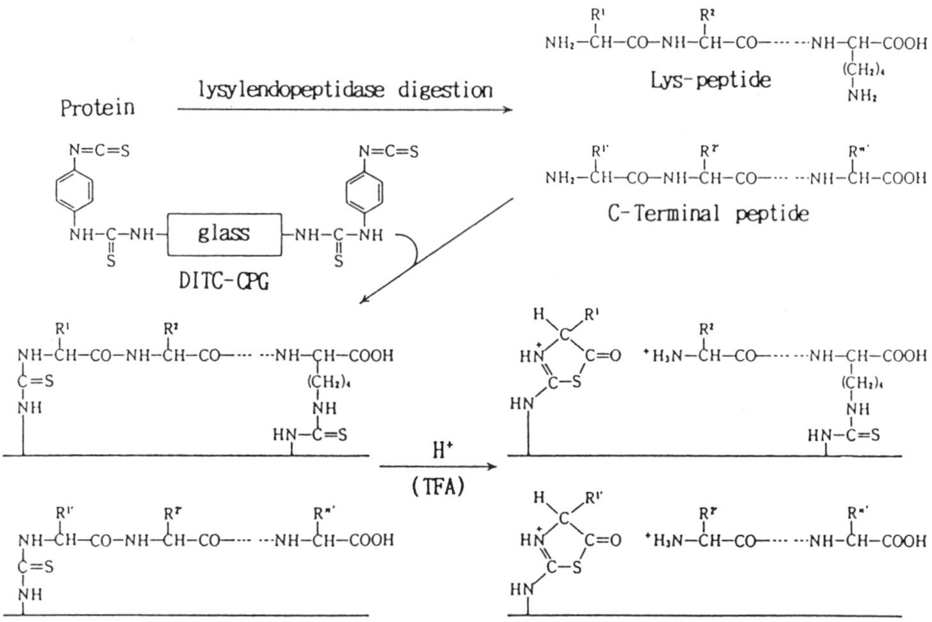

FIGURE 5. Principle of the isolation of the C-terminal peptide from a lysylendopeptidase digest of a protein by means of DITC-CPG.

amino acid residue of each peptide is cleaved as the anilinothiazolinone derivative, but the linkage of ε-amino groups with the support remains intact. Therefore, only the C-terminal peptide (after losing the first residue) is selectively released from the support. It can be purified, when necessary, by reversed-phase HPLC, and its sequence is then determined by the Edman method.

3.2. Isolation of a C-Terminal Peptide from a Lysylendopeptidase Digest of SCM-Hen Egg Lysozyme

Ouchi *et al.* (1988) described the isolation of a C-terminal peptide from a lysylendopeptidase digest of a reduced and S-carboxymethylated (SCM) derivative of hen egg lysozyme. The protein (2.0 nmol) was digested with lysylendopeptidase [enzyme:substrate = 1:300 (mol/mol)] in 0.05 ml of $0.05M$ N-ethylmorpholine–acetate buffer, pH 8.7, containing $5M$ urea at 37°C for 6 h. After the pH of the solution was adjusted to 10 by adding N-ethylmorpholine, 80 mg of DITC-CPG (binding capacity for Ala, 1.5 nmol/mg) was added. The suspension was then incubated with shaking under a stream of N_2 gas at 40°C for 1 h. About a 10-fold molar excess of DITC-CPG relative to the total amino groups of the protein should be added to achieve complete attachment of the peptide fragments to the support. The pore glass coupled with the peptides was washed on a glass filter with about 2 ml each of four different media in the following order: 1, 0.1% TFA; 2, 0.1% TFA + 50% acetonitrile–2-propanol (3:7, v/v); 3, 0.1% TFA + 80% acetonitrile–2-propanol (3:7, v/v); 4, methanol. The pore glass was then dried *in vacuo* at 40°C. A suitable volume of anhydrous TFA was then added, and the suspension was incubated under a stream of N_2 gas at 40°C for 15 min. The peptides released from the support were extracted with media 1–3.

The pooled extracts were subjected to reversed-phase HPLC. The elution profile of the peptides released from the pore glass (Fig. 6b) was compared with that of the whole protease digest (Fig. 6a). The former gave three peaks, denoted as I, II, and III. The results of amino acid composition analysis and sequence analysis suggest that peak II corresponds to Thr-Asp-Val-Gln-Ala-Trp-Ile-Arg-Gly-SCMCys-Arg-Leu-OH, originating from the C-terminal portion (nos. 118–129) of SCM-lysozyme. The yield was 36%. Peaks I and III did not contain any amino acids and were attributed to by-products due to the reagents used in this experiment. This method was also successfully employed in the selective isolation of the C-terminal peptide from a lysylendopeptidase digest of human albumin (MW 66,000) (Kondo, 1989). It should be noted that this method is not applicable to proteins with Lys at the C-terminus.

4. CONCLUSION

Two methods for the determination of C-terminal sequences of proteins have been described. They are based on the selective isolation of C-terminal peptide fragments from protease digests of the proteins by the use of immobilized anhydrotrypsin in the first method and DITC-CPG in the second as specific adsorbents for the peptides. Although these methods still have several minor problems to be solved as discussed above, their practical

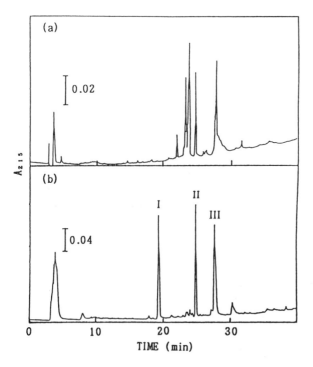

FIGURE 6. Reversed-phase HPLC on a Bakerbond Octyl column (0.46 × 25 cm) of the whole lysylendopeptidase digest of SCM-hen egg lysozyme (0.3 nmol) (a) and the extracts from DITC-CPG (2 nmol) (b). Elution was carried out with a linear gradient of acetonitrile concentration from 0 to 60% in 0.1% TFA at a flow rate of 1 ml/min.

use has been already established by successful applications to various proteins, including those with high molecular weights ($>$60,000). The second method is superior to the first in terms of the higher stability of the adsorbent used in this method under harsh conditions (i.e., use of strong detergents). On the other hand, the first method has the advantage of general utility; it is applicable even to proteins with Lys (or Arg) at the C-terminus. Another advantage of the first method is that the peptide fragments derived from internal parts of the protein, as well as the C-terminal fragment, can be recovered and used for further studies. In both methods, reversed-phase HPLC was necessary for the final purification of the required peptides, and less than one nanomole of the amounts have been used to date.

REFERENCES

Fong, S.L., and Hargrave, P. A., 1977, Preparation of carboxyl-terminal tryptic peptides from proteins by cleavage at arginine, *Int. J. Pept. Protein Res.* **10:**139–145.

Gibó, G., Furka, Á., Sebestyén, F., and Fehérvari, Á., 1986, Isolation of the C-terminal fragment of proteins after re-digestion of their esterified tryptic hydrolysates, *Int. J. Pept. Protein Res.* **27:**355–359.

Hawke, D. H., Lahm, H-W., Shively, J. E., and Todd, C. W., 1987, Microsequence analysis of peptides and proteins: Trimethylsilylisothiocyanate as a reagent for COOH-terminal sequence analysis, *Anal. Biochem.* **166:**298–307.

Hirabayashi, J., and Kasai, K., 1990, Specific isolation by anhydrotrypsin-agarose chromatography of a recombinant protein tagged with an affinity tail arginine at the C-terminus, *J. Mol. Recog.* **3:**204–207.

Horn, M. J., 1975, Amination of carboxy-terminal homoserine peptides as an aid in peptide separation, *Anal. Biochem.* **69:**583–589.

Ikenaka, T., Odani, S., Sakai, M., Nabeshima, Y., Sato, S., and Murao, S., 1974, Amino acid sequence of an alkaline protease inhibitor (*Streptomyces* subtilisin inhibitor) from *Streptomyces albogriseoulus* S-3253, *J. Biochem. (Tokyo)* **76:**1191–1209.

Ishii, S., Kumazaki, T., Fujitani, A., and Terasawa, K., 1988, Selective isolation of C-terminal peptides from proteolytic mixtures of proteins by affinity chromatography on immobilized anhydrotrypsin and an-hydrochymotrypsin, *Makromol. Chem., Macromol. Symp.* **17:**281–290.

Isobe, T., Ichimura, T., and Okuyama, T., 1986, Identification of the C-terminal portion of a protein by comparative peptide mapping, *Anal. Biochem.* **155:**135–140.

Kawasaki, H., Imajoh, S., and Suzuki, K., 1987, Separation of peptides on the basis of the difference in positive charge: Simultaneous isolation of C-terminal and blocked N-terminal peptides from tryptic digests, *J. Biochem. (Tokyo)* **102:**393–400.

Kondo, J., 1989, Specific isolation method for C-terminal peptides by using *p*-phenylenediisothiocyanate-controlled pore glass (DITC-CPG), *Mitsubishi Kasei R & D Reviews* **3:**110.

Kumazaki, T., Nakako, T., Arisaka, F., and Ishii, S., 1986, A novel method for selective isolation of C-terminal peptides from tryptic digests of proteins by immobilized anhydrotrypsin: Application to structural analyses of the tail sheath and tube proteins from bacteriophage T4, *Proteins: Struct., Funct., Genet.* **1:**100–107.

Kumazaki, T., Terasawa, K., and Ishii, S., 1987, Affinity chromatography on immobilized anhydrotrypsin: General utility for selective isolation of C-terminal peptides from protease digests of proteins, *J. Biochem. (Tokyo)* **102:**1539–1546.

Kumazaki, T., Fujitani, A., Terasawa, K., Shimura, K., Kasai, K., and Ishii, S., 1988, Immobilized an-hydrochymotrypsin as a biospecific adsorbent for the peptides produced by chymotryptic hydrolysis, *J. Biochem. (Tokyo)* **103:**297–301.

Kwon, S.-T., Terada, I., Matsuzawa, H., and Ohta, T., 1988, Nucleotide sequence of the gene for aqualysin I (a thermophilic alkaline serine protease) of *Thermus aquaticus* YT-1 and characteristics of the deduced primary structure of the enzyme, *Eur. J. Biochem.* **173:**491–497.

Long, G. L., Chandra, T., Woo, S. L. C., Davie, E. W., and Kurachi, K., 1984, Complete sequence of the cDNA for human α_1-antitrypsin and the gene for the S variant, *Biochemistry* **23:**4828–4837.

Ouchi, T., Kondo, J., and Hishida, T., 1988, Specific isolation method for C-terminal peptides, *Seikagaku* **60:**875 [in Japanese].

Yokosawa, H., and Ishii, S., 1976, The effective use of immobilized anhydrotrypsin for the isolation of biologically active peptides containing L-arginine residues in C-termini, *Biochem. Biophys. Res. Commun.* **72:**1443–1449.

Yokosawa, H., and Ishii, S., 1977, Anhydrotrypsin: New features in ligand interactions revealed by affinity chromatography and thionine replacement, *J. Biochem. (Tokyo)* **81:**647–656.

Yokosawa, H., and Ishii, S., 1979, Immobilized anhydrotrypsin as a biospecific affinity adsorbent for the peptides produced by trypsin-like proteases, *Anal. Biochem.* **98:**198–203.

9

Receptor-Affinity Chromatography (RAC)

Michele Nachman-Clewner, Cheryl Spence, and Pascal Bailon

1. INTRODUCTION

Affinity chromatography, which utilizes the ability of macromolecules in solution to bind specifically and reversibly to immobilized substrates and other ligands, is an ideal tool for the purification of biomolecules such as enzymes, hormones, receptors, and antibodies, among others. Since the discovery of monoclonal antibody-producing murine hybrid cell lines by Kohler and Milstein (1975), immunoaffinity chromatography involving antigen–antibody interactions has been increasingly used in the purification of biologically active molecules (Leonard *et al.*, 1984; Tarnowski *et al.*, 1986; Bailon *et al.*, 1987). An alternate affinity purification method, termed receptor-affinity chromatography, has been developed recently (Bailon *et al.*, 1987; Bailon and Weber, 1988). Receptor-affinity chromatography (RAC) takes advantage of the specificity and reversibility of receptor–ligand interactions.

Conceptually, only a fully active biomolecule in its native conformation should bind to its natural receptor with high avidity. This unique property of receptor–ligand interactions bestows several useful characteristics upon receptor-affinity chromatography. The receptor-affinity adsorbent exhibits selectivity and high avidity toward the fully active biomolecule. At the same time, the receptor–ligand complex formed during adsorption can be easily dissociated with mild, nonspecific reagents. The specificity exhibited in RAC is not necessarily characteristic of immunoaffinity chromatography, in which different molecular forms of antigen with varying degrees of biological activity and renaturation bind to the immunosorbent (Bailon *et al.*, 1987; Weber and Bailon, 1990). The aforementioned properties of ligand–receptor interactions make RAC an ideal tool for the downstream purification of recombinant proteins from microbial and mammalian sources.

A schematic illustration of receptor-affinity purification is given in Fig. 1. The receptor [e.g., the interleukin-2 receptor (IL-2R)] is chemically bonded to an inert polymer support and packed into a column. After proper equilibration, the crude feed stream containing the target protein [e.g., interleukin-2 (IL-2)] is passed through the column. The unadsorbed

Michele Nachman-Clewner • Department of Physiology, Cornell Medical School, New York, New York 10021. *Cheryl Spence and Pascal Bailon* • Protein Biochemistry, Molecular Sciences Department, Hoffmann-La Roche Inc., Nutley, New Jersey 07110.

Molecular Interactions in Bioseparations, edited by That T. Ngo. Plenum Press, New York, 1993.

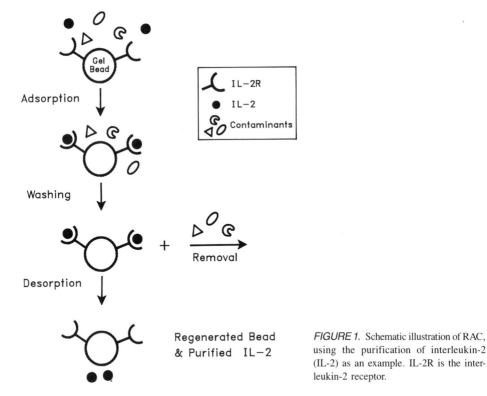

FIGURE 1. Schematic illustration of RAC, using the purification of interleukin-2 (IL-2) as an example. IL-2R is the interleukin-2 receptor.

contaminants are washed away, and the specifically adsorbed target protein is then eluted with mild desorbing agents. A regeneration step prepares the affinity sorbent for the next cycle of operation.

In this chapter we describe the utilization of this technology for the purification of recombinant human interleukin-2 (rIL-2) rIL-2 muteins, murine rIL-2, an interleukin-2–*Pseudomonas* exotoxin fusion protein (IL2-PE40), and a humanized monoclonal antibody to IL-2 receptor (anti-Tac-H), using a multipurpose IL-2R affinity adsorbent.

2. CONSTRUCTION OF RECEPTOR-AFFINITY ADSORBENT

2.1. Purification of Interleukin-2 Receptor

A soluble form of human interleukin-2 receptor (IL-2R), denoted IL-2RΔNae, has been engineered and expressed in Chinese hamster ovary (CHO) cells using gene-linked coamplification technology (Hakimi *et al.*, 1987). IL-2RΔNae lacks 28 amino acids at the carboxyl terminus and contains the naturally occurring N- and O-linked glycosylation sites. The deletions result in the removal of the presumptive transmembrane and cytoplasmic domains of the IL-2R, thus allowing it to be secreted into the medium by the transfected CHO cells.

IL-2RΔNae has been purified employing IL-2 ligand-affinity chromatography, as described by Weber *et al.*, 1988, according to the scheme shown in Fig. 2. The affinity-

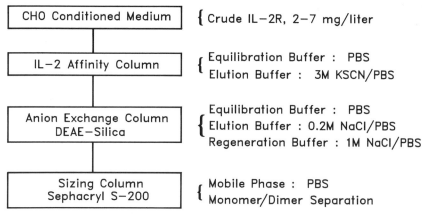

CHO Conditioned Medium
Crude IL−2R, 2−7 mg/liter
IL−2 Affinity Column
Equilibration Buffer : PBS
Elution Buffer : 3M KSCN/PBS
Anion Exchange Column DEAE−Silica
Equilibration Buffer : PBS
Elution Buffer : 0.2M NaCl/PBS
Regeneration Buffer : 1M NaCl/PBS
Sizing Column Sephacryl S−200
Mobile Phase : PBS
Monomer/Dimer Separation

FIGURE 2. IL-2RΔNae purification scheme. PBS, Phosphate-buffered saline.

purified material contained a monomer and a reducible dimer in a 3:1 ratio, as well as a strongly UV-absorbing nonproteinaceous contaminant, which was removed by the anion-exchange column. The monomeric and dimeric forms of the IL-2RΔNae were separated from each other by gel permeation chromatography in Sephacryl S-200 with phosphate-buffered saline (PBS), pH 7.4, as the mobile phase.

2.2. Immobilization of IL-2RΔNae

The monomeric form of IL-2RΔNae was immobilized to NuGel P-AF-poly-*N*-hydroxysuccinimide (PNHS, 500 Å, 40–60 μm, Separation Industries, Metuchen, New Jersey) at a coupling density of 1–2 mg/ml of gel. The coupling reaction is shown in Fig. 3.

Twenty-five grams of NuGel-PNHS, equivalent to a gel volume of 50 ml, was quickly washed three times with 50 ml of ice-cold water in a coarse sintered glass funnel. The gel was then mixed with 50 ml of IL-2R at a protein concentration of 2.0 mg/ml, in 0.1*M* potassium phosphate buffer, pH 7.0. The mixture was shaken overnight on a Labline orbital shaker at 4°C. The uncoupled IL-2R was collected by filtering the reaction mixture. The gel was washed with two volumes of PBS buffer. The filtrate and the washes were combined, and an aliquot (1–2 ml) was dialyzed against PBS buffer. The gel was then mixed with

NuGel−(CH2)3−O−CH2−CH2−NH−CH2−CO−ON
+ R−NH2
NuGel−(CH2)3−O−CH2−CH2−NH−CH2−CO−NHR
+ HON

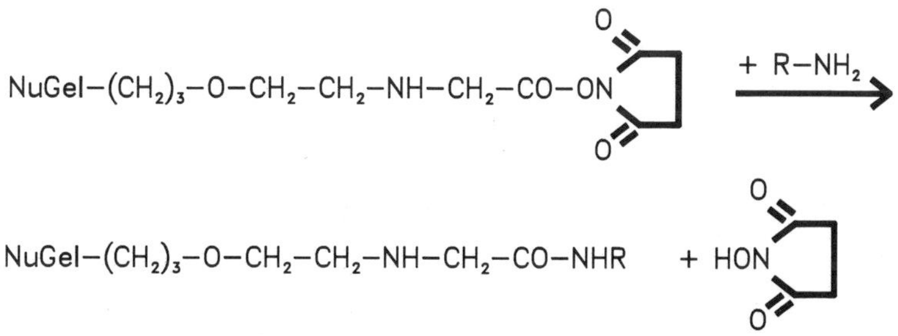

FIGURE 3. Protein coupling to NuGel P-AF-poly-*N*-hydroxysuccinimide.

150 ml of 0.1*M* ethanolamine-HCl, pH 7.0, for 1 hr, to neutralize any remaining *N*-hydroxy-succinimide groups. The IL-2R NuGel was washed three times with PBS and stored in the same buffer containing 0.02% sodium azide. The protein content of the dialyzed uncoupled IL-2R was determined using the A_{280} value of 1.65 for a 1-mg/ml IL-2R solution. Calculations based on (1) the difference between the starting amount of IL-2R (100 mg) and the amount of protein in the total volume of the uncoupled IL-2R (22.5 mg) and (2) the 50-ml gel volume yielded an IL-2R coupling density of 1.55 mg/ml of gel.

2.3. Determination of IL-2 Binding Capacity of the Receptor Column

A receptor column of known volume (0.5–1.0 ml) is saturated with rIL-2, purified or crude. After washing of the column to remove the unadsorbed materials, the bound rIL-2 is desorbed and the protein content determined. The binding capacity of the adsorbent is defined as the number of nanomoles of IL-2 bound per unit volume of gel. In Table 1 the IL-2 binding capacity of the receptor adsorbent is compared to that of an immunoadsorbent containing the monoclonal antibody 5B1. The >100% binding efficiency observed for the IL-2R suggested that the IL-2/IL-2R binding stoichiometry was >1. On a weight-to-weight basis, nine times more antibody was needed per unit volume of gel to recover the same amount of IL-2 as that retained by the receptor column.

In general, coupling conditions such as pH, activated group density on the matrix, and receptor coupling density affect the IL-2 binding capacity of the receptor adsorbents. The coupling efficiencies and binding capacities were optimal at a coupling pH of 7–8, with 20–40 μmol of activated groups per milliliter of affinity gel and a coupling density of 1–2 mg of IL-2R/ml of gel (Bailon *et al.*, 1991).

3. RECOMBINANT IL-2 PRODUCTION

A synthetic gene for IL-2 was constructed and introduced into *Escherichia coli* via the plasmid RR1/pRK 248 Cl$_{ts}$/pRC 233 (Ju *et al.*, 1987) and grown in appropriate medium in large fermentors. A CHO cell line transfected with the IL-2 gene was the source of mammalian glycosylated rIL-2.

TABLE 1
IL-2 Binding Capacities of Receptor and Immunoaffinity Adsorbents

| Adsorbent | Coupling density (mg/ml) | Binding Capacity | | | Binding efficiency[b] (%) |
| | | Expected[a] (nmol/ml) | Observed | | |
			(mg/ml)	(nmol/ml)	
Receptor	1.55	62	1.3	84	136
5B1 antibody	12.80	162	1.2	77	48

[a]Binding capacities were calculated taking the M_r of IL-2R and rIL-2 as 25 and 15.5 kDa, respectively. Two IL-2 binding sites (equivalents) were assumed for the antibody, and a 1:1 stoichiometry was assumed for IL-2R.
[b]The binding efficiency represents the percentage of receptor or antibody effective in capturing the IL-2, relative to the theoretically expected value.

4. IL-2 BIOASSAY

IL-2 bioactivity as determined by the IL-2-dependent proliferation of murine-cytolytic T lymphocytes line (CTLL) cells, as measured by the colorimetric analysis of lactic acid produced by the stimulated cells as an end product of glucose metabolism (Familletti and Wardwell, 1988). The quantity of IL-2 needed to effect a half-maximal response in the assay is defined as one unit of activity. A biologic response modifier protein (BRMP) reference reagent, human IL-2 (Jurkat), was used as a reference standard.

5.   RECEPTOR-AFFINITY PURIFICATION PROCEDURES

The general RAC purification scheme for the production of clinical grade rIL-2 from microbial and mammalian sources is given in Fig. 4.

5.1. Extraction, Solubilization, and Renaturation of rIL-2

rIL-2 is expressed in high concentrations in an insoluble form within the inclusion bodies of *E. coli* cells. The extraction of rIL-2 from inclusion bodies requires strong denaturants such as $7M$ guanidine-HCl. Under these conditions the solubilized rIL-2 is denatured. It is diluted 40-fold with PBS and allowed to age for 3–4 days, during which time refolding occurs. The clarified extract is then ready for application onto the receptor column. The conditioned CHO media are filtered through a 0.2-μm filter before being applied onto the receptor column.

5.2. Adsorption

The adsorption step is one of the most critical aspects of receptor-affinity chromatography. The buffer used to prepare the crude extract, as well as the flow rate, should be carefully chosen such that maximum adsorption takes place. In order to take advantage of the receptor column's ability to preferentially bind the fully renatured, soluble form of rIL-2 from a heterogeneous population, the column is operated at or above saturating conditions.

5.3. Washing and Elution

Immediately following adsorption, the receptor column is washed extensively with PBS to remove unadsorbed and nonspecifically adsorbed materials from within and surrounding the receptor beads.

The elution of the adsorbed rIL-2 from the receptor column is achieved by causing the dissociation of the receptor–ligand complex by treating it with the low-pH buffer, $0.2N$ acetic acid containing $0.2M$ sodium chloride.

5.4. Size-Exclusion Liquid Chromatography

The receptor-affinity purified rIL-2 is concentrated and subjected to gel permeation in a Sephadex G-50 superfine column equilibrated with $50mM$ sodium acetate containing 5 mg of mannitol/ml and $0.2M$ sodium chloride, pH 3.5. Gel permeation as a final step in the

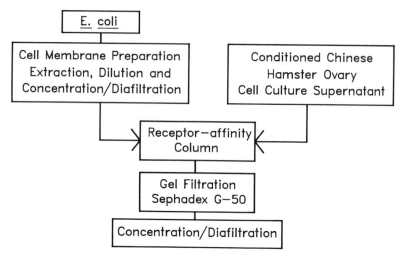

FIGURE 4. Receptor-affinity purification schemes.

purification scheme is a convenient way of preparing the rIL-2 free of high-molecular-weight contaminants such as oligomers, proteins of microbial origin, and pyrogens, as well as low-molecular-weight contaminants, such as fragments and salts. This step can also be used for the exchange of IL-2 into the final storage buffer. The gel permeation step is carried out under aseptic conditions.

5.5.   Concentration, Diafiltration, and Storage

The gel-filtered rIL-2 is concentrated to ~5mg/ml in an Amicon thin-channel concentrator using a 5000 M_r cutoff YM-5 membrane and stored in the acetate/mannitol buffer at 2–4°C. No loss of activity is observed for at least one year.

6.   GLYCOSYLATED MAMMALIAN rIL-2

The glycosylated mammalian rIL-2 was receptor-affinity-purified according to the same protocol as used for the microbial rIL-2.

The final recoveries for microbial and mammalian rIL-2 were 58% and 88%, respectively. Both forms of rIL-2 had similar specific activities of 2×10^7 U/mg.

7.   COMPARISON OF RECEPTOR-AFFINITY- AND IMMUNOAFFINITY-PURIFIED rIL-2

Conceptually, only a fully active biomolecule in its native conformation should bind to its receptor with high avidity from a heterogenous population. Therefore, the receptor-affinity-purified rIL-2 is expected to be biochemically and biologically more homogeneous than the immunoaffinity-purified material, which may also contain other molecular forms.

We tested this hypothesis by comparing the gel permeation profiles of receptor-affinity- and immunoaffinity-purified rIL-2. Unlike the receptor-affinity-purified rIL-2, the immunoaffinity-purified material contained significant amounts of oligomeric forms of rIL-2, which elute as a high-molecular-weight fraction (Bailon *et al.*, 1987). The specific activity of the high-molecular-weight fraction is only one-tenth that of the soluble monomeric form (Weber and Bailon, 1990). These results clearly indicate that in immunoaffinity chromatography several molecular forms of rIL-2 with varying degrees of biological activity and renaturation bind to the antibody column. In contrast, the receptor-affinity-purified rIL-2 contained essentially the soluble monomeric form.

8. RECEPTOR-AFFINITY PURIFICATION OF HUMAN rIL-2 ANALOGS

Employing recombinant DNA technology, four human rIL-2 analogs, a murine homolog, and an interleukin-2 *Pseudomonas* exotoxin fusion protein have been synthesized in *E. coli* or yeast (*Pichia pastoris*) and purified according to the receptor-affinity purification protocol described earlier. All purified rIL-2 analogs were subjected to sodium dodecyl sulfate-polyacrylamide gel electrophoresis (SDS-PAGE) analysis (Fig. 5).

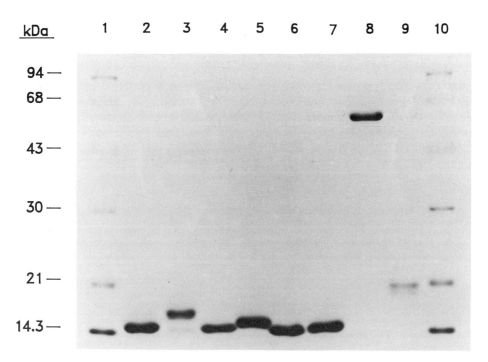

FIGURE 5. SDS-PAGE analysis of receptor-affinity purified rIL-2 and its analogs. Lanes 1 and 10: standard molecular weight marker proteins; lane 2, microbial rIL-2; lane 3, mammalian glycosylated rIL-2; lane 4, yeast rIL-2; lane 5, rIL-2 mutein Lys[20]; lane 6, rIL-2 mutein Des-Ala[1]-Ser[125]; lane 7, seleno-Met-rIL-2; lane 8, IL2-PE40; lane 9, murine rIL-2.

8.1. Human rIL-2 from Yeast (Pichia pastoris)

Human rIL-2 was cloned and expressed in *Pichia pastoris* and receptor-affinity purified. The yeast rIL-2 contains ~3% N-terminal methionine compared to ~90% for the rIL-2 produced in *E. coli*. Nevertheless, both the yeast and *E. coli* rIL-2s have similar specific activities and electrophoretic mobilities.

8.2. rIL-2 Lys²⁰ Mutein

Using site-directed mutagenesis, the N-terminal aspartic acid at position 20 (Asp^{20}) in human rIL-2 was substituted with a lysine residue (Collins *et al.*, 1988). This substitution caused a thousandfold reduction in bioactivity. rIL-2 Lys^{20} is one of the human rIL-2 analogs used in the study aimed at identifying the amino acid residues involved in the binding of IL-2 to its high-affinity receptor and to the individual receptor subunits (Collins *et al.*, 1988).

8.3. rIL-2 Des-Ala¹-Ser¹²⁵ Mutein

In another of the rIL-2 muteins that we purified, alanine at position 1 is deleted and the cysteine at position 125 is substituted with a serine residue. The purified soluble monomeric form of the mutant IL-2 has a specific activity of 1.67×10^7 U/mg. It contains ~50% N-terminal methionine compared to ~90% for the wild-type rIL-2.

8.4. Selenomethionine-rIL-2

The selenomethionine derivative of rIL-2 was engineered and expressed in *E. coli*, and the receptor-affinity purified material was used in X-ray crystallographic structure determination studies. The selenium atom in the rIL-2 provides an anomalous scattering element, which is very useful in X-ray crystallographic studies at multiple wavelengths. Amino acid sequence analysis indicated that ~99% of the methionine residues were converted to the selenomethionine form. The selenomethionine-rIL-2 retained full bioactivity.

8.5. Murine rIL-2

The murine rIL-2 was needed to conduct chronic studies in mouse models without the antigenicity problems associated with species differences. The purified murine rIL-2 shows a doublet on SDS-PAGE, consisting of 17- and 18-kDa bands (Fig. 5, lane 9), and it has a specific activity of 1×10^7 U/mg. The amino acid sequence analysis indicated the presence of an unusual string of 12 consecutive glutamine residues (positions 15–26).

8.6. Interleukin-2–Pseudomonas Exotoxin Fusion Protein

The interleukin-2–*Pseudomonas* exotoxin fusion protein (IL2-PE40) is a 54.4-kDa chimeric protein in which the cell recognition domain of *Pseudomonas* exotoxin is replaced by IL-2. This fusion protein is a potential cytotoxic agent for cells bearing IL-2 receptors. Consequently, IL2-PE40 may be a useful therapeutic agent in the treatment of hyperimmunity associated with autoimmune diseases, organ transplant rejection, allograft rejection, rheumatoid arthritis, and the like, where IL-2R-positive cells are thought to be involved in the aberrant regulation of the immune response. In order to fully explore the clinical utility of this molecule, sufficient quantities of highly pure IL2-PE40 are needed. IL2-PE40 was engineered and expressed in *E. coli* (Lorberboum-Galski *et al.*, 1988) and purified by the receptor-affinity method. The bioactivity of IL2-PE40 was determined by its ability to inhibit the IL-2-dependent proliferation of murine CTLL cells (Bailon *et al.*, 1988).

8.7. SDS-PAGE Analysis of RAC-Purified Molecules

The apparent homogeneity of the receptor-affinity-purified biomolecules is demonstrated by the SDS-PAGE profile in Fig. 5.

9. RECEPTOR-AFFINITY PURIFICATION OF HUMANIZED ANTI-Tac (ANTI-Tac-H)

Anti-Tac-H is a chimeric monoclonal antibody in which selected amino acids from within the variable region of a murine monoclonal antibody to the low-affinity p55 sub-unit of IL-2 receptor (denoted Tac, for T-cell-activating antigen) are engineered into a human IgG framework using recombinant DNA technology (Queen *et al.*, 1989). It is a potential immunosuppressant drug lacking the antigenicity problems associated with the murine monoclonal antibody. Anti-Tac-H is being investigated as a therapeutic agent in the treatment of autoimmune diseases as well as for the prevention of organ transplant rejection.

Anti-Tac-H was secreted by SP2/O cells (a murine myeloma cell line) which were transfected with the gene encoding the chimeric monoclonal antibody. It was purified from SP2/O conditioned media employing RAC. The purification procedures were the same as those described earlier for the purification of mammalian rIL-2. SDS-PAGE analysis (Fig. 6) as well as bioassays (IL-2R binding assay) demonstrated the high purity of the RAC-purified anti-Tac-H. SDS-PAGE analysis under reducing conditions (Fig. 6, lane 6) shows the heavy and light chains of the IgG molecule. Lane 8 in Fig. 6 shows the anti-Tac-H under nonreducing conditions. The heterogeneity seen here is probably due to posttranslational modifications such as glycosylation.

10. CONCLUDING REMARKS

Receptor-affinity chromatography has been demonstrated as a versatile technique for the purification of wild-type human rIL-2, its various analogs, a murine homolog, an interleukin-2–*Pseudomonas* exotoxin fusion protein, and a chimeric monoclonal antibody to IL-2R. Recent progress in biotechnology has led to the production of a soluble form of IL-2R, which has made possible the development of the first reported receptor-affinity chromatography system. With the advent of recombinant DNA technology, soluble receptors of other biomolecules such as interleukin-1, tumor necrosis factor, and γ-interferon have also been produced. Therefore, we believe that in the near future receptor-affinity chromatography will become a practical method for the purification of biomolecules.

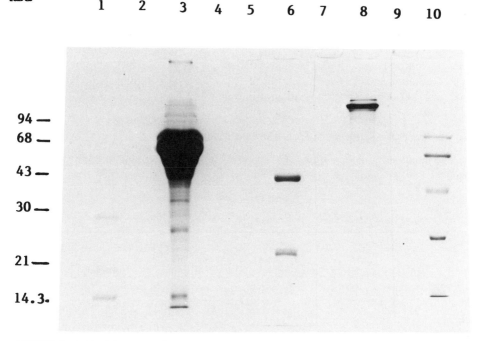

kDa

94 —
68 —
43 —
30 —
21 —
14.3.

FIGURE 6. SDS-PAGE analysis of receptor-affinity purified anti-Tac-H. Lanes 1 and 10: standard molecular weight marker proteins; lane 3, crude CHO media; lane 6, purified anti-Tac-H (reduced); lane 8, purified anti-Tac-H (nonreduced).

ACKNOWLEDGMENTS

We are grateful to P. Familletti, J. E. Fredericks, and D. Mueller for providing the *E. coli* cells and CHO cell culture supernatants. We thank J. E. Fredericks, P. Quinn, and J. Kondas for the IL-2, IL2-PE40, and anti-Tac-H bioassays, respectively. M. Miedel and Y.-C. Pan are acknowledged for performing the amino acid sequence analysis. We are grateful for G. Velicelebi and G. Davies (SIBIA Inc., San Diego, California) for engineering, cloning, and expressing IL-2 in *Pichia pastoris*, as well as for providing the cells. Finally, we thank Lisa Nieves for typing the manuscript.

REFERENCES

Bailon, P., and Weber, D. V., 1988, Receptor-affinity chromatography, *Nature (London)* **335:**839–840.

Bailon, P., Weber, D. V., Keeney, R. F., Fredericks, J. E., Smith, C., Familletti, P. C., and Smart, J. E., 1987, Receptor-affinity chromatography: A one-step purification for recombinant interleukin-2, *Bio/Technology* **5:** 1195–1198.

Bailon, P., Weber, D. V., Gately, M., Smart, J. E., Lorberboum-Galski, H., Fitzgerald, D., and Pastan, I., 1988, Purification and partial characterization of an interleukin 2-*Pseudomonas* exotoxin fusion protein, *Bio/Technology* **7:**1326–1329.

Bailon, P. S., Weber, D. V., and Smart, J. E., 1991, Practical aspects of receptor affinity chromatography, in:

Purification and Analysis of Recombinant Proteins (R. Seetharam and S. Sharma, eds.), Marcel Dekker, New York, pp. 267–283.

Collins, L., Tsien, W.-H., Seals, C., Hakimi, J., Weber, D., Bailon, P., Hoskins, J., Green, W. C., Toome, V., and Ju, G., 1988, Identification of specific residues of human interleukin 2 that affect binding to the 70-kDa subunit (p70) of the interleukin 2 receptor, *Proc. Natl. Acad. Sci. U.S.A.* **85:**7709–7713.

Familletti, P. C., and Wardwell, J. A., 1988, A novel colorimetric assay to determine cell growth in biological assays, *Bio/Technology* **6:**1169–1172.

Hakimi, J., Seals, C., Anderson, L. E., Podlaski, J., Lin, P., Danho, W., Jenson, J. C., Donadio, P. E., Familletti, P. C., Pan, Y.-C. E., Tsien, W.-H., Chizzonite, R. A., Csabo, L., Nelson, D. L., and Cullen, B. R., 1987, Biochemical and functional analysis of soluble human interleukin-2 receptor produced in rodent cells, *J. Biol. Chem.* **262:**17336–17341.

Ju, G., Collins, L., Kaffka, L., Tsien, W.-H., Chizzonite, R. A., Crowl, R., Bhatt, R., and Kilian, P., 1987, Structure–function analysis of human interleukin-2: Identification of amino acid residues required for biological activity, *J. Biol. Chem.* **262:**5723–5732.

Kohler, G., and Milstein, C., 1975, Continuous cultures of fused cells secreting antibody of predefined specificity, *Nature (London)* **256:**495–497.

Leonard, W. J., Depper, J. M., Crabtree, G. R., Rudikoff, S., Pumphry, J., Robb, R. J., Krönke, M., Svetlik, P. B., Peffer, N. J., Waldmann, T. A., and Greene, W. C., 1984, Molecular cloning and expression of cDNAs for the human interleukin-2 receptor, *Nature (London)* **311:**626–631.

Lorberboum-Galski, H., Fitzgerald, D., Chaudhary, V., Adhya, S., and Pastan, I., 1988, Cytotoxic activity of an interleukin 2-*Pseudomonas* exotoxin chimeric protein in *Escherichia coli*, *Proc. Natl. Acad. Sci. U.S.A.* **85:**1922–1926.

Queen, C., Schneider, W. P., Selick, H. E., Payne, P. W., Landolfi, N. F., Duncan, J. F., Avdalovic, N. M., Levitt, M., Junghans, R. P., and Waldmann, T. A., 1989, A humanized antibody that binds to IL-2 receptor, *Proc. Natl. Acad. Sci. U.S.A.* **86:**10029–10033.

Tarnowski, S. J., Roy, S. K., Liptak, R. A., Lee, D. K., and Ning, R. Y., 1986, Large-scale purification of recombinant human leukocyte interferons, *Methods Enzymol.* **119:**153–165.

Weber, D. V., and Bailon, P., 1990, Application of receptor-affinity chromatography to bioaffinity purification, *J. Chromatogr.* **510:**59–69.

Weber, D. V., Keeney, R. F., Familletti, P. C., and Bailon, P., 1988, Medium-scale ligand-affinity purification of two soluble forms of human interleukin-2 receptor, *J. Chromatogr.* **431:**55–63.

10

Immobilized Artificial Membrane Chromatography
Prediction of Drug Transport across Biological Barriers

Francisco M. Alvarez, Carey B. Bottom, Prashant Chikhale, and Charles Pidgeon

1. INTRODUCTION

The development of a simple, rapid method to predict drug transport across biological barriers has been a long-standing objective in the pharmaceutical sciences. Typical barriers to drug transport include the membranes in the intestinal tract, lungs, and eye, the blood–brain barrier, and many others. In addition to these "internal" barriers, the skin is an "external" barrier that molecules must penetrate if transdermal drug delivery is to be successful. The prediction of drug transport through any of these biological-membrane barriers is typically a lengthy process that requires substantial experimental effort; however, the experiments are frequently complicated by significant variability that makes the methods unreliable or useful for only a small class of compounds. This chapter describes a new approach for predicting drug transport through human skin. The method utilizes high-performance liquid chromatography (HPLC) columns packed with a stationary phase containing immobilized membrane lipids (i.e., phosphatidylcholine).

2. SKIN COMPOSITION AND TRANSDERMAL DRUG DELIVERY

The development of transdermal delivery systems is a particularly challenging problem, and there are no useful, simple models for predicting drug transport through the

Francisco M. Alvarez and Carey B. Bottom • Schering-Plough Research Institute, Kenilworth, New Jersey 07033. *Present address of F. M. A.*: Mylan Pharmaceuticals Inc., Morgantown, West Virginia 26504. *Present address of C. B. B.*: Chase Laboratory Co., Newark, New Jersey 07105. *Prashant Chikhale* • Department of Pharmaceutical Chemistry, The University of Kansas, Lawrence, Kansas 66045. *Charles Pidgeon* • Department of Medical Chemistry, Purdue University, West Lafayette, Indiana 47907.

Molecular Interactions in Bioseparations, edited by That T. Ngo. Plenum Press, New York, 1993.

skin. A recent review has clearly defined the state of the science and emphasizes that several artificial systems are worthy of further exploration (Houk and Guy, 1988). Clearly, the most desirable method would provide reproducible results rapidly. In addition, the method should permit the experimental data to be used to model the physical biochemical events associated with drug transport through the skin. Our current understanding of the chemistry responsible for drug penetration through the skin involves the skin lipids that reside inside the main diffusion paths accessible to drugs. The main diffusion paths (or transport channels) are in the intercellular spaces of the stratum corneum; this intercellular space comprises 10–30% of the total stratum corneum volume (Grayson and Elias, 1982) and is primarily lipid organized into lamellar structures (Elias, 1990).

3. COMPOSITION OF MAMMALIAN EPIDERMAL LIPIDS

The lipid content of the stratum corneum is predominantly sphingolipids with smaller amounts of cholesterol and fatty acids (Table 1, from Elias, 1990). Sphingolipids are a broad class of polar lipids, the most abundant of which contain the phosphatidylcholine head group at the 1 position of the sphingosine backbone and an amide-linked fatty acid at the 2 position. The intercellular lipid matrix of the skin creates two types of barriers to transport: (i) the lipid head groups provide a polar barrier for the transport of drugs, and (ii) the hydrocarbon fatty alkyl chains provide a nonpolar barrier. Thus, any method that is developed for predicting the transport of drugs through the skin should incorporate both polar and nonpolar interactions.

4. MODEL SYSTEMS FOR SKIN TRANSPORT STUDIES

Several synthetic membranes have been used to model the transport of drugs across the skin (Houk and Guy, 1988). In these experiments the synthetic membrane itself is substituted for mammalian skin in the flux measurements that are used to measure the skin permeability of drugs. A key supposition in these models is that the skin acts primarily as a passive diffusional barrier, with very little involvement of drug partitioning into the intercellular lipid pools. Consequently, for this method, the rate of diffusion is determined by only the drug concentration and the length of the diffusion pathway. The cellulose acetate

TABLE 1
Composition of Mammalian Epidermal Lipids[a]

Lipid type	Living layers (%)	Stratum corneum (%)
Phospholipids	40	Trace
Sphingolipids	10	35
Cholesterol	15	20
Triglycerides	25	Trace
Fatty acids	5	25
Other	5	10

[a]From Elias, 1990; reprinted with permission.

membrane is a good example of a synthetic membrane for diffusion studies. As model systems for skin, these acetate membranes give poor correlations between skin permeability and the octanol/water (o/w) partition coefficients of steroids; this limitation has been documented by several laboratories (Gary-Bobo *et al.*, 1969; Barry and El Eini, 1976; Barry and Brace, 1977). The main reason for the poor correlation is that this simple system completely neglects the role of drug partitioning during transport through the skin.

In addition to hydrophilic acetate membranes, an array of synthetic polymer membranes containing hydrophobic groups have been evaluated as potential models for the penetration of compounds through the skin. The hydrophobic groups are intended to provide functional groups for drug partitioning into lipids. Of these mixed-polarity membranes, dimethylpolysiloxane membranes have been used most extensively. The correlation of alkanol permeability through these membranes with octanol/water partition coefficient is good (Behl *et al.*, 1983). These model membranes, however, do not incorporate all of the structural features needed to evaluate interactions occurring during drug transport through the lipoidal intercellular channels.

In other instances, membrane laminates have also been used to mimic the skin barrier. Unfortunately, this model system is not useful because desorption of solutes out of the stratum corneum is linear with respect to the square root of time, whereas a linear relationship is obtained with laminates (Houk and Guy, 1988).

In addition to synthetic membrane models, a variety of skins have been used to model drug penetration through human skin. Examples of animal skins employed for this purpose include rat, mouse, rabbit, squirrel, and guinea pig skins and hairless variations of these animal skins (Wester and Maibach, 1985). Although animal skin is more readily available than human skin, the method of studying solute transport using any type of skin remains time-consuming and suffers from reproducibility problems. For instance, hairless mouse skin, frequently purported to be a good model, has been demonstrated to be more permeable than human skin and is therefore an inadequate model. Recent work by Itoh *et al.* (1990) has suggested the use of shed snake skin as a model membrane, and these authors proposed that the lipophilicity and the molecular weight response of snake skin more closely mimic those of human stratum corneum than do those of hairless mouse skin. In addition, the water evaporation rates of shed snake skin are comparable to those of human skin. Although snake skin appears to be a better model barrier than rat or mouse skin, permeability differences of some magnitude were observed between human and snake skin. It is likely that structural differences in the skin lipids and in the relative proportions of each lipid class in snake skin versus human skin account for the lack of optimum correlation.

In contrast to experiments with hydrophilic and mixed-polarity membranes, similar experiments using human epidermis as the membrane barrier demonstrated a significant correlation between permeability and the octanol/water partition coefficient (Flynn, 1985). It is clear from these studies that the role of partitioning into the lipid matrix of the intercellular space of the stratum corneum is critical for the percutaneous penetration of compounds. Model systems useful for predicting drug transport across the epidermis must include both partitioning of drugs into polar lipids (like sphingolipids) and tortuous paths of different lengths which both mechanisms play a major role in drug transport across the stratum corneum.

For any biological barrier, understanding the structure and composition of the main diffusion paths is critical for prediction of drug transport. As discussed above, the prevalent diffusion paths for transdermal drug delivery are the epidermal transport channels that are filled with lipid. Usually, channels are thought of as empty tubes that solvents can flow through. However, we emphasize that the transport channels in the skin are filled with lipids, and both the polar and nonpolar lipid components significantly influence the ability of a solute to penetrate the skin through these channels.

The octanol/water partition coefficient of a drug has been a physicochemical property used by pharmaceutical scientists for predicting absorption in the gastrointestinal tract. Obviously, this simple measurement only evaluates the partitioning processes associated with the *hydrocarbon* components of skin lipids and does not take into consideration the polar lipid head groups that form the first barrier to transport of drugs across virtually all biological membranes (Pidgeon, 1990). The octanol/water partition parameter has nevertheless been the tool of choice due to its simplicity and ease of determination.

The mechanism of drug penetration and subsequent absorption is clearly related to the difficulty of penetration and, more importantly, specific interactions with the membrane bilayer. The interactions are specific and occur at defined locations in the membrane bilayer. Recent experiments using small-angle X-ray diffraction have demonstrated the presence of specific sites at which drugs interact with membrane lipids (Mason *et al.*, 1989).

The membrane partition coefficient (K_m) is perhaps the most important parameter related to drug transport and absorption and of course is dependent on the physical and chemical nature of the biological membrane and the drug. Structural diversity in the membrane lipid composition has a significant influence on drug–membrane interactions and consequently on drug transport through membranes. As an example, Mason *et al.* (1991) have shown a significant decrease in the membrane partition coefficient for the drug nimodipine with a change in the membrane content from a 0.1:1 to a 0.6:1 cholesterol: phospholipid molar ratio. This study exemplifies the relevance of the membrane lipid composition to the membrane partition process and drug transport.

In summary, drug transport through the skin has two major components. One component is the length and tortuosity of the absorption path itself. Longer paths require more time for absorption. The second major component is the solubility and/or partitioning of drugs within the lipid matrix comprising the intercellular transport channels. Partitioning of drugs into lamellar structures of membrane lipids involves both nonpolar interactions with the hydrocarbon chains and polar interactions with the lipid head groups.

Methods used to study the polar and nonpolar interactions involve either mammalian skin(s), synthetic membranes, or physicochemical measurements such as octanol/water partition coefficients. The partitioning component can be modeled with octanol/water partitioning measurements whereas the tortuous paths can be modeled with synthetic membranes. Octanol/water partitioning measurements do not take into consideration the tortuosity of the diffusion path nor do these measurements takes into consideration drug–membrane interactions due to the lipid head groups. None of the existing physicochemical methods to study solute transport through skin incorporate both tortuous paths with different lengths and membrane lipid binding sites during the measurement. Although artificial membrane suspensions called liposomes can model both the polar and nonpolar binding components of drug partitioning into membranes, these artificial membranes do not have tortuous paths.

Since solute migration through a chromatography column is known to occur through numerous paths, a logical extension of the physicochemical methods described above is to immobilize membrane lipids on chromatography surfaces such as silica. Monolayers of immobilized lipids on chromatography particles would model both the hydrophobic and hydrophilic interactions of the stratum corneum intercellular lipid matrix and, when packed in HPLC columns, would provide tortuous paths for solute migration. The immobilized artificial membrane (IAM) column is clearly not optimally representative of tortuous pathways in biological membranes. However, chromatography columns containing immobilized membrane lipids are possible models for skin transport because they model the important components associated with drug transport (i.e., tortuous paths and lipids containing both hydrophilic and hydrophobic binding sites). HPLC C_{18} columns can model the tortuous paths and the hydrophobic contribution of drug partitioning in membranes, but they cannot model the hydrophilic drug–membrane interactions. This is because C_{18} columns contain only immobilized alkyl chains. Immobilized membrane lipids are expected to be superior to immobilized alkyl chains, at least with regard to their ability to model the transport of drugs across the skin or any other biological barrier.

We have synthesized chromatography particles containing immobilized membrane lipids and denoted the chromatographic method as immobilized artificial membrane (IAM) chromatography (for a recent review of IAM technology, see Pidgeon et al., 1992). In this chapter, we describe the first application of IAM technology for predicting the transport of solutes across a biological barrier, namely, the skin. In addition to predicting drug transport across mammalian skin, IAM technology is probably useful for drug design and delivery across any biological membrane (Alvarez et al., 1990). It is envisioned that columns with different polar lipid monolayers could be developed to provide specificity with regard to the biological membrane of interest.

5. IMMOBILIZED ARTIFICIAL MEMBRANE CHROMATOGRAPHY (IAM)

The IAM column consists of a monolayer of phosphatidylcholine linked to silica propylamine, and therefore this surface contains one-half of the membrane bilayer (Pidgeon and Venkataram, 1989). The IAM column is intended to mimic the lipid phase of cell membranes and thereby effect separation on the basis of membrane interactions (Pidgeon and Venkataram, 1989; Stevens et al., 1989; Pidgeon et al., 1992).

Since drug transport and absorption are based on binding interactions with membrane lipids, we postulated that IAM columns would be useful for predicting the transport of drugs through the skin. Preliminary results have demonstrated that the correlation between the drug's capacity factor on an IAM column and the permeability through human stratum corneum was excellent (Chikhale et al., 1989; Pidgeon et al., 1992). Herein, we describe in more detail the use of IAM columns for predicting drug transport. Results obtained using the IAM column are compared to those obtained with the widely used C_{18} columns and the octanol/buffer partition system.

6. EXPERIMENTAL METHODS

6.1. Skin Preparation for Transport Studies

Human skin from cadavers (thigh or abdominal region; Worth Regional Transplant Services, Lansing, Michigan) previously dermatomed was gently swirled in distilled water (preheated to 60°C) for about 2–3 min. The epidermal membrane (about 100 μm thick) was then carefully separated from the dermis in a glass trough containing distilled water at room temperature. The stratum corneum surface was then placed flat against a polyethylene plastic sheet, and the dermal side of the epidermal layer was covered with an absorbent paper saturated with 0.9% NaCl. This human epidermal membrane was used immediately or stored at 6°C and used within 24 h for the *in vitro* diffusion experiments.

6.2. Diffusion Experiments

Solute diffusion through human epidermal membranes was studied using side-by-side diffusion cells (Science Glass Company, Miami, Florida) containing a 1.77-cm² diffusional surface area. The diffusion cell was equipped with both a 5-ml donor reservoir and a 5-ml receiver reservoir. The donor reservoir contained a saturated solution of the drug in phosphate-buffered saline, pH 7.1, and the receiver reservoir contained nonbuffered saline. A typical preparation of donor solution was as follows. Excess drug was stirred in phosphate-buffered saline, pH 7.1, for 24 h at 32°C in an incubator. The suspension was filtered, and 5 ml of the filtrate was pipetted into the donor reservoir. All drugs were chemically stable in the aqueous donor vehicle. During the diffusion studies, the cells were unstirred. At appropriate time intervals, the entire contents of the receiver reservoir were collected and subjected to HPLC analysis, and the receiver reservoir was refilled with an equal volume of fresh saline.

The appropriate equations (based upon Fick's law of diffusion) are shown in Fig. 1 (Michaels *et al.*, 1975). These equations were used to obtain the permeability parameters from the *in vitro* diffusion experiments.

6.3. IAM Chromatography

A 12-μm immobilized artificial membrane phosphatidylcholine (IAM.PC) column (Regis Chemical Company, Morton Grove, Illinois), 15 cm × 4.6 mm i.d., was connected

1. $\quad J = (K_m \times D/\delta)C_d = K_p \times C_d$

2. $\quad D = \delta^2/6 \times t_{lag}$

3. $\quad K_m = C_m/C_v$

J = Flux of the drug across the membrane (mol/cm²/h)
K_m = Partition coefficient of the drug in the membrane
D = Diffusion coefficient of the drug in the membrane (cm²/h)
K_p = Permeability coefficient of the membrane for the drug (cm/h)
C_d = Concentration of the drug in the donor (mol/ml)
δ = Thickness of the membrane (cm)
t_{lag} = Lag-time for diffusion (h)
C_m = Concentration of the drug in the membrane (mol/ml)
C_v = Concentration of the drug in the vehicle (mol/ml)

FIGURE 1. Equations used for calculating skin permeability.

to a Waters HPLC system consisting of a 510 pump, a 712 Wisp automatic injector, a Lambda-Max model 480 LC spectrometer, and a 730 Data Module. A dilute solution of the drug (50–100 nmol/ml) in the mobile phase was injected into the column. The capacity factor, K', of the test compound was calculated using the equation $K' = (t_m - t_0)/t_0$, where t_m is the retention time in minutes of the test compound, and t_0 is the retention time of the practically unretained compound, citric acid. K'_{IAM} denotes that the capacity factor was measured on an IAM column. The test solutes were detected by UV absorbance measurements at $\lambda = 220$ nm (steroids and β-blockers) or by refractive index measurements (*n*-alcohols) using an LDC Refractomonitor III operating at 0.5 mA. The mobile-phase compositions used were 0, 10, and 30% acetonitrile in Dulbecco's phosphate-buffered saline 10× dilution (DPBS 10×) at pH 7.1 for the *n*-alcohols, β-blockers, and steroids, respectively. The flow rate was 1.0 ml/min for both the *n*-alcohols and β-blockers and 2.0 ml/min for the steroids. The column was washed with acetonitrile:water (20:80) at 1.0 ml/min for 1 hr, before and after analysis, and was preserved (stored) by washing it with acetonitrile at 1.0 ml/min for 1 hr.

6.4. C_{18} Column Chromatography

A Waters μ Bondapak 10-μm column, 15 cm × 4.6 mm i.d., was used with similar HPLC conditions as described above for the IAM column chromatography. Generally, mobile-phase compositions containing higher percentages of the organic modifier were required to elute the various compounds from the C_{18} stationary phase. Thus, the *n*-alcohols were eluted using acetonitrile:water (30:70), and the steroids were eluted with acetonitrile: DPBS 10× (50:50) as the mobile phase. The dead time, t_0, was measured with citric acid and the capacity factor, K'_{C18}, was calculated as described above for IAM chromatography.

7. CORRELATIONS TO SKIN PERMEABILITY

7.1. Linear Alcohols

The experimentally measured values for human skin permeation (K_p), skin–water partition coefficients (K_m), HPLC octadecylsilyl capacity factors (K'_{C18}), and HPLC IAM.PC capacity factors (K'_{IAM}) for the *n*-alcohols are presented in Table 2. The *n*-alcohols were chosen as a set of model solutes because of the known log-linear relationship between K_m and K_p (Fig. 2). Since K_m is log linear with K_p, the transport of *n*-alcohols through human skin is primarily governed by the partitioning processes in the stratum corneum (Scheuplein and Blank, 1971).

For the C_{18} columns, the relationship between the *n*-alcohols' capacity factor and their permeability across the human stratum corneum shows a biphasic curve (Fig. 3). Compared to the best fit line in Fig. 3, the lower-molecular-weight *n*-alcohols (propanol, butanol, and pentanol) are underestimated and the higher-molecular-weight *n*-alcohols (heptanol and octanol) are overestimated. For the low-molecular-weight *n*-alcohols, the alcohol functional group is a large volume fraction of the molecular volume, and consequently these molecules are polar. However, for the high-molecular-weight alcohols, the alcohol functional group is a small volume fraction of the total molecular volume, and consequently these molecules are relatively nonpolar. Thus, the size of the alcohol molecule itself determines if the OH

TABLE 2
Relative Retention of Linear Alcohols in the IAM.PC
and C_{18} HPLC Columns and Their Human Stratum
Corneum Partition and Permeability Coefficients

Alcohol	log K'_{IAM}	log $K'_{C_{18}}$	log $K_m{}^a$	log $K_p{}^b$
Methanol	−1.32	−2.19	−0.22	−3.00
Ethanol	−1.02	−1.89	−0.22	−3.00
Propanol	−0.94	−1.00	0.30	−2.85
Butanol	−0.54	−0.12	0.40	−2.60
Pentanol	−0.12	0.23	0.70	−2.22
Hexanol	0.38	0.59	1.00	−1.89
Heptanol	0.85	0.95	1.48	−1.50
Octanol	1.04	1.31	1.70	−1.28

[a]Data from Scheuplein (1965).
[b]K_p in units of cm/h; data from Scheuplein (1965).

functional group dominates the molecule's total polarity. Consequently, during the partitioning process, the alcohol functional group is very important for the low-molecular-weight alcohols and less important for the higher-molecular-weight alcohols. Polar solute–membrane interactions, important for the transport of hydrophilic solutes (such as the lower-molecular-weight n-alcohols), cannot be predicted by hydrophobic C_{18} stationary phases, as shown in Fig. 3. The C_{18} column can only measure nonspecific hydrophobic interactions. We note that when the octanol/water partition coefficients were used to predict

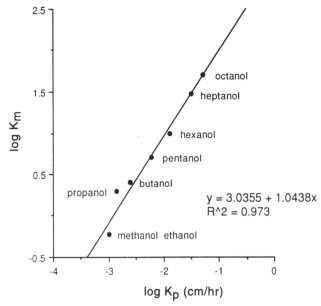

FIGURE 2. Relationship between stratum corneum permeability coefficients (K_p) of n-alcohols and their membrane partition coefficients (K_m) on a log–log scale.

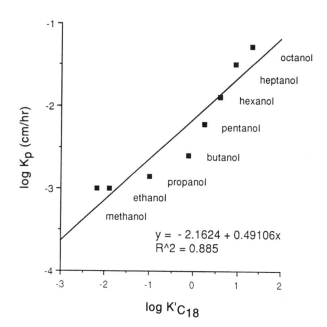

FIGURE 3. Relationship between relative retention of n-alcohols in the C_{18} column ($K'_{C_{18}}$) and their stratum corneum membrane permeability coefficients (K_p) on a log–log scale.

skin permeability coefficients of n-alcohols, a similar biphasic curve was obtained (not shown). In other words, as expected, C_{18} columns give similar results to those obtained from octanol/water partition measurements.

In contrast to these results for the C_{18} columns, the capacity factors for the n-alcohols eluting from the IAM column correlated very well with their membrane permeability coefficients (Fig. 4). In addition, the capacity factors of the n-alcohols on IAM.PC columns exhibited a log-linear relationship with membrane partition coefficients, K_m (not shown). These linear profiles indicate that the IAM.PC column accurately mimics the polar and nonpolar interactions that n-alcohols exhibit with the skin stratum corneum. In other words, the lower-molecular-weight n-alcohols exhibit polar interactions with the immobilized phospholipid head group, and the higher-molecular-weight alcohols exhibit both polar and nonpolar interactions with the immobilized phospholipid molecule. Thus, there is a smooth transition in the interaction of low- and high-molecular-weight alcohols with the IAM.PC column, and this resulted in the log-linear relationship shown in Fig. 4.

7.2. Steroids

The steroids were another group of model solutes that were evaluated. They are structurally more complex and bulkier than the n-alcohols, and their permeation across human skin has been reported to be a function of both their membrane partition coefficient, K_m, and the diffusion coefficient, D (Scheuplein and Blank, 1971). Steroids were thus used as model compounds to evaluate the ability of the IAM.PC stationary phase to predict the skin transport of drugs when both K_m and D significantly contribute to the transport process.

Table 3 lists the steroids that were evaluated. All the relevant permeation parameters

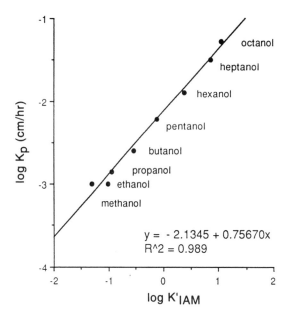

$$y = -2.1345 + 0.75670x$$
$$R^2 = 0.989$$

FIGURE 4. Relationship between relative retention of n-Alcohols in the IAM.PC column (K'_{IAM}) and their stratum corneum membrane permeability coefficients (K_p) on a log–log scale.

TABLE 3

Permeability of Human Epidermis to Steroids *In Vitro*
and Their Relative Retention in the IAM.PC and C_{18} HPLC Columns

Steroid	$\log J^a$	$\log K_p^b$	$\log K_m$	$\log D^c$	$\log K'_{IAM}$	$\log K'_{C_{18}}$
Prednisone	−4.14	−3.62	0.00	−5.62	−0.50	−0.13
Cortisone	−4.31	−4.14	−0.62	−5.52	−0.43	−0.10
Hydrocortisone	−4.19	−3.90	0.06	−5.96	−0.39	−0.14
Prednisolone	−4.77	−4.47	−0.34	−6.13	−0.38	−0.17
Dexamethasone	−4.41	−3.70	−0.29	−5.41	−0.20	0.01
Corticosterone	−3.85	−3.37	−0.12	−5.24	−0.17	0.11
Cortexolone	−3.81	−2.90	1.40	−6.30	−0.06	0.14
Spironolactone	−4.51	−3.12	−0.13	−4.99	0.07	0.57
Cortexone	−2.56	−1.61	1.59	−5.20	0.12	0.43
Testosterone	−3.11	−1.63	0.43	−4.06	0.13	0.40
Progesterone	−2.79	−1.44	2.17	−5.61	0.44	0.86
CH_3-Androstanolone	−4.18	−2.26	1.12	−5.38	0.47	0.71
Estradiol	−4.23	−2.28	1.08	−5.36	0.52	0.34
17α-Ethinyl-E_2	−3.45	−1.98	0.66	−4.64	0.63	0.44
Mestranol	−4.07	−1.97	1.99	−5.95	0.97	1.00

[a] J in units of $\mu mol/cm^2$ per hour.
[b] K_p in units of cm/h.
[c] D in units of cm^2/h.

from the diffusion studies and the capacity factors obtained using the IAM.PC and C_{18} columns are included in Table 3. The steroids varied widely in their lipophilicity and chemical structure, and when all the steroids shown in Fig. 5 were used, plots of the permeability coefficient ($\log K_p$) versus the capacity factor ($\log K'$) for the IAM.PC and the C_{18} columns showed comparable relationships (not shown).

When structurally similar analogs of the 4-pregnene steroids Δ^4-3-keto-steroids, Fig. 5) are used, the IAM.PC column gives a superior correlation compared to the C_{18} column. For the 4-pregnene series, the relationships between K_m and K_p (Fig. 6), $K'_{C_{18}}$ and K_p (Fig. 7), and K'_{IAM} and K_p (Fig. 8) are shown. The poor correlations in Fig. 6

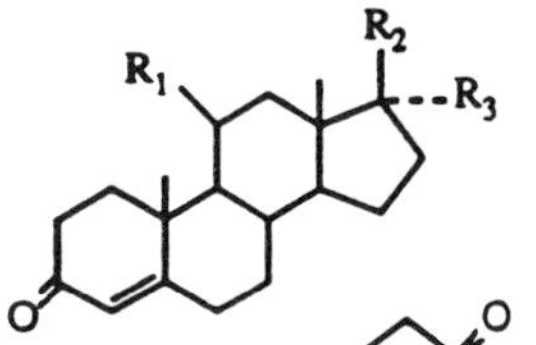

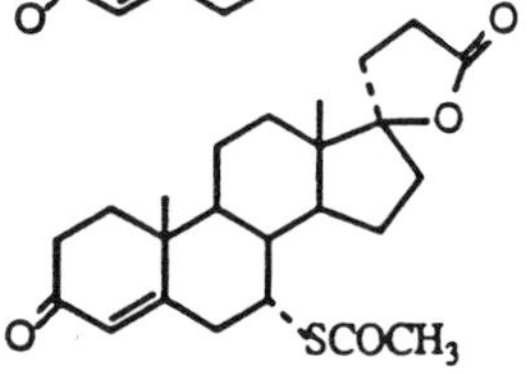

Δ^4-3-keto-steroid	R_1	R_2	R_3
Cortisone	O	$COCH_2OH$	OH
Hydrocortisone	OH	$COCH_2OH$	OH
Corticosterone	OH	$COCH_2OH$	H
Cortexolone	H	$COCH_2OH$	OH
Cortexone	H	$COCH_2OH$	H
Testosterone	H	OH	H
Progesterone	H	$COCH_3$	H

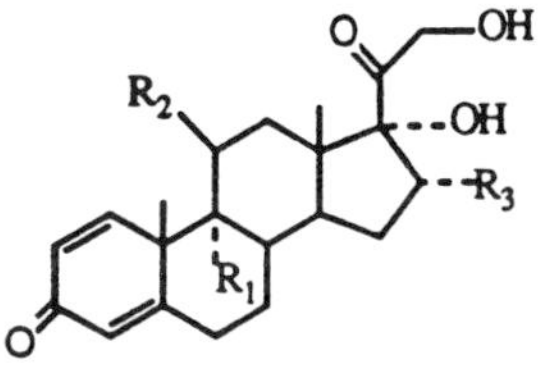

Spironolactone

$\Delta^{1,4}$-3-oxo-steroid	R_1	R_2	R_3
Prednisone	H	O	H
Prednisolone	H	OH	H
Dexamethasone	F	OH	CH_3

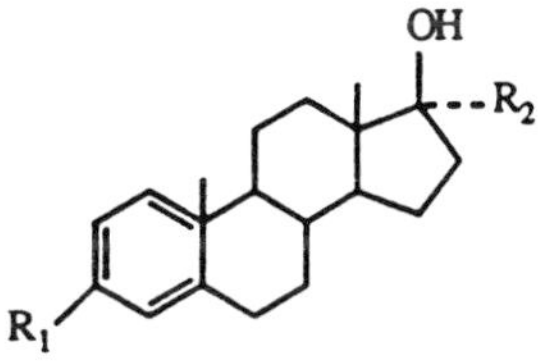

Estrogen	R_1	R_2
Estradiol	OH	H
17α-Ethinylestradiol	OH	C≡CH
Mestranol	OCH_3	C≡CH

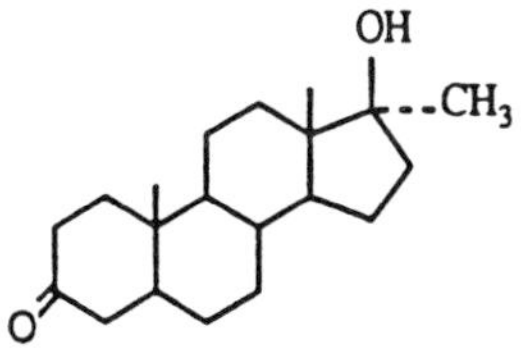

Methylandrostanolone

FIGURE 5. Chemical structures of the steroids.

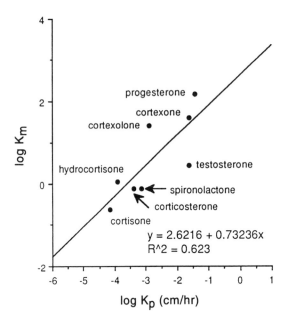

FIGURE 6. Relationship between stratum corneum coefficients (K_p) of 4-pregnene steroids and their membrane partition coefficients (K_m) on a log–log scale.

between K_m and K_p indicate that "partitioning" of the steroids in the human epidermis is not a good predictor of the "permeability" of the steroids through human skin. This observation that "partitioning" alone is not sufficient to accurately predict skin permeability of the 4-pregnene steroid analogs is supported by the results obtained using C_{18} columns. C_{18} columns can only measure partition effects, and the C_{18} stationary phase exhibited a poor

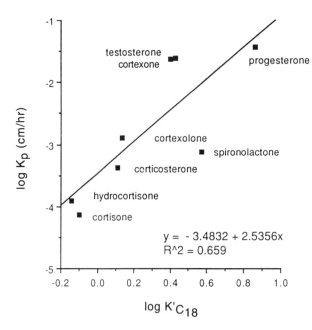

FIGURE 7. Relationship between relative retention of 4-pregnene steroids in the C_{18} column ($K'_{C_{18}}$) and their stratum corneum membrane permeability coefficients (K_p) on a log–log scale.

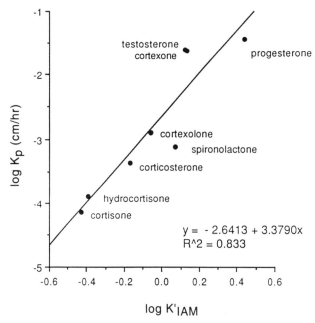

FIGURE 8. Relationship between relative retention of 4-pregnene steroids in the IAM.PC column (K'_{IAM}) and their stratum corneum membrane permeability coefficients (K_p) on a log–log scale.

correlation between $K'_{C_{18}}$ and K_p (Fig. 7). In contrast, the IAM.PC column gave a good correlation between the transport of 4-pregnene analogs and K'_{IAM} (Fig. 8). As stated above, binding of solutes to the phospholipid stationary phase occurs by both polar and nonpolar interactions; this is the reason for the better correlation using the IAM columns (Fig. 8) compared to that obtained using the C_{18} columns (Fig. 7).

7.3. β-Blockers

The last groups of compounds tested were heterogeneous with regard to molecular polarity. The mixture contained lipophilic, hydrophilic, and intermediate-polarity compounds. The compounds are listed in Table 4 along with their permeability characteristics. [Permeability data in Table 4 were obtained from Green *et al.* (1989).] The water-soluble β-blockers are structurally related cationic drugs previously characterized according to their lipophilicity (Woods and Robinson, 1981). Based on octanol/buffer partition coefficients ($K_{o/b}$), β-blockers are lipophilic (propranolol), hydrophilic (atenolol), and intermediate in behavior (oxprenolol and metoprolol). The other compounds listed in Table 4 are not structurally related to the β-blockers but represent examples of a neutral penetrant (caffeine), an anionic drug (sodium salicylate), and a cationic drug more basic than the β-blockers (naphazoline). A plot of log K'_{IAM} versus log K_p for all the compounds listed in Table 4 yielded a trend from low to high K_p, but the correlation was poor (not shown). However, an excellent correlation was observed when only the data for the structurally similar β-blockers were used (Fig. 9). The correlation of K_p with K'_{IAM} (Fig. 9) was significantly better than the correlation of K_p with $K_{o/b}$ (Fig. 10). It is interesting that

TABLE 4
IAM Relative Retention, Permeability
Coefficients, and Partition Coefficients
of β-Blockers and Miscellaneous Compounds

Compound	$\log K'_{IAM}$	$\log K_p{}^a$	$\log K_{o/b}{}^b$
Sodium salicylate	−0.48	−8.22	—
Caffeine	−0.42	−7.14	—
Atenolol	−0.03	−7.19	−2.52
Metoprolol	0.60	−6.63	−0.82
Propranolol	0.53	−6.47	0.73
Oxprenolol	0.87	−6.31	−0.29
Naphazoline	1.02	−7.04	—

^aK_p in units of cm/s; data from Green *et al.* (1989).
^{b}Data from Woods and Robinson (1981).

propranolol is more lipophilic than oxprenolol (Fig. 10), but oxprenolol is retained longer than propranolol on the IAM.PC column.

8. CONCLUSION AND DISCUSSION

The IAM.PC HPLC column was studied as a model system to predict the transport of solutes across human skin. The IAM.PC column differs from other chromatographic systems composed of phospholipids (Miyake *et al.*, 1987; Yang *et al.*, 1990) in that the lipid

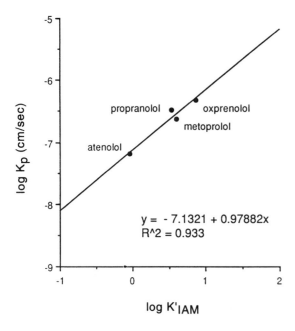

y = - 7.1321 + 0.97882x
R^2 = 0.933

FIGURE 9. Relationship between relative retention of β-Blockers in the IAM.PC column (K'_{IAM}) and their permeability (K_p) across excised human skin on a log–log scale.

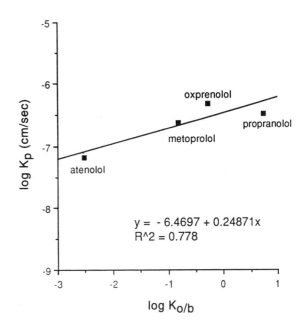

FIGURE 10. Relationship between octanol/buffer partition coefficients ($K_{o/b}$) of β-blockers and their permeability (K_p) across excised human skin on a log–log scale.

phase is covalently bound to the chromatographic support. Covalent immobilization of the phospholipid allows use of organic solvents in the mobile phase without destroying the stationary phase. Furthermore, the IAM.PC surface mimics one-half of a true bilayer in hydrocarbon thickness and contains the phosphatidylcholine head group. Sphingolipids are the most abundant lipid in the stratum corneum. On hydrolysis, sphingolipids yield one molecule of fatty acid and one molecule of sphingosine. The sphingosine moiety has a polar head group attached at the 1-hydroxyl position. In the IAM.PC column model, the phosphatidylcholine grouping mimics the polar head group of skin sphingolipids.

The phosphatidylcholine head group contains both a positive charge from a quaternary amine and a negative charge from phosphate. Titration of the IAM.PC surface showed that the surface was neutral with a pK_a of approximately 7 (Stevens *et al.*, 1989). Our studies have shown that the IAM.PC system is a better predictor of permeability (K_p) than the C_{18} reversed-phase column or the octanol–buffer partition systems for structurally similar compounds. Specific interactions with the lower alcohols (e.g., methanol, ethanol, and propanol) and the β-blocker oxprenolol were observed when comparing these partitioning systems versus the IAM column. In all cases these interactions in the IAM improved the correlation with K_p. The 4-pregnene steroids yielded a significantly better correlation between K_p and the capacity factor obtained from the IAM.PC column (K'_{IAM}) as compared to the membrane partition coefficient (K_m) or the capacity factor obtained from a C_{18} column ($K_{C_{18}}$). The above examples are indicative of specific interactions similar to those in affinity chromatography, involving polar interactions between the solute and some part or parts of the IAM surface.

A significant observation from this study is that for heterogeneous groups of compounds, comparable results are obtained from IAM.PC columns, C_{18} columns, and the octanol–buffer partition system. This may have been anticipated, given the complexity of the skin transport process. Many parameters (e.g, membrane partitioning, diffusion coefficient, diffusion path length, specific affinity interactions between lipids and drugs, etc.) contribute to the transport of any given solute, and the relative contribution, or importance, of each of these parameters depends on the drug and the biological barrier. It is likely that different parameters are important for different classes of drugs and different biological barriers. However, for drug development involving the preparation of homologous analogs, IAM columns are superior to all existing methods. For diverse groups of compounds and general screening of compounds, IAM columns prepared with phosphatidylcholine produce results comparable to those obtained with existing methods.

We note that solute flux measurements (using either mammalian skin or synthetic membranes) to measure K_p for skin transport are very time-consuming and require extensive experimental effort; consequently, only a few compounds can be evaluated within a month. However, physicochemical methods are rapid, and thus improvements in any physicochemical method that yields similar results to those provided by flux measurements would be a significant advancement. Improving the IAM method for studying drug transport requires an understanding of the lipids comprising the biological barrier. For instance, since sphingolipids and fatty acids are the major lipids in the stratum corneum, an improvement of the IAM method for predicting solute penetration through the skin would be to synthesize an IAM-sphingolipid or IAM-sphingolipid/fatty acid surface. In summary, we postulate that preparing IAM surfaces containing lipids that are similar to the biological barrier will provide methods to evaluate transport processes that are rapid and better than all existing methods.

ACKNOWLEDGMENTS

One of us (C.P.) is grateful for the research support provided from NSF (CTS 8908450), Schering-Plough, and Eli Lilly and Co. We thank Robert Markovich for preparing some of the figures and Susan Santos for typing the manuscript.

REFERENCES

Alvarez, F. M., Chikhale, P. J., Pidgeon, C., and Walton, L., 1990, Immobilized artificial membrane column: An *in vitro* model for drug–membrane interactions, presented at the American Association of Pharmaceutical Science (AAPS) Fifth Annual Meeting and Exposition, Las Vegas, Nevada, November 4–8, *Pharm. Res.* **7**(9) Supplement:S-3.

Barry, B. W., and Brace, A. R., 1977, Permeation of oestrone, oestradiol, oestriol and dexamethasone across cellulose acetate membrane, *J. Pharm. Pharmacol.* **29**:397–400.

Barry, B. W., and El Eini, D. I. D., 1976, Influence of non-ionic surfactants on permeation of hydrocortisone, dexamethasone, testosterone and progesterone across cellulose acetate membrane, *J. Pharm. Pharmacol.* **28**: 219–277.

Behl, C., Linn, E., Flynn, G., Pierson, C., Higuchi, W., and Ho, N., 1983, Permeation of skin and Eshar by antiseptics I: Baseline studies with phenol, *Pharm. Sci.* **72**:391–397.

Chikhale, P., Alvarez, F. M., Pidgeon, C., Walton, L., and Kuester, J., 1989, Simulation of human skin permeation by IAM column, presented at the American Association of Pharmaceutical Science (AAPS) Fourth Annual Meeting and Exposition, Atlanta, Georgia, October 22–26, *Pharm. Res.* **6**(9) Supplement:S-147.

Elias, P. M., 1990, The importance of epidermal lipids for the stratum corneum barrier, in: *Topical Drug Delivery Formulations*, Vol. 42 (D. W. Osborne and A. H. Amann, eds.), Marcel Dekker, New York, pp. 13–28.

Flynn, G., 1985, Mechanism of percutaneous absorption from physicochemical evidence, in: *Percutaneous Absorption* (R. Bronaugh and H. Maibach, eds.), Marcel Dekker, New York, pp. 17–42.

Gary-Bobo, C. M., DiPolo, R., and Solomon, A. K., 1969, Role of hydrogen-bonding in nonelectrolyte diffusion through dense artificial membranes, *J. Gen. Physiol.* **54:**369–382.

Graveson, S., and Elias, P., 1982, Isolation and lipid biochemical characterization of stratum corneum membrane complexes: Implications for the cutaneous permeability barrier, *J. Invest. Dermatol.* **78:**128–131.

Green, P. G., Hadgraft, J., and Ridout, G., 1989, Enhanced *in vitro* skin permeation of cationic drugs, *Pharm. Res.* **6:**628–632.

Houk, J., and Guy, R., 1988, Membrane models for skin penetration studies, *Chem. Rev.* **88:**455–471.

Itoh, T., Xia, J., Magavi, R., Nishihata, T., and Rytting, J., 1990, Use of shed snake skin as a model membrane for *in vitro* percutaneous penetration studies: Comparison with skin, *Pharm. Res.* **7:**1042–1047.

Mason, R., Campbell, S., Wang, S, and Herbette, L., 1989, A comparison of bilayer location and binding for the charged 1,4-dihydropyridine Ca^{2+} channel antagonist amlodipine with uncharged drugs of this class in cardiac and model membranes, *J. Mol. Pharmacol.* **36:**634–640.

Mason, R., Rhodes, D., and Herbette, L., 1991, Reevaluating equilibrium and kinetic binding parameters for lipophilic drugs based on a structural model for drug interaction with biological membranes, *J. Med. Chem.* **34:**869–875.

Michaels, A. S., Chandrasekaran, S. K., and Shaw, J. E., 1975, Drug permeation through human skin: Theory and *in vitro* experimental measurement, *Am. Inst. Chem. Eng. J.* **21:**985–996.

Miyake, K., Kitaura, F., and Mizuno, N., 1987, Phosphatidylcholine coated silica as a useful stationary phase for high performance liquid chromatographic determination of partition coefficients between octanol and water, *J. Chromatogr.* **389:**47–56.

Pidgeon, C., 1990, Solid phase membrane mimetics, *Enzyme Microb. Technol.* **2:**149–150.

Pidgeon, C., and Venkataram, U. V., 1989, Immobilized artificial membranes: Chromatography supports composed of membrane lipids, *Anal. Biochem.* **176:**36–47.

Pidgeon, C., Marcus, C., and Alvarez, F., 1992, Immobilized artificial membrane chromatography: Surface chemistry and applications, in: *Applications of Enzyme Biotechnology* (J. W. Kelly and T. O. Baldwin, eds.), Plenum Press, New York, pp. 601–620.

Scheuplein, R. J., 1965, Mechanism of percutaneous absorption. I. Routes of penetration and the influence of solubility, *J. Invest. Dermatol.* **45:**334–346.

Scheuplein, R. J., and Blank, I. H., 1971, Permeability of the skin, *Physiol. Rev.* **51:**700–747.

Stevens, J. M., Markovich, R. J., and Pidgeon, C., 1989, Characterization of immobilized artificial membrane HPLC columns using deoxynucleotides as model compounds, *BioChromatography* **4:**192–205.

Wester, R., and Maibach, H., 1985, *In vivo* animal models for percutaneous absorption, in: *Percutaneous Absorption*, Vol. 6 (R. L. Bronaugh and H. I. Maibach, eds.), Marcel Dekker, New York, pp. 251–266.

Woods, P. B., and Robinson, M. L., 1981, An investigation of the comparative liposolubilities of achenoceptor blocking agents, *J. Pharm. Pharmacol.* **33:**172–173.

Yang, Q., Wallsten, M., and Lundahl, P., 1990, Lipid-vesicle surface chromatography, *J. Chromatogr.* **506:** 379–389.

11

Affinity Chromatography Using Immobilized Antisense-Family Peptides

Irwin Chaiken

1. INTRODUCTION—ANTISENSE PEPTIDE RECOGNITION OF SENSE PEPTIDES

Antisense peptides are sequences of amino acids encoded in the antisense strand of DNA. These peptides normally are not expressed in cells. Nonetheless, as proposed originally by Mekler (1969), such peptides can bind to the corresponding sense peptides, the sequences encoded in the DNA complementary to the antisense strand (Fig. 1A). This was first shown experimentally by Bost *et al.* (1985) with adrenocorticotropin and the corresponding chemically synthesized antisense peptide. Subsequently, many other experimental observations of antisense peptide recognition of sense peptides have been reported [for reviews, see Brentani (1988), Blalock (1990), Chaiken (1992), and Tropsha *et al.* (1992); see also several recently published papers (Bajpai *et al.*, 1991; Gartner *et al.*, 1991; Fassina and Cassani, 1992; Fassina *et al.*, 1992a,b,c)].

Affinity chromatographic analysis of antisense peptides eluting on affinity supports containing immobilized sense peptides has proven to be a useful means to observe and characterize antisense peptide recognition (Shai *et al.*, 1987, 1989; Fassina *et al.*, 1989a,b; Chaiken, 1992; Tropsha *et al.*, 1992). A typical elution result is shown in Fig. 1B. This analysis for antisense RNase S-peptide shows that the sense–antisense peptide interaction occurs as a saturable binding phenomenon with a dissociation constant (K_d) in the micromolar range (judged from the extent of retardation of eluting antisense peptide), occurs in solution as well as on the solid phase (judged from competitive elution), and is reasonably selective (irrelevant antisense peptides are not retarded chromatographically on immobilized S-peptide, and antisense RNase S-peptide is not retarded on irrelevant affinity matrices). On the whole, affinity chromatographic analyses of antisense peptide interactions have shown that the recognition process of antisense peptides with immobilized sense peptides occurs with high (though not 100%) frequency for several systems examined, is selective when it occurs, and usually is of modest affinity.

While sense–antisense peptide recognition remains ill understood mechanistically, the repetitive experimental observation of this phenomenon has made it tempting to explore the

Irwin Chaiken • SmithKline Beecham, King of Prussia, Pennsylvania 19406.

Molecular Interactions in Bioseparations, edited by That T. Ngo. Plenum Press, New York, 1993.

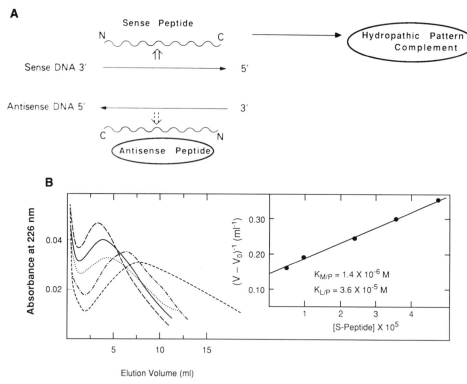

FIGURE 1. (A) Scheme of antisense-family peptides, showing the relationship of "antisense" and "hydropathic pattern complement" peptides to sense (native) amino acid sequence. The antisense peptides have sequences directly encoded in antisense DNA. Hydropathic pattern complement peptides have sequences based on sense peptide sequence, with residues in the former being hydrophilic in positions corresponding to residues in the sense peptide that are hydrophobic, and vice versa. [Adapted from Chaiken (1988).] (B) Competitive zonal elution high-performance liquid affinity chromatography (HPLAC) analysis of antisense S-peptide 20-mer binding to RNase S-peptide. Zones of antisense S-peptide (20-1), 7.6 nmol in 10 μl, were eluted on a 0.32-ml bed volume column of immobilized S-peptide, capacity of $9.1 \times 10^{-5} M$, in $0.2 M$ ammonium acetate, pH 5.7, containing the following molar concentrations of soluble S-peptide: ---, 0.47×10^{-5}; -··-, 0.95×10^{-5}; ···, 2.39×10^{-5}; —, 3.58×10^{-5}; --, 4.78×10^{-5}. Elution profiles are shown in the main figure; the inset shows the variation of elution volume V with soluble competitor concentration, plotted as $1/(V - V_0)$ vs. [S-peptide] as described before (Shai *et al.*, 1987). The calculated values of $K_{M/P}$ and $K_{L/P}$, the dissociation constants for the interactions of antisense peptide with matrix-bound and soluble (competing) native peptide, respectively, are given in the inset. [Adapted from Shai *et al.* (1987).]

use of antisense peptides as novel affinity ligands for biomolecular separation by immobilizing the antisense peptides. This chapter describes some of the results that have been obtained with this experimental approach and the opportunities in separation science that immobilized antisense peptides provide.

2. SEPARATIONS WITH IMMOBILIZED ANTISENSE PEPTIDES

The selectivity and modest affinity of antisense peptide recognition of sense peptides and proteins make immobilized antisense peptides attractive vehicles for peptide and protein separation. Antisense peptides can be synthesized corresponding to any native peptide or protein for which chromatographic separation is desired. The selectivity of antisense peptide recognition suggests that chromatographic resolution of peptides and proteins could be feasible even when these have closely related sequences, and the modest affinities expected make it likely that chromatographic elution from immobilized antisense peptides can be achieved by gentle, nonchaotropic conditions.

The feasibility of affinity chromatographic separation with immobilized antisense peptides was first demonstrated with the vasopressin system (Fassina *et al.*, 1989a). Arg^8-vasopressin (AVP) is encoded as the amino-terminal 9 residues of proAVP/NPII, the biosynthetic precursor of AVP and the associated neuropeptide carrier protein neurophysin II (BNPII in the bovine case). Based on the known DNA sequence of the precursor, antisense peptides were synthesized corresponding to the amino-terminal 12- and 20-residue sequences, each of which contains the AVP sequence (Fig. 2A). Both of these synthetic peptides bind to immobilized AVP with moderate and selective affinity (Fassina *et al.*, 1989a), and both can be immobilized to yield affinity supports that bind soluble AVP. As shown in Fig. 2B for the antisense (AS)-AVP(12-1) case, the affinity matrix can differentiate between AVP and the closely related peptide oxytocin (OT), even though these differ in only 2 out of the 9 residues (residues 3 and 8 are Ile and Leu for OT versus Phe and Arg for AVP). Interestingly, the selectivity of AVP versus OT binding to immobilized AS-AVP(12-1) is less than that found for the AVP receptor (for which affinity for AVP is between two and three orders of magnitude greater than that for OT) but greater than that for BNPII (for which the AVP and OT affinities are about equal). The selective affinity of immobilized AS-AVP(12-1) is sufficient to attain baseline chromatographic separation of AVP and OT (Fig. 2B). Strikingly, the affinity column also has been used to isolate AVP from crude acid extracts of posterior pituitary glands (Fig. 3).

Immobilized AS-AVP(12-1) has been used for chromatographic separation of AVP complexed to its cell receptor (Lu *et al.*, 1991). The fractionation of $[^3H]$-AVP–(rat liver membrane receptor) is shown in the right-hand panel of Fig. 4. This type of separation was predicted (Fig. 4, left-hand panel) on the basis of the hypothesis (Shai *et al.*, 1987, 1989; Fassina *et al.*, 1989a) that antisense and sense peptides interact with each other by multiple contacts all along their sequences and thus that sense peptide (or protein) when bound to its receptor may have enough structural elements remaining accessible to be able to bind to immobilized AS peptide in a three-way complex, (AS peptide)–(sense peptide)–(receptor). The data in Fig. 4 lend support to this prediction and hence provide evidence for the multicontact model of antisense–sense peptide recognition. In addition, it is worth noting that the receptor–AVP complex can be eluted from immobilized antisense peptide under mild, nonchaotropic conditions, because of the relatively weak affinity of the sense peptide for the antisense peptide, and yields a receptor that retains binding activity (Lu *et al.*, 1991). This suggests a possible means to isolate receptors of other peptides and proteins in a functionally intact state.

Immobilized antisense peptides have been shown to bind corresponding sense peptides and proteins in cases other than AVP. These include c-*raf* protein (Fassina *et al.*, 1989b), tumor necrosis factor (Fassina *et al.*, 1992a), endothelin (Fassina *et al.*, 1992c), big endothelin (Fassina *et al.*, 1992b), and interleukin-1β (Fassina and Cassani, 1992). In the latter four cases, the immobilized peptides actually were variant forms of antisense peptides with computer-generated sequences designed to be "hydropathic pattern complements" of the corresponding sense peptides (see Fig. 1A). These designed members of the antisense peptide family are discussed in the next section.

A

BOVINE ProAVP/NP (1-20) NH$_2$-Cys-Tyr-Phe-Gln-Asn-Cys-Pro-Arg-Gly-Gly-Lys-Arg-Ala-Met-Ser-Asp-Leu-Glu-Leu-Arg-COOH
(positions marked 1 ... 20)

AS[ProAVP/NP(12-1)] NH$_2$-Pro-Leu-Ala-Ala-Pro-Trp-Ala-Val-Leu-Glu-Val-Ala-COOH
(positions marked 12 ... 1)

AS[ProAVP/NP(20-1)] NH$_2$-Ser-Gln-Leu-Gln-Val-Gly-His-Gly-Pro-Leu-Ala-Ala-Pro-Trp-Ala-Val-Leu-Glu-Val-Ala-COOH
(positions marked 20 ... 1)

AVP NH$_2$-Cys-Tyr-Phe-Gln-Asn-Cys-Pro-Arg-Gly-CONH$_2$
(position marked 1)

Oxytocin NH$_2$-Cys-Tyr-Ile-Gln-Asn-Cys-Pro-Leu-Gly-CONH$_2$
(position marked 1)

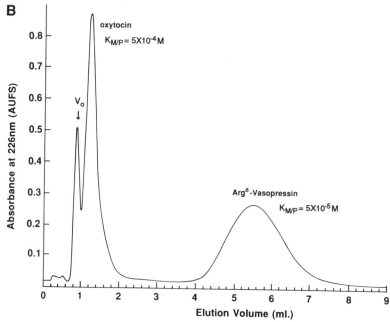

FIGURE 2. (A) Amino acid sequences of antisense peptides deduced from the corresponding complementary DNA of the amino-terminal 20 residues of proAVP/BNPII, the biosynthetic precursor of AVP and BNPII. The sequences of AVP and oxytocin (OT) are also shown. [Adapted from Fassina *et al.* (1989a).] (B) Separation of AVP and OT on AS[proAVP/BNPII(12-1)] immobilized on ACCELL (Waters-Millipore). The sample, containing 100 μg each of AVP and OT, and dissolved in 0.1M ammonium acetate, pH 5.5, was injected onto an immobilized 12-mer column (15 cm × 3 mm; amount of antisense peptide attached = 0.223 μmol), and elution (400 μl/min) was monitored at 226 nm, 1.0 AUFS (absorbance units full scale). Peak identities were verified by comparison with authentic peptide elution positions and by amino acid analysis. The affinity chromatographic peptide separation was carried out on a Beckman System Gold liquid chromatograph. [Adapted from Fassina *et al.* (1989a).]

3. PERFECTED HYDROPATHIC PATTERN PEPTIDES DESIGNED FROM SENSE PEPTIDE SEQUENCES

Early studies of the effects of sequence variation on antisense–sense peptide recognition showed that interaction affinity is affected by changes in hydrophilic and hydrophobic residues in the peptides (Shai *et al.*, 1989) and by changes in both ionic strength and content of organic solvents (Fassina *et al.*, 1989a). Model sequence-simplified antisense peptides, with Lys, Leu, Gly, and Ala placed in positions of all greatly hydrophilic residues (e.g., Lys, Glu, etc.), greatly hydrophobic residues (Leu, Val, etc.), Gly, and all other

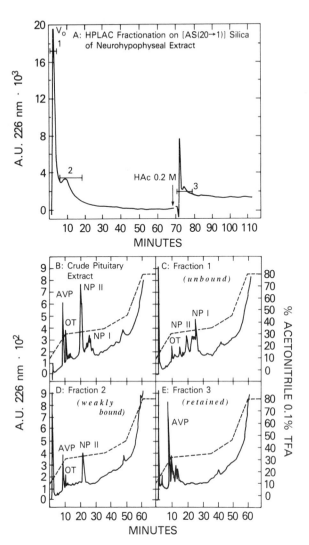

FIGURE 3. (A) High-performance liquid affinity chromatography (HPLAC) elution profile of crude pituitary extract applied to AS[proAVP/BNPII(20-1)]-ACCELL. The column (80 × 6.6 mm i.d.) was equilibrated at a flow rate of 1.0 ml/min with 50mM ammonium acetate, pH 5.7. Crude pituitary extract (50 μg) was injected in a 20-μl zone, and the effluent was monitored at 280 nm, 0.1 AUFS. After 70 min, the eluent was changed to 0.2M acetic acid. Fractions, collected as indicated, were frozen and lyophilized for subsequent high-performance liquid chromatography (HPLC) analysis. (B) Reversed-phase HPLC analysis of crude pituitary extract. A 50-μg amount of crude pituitary extract was injected onto the Axxiochrom RP C_{18} column (100 × 4.6 mm i.d.) equilibrated with 15% CH_3CN, 85% water, and 0.1% trifluoroacetic acid at a flow rate of 0.9 ml/min. Gradient as indicated in the figure. Effluent was monitored by UV absorbance at 226 nm, 0.1 AUFS. (C)–(E) Reversed-phase HPLC analysis of fractions 1, 2, and 3, respectively, of part (A). [Adapted from Fassina et al. (1989a).]

residues, respectively, act as mimics of direct-readout antisense peptides (Shai et al., 1989). At the same time, examination of the genetic code reveals a hydropathic pattern relationship between sense and antisense sequences (Mekler, 1969; Blalock and Smith, 1984; Tropsha et al., 1992). Thus, when a hydrophobic residue (e.g., Leu) is encoded in a trinucleotide codon of the sense strand of DNA, the corresponding trinucleotide anticodon of the antisense strand encodes a hydrophilic residue (e.g., Glu), and vice versa. From the amino acid pattern built into the genetic code, combined with experimental observation of the impact of hydropathic pattern sequence changes on antisense peptide recognition, it was proposed that sense–antisense peptide recognition is actually an amphipathic recognition process stabilized by the pattern of hydrophobic and hydrophilic residues along the peptide chains. Mechanistic visualization of how this could occur is shown in Fig. 5.

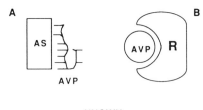

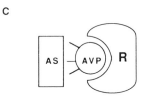

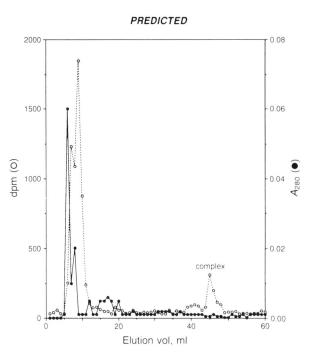

FIGURE 4. Top: Predicted binding of AS peptide to AVP–receptor complex. The three-way complex (C) was predicted based on the known selective binding of AVP to its AS peptide (A) (Fassina et al., 1989) and to its receptor (B) (Boer and Fahrenholz, 1985). AVP is visualized in (A) to interact with AS peptide by multipoint contacts along relatively elongated peptide chains (Shai et al., 1987; Chaiken, 1988; Tropsha et al., 1992). Receptor-bound AVP is visualized in (C) to have sufficient contact surface groups left exposed to provide enough of these multiple contacts to stabilize the interaction with AS peptide. [Adapted from Lu et al. (1991).] Bottom: Affinity capture of [^{3}H]-AVP–receptor complex on immobilized AS[proAVP/BNPII(20-1)]. Column bed was 2 ml. Elution was isocratic with 0.1M sodium acetate containing 0.005% maltoside (pH 5.4). Flow rate was 1 ml/min. DPM, Disintegrations per minute. [Adapted from Lu et al. (1991).]

The emerging view that antisense peptide interaction with sense peptide is stabilized by hydropathic pattern recognition suggests (Fig. 1A) that antisense peptide mimics could be synthesized not by direct readout of DNA but as "hydropathic complements" guided by the sequence of amino acids in the sense peptide. Such hydropathically opposite peptides have been synthesized and found not only to interact with sense peptides but to do so with higher affinity than antisense peptides directly read out from antisense DNA (Fassina et al., 1989b; Fassina and Cassani, 1992; Fassina et al., 1992a,b,c).

The higher affinity of hydropathically perfected "antisense" peptides makes these

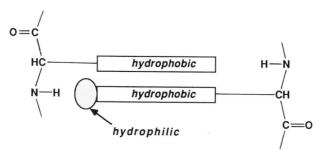

FIGURE 5. Scheme depicting the way in which hydropathic complementarity could occur between hydrophilic and hydrophobic amino acid residues encoded in opposite strands of DNA. The scheme illustrates the case when the hydrophilic side chain is a hydrogen-bond acceptor (e.g., Glu) and forms a hydrogen bond with the main-chain amide of the hydrophobic residue. A similar scheme could be drawn for the case in which the hydrophilic side chain is a hydrogen-bond donor (e.g., Lys) and forms a hydrogen bond with the main-chain carbonyl of the hydrophobic residue. [Adapted from Tropsha *et al.* (1992).]

potential replacements of direct-readout antisense peptides as immobilized ligands for affinity chromatography. This idea has been tested (Fassina *et al.*, 1989b; Fassina and Cassani, 1992; Fassina *et al.*, 1992a,b,c) as discussed in the previous section. The result for the case of interleukin-1β (IL-1β) is shown in Fig. 6. In this case, an affinity matrix containing the hydropathic complement peptide corresponding to residues 204–215 of IL-1β can be used to fractionate recombinant Il-1β (Fig. 6). The affinity matrix also can bind selectively and fractionate the synthetic peptide corresponding to residues 204–215 of IL-1β (Fassina and Cassani, 1992).

4. MULTIMOLECULAR SEPARATIONS USING IMMOBILIZED ANTISENSE-FAMILY PEPTIDES

Immobilized antisense-family peptides, both from direct readout of antisense DNA and hydropathically perfected based on sense peptide (or protein) sequence, may be usable to fractionate more than one peptide and/or protein sequence and even nonproteinaceous molecules. This expectation derives from the hypothesis that antisense peptide recognition of sense peptides is stabilized by hydropathic pattern complementarity. Hydropathic patterns are likely to be less unique than literal sequence and also could be similar to patterns in nonpeptides. Thus, an immobilized ligand that is interacting by hydropathic pattern recognition likely would bind several different peptides and perhaps nonpeptide molecules, hence the multiple-molecule recognition potential of immobilized antisense-family peptides. If the several molecular types that interact with a single immobilized antisense peptide bind with different affinities, a single chromatographic elution could lead to fractionation of several molecular interactors simultaneously. In the case of immobilized antisense vasopressin peptide, the affinity matrix actually interacts differentially with several members of the oxytocin/vasopressin peptide family (Fassina *et al.*, 1989a), and, as shown in Fig. 3, a larger vasopressin/neurophysin-related antisense peptide, immobilized on an affinity support, was used to separate several neurophysins and neuropeptides in a single elution run (Fassina *et al.*, 1989a). Peptide separations from

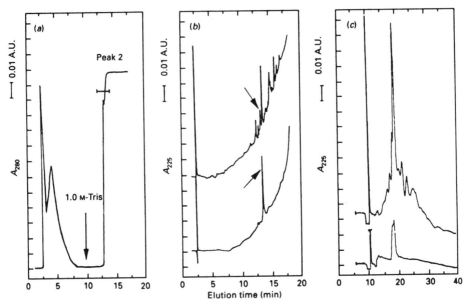

FIGURE 6. Affinity purification of recombinant IL-1β from crude *E. coli* lysate. (a) HPLAC profile of crude bacterial lysate containing recombinant human IL-1β eluted from an affinity column, containing a 12-residue hydropathic pattern complement peptide corresponding to IL-1β residues 204–215, equilibrated with 50mM Tris, pH 6.8. Bound material, denoted as peak 2, was collected for reversed-phase HPLC and high-performance electrophoresis chromatography (HPEC) analysis. (b) Reversed-phased HPLC analysis of crude *E. coli* lysate (top) and HPLAC-purified fraction 2 (bottom). IL-1β is indicated by the arrows. (c) HPEC analysis of crude bacterial lysate (top) and purified HPLAC fraction 2 (bottom). [Reproduced from Fassina and Cassani (1992) with permission.]

complex mixtures have been reported recently in the cases of tumor necrosis factor (Fassina *et al.*, 1992a), IL-1β (Fassina and Cassani, 1992), and endothelin (Fassina *et al.*, 1992b).

5. CONCLUSIONS

Antisense peptide recognition of sense peptides presents a new opportunity for affinity chromatographic separation of peptides, proteins, and other molecular species. The selectivity and moderate affinity characteristics expected for immobilized antisense-family peptides are ideal properties in chromatography. Such properties are the hallmark of ion-exchange chromatography, for example, and could enable immobilized antisense-family peptides to be used as general ligand affinity supports to achieve separations of multiple molecular species in a single elution. D-Antisense as well as L-antisense peptides recognize sense peptides (Fassina *et al.*, 1989b). Thus, affinity supports can be made with D-amino acids to ensure a greater resistance of an antisense peptide affinity matrix to proteases. Since antisense peptide recognition has been found to be a frequently occurring phenomenon, affinity chromatography using immobilized antisense-family peptides should be useful to solve separation problems for many peptide and protein systems in biology and biotechnology.

REFERENCES

Bajpai, A., Hooper, K. P., and Ebner, K. E., 1991, Interaction of antisense peptides with prolactin, *Biochem. Biophys. Res. Commun.* **180:**1312–1317.

Blalock, J. E., 1990, Complementarity of peptides specified by 'sense' and 'antisense' strands of DNA, *Trends Biotechnol.* **8:**140–144.

Blalock, J. E., and Smith, E. M., 1984, Hydropathic anti-complementarity of amino acids based on the genetic code, *Biochem. Biophys. Res. Commun.* **131:**203–207.

Boer, R., and Fahrenholz, F., 1985, Photoaffinity labelling of the V_1 vasopressin receptor in plasma membranes from rat liver, *J. Biol. Chem.* **260:**15051–15054.

Bost, K. L., Smith, E. M., and Blalock, J. E., 1985, Similarities between the corticotropin (ACTH) receptor and a peptide encoded by an RNA that is complementary to ACTH mRNA, *Proc. Natl. Acad. Sci. U.S.A.* **82:**1372–1375.

Brentani, R. B., 1988, Biological implications of complementary hydropathy of amino acids, *J. Theor. Biol.* **135:**495–499.

Chaiken, I., 1988, The design of peptide and protein recognition mimics using ideas from sequence simplification and antisense peptides, in: *Molecular Mimicry in Health and Disease* (A. Lernmark, T. Dyrberg, L. Terenius, and B. Hokfelt, eds.), Elsevier, Amsterdam, pp. 351–367.

Chaiken, I., 1992, Interactions and uses of antisense peptides in affinity technology, *J. Chromatogr.* **597:**29–36.

Fassina, G., and Cassani, G., 1992, Design and recognition properties of a hydropathically complementary peptide to human interleukin 1 beta, *Biochem. J.* **282:**773–779.

Fassina, G., Zamai, M., Brigham-Burke, M., and Chaiken, I. M., 1989a, Recognition properties of antisense peptides to Arg⁸-vasopressin/bovine neurophysin II biosynthetic precursor sequences, *Biochemistry* **28:**8811–8818.

Fassina, G., Roller, P. P., Olson, A. D., Thorgeirsson, S. S., and Omichinski, J. G., 1989b, Recognition properties of peptides hydropathically complementary to residues 356–375 of the c-*raf* protein, *J. Biol. Chem.* **264:**11252–11257.

Fassina, G., Cassani, G., and Corti, A., 1992a, Binding of human tumor necrosis factor alpha to multimeric complementary peptides, *Arch. Biochem. Biophys.* **196:**137–143.

Fassina, G., Consonni, R., Zetta, L., and Cassani, G., 1992b, Design of a hydropathically complementary peptide for Big Endothelin affinity purification, *Int. J. Pept. Protein Res.* **39:**540–548.

Fassina, G., Corti, A., and Cassani, G., 1992c, Affinity enhancement of complementary peptide recognition, *Int. J. Pept. Protein Res.* **39:**549–556.

Gartner, T. K., Loudon, R., and Taylor, D. B., 1991, The peptides APPLHK, EHIPA and GAPL are hydropathically equivalent peptide mimics of a fibrinogen binding domain of glycoprotein IIb/IIIa, *Biochem. Biophys. Res. Commun.* **180:**1446–1452.

Lu, F. X., Aiyar, N., and Chaiken, I., 1991, Affinity capture of Arg⁸-vasopressin-receptor complex using immobilized antisense peptide, *Proc. Natl. Acad. Sci. U.S.A.* **88:**3637–3641.

Mekler, L. B., 1969, Specific selective interaction between amino acid residues of the polypeptide chains, *Biofizika* **14:**581–584 [in Russian]; Engl. version: *Biophys. USSR* **14:**613–617 (1970).

Shai, Y., Flashner, M., and Chaiken, I. M., 1987, Antisense peptide recognition of sense peptides: Direct quantitative characterization with the ribonuclease S-peptide system using analytical high-performance affinity chromatography, *Biochemistry* **26:**669–675.

Shai, Y., Brunck, T. K., and Chaiken, I. M., 1989, Antisense peptide recognition of sense peptides: Sequence simplification and evaluation of forces underlying the interaction, *Biochemistry* **28:**8804–8811.

Tropsha, A., Kizer, J. S., and Chaiken, I. M., 1992, Making sense from antisense: A review of experimental data and developing ideas on sense–antisense peptide recognition, *J. Mol. Recog.* **5:**43–54.

12

Chromatographic Resolution of Chiral Compounds by Means of Immobilized Proteins

Stig Allenmark and Shalini Andersson

1. BACKGROUND

It is well known that biological recognition phenomena are intimately associated with the properties that characterize large complex molecules. A common feature of such molecules is their ability to bind other molecules via multiple, simultaneous, noncovalent interactions, which sometimes leads to very high binding constants—a classical example being the avidin–biotin complex ($K_b = 10^{15}$ M^{-1}). The selectivity achieved by such multiple interactions is often high and forms the basis of affinity chromatography.

Although enantioselectivity in protein–chiral ligand interactions was observed as early as in the 1950s, the first demonstration of an optical resolution by an affinity chromatographic technique was not made until the early 1970s (Stewart and Doherty, 1973) with the use of immobilized bovine serum albumin (BSA). However, the wide applicability of this principle was not shown until several years later (Allenmark *et al.*, 1982; Allenmark and Bomgren, 1982). Transfer of the technology to include "high-performance" silica columns (Allenmark *et al.*, 1983) as well as other proteins (Hermansson, 1983; Miwa *et al.*, 1987a, 1988; Erlandsson *et al.*, 1990) led to a rapid expansion of the application of protein columns for direct chromatographic optical resolution.

The general principle of the separation of a pair of enantiomers on a protein column is briefly outlined in Fig. 1. The two enantiomers experience a number of binding locations in the protein during their passage through the column. Since some of these will exert a stronger binding of one of the enantiomers (i.e., discriminate between the antipodes), the net result will be a difference in migration rates, leading to resolution. Regulation of the migration rates, affecting k' and α values,* is readily achieved via the mobile-phase system,

---

*The capacity ratio k' is defined as $k' = (t_R - t_0)/t_0$, where t_R and t_0 are the retention times for the analyte and an unretained solute, respectively. The separation factor α is the k'-ratio of two components (here two enantiomers); $\alpha = k'_2/k'_1 = (t_R(2) - t_0)/(t_R(1) - t_0)$, where 1 and 2 denote the first and last eluted enantiomer, respectively.

Stig Allenmark • Department of Organic Chemistry, University of Gothenburg, S-41296 Gothenburg, Sweden.
Shalini Andersson • IFM/Department of Chemistry, University of Linköping, S-58183 Linköping, Sweden.

Molecular Interactions in Bioseparations, edited by That T. Ngo. Plenum Press, New York, 1993.

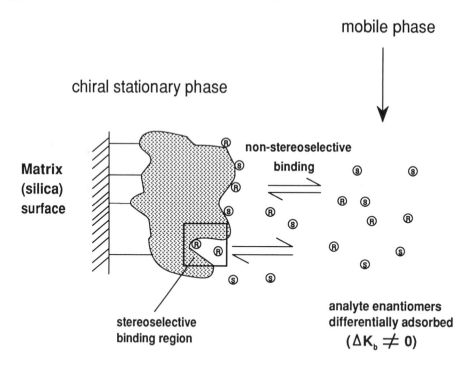

K_b and retention affected by pH, organic cosolvent, ionic strength, charged organic modifiers etc., causing changes in hydrophobic and charge interactions

FIGURE 1. Chromatographic enantioselectivity via a protein as stationary phase.

preferentially by changes in the contributions to hydrophobic and charge interactions (Fig. 1).

2. SOME GENERAL ASPECTS OF PROTEIN IMMOBILIZATION AND CROSS-LINKING

Since only a small part of the protein molecule is involved in enantioselection, it is understandable that the loading capacity of a protein column is very limited and that the protein content of a column should be made as high as possible. To achieve this, a variety of immobilization techniques and chromatographic support materials have been used. Apart from simple physical adsorption of the native protein to the support (Erlandsson et al., 1986), immobilization by covalent bonding is generally preferred. This has been done in a number of ways, leading to more or less cross-linking and/or multiple-point attachment of the protein. Theoretically, loss of conformational freedom in the protein should lead to increased resistance to denaturation by the mobile phase used; however, it might also impede binding interactions involving allosteric effects. This highly complex, but very important, problem area deserves further attention in the future.

3. PROTEINS USED AS CHIRAL STATIONARY PHASES

A variety of proteins, many containing a carbohydrate moiety, have been immobilized on chromatographic supports, and a majority of these proteins are acidic, as can be seen in Table 1. Interestingly, one protein (CBH I) is of microbial origin and can be produced extracellularly in large quantities.

Proteins have mainly been immobilized to functionalized silica. In most cases, terminally substituted alkyl trialkoxysilanes are used to introduce reactive end groups that can be further activated and used for protein coupling; the most useful end groups are shown in Table 2.

The instability of silica gel at alkaline pH has led to the immobilization of proteins to polymers that are stable over a wide pH range (Stewart and Doherty, 1973; Simek and Vespalec, 1989; Miwa et al., 1991). This permits chromatographic separations of basic chiral compounds to be carried out at their pK_a. However, the major drawback of such polymeric supports has been the low column efficiency, and they have therefore not been used extensively as support materials for binding proteins. Recently, Miwa et al. (1991) have reported the use of an amino residue-conjugated polymer for the immobilization of ovomucoid, which gives *almost* the same column efficiency as silica gel.

4. SOME MECHANISTIC ASPECTS OF ENANTIOSELECTIVE ADSORPTION TO PROTEINS

The ability of proteins to bind various ligands has been long known and is frequently applied in affinity chromatography (Müller and Wollert, 1975; Sudlow et al., 1976; Sjöholm et al., 1979; Lagercrantz et al., 1981; Fitos et al., 1983). Although the details of the type of stereoselective interactions involved, at the molecular level, are far from clear, it is

TABLE 1

Some Physical and Chemical Properties of Proteins Used as Chiral Stationary Phases

Protein	Source	Molecular weight	pI	Carbohydrate content (g/100 g of protein)
Avidin	Hen egg white	70,000	10	20.5
Cellulase (CBH I)	*Trichoderma reesei*	60,000–70,000	3.9	6
α-Chymotrypsin	Bovine pancreas	25,000	8.1–8.6	—
Orosomucoid (α₁-acid glycoprotein, AGP)	Human plasma	44,000	2.7	41.4
Ovomucoid	Hen egg white	28,000	4.5	17.4–33.5
Bovine serum albumin (BSA)	Bovine plasma	67,000	4.7	—
Human serum albumin (HSA)	Human plasma	68,000	4.7	—
Ovalbumin	Hen egg white	45,000	4.6	3.2

TABLE 2
Various Routes to Protein-Based Chiral Sorbents from Silica

Functionalized silica	Reactive coupling group[a]	Protein immobilized	References
Silica	Not published	Bovine serum albumin	Allenmark et al., 1983
	None	Bovine serum albumin	Erlandsson et al., 1986
Diol silica	Aldehyde	Orosomucoid	Hermansson, 1983
		Cellulase	Erlandsson et al., 1990
	Im$_2$CO	Human serum albumin	Domenici et al., 1990a
Silica-bound hydro-philic polymer	Aldehyde	α-Chymotrypsin	Wainer et al., 1988
3-Aminopropylsilica	Aldehyde	Bovine serum albumin	Aubel and Rogers, 1987; Thompson et al., 1989; Andersson et al., 1992
		Human serum albumin	Andersson and Allenmark, 1992
	DSC	Ovomucoid	Miwa et al., 1987a
		Avidin	Miwa et al., 1988
		Bovine serum albumin	Andersson et al., 1992
	DSS	Ovomucoid	Oda et al., 1991
		Avidin	Oda et al., 1991
		Ovalbumin	Oda et al., 1991

[a]Abbreviations: Im$_2$CO, 1,1'-carbonyldiimidazole; DSC, N,N'-disuccinimidyl carbonate; DSS, N,N'-disuccinimidyl suberate.

generally considered that the dominating contributions to the binding of a ligand to a protein are electrostatic and hydrophobic interactions. However, hydrogen bonding, charge-transfer, and dipole–dipole interactions may also be of some importance in the formation of reversible diastereomeric complexes between a protein and a chiral ligand. The stability of these complexes will be dependent upon both the binding and the repulsive interactions involved, and a small difference in the stability of the two diastereomeric complexes ($\Delta\Delta G = 0.235$ kJ gives $\alpha = 1.10$) is often sufficient for a chromatographic separation of enantiomers.

Although the mechanism of chiral recognition by protein-based chiral stationary phases is highly complex, certain empirical guidelines for the regulation of the overall retention of an analyte have been deduced. Since quite a number of mobile-phase parameters affect the retention and resolution of a chiral compound, optimization of resolution can be considered as a multivariate analysis problem and may be treated by chemometric methods (Glajch and Snyder, 1990), including a modified sequential simplex technique (Noctor et al., 1989).

It is generally assumed that upon titration of a protein with a certain ligand, the latter will first bind to the sites of high affinity and subsequently to other sites of decreasing affinity. In chromatography, this means that the retention exerted by an immobilized protein will be dependent upon the amount of analyte injected, that is, the column load. Since different types of binding sites will cause different enantioselectivities, constant capacity ratios for a pair of enantiomers can only be obtained as long as the column load is kept low enough not to saturate the sites involved in the primary binding process. Therefore, the capacity of all protein-based chiral sorbents is generally low.

Recently, this problem was addressed via detailed studies of chromatographic band spreading and adsorption isotherms at different temperatures and mobile-phase compositions (Jacobson et al., 1990a,b, 1991).

Various attempts to localize the binding region primarily responsible for enantioselection have been made. By enzymatic limited degradation of BSA and immobilization of an N-terminal 38-kDa fragment (Erlandsson and Nilsson, 1989; Andersson *et al.*, 1990), it was found that the binding region for oxazepam was preserved in this fragment, whereas this was not the case for many other analytes.

Chemical modification of amino acid residues within a protein has also been used as a tool to change the enantioselectivity properties and as a mechanistic probe. Recently, Noctor and Wainer (1992) reported a significant change in retention and resolution properties after *in situ* acetylation of a human serum albumin (HSA) phase (Fig. 2). It was assumed that the effect was mainly caused by acetylation of a tyrosine residue present in the binding region.

The simplest way, however, to study binding interactions between enantiomers and a protein phase is to use additives to the mobile phase that may compete with, or in other ways affect, the binding process. Thus, any additive which competes with the analyte for a certain binding site will reduce the retention (lower k') of that analyte. Such competitive displacement effects have long been studied (Lagercrantz *et al.*, 1979; Hermansson and Eriksson, 1986; Andersson and Allenmark, 1989; Noctor *et al.*, 1991) and are useful probes to elucidate whether two enantiomers bind to the same site or not. Figure 3 shows the effect of ibuprofen added to the mobile phase on the retention of oxazepam enantiomers by HSA.

Although enantiomers often bind to the same site, this is not always the case. Fitos *et al.* (1986, 1990) were the first to report on allosteric interactions caused by (S)-warfarin on the binding of the enantiomers of lorazepam and other benzodiazepinones to HSA-Sepharose. Later, the same effect was shown chromatographically with an HSA column

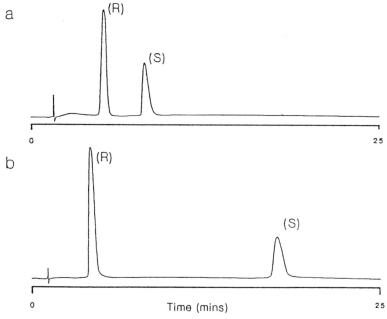

FIGURE 2. Resolution of racemic oxazepam hemisuccinate on unmodified HSA (a) and on acetylated HSA (b). [Reproduced with permission from Noctor and Wainer (1992). ©1992 Academic Press.]

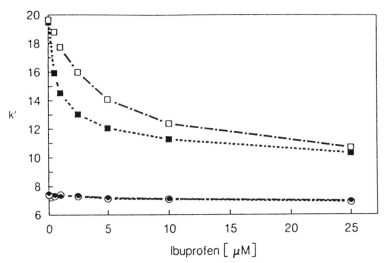

FIGURE 3. Effect of competition by ibuprofen on the retention of oxazepam enantiomers. □, (*S*)-oxazepam; ■, (*S*)-oxazepam + (*R*)-ibuprofen as modifier; ○, (*R*)-oxazepam; ●, (*R*)-oxazepam + (*R*)-ibuprofen as modifier. [Reproduced with permission from Domenici *et al.* (1990b). Copyright 1990 Wiley-Liss Inc.]

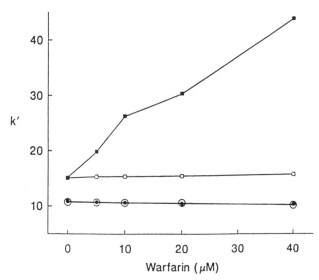

FIGURE 4. Illustration of the effect of allosteric interaction via a mobile-phase additive. Increased *k'* of (*S*)-lorazepam is observed with the addition of (*S*)-warfarin as a modifier. ■, (*S*)-lorazepam + (*S*)-warfarin as modifier; □, (*S*)-lorazepam + (*R*)-warfarin as modifier; ●, (*R*)-lorazepam + (*S*)-warfarin as modifier; ○, (*R*)-lorazepam + (*R*)-warfarin as modifier. [Reproduced with permission from Domenici *et al.* (1991) and from the copyright owner, the American Pharmaceutical Association.]

(Domenici *et al.*, 1991). A similar effect was also observed by Hermansson and Eriksson (1986), who found a dramatic increase in k_2' of naproxen during addition of N,N-dimethyl-octylamine to the mobile phase. These effects are consistent with the assumption of different binding sites for the enantiomers. Figure 4 shows the effect of added warfarin on the chromatographic retention by HSA of the enantiomers of lorazepam.

5. SELECTED EXAMPLES OF APPLICATIONS OF THE TECHNIQUE IN BIOTECHNOLOGY AND THE LIFE SCIENCES

The use of direct liquid-chromatographic optical resolution with protein columns and aqueous mobile-phases systems has opened unique possibilities to determine the enantiomeric composition of very small amounts of analyte. Since numerous drugs have been resolved into enantiomers on protein columns (Miwa *et al.*, 1987b, 1988; Allenmark, 1988; Jadaud and Wainer, 1989; Allenmark and Andersson, 1989a; Enquist and Hermansson, 1990; Domenici *et al.*, 1990b; Marle *et al.*, 1991; Oda *et al.*, 1991), enantioselective pharmacokinetic drug monitoring (Oda *et al.*, 1991; Iredale and Wainer, 1992) is one of the application areas under rapid expansion. Others include determination of enzyme stereoselectivity (Allenmark *et al.*, 1986; Allenmark and Andersson, 1989b; Allenmark and Ohlsson, 1992) and chiral amino acid analysis (Allenmark and Andersson, 1991).

ACKNOWLEDGMENTS

Financial support from the Swedish Natural Science Research Council, the National Swedish Board for Technical Development, EKA Nobel AB, and Astra Hässle AB is gratefully acknowledged.

REFERENCES

Allenmark, S. G., 1988, *Chromatographic Enantioseparation—Methods and Applications*, Horwood/Wiley, Chichester, New York.

Allenmark, S., and Andersson, S., 1989a. Optical resolution of some biologically active compounds by chiral liquid chromatography on BSA-silica (Resolvosil) columns, *Chirality* **1**:154–160.

Allenmark, S., and Andersson, S., 1989b, Chiral liquid chromatographic monitoring of asymmetric carbonyl reduction by some yeast organisms, *Enzyme Microb. Technol.* **11**:177–179.

Allenmark, S., and Andersson, S., 1991, Chiral amino acid microanalysis by direct optical resolution of fluorescent derivatives on BSA-based (Resolvosil) columns, *Chromatographia* **31**:429–433.

Allenmark, S., and Bomgren, B., 1982, Direct liquid chromatographic separation of enantiomers on immobilized protein stationary phases. II. Optical resolution of a sulphoxide, a sulphoximine and a benzoylamino acid. *J. Chromatogr.* **252**:297–300.

Allenmark, S., and Ohlsson, A., 1992, Enantioselectivity of lipase-catalyzed hydrolysis of some 2-chloroethyl 2-arylpropanoates studied by chiral reversed-phase liquid chromatography, *Chirality* **4**:98–102.

Allenmark, S., Bomgren, B., and Borén, H., 1982, Direct resolution of enantiomers by liquid affinity chromatography on albumin-agarose under isocratic conditions, *J. Chromatogr.* **237**:473–477.

Allenmark, S., Bomgren, B., and Borén, H., 1983, Direct liquid chromatographic separation of enantiomers on immobilized protein stationary phases. III. Optical resolution of a series of N-aroyl D,L-amino acids by high-performance liquid chromatography on bovine serum albumin covalently bound to silica, *J. Chromatogr.* **264**:63–68.

Allenmark, S., Bomgren, B., and Borén, H., 1986, Enantioselective microbial degradation of some racemates studied by chiral liquid chromatography, *Enzyme Microb. Technol.* **8**:404–408.

Andersson, S., and Allenmark, S., 1989, Influence of amphiphilic mobile phase additives upon the direct liquid chromatographic optical resolution by means of BSA-based chiral sorbents, *J. Liq. Chromatogr.* **12**:345–357.

Andersson, S., and Allenmark, S., 1992, Some mechanistic aspects on chiral discrimination of organic acids by immobilized bovine serum albumin (BSA), *Chirality* **4**:24–29.

Andersson, S., Allenmark, S., Erlandsson, P., and Nilsson, S., 1990, Direct liquid chromatographic separation of enantiomers on immobilized protein stationary phases. VIII. A comparison of a series of sorbents based on bovine serum albumin (BSA) and its fragments, *J. Chromatogr.* **498**:81–91.

Andersson, S., Thompson, R. A., and Allenmark, S. G., 1992, Direct liquid chromatographic separation of enantiomers on immobilized protein stationary phases IX. Influence of the crosslinking reagent on the retentive and enantioselective properties of chiral sorbents based on bovine serum albumin, *J. Chromatogr.* **591**:65–73.

Aubel, M., and Rogers, L. B., 1987, Effect of pretreatment on the enantioselectivity of silica-bound bovine serum albumin used as high-performance liquid chromatographic stationary phases, *J. Chromatogr.* **392**:415–420.

Domenici, E., Bertucci, C., Salvadori, P., Felix, G., Cahagne, I., Motellier, S., and Wainer, I. W., 1990a, Synthesis and chromatographic properties of an HPLC chiral stationary phase based upon human serum albumin, *Chromatographia* **29**:170–176.

Domenici, E., Bertucci, C., Salvadori, P., Motellier, S., and Wainer, I. W., 1990b, Immobilized serum albumin: Rapid HPLC probe of stereoselective protein-binding interactions, *Chirality* **2**:263–268.

Domenici, E., Bertucci, C., Salvadori, P., and Wainer, I. W., 1991, Use of a human serum albumin-based high-performance liquid chromatography chiral stationary phase for the investigation of protein binding: Detection of the allosteric interaction between the warfarin and benzodiazepine binding sites. *J. Pharm. Sci.* **80**:164–166.

Enquist, M., and Hermansson, J., 1990, Influence of uncharged mobile phase additives on retention and enantioselectivity of chiral drugs using an α_1-acid glycoprotein column, *J. Chromatogr.* **519**:271–283.

Erlandsson, P., and Nilsson, S., 1989, Use of fragment of bovine serum albumin as a chiral stationary phase in liquid chromatography, *J. Chromatogr.* **482**:35–51.

Erlandsson, P., Hansson, L., and Isaksson, R., 1986, Direct analytical and preparative resolution of enantiomers using albumin adsorbed to silica as a stationary phase, *J. Chromatogr.* **370**:475–483.

Erlandsson, P., Marle, I., Hansson, L., Isaksson, R., Pettersson, C., and Pettersson, G., 1990, Immobilized cellulase (CBH I) as a chiral stationary phase for direct resolution of enantiomers, *J. Am. Chem. Soc.* **112**: 4573–4574.

Fitos, I., Simonyi, M., Tegyey, Z., Ötvös, L., Kajtár, J., and Kajtár, M., 1983, Resolution by affinity chromatography: Stereoselective binding of racemic oxazepam esters to human serum albumin, *J. Chromatogr.* **259**:494–498.

Fitos, I., Tegyey, Z., Simonyi, M., Sjöholm, I., Larsson, T., and Lagercrantz, C., 1986, Stereoselective binding of 3-acetoxy-, and 3-hydroxy-1,4-benzodiazepine-2-ones to human serum albumin, *Biochem. Pharmacol.* **35**(2):263–269.

Fitos, I., Visy, J., Magyar, A., Kajtár, J., and Simonyi, M., 1990, Stereoselective effect of warfarin and bilirubin on the binding of 5-(*o*-chlorophenyl)-1,3-dihydro-3-methyl-7-nitro-2*H*-1,4-benzodiazepin-2-one enantiomers to human serum albumin, *Chirality* **2**:161–166.

Glajch, J. L., and Snyder, L. R. (eds.), 1990, *Computer-Assisted Method Development for High-Performance Liquid Chromatography*, Elsevier, Amsterdam.

Hermansson, J., 1983, Direct liquid chromatographic resolution of racemic drugs using α_1-acid glycoprotein as the chiral stationary phase, *J. Chromatogr.* **269**:71–80.

Hermansson, J., and Eriksson, M., 1986, Direct liquid chromatographic resolution of acidic drugs using a chiral α_1-acid glycoprotein column (Enantiopac), *J. Liq. Chromatogr.* **9**:621–639.

Iredale, J., and Wainer, I. W., 1992, The determination of hydroxychloroquine and its major metabolites in plasma using sequential achiral/chiral high performance liquid chromatography, *J. Chromatogr.* **573**:253–258.

Jacobson, S., Golshan-Shirazi, S., and Guiochon, G., 1990a, Chromatographic band profiles and band separation of enantiomers at high concentration, *J. Am. Chem. Soc.* **112**:6492–6498.

Jacobson, S., Golshan-Shirazi, S., and Guiochon, G., 1990b, Measurements of the heats of adsorption of chiral isomers on an enantioselective stationary phase, *J. Chromatogr.* **522**:23–36.

Jacobson, S., Golshan-Shirazi, S., and Guiochon, G., 1991, Influence of the mobile phase composition on the adsorption isotherms of an amino-acid derivative on immobilized bovine serum albumin, *Chromatographia* **31**:323–328.

Jadaud, P., and Wainer, I. W., 1989, Stereochemical recognition of enantiomeric and diastereomeric dipeptides by high-performance liquid chromatography on a chiral stationary phase based upon immobilized α-chymotrypsin, *J. Chromatogr.* **476:**165–174.

Lagercrantz, C., Larsson, T., and Karlsson, H., 1979, Binding of some fatty acids and drugs to immobilized bovine serum albumin studied by column affinity chromatography, *Anal. Biochem.* **99:**352–364.

Lagercrantz, C., Larsson, T., and Denfors, I., 1981, Stereoselective binding of the enantiomers of warfarin and tryptophan to serum albumin from some different species studied by affinity chromatography on columns of immobilized serum albumin, *Comp. Biochem. Physiol.* **69C:**375–378.

Marle, I., Erlandsson, P., Hansson, L., Isaksson, R., Pettersson, C., and Pettersson, G., 1991, Separation of enantiomers using cellulase (CBH I) silica as a chiral stationary phase, *J. Chromatogr.* **586:**233–248.

Miwa, T., Ichikawa, M., Tsuno, M., Hattori, T., Miyakawa, T., Kayano, M., and Miyake, Y., 1987a, Direct liquid chromatographic resolution of racemic compounds. Use of ovomucoid as a column ligand, *Chem. Pharm. Bull.* **35:**682–686.

Miwa, T., Miyakawa, T., Kayano, M., and Miyake, Y., 1987b, Application of an ovomucoid-conjugated column for the optical resolution of some pharmaceutically important compounds, *J. Chromatogr.* **408:**316–322.

Miwa, T., Miyakawa, T., and Miyake, Y., 1988, Characteristics of an avidin-conjugated column in direct liquid chromatographic resolution of racemic compounds, *J. Chromatogr.* **457:**227–233.

Miwa, T., Sakashita, S., Ozawa, H., Haginaka, J., Asakawa, N., and Miyake, Y., 1991, Application of an ovomucoid-conjugated polymer column for the enantiospecific determination of chlorprenaline concentrations in plasma, *J. Chromatogr.* **566:**163–171.

Müller, W. E., and Wollert, U., 1975, High stereospecificity of the benzodiazepine binding site on human serum albumin, *Mol. Pharmacol.* **11:**52–60.

Noctor, T. A. G., and Wainer, I. W., 1992, The *in situ* acetylation of an immobilized human serum albumin chiral stationary phase for high performance liquid chromatography in the examination of drug protein-binding phenomena, *Pharm. Res.* **9:**480–484.

Noctor, T. A. G., Fell, A. F., and Kaye, B., 1989, Optimization, in: *Chiral Liquid Chromatography* (W. J. Lough, ed.), Blackie, Glasgow, pp. 235–243.

Noctor, T. A. G., Félix, G., and Wainer, I. W., 1991, Stereochemical resolution of enantiomeric 2-aryl propionic acid non-steroidal anti-inflammatory drugs on a human serum albumin based high-performance liquid chromatographic chiral stationary phase, *Chromatographia* **31:**55–59.

Oda, Y., Asakawa, N., Abe, S., Yoshida, Y., and Sato, T., 1991, Avidin protein conjugated column for direct injection analysis of drug enantiomers in plasma by high-performance liquid chromatography, *J. Chromatogr.* **572:**133–141.

Simek, Z., and Vespalec, R., 1989, Bovine serum albumin bonded to hydroxyethyl methacrylate polymer for chiral separations, *J. High Resolut. Chromatogr. Chromatogr. Commun.* **12:**61–62.

Sjöholm, I., Ekman, B., Kober, A., Ljungstadt-Pahlman, I., Seiving, B., and Sjödin, T., 1979, Binding of drugs to human serum albumin, *Mol. Pharmacol.* **16:**767–777.

Stewart, K. K., and Doherty, R. F., 1973, Resolution of D,L-tryptophan by affinity chromatography on bovine-serum albumin-agarose columns, *Proc. Natl. Acad. Sci. U.S.A.* **70:**2850–2852.

Sudlow, G., Birkett, D. J., and Wade, D. N., 1976, Further characterization of specific binding sites on human serum albumin, *Mol. Pharmacol.* **12:**1052–1061.

Thompson, R. A., Andersson, S., and Allenmark, S., 1989, Direct liquid chromatographic separation of enantiomers on immobilized protein stationary phases. VII. Sorbents obtained by entrapment of crosslinked BSA in silica, *J. Chromatogr.* **465:**263–270.

Wainer, I. W., Jadaud, P., Schombaum, G. R., Kadodkar, S. V., and Henry, M. P., 1988, Enzymes as HPLC stationary phases for chiral resolutions: Initial investigations with α-chymotrypsin, *Chromatographia* **25:** 903–907.

13

Chromatography with Cyclodextrin-Based Stationary Phases

Karen L. Williams and Apryll M. Stalcup

1. INTRODUCTION

A significant number of the organic components of foods, flavorings, pharmaceuticals, and pesticides are chiral molecules (Ariëns, 1988). The role of stereochemistry in biological processes was first recognized more than 100 years ago by Pasteur and van't Hoff–Le Bel (Pasteur, 1901). Much of the stereoselectivity of these chemicals has remained shrouded in mystery, however, because techniques for chiral separations were not well developed. The fact that enantiomers exhibit identical physical and chemical properties in an achiral environment renders their separation extremely difficult.

The importance of chiral separations has become readily apparent in the pharmaceutical industry. Cost factors often preclude the synthesis of the desired enantiomer. Therefore, synthetically derived chiral pharmaceuticals are often administered as the racemic mixture (Ariëns, 1989) despite the fact that, in some cases, only one enantiomer, or eutomer, has the desired therapeutic effect. The other enantiomer, or distomer, may be responsible for undesirable side effects or compete for active sites (Drayer, 1986). The U.S. Food and Drug Administration (FDA) is currently formulating guidelines for drugs with chiral centers (DeCamp, 1989). Future chiral drugs will probably be required to undergo bioefficacy and biotoxicity studies on both the pure enantiomers as well as the racemic mixture before approval.

Numerous techniques have been employed to achieve the separation of optical isomers. Perhaps the most impressive gains have been made in the field of high-performance liquid chromatography (HPLC) (Armstrong, 1987). Although impressive gains have also recently been made in capillary gas-chromatographic methods, the focus of the work presented here will be on liquid-chromatographic methods using chiral stationary phases (CSP).

Various types of CSP have been developed, including the Pirkle-type (Pirkle *et al.*, 1980), cellulosic (Lindner and Mannschreck, 1980), protein-based (Domenici *et al.*, 1990), and cyclodextrin (Armstrong and DeMond, 1984) CSP. In each case, slight differences in the associations of the enantiomers with the chiral ligand of the stationary phase cause

Karen L. Williams and Apryll M. Stalcup • Department of Chemistry, University of Hawaii at Manoa, Honolulu, Hawaii 96822.

Molecular Interactions in Bioseparations, edited by That T. Ngo. Plenum Press, New York, 1993.

one enantiomer to be retained preferentially. Although each CSP is able to separate a large number of enantiomers, the application of each CSP is typically limited to structurally related chiral compounds.

The cyclodextrin (CD) bonded phases have proven to be applicable to a wide variety of compounds. Both native and derivatized CD CSP have been successful in separating a wide variety of enantiomeric pairs and will be discussed in subsequent sections. This chapter is not intended as an exhaustive review of chiral separations using CD-based CSP but merely to highlight and discuss some of the more important recent developments.

2. NATIVE CYCLODEXTRIN CSP

Native CD are chiral, cyclic molecules composed of glucose units bonded through α-(1,4)-linkages (Bender and Komiyama, 1978). The CD most commonly used for bonded stationary phases are comprised of six (α), seven (β), or eight (γ) glucose units that form a truncated cone shape (Ward and Armstrong, 1988). A structural diagram is shown in Fig. 1. The secondary hydroxyl groups of the glucose molecules line the mouth of the CD, while the primary hydroxyls form the narrow end of the cone. The inner cavity of the CD is relatively hydrophobic in comparison to the hydrophilic mouth of the CD. The cavity of the CD varies in diameter from 0.45 to 0.60 nm for α-CD and from 0.8 to 1.0 nm for γ-CD (Hinze, 1981).

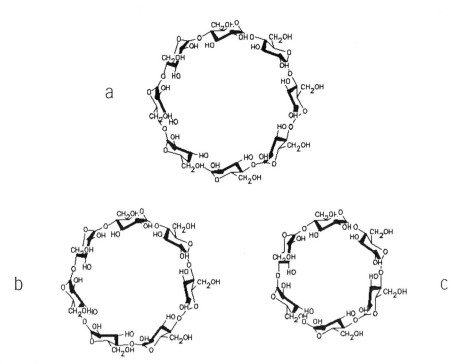

FIGURE 1. Schematic showing the structures and relative sizes of the three most common cyclodextrins: (1) γ-cyclodextrin (8 glucose units); (b) β-cyclodextrin (7 glucose units); (c) α-cyclodextrin (6 glucose units).

The ability of CDs to form inclusion complexes encouraged the development of CD-based CSP for HPLC. However, the first attempts to bind CDs to silica through amine (Fujimura *et al.*, 1983) or amide linkages (Kawaguchi *et al.*, 1983) were not highly successful. The resulting phases were hydrolytically unstable, and the nitrogen-containing linkages affected the selectivity of some compounds through the addition of nonstereo-specific interactions (Feitsma *et al.*, 1985).

Bonded phases without nitrogen linkages were developed by Armstrong (1985). A spacer is first attached to the silica. Subsequently, the CD is linked to the spacer through one of its primary hydroxyl groups. The linkage is stable in aqueous and hydro-organic mobile-phase systems and does not introduce additional nonstereospecific interaction sites. However, the bulky nature of the CD results in less than complete coverage of the spacer moiety. In fact, for β-CD, the concentration of the spacer is approximately an order of magnitude larger than the concentration of the CD (Stalcup and Williams, 1992).

In the presence of polar solvents, the hydrophobic character of the CD cavity encourages the nonpolar portion of a compound to reside within the cavity. For this reason, most separations accomplished with native CD have been performed in the reversed-phase mode. The size and geometry of the guest molecule appear to play an important role in the retention mechanisms of CD stationary phases (Armstrong *et al.*, 1986). For β-CD, the cavity can easily accommodate solutes comparable in size to naphthalene. Chiral recognition is improved if the compound consists of two or more ring moieties, at least one of which is aromatic (Han *et al.*, 1988). In addition, enantioselectivity is enhanced when the substituents on the chiral center of the enantiomer are capable of forming hydrogen bonds with the secondary hydroxyls along the mouth of the CD (Maguire, 1987). Compounds which do not fit tightly within the cavity exhibit lower binding constants and often show reduced selectivity (Armstrong and Li, 1987).

Native CD stationary phases have been applied to the separation of numerous types of compounds (Table 1). Another exciting application of β-CD involves the determination of enantiomers in human serum by direct injection (Stalcup and Williams, 1992). The excess concentration of the spacer moieties noted previously results in a stationary phase with both diol and CD functionalities. Diol phases have proven to be compatible with the analysis of proteins by HPLC (Schmidt *et al.*, 1980). Thus, the diol functionality of the β-CD phase permits the proteinaceous components of the serum to elute with the void volume. The drug enantiomers are retained by the CD moiety and are then resolved, as illustrated in Fig. 2. As a result, human serum samples can be directly injected onto the chromatographic column without pretreatment. The benefits of this method include reduced analysis time and reduced risk of perturbation of the original enantiomeric composition in the blood sample.

3. HYDROXYPROPYL-β-CYCLODEXTRIN CSP

Despite the many successes of the native CD phases, not all chiral compounds can be resolved on these columns because the solutes do not meet the criteria for chiral recognition. As mentioned previously, one of the important factors in chiral recognition is hydrogen bonding between the solute and the secondary hydroxyl groups along the mouth of the CD. Reacting β-CD with propylene oxide introduces an additional stereogenic interaction site and extends the region of interaction with the CD. (*R*,*S*)-2- and (*S*)-2-hydroxypropyl-β-CD

TABLE 1
Enantiomeric Separations Obtained for Selected Compounds
in the Reversed-Phase Mode on the Native β-Cyclodextrin Bonded Phase

Parent compound	Structure	Column (cm)	Mobile phase[a]	k'	α	R_s	Reference[b]
Chlorpheniramine maleate	CH₂CH₂N(CH₃)₂ ...	2 × 25	A	5.11	1.07	0.84	1
Chlorthalidone	... SO₂NH₂ ... Cl ...	2 × 25	A	1.93	1.28	2.02	1
Hexobarbital	... CH₃ ... OH ...	25	B	6.08	1.15	1.39	1

Mephenytoin	25	C	0.48	1.33	1.83	2
Mephobarbital	25	B	6.27	1.12	1.10	1
Propranolol	2 × 25	D	2.78	1.04	1.40	2

[a]A, 10% acetonitrile/buffer (100mM phosphate, pH 4.2); B, 5% acetonitrile/buffer (100mM phosphate, pH 6.9); C, 40% methanol/buffer (1% triethylammonium acetate, pH 4.1); D, 25% methanol/buffer (1% triethylammonium acetate, pH 4.1)
[b]1, Stalcup and Williams, 1992; 2, Armstrong et al., 1986.

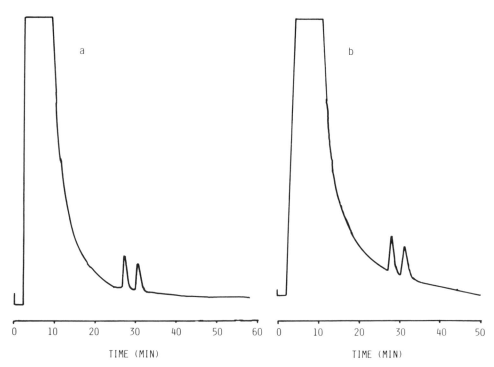

FIGURE 2. Chromatographic separation of hexobarbital enantiomers on a 25 cm × 4.6 mm i.d. Cyclobond I (β-cyclodextrin) column. Experimental conditions are given in Table 1. (a) After 20 serum injections; (b) after 60 serum injections. Sample spike was 50 μg of hexobarbital/ml of serum.

have been used to prepare bonded phases for HPLC (Stalcup *et al.*, 1990). The hydroxy-propyl groups appear to play different roles in chiral recognition, depending on the solute. In contrast to the rigid hydroxyl groups of the native CD, the OH functionality of the hydroxypropyl group is free to rotate and may allow hydrogen bonding interactions that were not previously possible with the native CD. Evidence that, for some compounds, the optical purity of the substituent does not play a role is supplied by the fact that chiral separations for chlorthalidone and 1,2,3,4-tetrahydro-1-napthol are obtained on the racemic column, as shown in Table 2.

However, the hydroxypropyl groups may also introduce steric interaction effects by partially occluding the mouth of the CD. Thus, one enantiomer may be hindered in forming an inclusion complex, especially if the substituents on the chiral center are bulky. This type of interaction is illustrated by the separation of nisoldipine (Table 2).

In addition, for some analytes, the enantioselectivity originates from the additional stereogenic center introduced by the hydroxypropyl group. Compounds which are too bulky to be fully included into the cavity can be resolved on the optically pure column but not on the racemic column. In this case, even though the chiral center is outside the cavity, interaction with the optically pure hydroxypropyl groups permits stereoselective hydrogen bonding. The resolution of 3,3′,5-triiodo-D,L-thyronine (Table 2) exemplifies this situation.

4. MISCELLANEOUS DERIVATIZED β-CYCLODEXTRIN CSP

Until recently, CD-based CSP have been used almost exclusively in the reversed-phase mode because the mobile-phase conditions favor the formation of inclusion complexes. Under normal-phase conditions, the apolar solvent probably occupies the CD cavity, and any solute interactions occur at the mouth of the CD. Recently, derivatized CD phases that can be used in the normal-phase mode have been developed. These phases appear to be analogous to the derivatized cellulosic CSP. The addition of carbonyl or aromatic functionalities onto the native CD produces CSP with π–π and hydrogen bonding interactions similar to Pirkle-type phases. The derivatization can be used to incorporate an additional stereogenic center, which has been shown not only to enhance chiral selectivity, but also to influence the elution order of some enantiomers. Significantly, normal-phase chiral separations comparable to those achieved in the reversed-phase mode have also been achieved on these derivatized CD bonded CSP.

Normal-phase CD CSP have been prepared from acetic anhydride, (R)-, (S)-, and (R,S)-1-(naphthyl)ethyl isocyanate, 2,6-dimethylphenyl isocyanate, and p-toluoyl chloride derivatives of β-CD. The degree of substitution is varied by changing the amount of the derivatizing agent and the reaction time. Except for the acetylated CD phase, an appreciable number of hydroxyl groups remain for hydrogen bonding (Armstrong et al., 1990). Although most of the derivatized phases exhibit unique selectivities, as illustrated in Table 3, the naphthylethyl isocyanate-derivatized CD produces a CSP comparable in selectivity to a Pirkle-type naphthylvaline column (Stalcup et al., 1991). In addition, the configuration of the naphthylethyl substituent allows flexibility in enantiomeric elution order for some compounds.

Although the mechanism is not completely understood, chiral recognition for the naphthylethyl-derivatized CD probably arises from a combination of interactions with the CD and the chiral naphthylethyl carbamate substituents. The role of the CD in chiral recognition can be deduced from the selectivity of the racemic column. For some derivatized (3,5-dinitrobenzoyl) amino acids (e.g., DL-norleucine, DL-tryptophan), no separation is obtained on the racemic column (Table 3). Hence, the enantioselectivity is dictated by the configuration of the naphthylethyl substituent. In addition, reversal of enantiomeric elution order is obtained for some solutes on the R-substituted CD column versus the S-substituted column. For the native CD phases, elution order is often difficult to predict and is not generally reversible.

In some instances, the enantioselectivity appear to be governed by the CD. For derivatized DL-phenylalanine, DL-isoleucine, and DL-valine, the elution order and selectivity are independent of the configuration of the naphthylethyl substituents (Table 3). Comparable selectivities are obtained on the optically pure and the racemic columns.

A third chiral recognition mechanism involves contributions from both the CD and the chiral substituents. The interaction between the CD and one configuration of the naphthylethyl substituent can be complementary (synergistic), resulting in improved chiral recognition relative to that on the racemic column and on the oppositely configured column, or competitive (antagonistic), leading to reduced selectivity. Elution order is more difficult to predict for this mode of chiral recognition. Examples of some of the normal-phase separations obtained on the naphthylethyl-derivatized CD, as well as some of the other derivatized CD columns, are illustrated in Table 3.

Although the (R)- and (S)-naphthylethyl carbamate phases were originally developed

TABLE 2

Enantiomeric Separations for Selected Compounds Obtained in the Reversed-Phase Mode
on the Hydroxypropyl-β-Cyclodextrin Bonded Phase[a]

Parent compound	Structure	Column[b]	Mobile phase[c]	k'	α	R_s
(±)-1,1′-Bi-2-naphthol		S	A	4.40	1.08	0.6
Chlorthalidone		S	B	1.33	1.38	2.2
		RAC	B	1.50	1.31	2.8
Nisoldipine		S	C	4.81	1.13	1.04

Compound		Config	Mobile phase	k	α	R_s
SQ 31 236	(3-nitrophenyl dihydropyrimidine-2-thione; NCO$_2$CH$_2$HC(CH$_3$)$_2$; CH$_2$–N(CH$_3$)(CH$_2$)$_2$O$_2$C; H$_3$C)	S	D	2.00	1.21	2.00
		RAC	E	10.86	1.28	1.25
SQ 31 579	(2-trifluoromethylphenyl dihydropyrimidine-2-thione; F$_3$C; CH$_3$CH$_2$O$_2$C; H$_3$C; N-benzylpiperidin-4-yl carbamate)	S	B	5.08	1.14	0.55
(±)-1,2,3,4-tetrahydro-1-naphthol	(OH)	S	F	2.36	1.08	0.7
		RAC	F	1.92	1.08	0.5
3,3',5-triiodo-D,L-thyronine	(HO– –O– –CH$_2$–CHCOOH; NH$_3$; I)	S	D	7.18	1.05	0.65

[a]Data taken from Stalcup et al. (1990).

[b]S, (S)-2-Hydroxypropyl-β-cyclodextrin; RAC, (R,S)-2-hydroxypropyl-β-cyclodextrin.

[c]A, 80% methanol/buffer (1% triethylammonium acetate, pH 4.1); B, 5% acetonitrile/buffer (1% triethylammonium acetate, pH 4.1); C, 10% acetonitrile/buffer (1% triethylammonium acetate, pH 4.1); D, 15% acetonitrile/buffer (1% triethylammonium acetate, pH 4.1); E, 20% acetonitrile/buffer (1% triethylammonium acetate, pH 4.1); F, 100% buffer (1% triethylammonium acetate, pH 4.1).

TABLE 3
Enantiomeric Separations Obtained for Selected Compounds
in the Normal Phase Mode on Derivatized β-Cyclodextrin Bonded Phases

Parent compound	Structure	Derivatizing agents[c]	Column[a]	Mobile phase[b]	k'^d	α	Reference[e]
(±)-Glutethimide		—	DMP	A	14.3	1.10	1
DL-Isoleucine		DNB	RN	B	3.03^L	1.10	2
		DNB	SN	B	2.49^L	1.11	2
		DNB	RSN	B	3.61^L	1.11	2
DL-Norleucine		DNB	RN	B	3.11^D	1.12	2
		DNB	SN	B	2.49^L	1.08	2
		DNB	RSN	B	3.92	1.00	2
(±)-Phensuximide		—	DMP	A	11.2	1.30	1
		—	RN	A	3.38	1.07	1

Compound	Structure						
DL-Phenylalanine		DNB	RN	C	7.41^L	1.06	2
		DNB	SN	C	7.31^L	1.10	2
		DNB	RSN	C	9.68^L	1.11	2
DL-Tryptophan		DNB	RN	C	7.91^D	1.16	2
		DNB	SN	C	7.21^L	1.22	2
		DNB	RSN	C	9.89	1.00	2
DL-Valine		DNB	RN	B	3.16^L	1.05	2
		DNB	SN	B	2.70^L	1.05	2
		DNB	RSN	B	3.71^L	1.08	2

[a]DMP, 2,6-Dimethylphenylcarbamoylated β-cyclodextrin; RN, (R)-$(-)$-1-(1-naphthyl)ethylcarbamoylated β-cyclodextrin; SN, (S)-$(+)$-1-(1-naphthyl)ethylcarbamoylated β-cyclodextrin; RSN, (R,S)-$(\pm)$-1-(1-naphthyl)ethylcarbamoylated β-cyclodextrin.
[b]A, 10% isopropanol/n-hexane; B, 50% acetonitrile/ethanol C, 1% acetic acid/methanol.
[c]DNB, 3,5-Dinitrobenzamide.
[d]Configuration of the first eluting enantiomer indicated as a superscript, when known.
[e]1, Armstrong et al., 1990; 2, Stalcup et al., 1991.

for normal-phase applications, they have proven to be a multimodal CSP (Armstrong *et al.*, 1991). Because the derivatized CD is bonded to the silica rather than adsorbed, the phases do not suffer the same mobile-phase limitations as the derivatized cellulosic CSP. An entirely different group of enantiomers can be resolved by utilizing the derivatized CD under reversed-phase conditions.

The separation mechanism in the reversed-phase mode appears to be completely different than under normal-phase conditions. This assumption is supported by the fact that compounds which could be separated in the reversed-phase mode could not be separated on the same column when used in the normal-phase mode. In addition, solutes which had previously been resolved in the normal-phase mode were not separated in the reversed-phase mode. However, the configuration of the naphthylethyl substituent does play an important role in enantioselectivity under reversed-phase conditions. Table 4 lists some

TABLE 4

Enantiomeric Separation Obtained for Selected Compounds
in the Reversed-Phase Mode on Derivatized Cyclodextrin Bonded Phases[a]

Parent compound	Structure	Column[b]	Mobile phase[c]	k'	α
Dyfonate		RN	A	12.0	1.02
Flurbiprofen		SN	B	25.0	1.04
Ibuprofen		RN SN	C D	3.4 2.8	1.13 1.11
Lorazepam		SN	B	5.0	1.04
Oxazepam		SN	A	1.2	1.16

[a]Data taken from Armstrong *et al.* (1991).
[b]RN, (*R*)-(−)-1-(1-Naphthyl)ethylcarbamoylated β-cyclodextrin; SN, (*S*)-(+)-1-(1-naphthyl)ethylcarbamoylated β-cyclodextrin.
[c]A, 30% acetonitrile/buffer (1% triethylammonium acetate, pH 4.5); B, 20% acetonitrile/buffer (1% triethylammonium acetate, pH 4.5); C, 35% acetonitrile/buffer (1% triethylammonium acetate, pH 7.1); D, 40% acetonitrile/buffer (1% triethylammonium acetate, pH 7.1).

of the biologically important racemates that have been separated using reversed-phase solvent conditions.

5. CONCLUSION

The development of native and derivatized CD CSP has been an important advance in the area of chiral separations. The utilization of native CD columns for the direct injection of human serum offers simplified methodology for the study of drug pharmacokinetics. In addition, the CD can be derivatized, providing flexibility in the design of CSP with the desired functionality.

REFERENCES

Ariëns, E. J., 1988, Stereoselectivity of bioactive agents, in: *Stereoselectivity of Pesticides* (E. J. Ariëns, J. J. S. van Rensen, and W. Welling, eds.), Elsevier, Amsterdam, pp. 39–108.

Ariëns, E. J., 1989, Racemates—an impediment in the use of drugs and agrochemicals, in: *Chiral Separations by HPLC.* (A. M. Krstulović, ed.), John Wiley & Sons, New York, pp. 31–68.

Armstrong, D. W., 1985, Bonded phase material for chromatographic separations, U.S. Patent 4,549,499.

Armstrong, D. W., 1987, Optical isomer separation by liquid chromatography, *Anal. Chem.* **59**(2):84A–91A.

Armstrong, D. W., and DeMond, W., 1984, Cyclodextrin bonded phases for the liquid chromatographic separation of optical, geometrical, and structural isomers, *J. Chromatogr. Sci.* **22**:411–415.

Armstrong, D. W., and Li, W., 1987, Optimization of liquid chromatographic separations on cyclodextrin-bonded phases, *Chromatogr. Forum* **2**:43–48.

Armstrong, D. W., Ward, T. J., Armstrong, R. D., and Beesley, T. E., 1986, Separation of drug stereoisomers by the formation of β-cyclodextrin inclusion complexes, *Science* **232**:1132–1135.

Armstrong, D. W., Stalcup, A. M., Hilton, M. L., Duncan, J. D., Faulkner, J. R., Jr., and Chang, S.-C., 1990, Derivatized cyclodextrins for normal-phase liquid chromatographic separation of enantiomers, *Anal. Chem.* **62**:1610–1615.

Armstrong, D. W., Chang, C.-D., and Lee, S. H., 1991, (*R*)- and (*S*)-Naphthylethylcarbamate-substituted β-cyclodextrin bonded stationary phases for the reversed-phase liquid chromatographic separation of enantiomers, *J. Chromatogr.* **539**:83–90.

Bender, M. L., and Komiyama, M., 1978, *Cyclodextrin Chemistry*, Springer-Verlag, Berlin.

DeCamp, W. H., 1989, The FDA perspective on the development of stereoisomers, *Chirality* **1**(1):2–6.

Domenici, E., Bertucci, C., Salvadori, P., Félix, G., Cahagne, I., Motellier, S., and Wainer, I. W., 1990, Synthesis and chromatographic properties of an HPLC chiral stationary phase based upon human serum albumin, *Chromatographia* **29**(3/4):170–176.

Drayer, D. E., 1986, Pharmacodynamic and pharmacokinetic differences between drug enantiomers in humans: An overview, *Clin. Pharmacol. Ther.* **40**(2):125–133.

Feitsma, K. G., Bosman, J., Drenth, B.F. H., and DeZeeuw, R. A., 1985, A study of the separation of enantiomers of some aromatic carboxylic acids by high-performance liquid chromatography on a β-cyclodextrin-bonded stationary phase, *J. Chromatogr.* **333**:59–68.

Fujimura, K., Ueda, T., and Ando, T., 1983, Retention behavior of some aromatic compounds on chemically bonded cyclodextrin silica stationary phase in liquid chromatography, *Anal. Chem.* **55**:446–450.

Han, S. M., Han, Y. I., and Armstrong, D. W., 1988, Structural factors affecting chiral recognition and separation on β-cyclodextrin bonded phases, *J. Chromatogr.* **441**:376–381.

Hinze, W. L., 1981, Applications of cyclodextrins in chromatographic separations and purification methods, *Sep. Purif. Methods* **10**(2):159–237.

Kawaguchi, Y., Tanaka, M., Nakae, M., Funazo, K., and Shono, T., 1983, Chemically bonded cyclodextrin stationary phases for liquid chromatographic separation of aromatic compounds, *Anal. Chem.* **55**:1852–1857.

Lindner, K. R., and Mannschreck, A., 1980, Separation of enantiomers by high-performance liquid chromatography on triacetylcellulose, *J. Chromatogr.* **193**:308–310.

Maguire, J. H., 1987, Some structural requirements for resolution of hydantoin enantiomers with a β-cyclodextrin liquid chromatography column, *J. Chromatogr.* **387:**453–458.

Pasteur, L., 1901, On the asymmetry of naturally occurring organic compounds, the foundations of stereochemistry, in: *Memoirs by Pasteur, Van't Hoff, Le Bel, and Wislicenus* (G. M. Richardson, ed.), American Book Co., New York, pp. 1–33.

Pirkle, W. H., House, D. W., and Finn, J. M., 1980, Broad spectrum resolution of optical isomers using chiral high-performance liquid chromatographic bonded phases, *J. Chromatogr.* **192:**143–158.

Schmidt, D. E., Jr., Giese, R. W., Conron, D., and Karger, B. L., 1980, High performance liquid chromatography of proteins on a diol-bonded silica gel stationary phase, *Anal. Chem.* **52:**177–182.

Souter, R. W., 1985, *Chromatographic Separations of Stereoisomers*, CRC Press, Boca Raton, Florida.

Stalcup, A. M., and Williams, K. L., 1992, Determination of enantiomers in human serum by direct injection onto a β- cyclodextrin HPLC bonded phase, *J. Liq. Chromatogr.* **15:**129–137.

Stalcup, A. M., Chang, S.-C., Armstrong, D. W., and Pitha, J., 1990, (*S*)-2-Hydroxypropyl-β-cyclodextrin, a new chiral stationary phase for reversed-phase liquid chromatography, *J. Chromatogr.* **513:**181–194.

Stalcup, A. M., Chang, S.-C., and Armstrong, D. W., 1991, Effect of the configuration of the substituents of derivatized-β-cyclodextrin bonded phases on enantioselectivity in normal phase liquid chromatography, *J. Chromatogr.* **540:**113–128.

Ward, T. J., and Armstrong, D. W., 1988, Cyclodextrin-stationary phases, in: *Chromatographic Chiral Separations* (M. Zief and L. J. Crane, eds.), Marcel Dekker, New York, pp. 131–163.

Immunoaffinity Separation

Suitable Antibodies as Ligands in Affinity Chromatography of Biomolecules

Eizo Sada and Shigeo Katoh

1. INTRODUCTION

Immunoaffinity chromatography is especially useful for purification of biological materials because of the highly specific molecular recognition between antigen and antibody. Furthermore, antibodies can generally be raised against many biological materials foreign to the host. The high affinity of the immunointeraction, however, requires that extreme elution conditions, such as low pH (pH 2–3) or the addition of denaturing agents, be employed; under such conditions, many bioactive materials are denatured. Thus, immunoaffinity chromatography has two conflicting requirements: a high affinity for the adsorption step and a weak affinity for the elution step. Therefore, both the selection of antibody ligands having these characteristics and the regulation of their affinity by environmental factors are essential for effective utilization of immunoaffinity chromatography in downstream processes.

Since an antibody directed against a specific antigenic site has a complementary structure to that of this site, it is natural to expect the existence of a relationship between the equilibrium characteristics of antigen–antibody binding and the features of antigenic sites. This point will be treated in this chapter by considering haptens and peptides of different characteristics, and guidelines will be suggested for the selection of antigenic determinants in order to obtain antibodies with suitable adsorption behavior for immunoaffinity chromatography.

2. ADSORPTION EQUILIBRIUM OF ANTIHAPTEN AND ANTIPEPTIDE ANTIBODIES

2.1. Preparation of Antibodies

The antigens used to obtain antihapten antibodies were prepared by coupling diazotized *p*-aminobenzoic acid and 2.4-dinitroaniline to bovine serum albumin (BSA). The

Eizo Sada and Shigeo Katoh • Chemical Engineering Department, Kyoto University, Kyoto 606, Japan.

Molecular Interactions in Bioseparations, edited by That T. Ngo. Plenum Press, New York, 1993.

peptide antigens used for immunization are shown in Table 1. P1 (corresponding to residues 112–121 of sperm whale myoglobin) contains three histidine residues, which have a unique pK_a value of 6.0 and thus are positively charged at low pHs. P2 (residues 148–158 of tobacco mosaic virus protein) has no charged residues except for the carboxyl and amino terminals; in contrast, P3 contains many charged residues. In HA-119, an essential histidine residue in P1 has been replaced by alanine, as shown in Table 1, to clarify the effect of the histidine residue on the binding between the peptide and antibody. These peptides were coupled to keyhole limpet hemocyanin (KLH) with glutaraldehyde according to the method of Mariani et al. (1987).

The antisera against the haptens and peptides were raised in rabbits. A mixture of equal volumes of conjugate and complete Freund's adjuvant was injected into rabbits. Booster injections were repeated twice at 10-day intervals. Specific antibodies were purified from pooled antisera by affinity chromatography and coupled to Sepharose 4B to obtain immunoadsorbents.

2.2. Measurement of Adsorption Equilibrium

The immunoadsorbents were packed in a column, and a solution of haptens or peptides adjusted to the desired ionic strength and pH was applied to the column. The absorbance at 280 or 215 nm of the effluent solution was measured continuously, and the total amount of adsorbed antigens was obtained by numerical integration of the breakthrough curves, with the assumption that the total void fraction of the packed column was 0.96. The average association constant K_0 and the heterogeneity index σ for each antibody were estimated by assuming that the heterogeneity of the free energy of antigen–antibody binding can be described by the normal distribution function.

3. EQUILIBRIUM CHARACTERISTICS OF HAPTENS WITH AND WITHOUT A CHARGED GROUP

The equilibrium behavior of antihapten antibodies was measured (Sada et al., 1988). Figures 1 and 2 show the effects of pH and ionic strength on the amounts of 2.4-dinitroaniline and p-aminobenzoic acid adsorbed by antihapten antibodies at a hapten concentration of $7.6 \times 10^{-7}M$. The adsorption capacity of anti-dinitrophenyl (DNP) antibody was little affected by pH and ionic strength, which shows that the interaction between the unionizable hapten and antibody is stable to changes in environmental

TABLE 1
Peptides Used for Immunization and Values of K_0

Peptide	Sequence	$K_0(M^{-1})$
P1	Ile-His-Val-Leu-His-Ser-Arg-His-Pro-Gly	1.3×10^6
P2	Ser-Gly-Leu-Val-Trp-Thr-Ser-Gly-Pro-Ala-Thr	1.3×10^6
P3	Ile-Lys-Lys-Lys-Thr-Glu-Arg-Glu-Asp-Leu	2.0×10^6
HA-119	Ile-His-Val-Leu-His-Ser-Arg-*Ala*-Pro-Gly	3.0×10^5

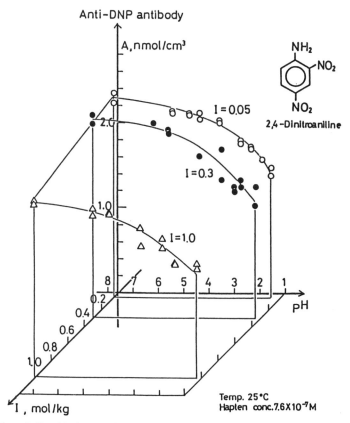

FIGURE 1. Effect of pH and ionic strength on adsorption capacity of anti-DNP antibody for 2,4-dinitroaniline.

conditions. In contrast, as shown in Fig. 2, the adsorption capacity of anti-p-azobenzoate antibody drastically decreased with increase in ionic strength and with decrease in pH, particularly near the pK_a of p-aminobenzoic acid. Thus, the interaction between the charged hapten and antibody depends strongly on pH and ionic strength, because the main force contributing to this interaction is an electrostatic one.

Figure 3 shows the effect of pH on the adsorption capacities of the anti-p-azobenzoate antibody for a series of para-substituted benzoic acids. The adsorption capacities for the acids with higher pK_a values decreased in the higher pH ranges near their pK_a values, shown in Table 2. These results indicate that the charged states of haptens strongly affect the interaction between haptens and antibodies.

4. EQUILIBRIUM CHARACTERISTICS OF PEPTIDES OF DIFFERENT CHARACTERISTICS

Recently, it has become clear that short synthetic peptides corresponding to parts of a protein can elicit antibodies that are reactive with the native protein. Therefore, this peptide

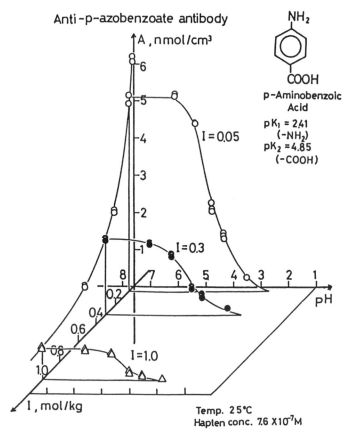

FIGURE 2. Effect of pH and ionic strength on adsorption capacity of anti-*p*-azobenzoate antibody for *p*-aminobenzoic acid.

immunization may be an effective method to prepare antibodies suitable for immunoaffinity chromatography if the relationships between the amino acid composition of peptides and the dependence of adsorption equilibria on pH and ionic strength are clarified.

The adsorption equilibria between antibodies and peptides of different characteristics, shown in Table 1, were measured (Kondo *et al.*, 1990). Figure 4 compares the effect of ionic strength adjusted by addition of NaCl on the adsorption capacity of anti-P1, anti-P2, and anti-P3 antibodies at pH 7.6. The adsorption capacity of anti-P3 antibody decreased extensively with increase in ionic strength. This is probably because electrostatic forces are a dominant factor in stabilizing the interaction between P3 and anti-P3 antibody, because P3 has seven charged amino acids. P1 and P2 have few charged residues at neutral pH, and hence the adsorption capacities of anti-P_1 and anti-P_2 were little affected by ionic strength.

Figure 5 shows the pH dependence of the adsorption capacities of anti-P1, anti-P2, and anti-P3 antibodies at an ionic strength of 0.3 mol/kg. The decrease in the adsorption capacity of anti-P1 antibody in the mildly acidic pH range was caused by the ionization of histidine residues below pH 6, because P1 contains three histidine residues. On the other

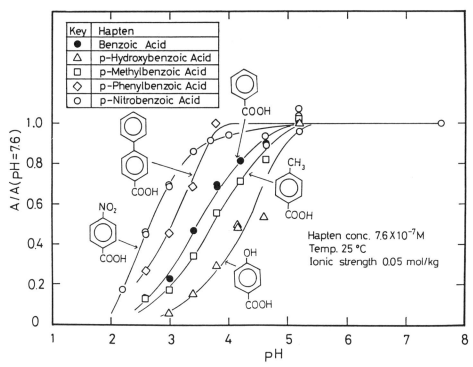

FIGURE 3. pH dependence of adsorption capacity of anti-*p*-azobenzoate antibody for para-substituted benzoic acids.

hand, anti-P2 antibody bound its corresponding peptide over a wider pH range than the other two antibodies. This is because P2 does not contain any charged residues.

To confirm the effect of charged residues on the adsorption equilibrium between peptides and antipeptide antibodies, the additional peptide, HA-119, was investigated. In this peptide, an essential histidine residue for binding of P1 to anti-P1 antibody has been replaced by alanine, which is not charged at neutral pH (Sada *et al.*, 1990). In Fig. 6, the adsorption capacity of anti-HA-119 for HA-119 is compared with that of anti-P1 for P1. The stronger dependence of the adsorption capacity of anti-P1 on pH demonstrates the effect of ionization of the histidine residue on the adsorption equilibrium.

TABLE 2

pK_a Values of para-Substituted
Benzoic Acids in Water

Substituent	pK_a	Substituent	pK_a
—NH$_2$	4.86	—C$_6$H$_5$	4.21
—OH	4.57	—H	4.20
—CH$_3$	4.37	—NO$_2$	3.42

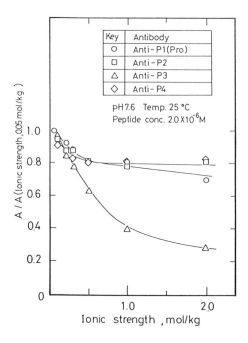

FIGURE 4. Effect of ionic strength on adsorption capacity of antipeptide antibodies for peptides.

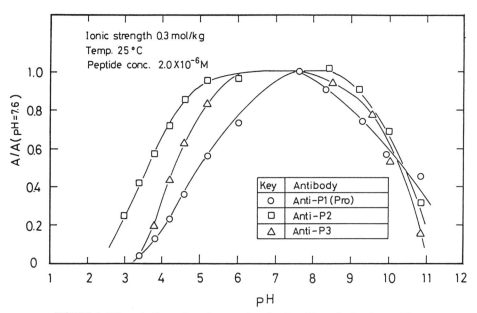

FIGURE 5. Effect of pH on adsorption capacity of antipeptide antibodies for peptides.

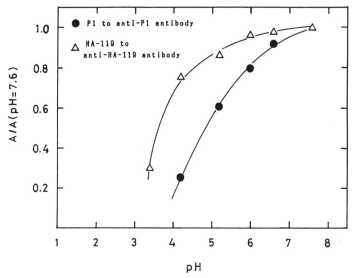

FIGURE 6. Comparison of pH dependence of adsorption capacities of anti-P1 and anti-HA-119 antibodies.

These results indicate that charged states of amino acid residues essential for binding have strong effects on the dependence of adsorption behavior on pH and ionic strength and might be useful as a selection criterion for peptide sequences from proteins for immunization in order to raise antibodies suitable as immunoaffinity ligands, that is, antibodies whose binding affinity is sensitive to pH or ionic strength.

REFERENCES

Kondo, A., Takamatsu, H., Katoh, S., and Sada, S., 1990, Adsorption equilibrium in immunoaffinity chromatography with antibodies to synthetic peptides, *Biotechnol. Bioeng.* **35:**146–151.

Mariani, M., Bracci, L., Presentini, R., Nucci, D., Neri, P., and Antoni, G., 1987, Immunogenicity of a free synthetic peptide: carrier-conjugation enhances antibody affinity for the native protein, *Mol. Immunol.* **24:**297–303.

Sada, E., Katoh, S., Miyoshi, H., Yamanaka, K., and Kondo, A., 1988, Effects of charged groups in haptens on adsorption equilibrium of hapten antibody, *Biotechnol. Bioeng.* **32:**467–474.

Sada, E., Katoh, S., and Sohma, Y., 1990, Effects of histidine residues on adsorption equilibrium of peptide antibodies, *J. Immunol. Methods* **130:**33–37.

15

Immunoaffinity Purification of Organelles

Peter J. Richardson and J. Paul Luzio

1. INTRODUCTION

The vast majority of subcellular fractionation techniques exploit physical differences between organelles, that is, size, density, charge or hydrophobicity. The most widely used methods are density-gradient centrifugation, free-flow electrophoresis, and polymer-phase partitioning. These approaches are of limited use when organelles with similar physical properties are to be fractionated. In such situations, procedures based on biological differences are required. Of these, methods based on immunological techniques have probably the greatest potential since they rely solely on the presence of specific antigens on the organelles of interest. In 1975, de Duve postulated that each biochemical marker is restricted to a single subcellular site (e.g., cytochrome oxidase in mitochondria). This remains the basis on which the purity of many isolated organelles is assessed, although it has become apparent that subcellular organelles previously thought to be homogeneous may differ in their composition (Reijnierse *et al.*, 1975), while formerly "specific" markers may be found in a variety of subcellular compartments, an example being 5′-nucleotidase (Stanley *et al.*, 1982). Consequently, organelles are now frequently identified by their function (e.g., transport vesicle, synaptic vesicle, transcytotic carrier vesicle) as much as by their composition and appearance under electron microscopes. Since the function of organelles will be reflected in their composition, it is now generally accepted that each functionally distinct organelle has a distinct antigenic pattern. Immunological techniques ought therefore to be capable of isolating any defined organelle, with the proviso that a specific antigen is expressed on the organelle's surface. The latter is unlikely to be a serious limitation since there are few organelle functions that are not reflected in the surface composition. This chapter is therefore concerned with the development and application of immunological techniques for subcellular fractionation. The emphasis is on those techniques involving the antibody-mediated adsorption of specific organelles to solid matrices, that is, immunoaffinity purification. The applicability of this approach to the fractionation of two complex heterogeneous systems (brain homogenates and intracellular membrane traffic pathways) is demonstrated.

Peter J. Richardson • Department of Pharmacology, University of Cambridge, Cambridge CB2 1QJ, England. *J. Paul Luzio* • Department of Clinical Biochemistry, University of Cambridge, Addenbrooke's Hospital, Cambridge CB2 2QR, England.

Molecular Interactions in Bioseparations, edited by That T. Ngo. Plenum Press, New York, 1993.

The most common method used for separating organelles is still centrifugation, in which membranes and particles are separated on the basis of differences in their size, shape, and buoyant density. Centrifugation, especially on density gradients, is a versatile technique but is limited by the inherent differences in organelle densities. For example, centrifugation of homogenates of mammalian brain revealed the presence of sealed nerve terminals (Gray and Whittaker, 1962) and synaptic vesicles (Whittaker *et al.*,1964), but, despite some partial success, this approach has been unable to fractionate these organelles into transmitter-specific types. Similarly, centrifugation has proved of little use in the separation of the individual organelles involved in intracellular membrane traffic, largely because of the similar densities of the vesicles involved. There have been a number of antibody-based innovations which have been used to increase the selectivity of physical separation methods. For example, polymer two-phase separations, which are based on differences in the hydrophobicity of membranes (Albertson, 1986), can be used in conjunction with antibodies for cell separations (Sharp *et al.*, 1986).

Increasingly, immunoaffinity techniques are being used in conjunction with centrifugation to isolate specific organelles from such complex mixtures. The initial partial purification of the organelles by differential and/or density-gradient centrifugation often enhances the purity of the immunoisolated product. This increase in purity is however at the expense of increasing the time between homogenization and the final fractionation.

Fluorescence-activated cell sorting (FACS) is a technique which separates fluorescent particles (cells organelles, chromosomes, etc.) from those lacking a fluorescent chromophore. Because each particle is sorted separately, this technique is not suitable for preparative subcellular fractionation nor for the separation of labile organelles, due to the long separation times involved. However, it is ideal for analytical purposes, as nonspecific contamination of the product is minimized (see below). The use of appropriate specific antibodies with FACS has, for example, permitted the separation of plasma membrane domains from epithelial cells (Gorvel and Maroux, 1987).

In common with the FACS procedure, immunoaffinity techniques have the great advantage of not exposing the organelles to unusual or hyperosmotic media (cf. centrifugation). They are extremely rapid, which may be essential when unstable or labile organelles are being isolated (Burger *et al.*, 1989), and they can be used in both analytical or preparative procedures and when only small amounts of material are available (e.g., tissue culture cells; Howell *et al.*, 1989). In 1972 Hales discussed the possibility of applying immunoaffinity procedures to subcellular fractionation, although it was only during the 1980s that the approach gained more widespread use. One reason for this may have been the perceived difficulty in eluting the purified organelles from the solid phase, given the extreme conditions required to break antibody–antigen bonds (e.g., pH 2.2, 4*M* salt, etc.). After the early reported successes of the approach (Luzio *et al.*, 1974, 1976; Kawajiri *et al.*, 1977), it became apparent that most membrane and organelle properties can be studied while the organelle is still attached to the solid phase (see, e.g., Richardson and Brown, 1987). One other major hindrance to the widespread use of the technique has been the lack of suitable organelle-specific antibodies, although, since the advent of monoclonal antibodies (Kohler and Milstein, 1975), it has become possible to isolate antibodies (almost) on demand. More recently, immunoaffinity procedures have become part of the standard methodology of subcellular fractionation. Indeed, five years ago immunoaffinity was usually referenced in the "key words" of publications; nowadays it is seldom mentioned.

Historically, there have been two approaches to the immunoaffinity purification of organelles—the direct and indirect methods. The principles behind these methods are shown in Fig. 1. Briefly, the direct method involves the covalent linking of an anti-membrane-component antibody to the solid matrix, followed by incubation of the hetero-geneous organelle preparation with this matrix. Successful applications have been reported (Kawajiri *et al.*,1977; Pontremoli *et al.*, 1984; Burger *et al.*, 1989), but in general this approach has not been widely used. The indirect method involves the covalent coupling of an antibody-binding reagent to the matrix, which is then incubated with the organelle preparation already sensitized with the specific antibody. This method has been very successful, perhaps because it provides a long, flexible spacer arm between the membrane particle and the matrix. It also has the added advantage that a panel of specific anti-bodies can be used to isolate a number of organelles with a single matrix.

The success of immunoaffinity-based fractionation procedures is determined by the quality of both the primary antibody, which recognizes the organelle of interest, and the solid phase. Obviously, the primary antibody should only bind to the organelle of interest and should be of the highest possible affinity. Similarly, the solid phase should bind the primary antibody efficiently, while exhibiting minimal binding to organelles lacking the primary antibody. In addition, the separation of the solid phase from the unbound organelles

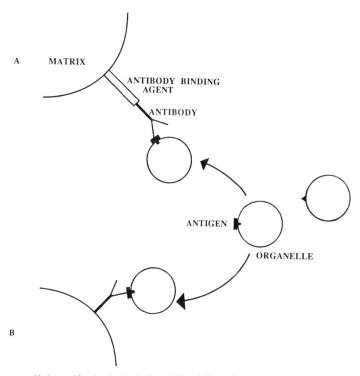

FIGURE 1. Immunoaffinity purification by the indirect (A) and direct (B) techniques. In the indirect technique an antibody-binding agent coupled to a solid matrix is used to capture the organelle bearing antibodies which recognize the specific antigen. These antibodies can be bound to either the organelle or to the antibody binding agent prior to organelle capture (see text). In the direct technique the antibody is (covalently) linked to the matrix.

should be rapid and complete. The design and required properties of antibodies and solid phases are discussed in the next two sections.

2. PRIMARY ANTIBODY

As mentioned above, the primary antibody should be specific to the organelle of interest. The antigen recognized should ideally be present at concentrations greater than 50 molecules/μm^2 (Devaney and Howell, 1985; Gruenberg and Howell, 1985) on the surface of the organelle. Polyclonal antisera are rarely of sufficient specificity, although there are a number of different ways in which they can be used, depending on the nature of the organelle being purified and the original source of the antigen. For instance, plasma membrane vesicles have been isolated from cell homogenates after incubation of intact cells with polyclonal antiserum at low temperatures (0–4°C). Under these conditions, little endocytosis of bound antibody occurs (Luzio and Stanley, 1983), and the unbound antibodies can be removed prior to homogenization. Given the high affinity of most antibodies, little redistribution of bound antibody subsequently occurs when the cell is disrupted. Consequently, even though one of the recognized antigens may be present in non-plasma membrane compartments, pure plasma membrane vesicles can be isolated (Luzio et al., 1976). In a similar manner, polyclonal antisera have been used to fractionate brain nerve terminals according to their neurotransmitter content. In one case, antisera raised against a number of soluble transmitter-synthesizing enzymes have been shown to recognize epitopes expressed primarily on the corresponding nerve terminal surfaces (Docherty et al., 1987). An alternative approach has been to use antibodies raised against antigens whose subcellular distribution has been well characterized. For example, antibodies raised against the cytoplasmic or exoplasmic domains of a vesicular stomatitis virus can be used to isolate endosomes and plasma membranes from cultured cells infected with the virus (Gruenberg and Howell, 1986). In another example, antisera raised against the synaptic membranes of the *Torpedo* electromotor synapse recognize two gangliosides expressed solely on the surface of cholinergic neurons in mammalian brain (Obrocki and Borroni, 1988). These gangliosides bear the only epitopes conserved between the *Torpedo* cholinergic neuron and mammalian cells (Jones et al., 1981). Consequently, these antisera can be used in the immunoaffinity purification of cholinergic nerve terminals and plasma membranes (Richardson et al., 1984).

However, even in such situations as described above, polyclonal antisera suffer from one other drawback: in any given antiserum the proportion of antibodies recognizing the immunizing antigen seldom exceeds 5%. Consequently, after incubation of the organelle with the primary antibody, a large excess of unwanted immunoglobulin must be removed since it would otherwise compete for the binding sites on the solid matrix. This is particularly important when isolating low-density membranes which cannot be centrifuged (and so washed) rapidly. In order to overcome this, the antibodies can be affinity purified if sufficient antigen is available (Brown et al., 1990). Indeed, it may be that affinity-purified polyclonal antibodies are the best available source of primary antibodies.

Many recent immunoaffinity purification protocols have utilized monoclonal antibodies (Table 1). Monoclonal antibodies have two great advantages in that they can be isolated when crude subcellular fractions are used as immunogens and they can be used in small amounts compared to polyclonal sera. Therefore, in theory, a monoclonal antibody

TABLE 1
Immunoaffinity-Purified Organelles

Organelle	Primary antibody	Solid phase	Reference
Plasma membrane	Polyclonal	Cellulose	Luzio et al., 1974
	Polyclonal	Cellulose	Luzio et al., 1976
	Polyclonal	Cellulose	Luzio, 1977
	Polyclonal	Cellulose	Westwood et al., 1979
	Polyclonal	Polyacrylamide	Miljanich et al., 1982
	Polyclonal	S. aureus[a]	Roman and Hubbard, 1984
	Monoclonal, polyclonal	Polyacrylamide, S. aureus	Gruenberg and Howell, 1985
	Monoclonal	Cellulose	Devaney and Howell, 1985
Nerve terminals	Polyclonal	Cellulose	Richardson et al., 1984
	Polyclonal	Polyacrylamide	Docherty et al., 1987
	Polyclonal	Cellulose	Richardson et al., 1987
	Polyclonal	Cellulose	Brown et al., 1990
Synaptic and secretory vesicles	Monoclonal	Polyacrylamide	Matthew et al., 1981
	Monoclonal	Polyacrylamide	Floor and Leeman, 1985
	Monoclonal	Polyacrylamide, polystyrene[b]	Lowe et al., 1988
	Monoclonal	Polyacrylamide	Volknandt et al., 1988
	Monoclonal	Polyacrylamide	Floor and Feist, 1989
	Monoclonal	Methacrylate	Burger et al., 1989
Golgi compartments	Polyclonal	Polyacrylamide	Ito and Palade, 1978
	Monoclonal	Cellulose, polystyrene	deCurtis et al., 1988
Endosomes	Polyclonal	S. aureus	Mueller and Hubbard, 1986
	Monoclonal	Polyacrylamide	Gruenberg and Howell, 1986
	Monoclonal	Polyacrylamide, polystyrene	Gruenberg and Howell, 1987
Glucose transporter vesicles	Polyclonal	S. aureus	Biher and Lienhard, 1986
	Monoclonal	Sepharose	Zorzano et al., 1989
Transport vesicles	Monoclonal	Cellulose, polystyrene	deCurtis and Simons, 1989
	Monoclonal	Polyacrylamide, polystyrene	Gruenberg et al., 1989
	Monoclonal	Cellulose	Wandinger-Ness et al., 1990
	Monoclonal	Polystyrene	Salamero et al., 1990
	Monoclonal	Sepharose	Sztul et al., 1991
Calthrin-coated vesicles	Polyclonal	S. aureus	Merisko et al., 1982
	Monoclonal	S. aureus	Pfeffer and Kelly, 1985
Lysosomes	Polyclonal	Polyacrylamide	Debanne et al., 1984
	Monoclonal	Sepharose	Pontremoli et al., 1984
Microsomes	Polyclonal	Sepharose	Kawajiri et al., 1977

[a]S. aureus: Fixed bacterial cells bearing protein A.
[b]In all the situations where polystyrene supports have been quoted, magnetic separation techniques have been used.

specific to any given organelle could be isolated. However, the design of suitable screens for such antibodies may be difficult if the specific antigenic determinant is unknown, and the final demonstration of organelle specificity of an antibody requires immunoelectron-microscopy. Consequently, many of the successful purifications reported to date have used antibodies directed against known antigens (Sztul *et al.*, 1991; Burger *et al.*, 1989). Until recently, it has been difficult to define appropriate antigens for many organelles. However, as more antigens are identified, recombinant DNA techniques are being used to define the cytoplasmic tails of these antigens, and antibodies are being raised to the corresponding synthetic peptides or proteins expressed in bacterial expression vectors. Because mono-clonal antibodies can be used in small amounts, it is seldom necessary to wash away primary antibody unbound to the organelles.

When using either affinity-purified polyclonal or monoclonal antibodies, the primary antibody can be preadsorbed onto the solid matrix prior to mixing with the organelle mixture. This can greatly reduce the time required to isolate the desired organelle, as washing of the sensitized organelle is not required. This also serves to reduce the effects of primary antibody competing with organelle-bound antibody for the solid-phase binding sites (see, e.g., Volknandt *et al.*, 1988). If a choice between affinity-purified polyclonal or monoclonal antibodies is available, it is likely that the former would be more appropriate for use as the primary antibody. It is conceivable that the recognition of only one antigenic site by the monoclonal antibody may reduce its efficiency as compared to that of a group of antibodies recognizing a number of different sites. This, however, has not been rigorously examined.

3. ANTIBODY BINDING AGENT

The antibody binding agent should be readily available, should usually be of high affinity, and should confer on the immunoadsorbent a high binding capacity.

A number of different agents are available for covalent linking to the solid matrix. These include protein A, protein G, anti-immunoglobulin antibodies as second (or linking) antibodies, and avidin. Protein A (from *Staphylococcus aureus*) binds to the F_c region of many immunoglobulin subclasses (Kronvall *et al.*, 1970). Most of those immunoglobulins not recognized by protein A with high affinity (e.g., mouse IgG1, human IgG3, and most sheep and bovine IgGs) can be detected with protein G from *Streptococcus* spp., the exceptions being rat, chicken, and human IgA and IgD. Even these immunoglobulins can be bound by the use of another antibody, for example, protein A, covalently linked to the matrix and bound to an anti-immunoglobulin antibody which in turn will recognize the primary antibody (Schneider *et al.*, 1982). However, when anti-immunoglobulin antibodies are used, it is more usual to couple them directly to the matrix. As with the primary antibody, secondary antibodies may be either affinity-purified polyclonal or monoclonal antibodies. By the same reasoning outlined above, it is likely that affinity-purified polyclonal antibodies are more suitable since they recognize more antigenic sites. Indeed, most commercially available immunoadsorbents are of this type. Monoclonal second antibodies have one great advantage in that they are available in almost unlimited quantities from one hybridoma cell line.

Avidin has great potential as an agent for binding biotinylated antibodies since the binding reaction is of very high affinity ($K_a = 10^{15}$ M^{-1}). Immobilized avidin could be used as a "universal" immunoadsorbent capable of binding a panel of biotinylated antibodies. The use of a commercially available disulfide derivative of biotin would also permit the elution of bound organelle under relatively mild reducing conditions (Shimkus *et al.*, 1985).

The choice of antibody binding agent is a major factor in determining the binding capacity of an immunoadsorbent. For instance, polyclonal antibodies (not affinity purified) on a cellulose matrix had a binding capacity of approximately 1 μg primary antibody per mg matrix (Luzio *et al.*, 1976). Substitution of a monoclonal antibody for the polyclonal antibodies raised this to 250 μg/mg (Richardson *et al.*, 1984). The antibody binding agents discussed are all suitable for use on immunoadsorbents and are readily available so the major determinant of the immunoadsorbent characteristics is the choice of the solid support.

4. THE SOLID SUPPORT

A variety of solid supports have been used for immunoaffinity purification, including bacterial cells bearing protein A (Roman and Hubbard, 1984), Sepharose (Ghetie *et al.*, 1978), polyacrylamide (Matthew *et al.*, 1981), polystyrene (de Curtis *et al.*, 1988), and cellulose (Luzio *et al.*, 1974). The optimum characteristics for a solid support are high binding capacity, low nonspecific binding, low porosity, flexibility, and ease of separation from unbound organelles.

A high binding capacity of the final immunoadsorbent serves a number of useful purposes. For instance, it reduces the competition by free primary antibody for the binding sites on the solid phase. After sensitization of the organelle and subsequent washing, it is likely that such free antibody will still be present. This would compete effectively with organelle-bound antibody for the immunoadsorbent binding sites, should the number of these sites be limiting. Similarly, the binding of the sensitized organelle to the solid support will be assisted by the presence of binding sites covering the surfaces of both. Finally, the amount of nonspecific binding is to some extent proportional to the amount of solid phase present (Miljanich *et al.*, 1982). Therefore, small amounts of high-capacity immunoadsorbent should be used wherever possible. This is particularly important when polyclonal primary antibodies are used since more immunoglobulin tends to be used than with monoclonal antibodies.

Although the antibody binding agents discussed in the previous section have different intrinsic binding capacities (i.e., protein A and avidin can bind 4 moles/mole, while the linking IgG antibodies can only bind 2 moles/mole), the major determinant of the binding capacity of the immunoadsorbent is the nature of the matrix. For example, cellulose (Luzio *et al.*, 1974; Richardson *et al.*, 1984), when activated by diazotization (Hales and Woodhead, 1980), has one reactive group every 41 glucose residues, corresponding to a binding capacity of approximately 150 μequiv/g (Addison, 1971). This is ten times greater than the binding capacity of cyanogen bromide-activated Sepharose. Polyacrylamide and polystyrene beads have intermediate binding capacities and, with cellulose, are now the most widely used solid supports.

Nonspecific binding of organelle to the solid support is the major limiting factor in immunoaffinity purification. There appears to be little intrinsic difference between cellulose, polyacrylamide, and Sepharose in terms of nonspecific binding. However, when low-binding-capacity immunoadsorbents are used, nonspecific binding tends to increase with the increasing amounts of solid phase required. Nonporous supports are used to prevent access of contaminating molecules to the core of the matrix, since excessive washing would be required to remove them. In addition, the coupling of the antibody binding agent to the inside of the matrix would serve no useful purpose since the binding sites would be inaccessible to the organelles. It is thought that a flexible solid support may assist the binding of organelles. Cellulose is particularly flexible (Krassig, 1985) as compared to polyacrylamide or polystyrene.

The ease, and completeness, of separation of bound from unbound organelles has determined the choice of solid support in many cases. Currently, two major methods are used—centrifugation and magnetism. Centrifugation methods rely on the fact that most solid supports are denser than the organelles being isolated. They are used in most polyacrylamide-mediated immunoadsorption protocols and all of those using cellulose or bacterial cells. Rapid separations (a few minutes) at low g forces ($10,000 \times g$ or less) are routinely obtained. However, in those cases where contaminating organelles may be sedimented, other procedures are required. For instance, in the isolation of cholinergic nerve terminals, unacceptable quantities of unbound nerve terminals co-sedimented with the immunoadsorbent at $10,000 \times g$ in saline media. To prevent this, the density of the medium was increased by the inclusion of 45% (v/v) Percoll. This resulted in the complete separation of bound from unbound nerve terminals (Richardson et al., 1984). One other complication may also be encountered in situations where high-capacity immunoadsorbents are used in the presence of an excess of organelle. When very high binding of organelle to the solid support occurs, the density of the support tends toward that of the organelle, thus destroying the separation protocol. This can be avoided by reducing the organelle/solid support ratio or by using magnetic separation.

The recent advent of monosized magnetic polystyrene beads has led to a great increase in the use of magnetic separation techniques (see Ugelstad et al., this volume, Chapter 16). A variety of other magnetic particles have been used, including polyacrylamide-agarose (Docherty et al., 1987), dextran (Molday and Molday, 1984), and polyacrolein (Howell et al., 1985). The original hope of using magnetic particles in continuous-flow systems, with the particles kept in suspension by a magnetic field, has not been realized. Such an approach requires very strong magnetic fields (Molday and Molday, 1984) or specifically designed "magnetic bottles" (Howell et al., 1985). Consequently, most magnetic separation techniques are used in batch processes, the magnetic particles being separated using a small hand-held magnet.

5.   RECOMMENDATIONS

The primary antibody should be monoclonal or a polyclonal that has been affinity purified on the antigen being recognized. The minimum amount of antibody that is consistent with a reasonable yield should be used. In those cases where low-density or labile organelles are to be isolated, the primary antibody should be prebound to the solid phase. If only unpurified polyclonal primary antibodies are available, then a high-capacity immuno-adsorbent must be used (i.e., cellulose). Where possible and especially when relatively dense organelles are present, magnetic separation is recommended.

Obviously, in order to assess the efficiency and yield of the separation protocol, suitable markers should be measured. When assessing nonspecific binding, markers for contaminating organelles should be measured as well as the yield of the desired organelle in the absence and presence of the primary antibody. The efficiency of the separation can be most easily expressed as the ratio of the percent yield of the desired component to the percent yield of the contaminating organelle. This may be problematic as the organelle markers may not be absolutely specific, the recognized antigen may not be organelle specific, or different organelles may give different degrees of nonspecific binding (Luzio *et al.*, 1976). However, the ratio of the percent yield of the desired component in the presence and absence of the primary antibody will given an indication of the nonspecific binding. From this, it can also be deduced whether the primary antibody recognizes more than one organelle.

In many situations immunoaffinity protocols do not "work" the first time. In our experience, it is worth checking primary antibody binding to organelle and immunoadsorbent independently and to vary the amounts of antibody and immunoadsorbent used until the optimum conditions are obtained. The immunoadsorbent requires washing after binding the organelle. Washing procedures should be mild to prevent organelle disruption, while the use of a small amount of immunoadsorbent reduces the number of washes required.

6. APPLICATIONS OF IMMUNOAFFINITY-PURIFIED SUBCELLULAR FRACTIONATION

As mentioned earlier, immunoaffinity protocols are likely to be most useful for the separation of functionally distinct but physically similar organelles from tissue homogenates. Some of the applications are listed in Table 1. The large number of plasma membrane isolations reflects the development of these procedures and, in the absence of monoclonal antibodies, the ease with which polyclonal antibodies could be "affinity-purified" on the surface of intact cells (Luzio *et al.*, 1976). Subsequently, more specific primary antibodies have permitted the examination of intracellular membrane traffic pathways, as well as the subfractionation of the complex homogenates of mammalian brain.

In the brain these procedures have been employed for a number of purposes which exemplify their usefulness. For example, Docherty *et al.* (1987) have discovered a series of apparently neurotransmitter-specific plasma membrane proteins that have epitopes in common with the corresponding cytoplasmic transmitter-synthesizing enzymes. This has permitted the isolation of transmitter-specific nerve terminals using primary antibodies raised against the synthetic enzymes. Using a different approach, immunoaffinity-purified nerve terminals containing acetylcholine (Richardson *et al.*, 1984) have been used to investigate certain specific aspects of cholinergic neurobiology. These include the demonstration of the release of ATP from these terminals, and its subsequent extracellular degradation to adenosine, which feedback-regulated acetylcholine release (Richardson and Brown, 1987; Richardson *et al.*, 1987; Brown *et al.*, 1990). This would not have proved possible without prior purification of these nerve terminals since ATP and its metabolites are distributed throughout all of the nerve terminals of the brain. As with the experiments of Docherty *et al.* (1987), it has been possible to demonstrate the presence of specific molecules on (or in) the cholinergic nerve terminal. For instance, the A_2 adenosine receptor has been shown to be restricted to these nerves (Fig. 2). By varying the amount of primary

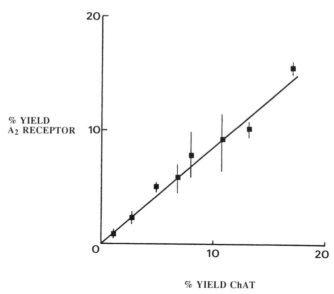

% YIELD ChAT

FIGURE 2. Copurification of the adenosine A_2 receptor with rat striatal cholinergic nerve terminals. Choline acetyltransferase (ChAT) activity was used as a measure of the cholinergic nerve terminals while the A_2 receptor was measured by N-ethylcarboxamidoadenosine stimulation of adenylate cyclase. The yield of nerve terminals was varied by adjusting the amount of primary antibody used. [Redrawn from Brown et al. (1990).]

antibody used, the yield of cholinergic nerve terminal was controlled and correlated with the yield of the A_2 receptor (Brown et al., 1990). It can be seen that the maximum yield obtained was approximately 20%. Under the conditions used, nonspecific binding of other (noncholinergic) nerve terminals increased to unacceptable values (greater than 2%) at higher yields.

Immunoaffinity procedures have also been used to analyze the protein and transmitter content of synaptic vesicles. For instance, Floor and Feist (1989) demonstrated the presence of the vesicle proteins SV2, synaptophysin, and p65 on all the light synaptic vesicle populations derived from mammalian brain (but see Volknandt et al., 1988). Burger et al. (1989) were able to demonstrate the presence of the excitatory transmitter glutamate in immunoisolated synaptic vesicles. This was made possible partly by the speed of the affinity purification procedure (less than 2 h) and was the first convincing demonstration that this transmitter was indeed stored in vesicles. In a different approach, Pfeffer and Kelly (1985) were able to demonstrate the localization of synaptic vesicle proteins (SV2) in coated vesicles, suggesting that the vesicle membrane may be retrieved from the synaptic plasma membrane via clathrin-coated pits. Using immunoaffinity procedures, the same group demonstrated the presence of three synaptic vesicle antigens (p38, p65, and SV2) on chromaffin granule membranes (Lowe et al., 1988).

The study of intracellular membrane traffic pathways is bedeviled by the difficulty of separating physically similar smooth membrane fractions and by the lack of defined endogenous markers for many of these fractions. Although immunoaffinity techniques offer a solution to the former problem, they are themselves dependent on the presence of specific antigens in the membrane of interest. Such antigens should at best be unique, though it is often possible to isolate a compartment containing a high concentration of an antigen that is present at lower concentration in other membranes. A further refinement is to use immunoaffinity isolation after a centrifugation step so that contaminating membranes containing the antigen of interest at high concentration are removed before the immuno-affinity step.

A good example of the use of antibodies to an endogenous membrane protein to isolate membrane-bound compartments on membrane traffic pathways is provided by the immuno-isolation of hepatocyte subcellular fractions using antibodies to the cytoplasmic tail of the polymeric IgA receptor (pIgAR) Salamero et al., 1990; Sztul et al., 1991). This receptor is synthesized on endoplasmic reticulum (ER)-bound polysomes and inserted into the ER membrane before passing the Golgi complex. Here it is terminally glycosylated before being routed to the basolateral cell surface (blood sinusoidal surface). From this surface the receptor is endocytosed, with or without bound ligand, by the coated pit route and enters the early, peripheral endosome compartment in common with other receptors and ligands entering the cell. In contrast to these other molecules, the pIgAR, with or without bound ligand, is sorted into a transcytic route (Geuze et al., 1984). It appears to pass across the cell in a transcytic carrier vesicle which fuses with the apical cell surface (bile canalicular surface), where the receptor is proteolytically cleaved, releasing a receptor fragment (secretory component) or a fragment bound to pIgA (secretory IgA) into the bile. The receptor is a type I membrane protein with a single transmembrane spanning domain and a carboxy-terminal cytoplasmic tail. There is currently much interest in tail modifications that may result in the provision of targeting signals and also interest in determining the sites in the membrane traffic pathway where such modifications occur (Breitfeld et al., 1989; Casanova et al., 1990). The pIgAR is present at high concentration in the Golgi complex and in transcytic carrier vesicles. Antibodies to the cytoplasmic tail have been used in the immunoaffinity purification of both, starting in each case with fractions partially purified by conventional means (Salamero et al., 1990; Sztul et al., 1991). The immobilized Golgi complexes have been used to establish cell-free assays to determine conditions to generate a population of vesicles from the trans-Golgi network that contain the pIgAR and >70% of newly synthesized, sialylated secretory proteins. Such vesicles mediate the constitutive pathway of secretion to the basolateral plasma membrane (Salamero et al., 1990). The purification of transcytic carrier vesicles should similarly make it possible to establish cell-free assays and define the conditions for transcytic carrier vesicle fusion with the apical plasma membrane. Proteolytic cleavage of the pIgAR could be used as an assay for the fusion event.

Whereas endogenous proteins such as the pIgAR may be used as target antigens for immunoaffinity isolation of intracellular membrane fractions, an alternative approach is to introduce an exogenous antigen into the membrane traffic pathway of interest. The most successful example of this approach has been the use of implanted vesicular stomatitis virus (VSV) G-protein to isolate endosome compartments from cultured cells (Gruenberg and Howell, 1987). Implantation of VSV G-protein is carried out at the cell surface at 4°C. Subsequent warming of the cells results in endocytosis, and the cytoplasmic tail of VSV G-protein can then be used as antigen to isolate endocytic fractions. These fractions may be time-defined in that they can be prepared at different times after internalization. The fractions may be used in cell-free assays to identify conditions for interaction and fusion with the same and other endocytic compartments. In fact, fusion activity is maximal within 5 minutes of internalization and decreases as markers are chased in vivo to later stages of the pathway (Gruenberg and Howell, 1987, 1989). Experiments of this sort have led to the concept of "carrier vesicles" being involved in transfer of material from one compartment of the endocytic pathway to another, later compartment (Griffiths and Gruenberg, 1991). However, this is a controversial area, and other workers support the alternative hypothesis of maturation of early endosomes to become late endosomes (Murphy, 1991). Such difficulties may only be resolved by identifying unique membrane proteins in particular compartments if they exist. Recently, strategies involving the use of expression cDNA libraries have been developed in order to identify and characterize integral membrane antigens located at particular sites on membrane traffic pathways (Brake *et al.*, 1990; Luzio *et al.*, 1990). The identification of such antigens should provide a new impetus to the immunoisolation of intracellular organelles.

7. CONCLUSION

In recent years immunoaffinity procedures have become an increasing part of the repertoire of cell biologists. With the advent of commercially available good-quality immunoadsorbents from a variety of companies (e.g., Dynal, Pharmacia, Bio-Rad, Pierce, etc.) and the widespread use of monoclonal antibodies, the immunoisolation of organelles has become a standard technique. As more specific primary antibodies become available, it is likely that these procedures will be more widely used both as an adjunct to subcellular fractionation and to assess the distribution or localization of specific antigens.

ACKNOWLEDGMENTS

Experimental work by the authors reported in this chapter was supported by grants from the Medical Research Council, the British Diabetic Association, the Beit Foundation, St John's College Cambridge, and the Muscular Dystrophy Group of Great Britain and Northern Ireland.

REFERENCES

Addison, G. M., 1971, Preparation and Properties of Labelled Antibodies, Ph.D. Thesis, University of Cambridge.

Albertson, P.-A., 1986, *Partition of Cell Particles and Macromolecules*, John Wiley & Sons, New York.

Biher, J. W., and Lienhard, G. E., 1986, Isolation of vesicles containing insulin responsive intracellular glucose transports from 3T3-Li adipocytes, *J. Biol. Chem.* **261**:16180–16184.

Brake, B., Braghetta, P., Banting, G., Luzio, J. P., and Stanley, K. K., 1990, A new recombinant DNA strategy for the molecular cloning of rare membrane proteins, *Biochem. J.* **267**:631–637.

Breitfeld, P. P., Casanova, J. E., Simister, N. E., Ross, S. A., McKinnon, W. C., and Mostov, K. E., 1989, Sorting signals, *Curr. Opin. Cell Biol.* **1**:617–623.

Brown, S. J., James, S., Reddington, M., and Richardson, P. J., 1990, Both A_1 and A_{2a} receptors regulate striatal acetylcholine release, *J. Neurochem.* **55**:31–38.

Burger, P. M., Mehl, E., Cameron, P. L., Maycox, P. R., Baumert, M., Lottspeich, F., DeCamilli, P., and Jahn, R., 1989, Synaptic vesicles immunoisolated from rat cerebral cortex contain high levels of glutamate, *Neuron* **3**: 715–720.

Casanova, J. E., Breitfeld, P. P., Ross, S. A., and Mostov, K. E., 1990, Phosphorylation of the polymeric immunoglobulin receptor required for its efficient transcytosis, *Science* **248**:742–745.

Debanne, M. T., Bolyos, M., Gauldie, J., and Regoeczi, E., 1984, Two populations of prelysosomal structures transporting asialoglycoproteins in rat liver, *Proc. Natl. Acad. Sci. U.S.A.* **81:**2995–2999.

deCurtis, I., and Simons, K., 1989, Isolation of exocytic carrier vesicles from BHK cells, *Cell* **58:**719–727.

deCurtis, I., Howell, K. E., and Simons, K., 1988, Isolation of a fraction enriched in the trans-Golgi network from BHK cells, *Exp. Cell Res.* **175:**248–265.

de Duve, C., 1975, Exploring cells with a centrifuge, *Science* **189:**186–194.

Devaney, E., and Howell, K. E., 1985, Immunoisolation of a plasma membrane fraction from the Fao cell, *EMBO J.* **4:**3123–3130.

Docherty, M., Bradford, H. F., and Wu, J.-Y., 1987, Corelease of glutamate and aspartate from cholinergic and GABAergic synaptosomes, *Nature (London)* **330:**64–66.

Floor, E., and Feist, B. E., 1989, Most synaptic vesicles isolated from rat brain carry three membrane proteins, SV2, synaptophysin and p65, *J. Neurochem.* **52:**1433–1437.

Floor, E., and Leeman, S. E., 1985, Evidence that large synaptic vesicles containing Substance P and small synaptic vesicles have a surface antigen in common in rat, *Neurosci. Lett.* **60:**231–237.

Geuze, H. J., Slot, J. W., Strous, G. J. A. M., Peppard, J., von Figura, K., Hasalik, A., and Schwartz, A. L., 1984, Intracellular receptor sorting during endocytosis: Comparative immunoelectronmicroscopy of multiple receptors in rat liver, *Cell* **37:**195–204.

Ghetie, V., Mota, A., and Sjoquist, J., 1978, Separation of cells by affinity chromatography on SpA-Sepharose 6MB, *J. Immunol. Methods* **21:**133–141.

Gorvel, J.-P., and Maroux, S., 1988, Characterization of intestinal membrane vesicles with flow cytometry, in: *Cell Free Analysis of Membrane Traffic* (J. Moore, ed.), Alan R. Liss, New York, pp. 195–210.

Gray, E. G., and Whittaker, V. P., 1962, The isolation of nerve endings from brain: An electronmicroscopic study of all fragments derived by homogenization and centrifugation, *J. Anat.* **96:**79–87.

Griffiths, G., and Gruenberg, J., 1991, The arguments for pre-existing early and late endosomes, *Trends Cell Biol.* **1:**5–9.

Gruenberg, J., and Howell, K. E., 1985, Immunoisolation of vesicles using antigenic sites either located on the cytoplasmic or exoplasmic domain of an implanted viral protein. A quantitative analysis, *Eur. J. Cell Biol.* **38:**312–321.

Gruenberg, J., and Howell, K. E., 1986, Reconstitution of vesicle fusions occurring in endocytosis with a cell-free system, *EMBO J.* **5:**3091–3101.

Gruenberg, J., and Howell, K. E., 1987, An internalized transmembrane protein resides in a fusion-competent endosome for less than 5 minutes, *Proc. Natl. Acad. Sci. U.S.A.* **84:**5758–5762.

Gruenberg, J., and Howell, K. E., 1989, Membrane traffic in endocytosis: Insights from cell-free assays, *Annu. Rev. Cell Biol.* **5:**453–481.

Gruenberg, J., Griffiths, G., and Howell, K. E., 1989, Characterization of the early endosome and putative endocytic carrier vesicles *in vivo* with an assay of vesicle fusion *in vitro*, *J. Cell Biol.* **108:**1301–1316.

Hales, C. N., 1972, Immunological techniques in diabetes research, *Diabetologia* **8:**229–235.

Hales, C. N., and Woodhead, J. S., 1980, Labelled antibodies and their use in the immunoradiometric assay, *Methods Enzymol.* **70:**334–355.

Howell, K. E., Ansorge, W., and Gruenberg, J., 1985, Immunoisolation system using beads maintained in free flow within a magnetic field, in: *Microspheres: Medical and Biological Applications* (A. Rembaum and Z. Tokes, eds.), CRC Press, Boca Raton, Florida, pp. 33–52.

Howell, K. E., Devaney, E., and Gruenberg, J., 1989, Subcellular fractionation of tissue culture cells, *Trends Biochem. Soc.* **14:**44–47.

Ito, A., and Palade, G. E., 1978, Presence of NADPH-cytochrome P-450 reductase in rat liver Golgi membranes. Evidence obtained by immunoadsorption method, *J. Cell Biol.* **79:**590–597.

Jones, R. T., Walker, J. H., Richardson, P. F., Fox, G. W., and Whittaker, V. P., 1981, Immunohistochemical localization of cholinergic nerve terminals, *Cell Tissue Res.* **218:**355–373.

Kawajiri, K., Ito, A., and Omura, T., 1977, Subfractionation of rat liver microsomes by immunoprecipitation and immunoadsorption methods, *J. Biochem.* **81:**779–789.

Kohler, G., and Milstein, C., 1975, Continuous cultures of fused cells secreting antibody of predefined specificity, *Nature (London)* **256:**495–497.

Krassig, H., 1985, Structure of cellulose and its relation to properties of cellulose fibres, in: *Cellulose and Its Derivatives: Chemistry, Biochemistry and Applications* (J. F. Kennedy, G. O. Phillips, D. J. Wedlock, and P. A. Williams, eds.,), Ellis Horwood, Chichester, U.K., pp. 3–25.

Kronvall, G., Seal, V. S., Finstad, J., and Williams, R. C., 1970, Phylogenetic insight into evolution of mammalian F_c fragment of γG globulin using staphylococcal protein A, *J. Immunol.* **104:**140–147.

Lowe, A. W., Madeddie, L., and Kelly, R. B., 1988, Endocrine secretory granules and neuronal synaptic vesicles have three integral membranes in common, *J. Cell Biol.* **106**:51–59.

Luzio, J. P., 1977, Immunological approaches to the study of membrane features in adipocytes, in: *Methodological Surveys in Biochemistry*, Vol. 6 (E. Reid, ed.), Ellis Horwood, Chichester, U.K., pp. 131–142.

Luzio, J. P., and Stanley, K. K., 1983, The isolation of endosome derived vesicles from rat hepatocytes, *Biochem. J.* **216**:27–36.

Luzio, J. P., Newby, A. C., and Hales, C. N., 1974, Immunological isolation of rat fat cell plasma membranes, *Biochem. Soc. Trans.* **2**:1385–1386.

Luzio, J. P., Newby, A. C., and Hales, C. N., 1976, A rapid immunological procedure for the isolation of hormonally sensitive rat fat cell plasma membrane, *Biochem. J.* **154**:11–21.

Luzio, J. P., Brake, B., Banting, G., Howell, K. E., Braghetta, P., and Stanley, K. K., 1990, Identification, sequencing and expression of an integral membrane protein of the trans-Golgi network (TGN38), *Biochem. J.* **270**:97–102.

Matthew, W. D., Tsavaler, L., and Reichardt, L. F., 1981, Identification of a synaptic vesicle-specific membrane protein with a wide distribution in neuronal and neurosecretory tissue, *J. Cell Biol.* **91**:257–269.

Merisko, E. M., Farquhar, M. G., and Palade, G. E., 1982, Coated vesicle isolation by immunoadsorption on *Staphylococcus aureus* cells, *J. Cell Biol.* **92**:846–857.

Miljanich, G. P., Brasier, A. R., and Kelly, R. B., 1982, Partial purification of presynaptic plasma membrane by immunoadsorption, *J. Cell Biol.* **84**:88–96.

Molday, R. A., and Molday, L. L., 1984, Separation of cells labelled with immunospecific iron dextran microspheres using high gradient magnetic chromatography, *FEBS Lett.* **170**:232–238.

Mueller, S. C., and Hubbard, A. L., 1986, Receptor-mediated endocytosis of asialoglycoproteins by rat hepatocytes: Receptor-positive and receptor-negative endosomes, *J. Cell Biol.* **102**:932–942.

Murphy, R. F., 1991, Maturation models for endosome and lysosome biogenesis, *Trends Cell Biol.* **1**:77–82.

Obrocki, J., and Borroni, E., 1988, Immunocytochemical evaluation of a putative cholinergic specific ganglioside antigen (Chol-1) in the central nervous system of the rat, *Exp. Brain Res.* **72**:71–82.

Pfeffer, S. R., and Kelly, R. B., 1985, The subpopulation of brain coated vesicles that carries synaptic vesicle proteins contains two unique polypeptides, *Cell* **40**:949–957.

Pontremoli, S., Diamine, G., Michetti, M., Salamino, F., Sparatore, B., and Honecker, B. L., 1984, Binding of monoclonal antibody to Cathepsin M located on the external surface of rabbit lysosomes, *Arch. Biochem. Biophys.* **233**:267–271.

Reijnierse, L. A., Velstra, H., and Van den Berg, C. J., 1975, Subcellular localization of γ-aminobutyrate transaminase and glutamate dehydrogenase in adult rat brain, *Biochem. J.* **152**:469–475.

Richardson, P. J., and Brown, S. J., 1987, ATP release from affinity purified cholinergic nerve terminals, *J. Neurochem.* **48**:622–630.

Richardson, P. J., Siddle, K., and Luzio, J. P., 1984, Immunoaffinity purification of intact metabolically active, cholinergic nerve terminals from mammalian brain, *Biochem. J.* **219**:647–654.

Richardson, P. J., Brown, S. J., Bailyes, E. M., and Luzio, J. P., 1987, Ectoenzymes control adenosine modulation of immunoisolated cholinergic synapses, *Nature (London)* **327**:232–234.

Roman, L. R., and Hubbard, A. L., 1984, A domain specific marker for the hepatocyte plasma membrane. III. Isolation of bile canalicular membrane by immunoadsorption, *J. Cell Biol.* **98**:1497–1504.

Salamero, J., Sztul, E. S., and Howell, K. E., 1990, Exocytic transport vesicles generated *in vitro* from the trans-Golgi network carry secretory and plasma membrane proteins, *Proc. Natl. Acad. Sci. U.S.A.* **87**:7717–7721.

Shimkus, M., Levy, T., and Herman, T., 1985, A chemically cleavable biotinylated nucleotide: Usefulness in the recovery of protein–DNA complexes from avidin affinity columns, *Proc. Natl. Acad. Sci. U.S.A.* **82**:2593–2597.

Schneider, C., Newman, R. A., Sutherland, D. R., Asser, V., and Greaves, M. F., 1982, A one step purification of membrane proteins using a high efficiency immunomatrix, *J. Biol. Chem.* **257**:10766–10769.

Sharp, K. A., Yalpani, M., Howard, S. J., and Brooks, D. E., 1986, Synthesis and application of a polyethylene-glycol-antibody affinity ligand for cell separations in aqueous two phase systems, *Anal. Biochem.* **154**:110–117.

Stanley, K. K., Newby, A. C., and Luzio, J. P., 1982, What do ectoenzymes do? *Trends Biochem. Sci.* **7**:145–147.

Sztul, E., Kaplin, A., Saucan, L., and Palade, G., 1991, Protein traffic between distinct plasma membrane domains: Isolation and characterization of vesicular carriers involved in transcytosis, *Cell* **64**:81–89.

Ugelstadt, J., Soderberg, L., Berge, A, and Bergstrom, J., 1983, Monodisperse polymer particles—a step forward for chromatography, *Nature (London)* **303**:95–96.

Volknandt, W., Henkel, A., and Zimmermann, H., 1988, Heterogeneous distribution of synaptophysin and protein 65 in synaptic vesicles isolated from rat cerebral cortex, *Neurochem. Int.* **12:**337–345.

Wandinger-Ness, A., Bennett, M. K., Antony, C., and Simons, K., 1990, Distinct transport vesicles mediate the delivery of plasma membrane proteins to the apical and basolateral domains of MDCK cells, *J. Cell Biol.* **111:** 987–1000.

Westwood, S. A., Luzio, J. P., Flockhart, D. A., and Siddle, K., 1979, Investigation of the subcellular distribution of cyclic AMP phosphodiesterase in rat hepatocytes, using a rapid immunological procedure for the isolation of plasma membrane, *Biochim. Biophys. Acta* **583:**454–466.

Whittaker, V. P., Michaelson, I. A., and Kirkland, R. J. A., 1964, The separation of synaptic vesicles from disrupted nerve ending particles, *Biochem. J.* **90:**293–303.

Zorzano, A., Wilkinson, W., Kotliar, N., Thoidis, G., Wadzinski, B. E., Ruoho, A. E., and Pilch, P. F., 1989, Insulin-regulated glucose uptake in rat adipocytes is mediated by two transporter isoforms present in at least two vesicle populations, *J. Biol. Chem.* **264:**12358–12363.

16

Immunoaffinity Separation of Cells Using Monosized Magnetic Polymer Beads

John Ugelstad, Ørjan Olsvik, Ruth Schmid, Arvid Berge,
Steinar Funderud, and Kjell Nustad

1. INTRODUCTION

A widespread biomedical use of magnetic particles has recently been initiated as a result of the preparation of a new generation of magnetic particles by Ugelstad *et al.* (1980). Originally, these particles were used for removal of tumor cells from bone marrow, and selective isolation of cells remains a very important area of application. More recently, their area of application has been extended to include prokaryotic cells, viruses, and subcellular components. A new and rapidly increasing area of application for magnetic beads is in different fields of DNA technology; this is discussed in Chapter 31 of this volume.

Reviews of the preparation of the different types of magnetic particles (Platsoucas, 1987) and use of different particles in cell separation have recently been published (Padmanabhan *et al.*, 1989; Kemshead *et al.*, 1990; Trickett, 1990).

The magnetic beads to be used in cell separation should fulfill the following criteria:

1. They should be chemically stable, and they should not aggregate in the media used in cell separation.
2. They should show very little magnetic remanence after having been subject to the magnetic field.
3. They should not bind to cells nonspecifically.
4. There should be very little leakage of antibody from the particles during storage.
5. They should allow a fast and complete magnetic separation of the cells labeled with particles and the excess of particles from the unlabeled cells.
6. They should be of a size which minimizes phagocytosis.

John Ugelstad, Ruth Schmid, and Arvid Berge • SINTEF, Applied Chemistry, N-7034 Trondheim, Norway. *Steinar Funderud and Kjell Nustad* • The Norwegian Radium Hospital, Montebello, N-0310 Oslo 3, Norway. *Ørjan Olsvik* • Enteric Diseases Branch, Division of Bacterial and Mycotic Diseases, Centers for Disease Control, Atlanta, Georgia 30333, and Department of Microbiology and Immunology, Norwegian College of Veterinary Medicine, Oslo 1, Norway.

Molecular Interactions in Bioseparations, edited by That T. Ngo. Plenum Press, New York, 1993.

2. MONOSIZED SUPERPARAMAGNETIC POLYMER PARTICLES (DYNABEADS)

The particles that best meet the combined set of requirements given above are the monosized superparamagnetic polymer particles developed by Ugelstad *et al.* (1980), which are now marketed as "Dynabeads." These particles are based on monosized polymer particles prepared by the method of activated swelling, which allows an easy preparation of polymer particles in sizes from 1 to 100 μm with a very narrow size distribution. The superparamagnetic particles made on the basis of these particles are prepared by the reaction of Fe^{2+} from an aqueous solution with preformed oxidizing groups covalently bound to the polymer throughout the entire volume of the particles. The magnetic iron oxides in the form of γ-Fe_2O_3 or Fe_3O_4 are present as very small grains evenly distributed throughout the whole volume of the particles, which ensures that the particles become superparamagnetic (Berge *et al.*, 1990). The particles may be produced with different chemical structures on the surface and especially with groups that allow covalent binding of the different ligands one wants to attach to the particles. The groups on the particles include —OH, —NH_2, —COOH, —SH, and aldehyde groups.

The particles which are available commercially at present are the M-450 and the M-280 particles shown in Figs. 1 and 2. The properties of these particles are given in Table 1. Of these particles, the M-450 particles are the most widely used for cell separation while the M-280 particles up to now have been mainly applied in diagnostics and in DNA technology.

The present M-450 particles contain reactive epoxy compounds on the surface and need no further activation. They will react with —NH_2 and —SH groups in the ligands. The

FIGURE 1. Scanning electron micrograph of M-450 particles.

FIGURE 2. Scanning electron micrograph of M-280 particles.

M-280 particles have short-chain glycol groups, $-(CH_2-CH_2O)_{2-4}H$, on the surface and are preferentially activated with sulfonyl chlorides (Mosbach and Nilsson, 1982). The sulfonic acid esters formed in the first activating step, $-(CH_2-CH_2O)_{2-4}S(O)_2R$, may react with $-NH_2$ groups or with $-SH$ groups in the ligands to form stable $-CH_2-NH-L$ or $-CH_2-S-L$ bonds between particles and ligands.

$$-(CH_2-CH_2O)_{2-4}S(O)_2R + NH_2-L \rightarrow -(OCH_2-CH_2)_{2-4}-NH-L$$
$$-(CH_2-CH_2O)_{2-4}S(O)_2R + HS-L \rightarrow -(OCH_2-CH_2)_{2-4}-S-L$$

In systems with IgG and IgM antibodies as ligands, the ligands are very rapidly attached to the particles by physical adsorption, which is followed by a relatively slow chemical reaction (Ugelstad *et al.*, 1991). The latter is much faster with tresyl chloride than with tosyl chloride (and epoxy groups).

The M-450 particles are very stable and maintain their binding capacity for more than a year when stored in water at 4°C. The sulfonyl esters on the particles resulting from

TABLE 1
Properties of the M-450 and M-280 Particles

Particle type	Diameter (μm)	Surface area (m²/g)	Fe content (wt %)	Density (g/cm³)	Particles per mg
M-450	4.5	3–5	22	1.5	1.4×10^7
M-280	2.8	4–6	12	1.3	6.7×10^7

activation with sulfonyl chlorides have been found to be very stable; particles with tosyl esters are especially stable, losing less than 10% of their activity for covalent coupling of antibodies after one year of storage when stored in 75% ethanol–water. Activation with sulfonyl chlorides has also turned out to be very efficient for binding of streptavidin for subsequent reaction with biotinylated ligands.

3. IMMUNOMAGNETIC CELL SEPARATION OF EUKARYOTIC CELLS

Immunomagnetic separation of cells by means of the Dynabeads is described in a steadily increasing number of scientific papers. Recently, several review articles dealing with immunomagnetic cell separation with the Dynabeads have appeared (Ugelstad et al., 1988; Lea et al., 1988; Padmanabhan et al., 1989; Kemshead et al., 1990). A detailed description of the procedures to be used for fractionation of lymphocytes has been given (Funderud et al., 1987). The guidelines given in the latter paper are valid for most cell separations carried out with Dynabeads.

The immunomagnetic separation of cells may be carried out in two different ways:

1. Indirect method: cells are first incubated with mouse monoclonal antibodies and are subsequently rosetted with anti-mouse IgG-coated beads.
2. Direct method: the appropriate antibodies are attached directly to the beads. This method may involve a direct coupling of the monoclonal antibody (mAb) to the beads or the mAb is coupled to the beads through a spacer arm in the form of a polyclonal antibody directed against the Fc part of the mAb.

The methods of choice for some particular purposes are discussed below.

3.1. Removal of Cancer Cells from Bone Marrow

The immunomagnetic separation procedure for the separation of tumor cells from bone marrow for autologous bone marrow transplantation was initiated in 1979 by Kemshead, Rembaum, and Ugelstad, who used magnetic beads to purge bone marrow of neuroblasts. The first paper discussing the preliminary results of these investigations was published in 1984 (Treleaven et al., 1984). Since then, the technique has been extended to other tumors, and the method has been modified both as regards the beads applied and the method of performing the purging procedure. Bone marrow purging in connection with autologous bone marrow transplantation requires a combination of a high efficiency in removal of tumor cells and high recovery of hemopoietic stem cells.

For a complete list of the types of tumors where the efficiency of purging with Dynabeads has been evaluated, the reader is referred to the review papers cited above. More recent work includes the following tumor types: neuroblastoma (Combaret et al., 1989; Lopez et al., 1989; Kemshead et al., 1990), lymphoma (Kvalheim et al., 1988b, 1989; Geisler et al., 1989), leukemia (Janssen et al., 1989, 1990; Morgan et al., 1989; Murakami et al., 1990; Trickett, 1990), myeloma (Shimazaki et al., 1990), breast cancer (Shpall et al., 1990), and small cell lung cancer (Elias et al., 1990).

In all cases model experiments with marrow contaminated with tumor cells have been carried out. In some cases, such as neuroblastoma, lymphoma, and leukemia, purging with Dynabeads has been used clinically. The purging procedure has in most cases involved purging of the marrow by the indirect method. In this case the cell suspension is most often sensitized with a cocktail of monoclonal IgG antibodies before mixing with the magnetic beads. Prior sensitization of the target cells will ensure a proper orientation of the mAbs and an optimal number of interaction possibilities between particles and cells. The efficiency of purging in removal of target cells will depend on the antigen expression, on the specificity and avidity of the monoclonal antibodies, and on the choice of the secondary anti-mouse antibody. There are, in addition, some important considerations which should be taken into account as regards the amount and the state of the secondary anti-mouse antibody used to sensitize the beads, the ratio of beads to target cells, and the concentration of particles and cells in the suspension during the incubation of the cells with the beads. With the M-450 beads, the concentration of the beads in the incubation mixture should preferentially be in the range of 10^8 beads/ml, corresponding to 7 mg of particles/ml. This concentration of beads should be applied regardless of the percent of target cells in the cell mixture. A too low concentration of beads may lead to an insufficient rosetting of the cells within the time of the incubation period. The ratio of beads to target cells will vary depending on the population of target cells. With 10% target cells in the cell suspension, the optimal ratio of the number of beads to the number of target cells has been found to be between 10:1 and 50:1, which in this case would mean a ratio of number of beads to total number of cells of between 1:1 and 5:1. With a lower percentage of target cells, this ratio of beads to total number of cells should be maintained. Thus, in the case where the purging process is carried out twice, the same ratio of beads to total cells should be applied in the second purging process.

The efficiency of the purging process obtained with the Dynabeads varies somewhat for different types of target cells. Model experiments with different types of tumor cells added to the bone marrow have clearly indicated that it is beneficial to repeat the purging procedure once. The additional purging reduces the tumor cells an additional 1–2 logs. With the indirect method and with initially 10% target cells in the bone marrow, two times purging gives 3 to more than 5 logs depletion of tumor cells. The variation in efficiency of the purging is not the result of an irreproducible purging process but reflects the difference in the type of tumor cells to be removed and the difference in the efficiency of the antibodies which are applied.

It may be noted that Kvalheim *et al.* (1989) reported that the M-280 Dynabeads tend to be somewhat more efficient than the M-450 beads in purging of bone marrow for removal of B-lymphoma cells. Kvalheim *et al.* (1988a) found that they could achieve an extremely efficient removal of B-lymphoma (Rael) cells from bone marrow by applying a monoclonal IgM antibody (AB-4) directly attached to the M-450 beads. Model experiments with 10% contamination with B-lymphoma cells gave a 6 logs depletion of cancer cells. AB-4 is an HLA-DR-specific antibody that binds to most non-Hodgkin's lymphomas and leukemias. It has been found that bone marrow stem cells express HLA-DR antigens, and therefore treatment with magnetic beads sensitized with anti-HLA-DR antibodies might have been expected to severely deplete the stem cells and result in graft failure. Kvalheim *et al.* found that removal of cells carrying the Ab-4 epitope only slightly influenced the growth of the bone marrow cells as measured by stem cell assays.

Quite recently, some conflicting data have appeared from experiments with an alternative type of beads, where the purging procedure has been optimized for these beads and the efficiency of these beads has been compared with that of Dynabeads under nonoptimal conditions (Biéva *et al.*, 1989). Kemshead *et al.* (1990) compared the different types of beads for removal of tumor cells (including the beads recommended by Biéva *et al.*), applying the indirect method and varying the amounts of polyclonal IgG and the incubation conditions as well as the ratio of beads to tumor cells. They concluded that the Dynabeads are superior as regards both efficiency in removal of tumor cells and low nonspecific binding to irrelevant cells. Unlike Biéva *et al.*, Kemshead *et al.* make the obvious reservation that the possibility exists that they may not have found optimal conditions for all the different types of beads.

3.2. Depletion of Normal T Cells in Allografts

Elimination of graft versus host disease (GVHD) in allogeneic bone marrow transplantation requires an effective removal of T cells from the bone marrow of the donor.

Vartdal *et al.* (1987b) applied a variation of the direct method for T-cell depletion. The beads were coated with rabbit anti-mouse IgG. They were subsequently coated with T-cell-specific mAbs by incubation with a mixture of $CD2^+$ and $CD3^+$ antibodies. A 3 logs depletion of T cells was reported. Similar results have recently been obtained by Frame *et al.* (1989), Gee *et al.* (1989), and Knobloch *et al.* (1990).

A complete T-cell depletion in connection with allogeneic bone marrow transplantation for leukemia patients is complicated by the fact that leukemia patients who survive GVHD are reported to survive free of disease longer than patients without GVHD. This may reflect an antileukemic effect associated with GVHD, which is referred to as the graft versus leukemia effect. A better understanding of the precise role of T lymphocytes in mediating GVHD, graft failure, and graft versus leukemia effect in allogeneic bone marrow transplantation is needed. The aim of such studies is to define which subsets of T cells can be removed without impairing the graft versus leukemia effect and graft failure. Champlin *et al.* (1990) have recently reported results of depletion of $CD8^+$ T lymphocytes from the allograft and found that the graft versus leukemia effect was retained. Immunomagnetic cell separation, where one may selectively remove subsets of T cells, may possibly be of help in addressing the problems of T-cell depletion connected with allogeneic bone marrow transplantation.

3.3. Isolation of Stem Cells

Treatment with marrow stem cells is a possible alternative strategy to overcome problems of GVHD in allogeneic bone marrow transplantation and cancer relapse in autologous bone marrow transplantation. In this case, one attempts to isolate in pure form hemopoietic stem cells which are able to reconstitute the hemopoietic system of the recipient.

Cottler-Fox *et al.* (1990) prepared purified hemopoietic precursor cells by a negative selection technique using monoclonal antibodies against CD3, CD4, CD14, and CD16 and Dynabeads coated with sheep anti-mouse IgG to produce a mononuclear marrow cell fraction that comprised no more than 2% of the initial cell population yet contained all the hemopoietic progenitor cells originally present.

Positive isolation of stem cells for grafting represents the most promising alternative (Civin *et al.*, 1990). A selection process such as this would not need to be 100% effective to provide a tumor-free graft.

Most work on isolation of stem cells up to now has been concentrated on the use of a $CD34^+$ monoclonal antibody. The CD34 antigen is present on 1–4% of human bone marrow cells which have been shown to include hemopoietic stem cells. Civin *et al.* (1990) isolated cells with the use of Dynabeads covered with sheep anti-mouse antibody, where the cells initially were covered with CD34$^+$ monoclonal antibody. They were able to detach the beads with the use of the enzyme chymopapain.

Quite recently, Smeland *et al.* (1991) developed a new and very promising method for isolation of stem cells. The monoclonal CD34$^+$ antibody BI3C5 was coupled directly to the beads. After isolation of the CD34$^+$ cells with beads attached to them, the beads were detached from the cells by an anti-Fab antiserum. The recovery of the CD34$^+$ cells was 40–70% and the purity was >96%.

3.4. Quantification of Lymphocytes in Peripheral Blood

M-450 beads are being increasingly used for isolation of cell subsets directly from blood (Brinchmann *et al.*, 1988, 1989a,b, 1990). IgM mAbs specific for T cells (CD4, CD8) and B cells (CD19) have been attached directly to the beads. An IgG1 mAb specific for CD2 is coated as a second layer onto beads covered with a first layer of sheep anti-mouse polyclonal antibody. After isolation of the rosetted target cells, the cells are lysed with a detergent, and the number of cells counted in a hemocytometer after staining. The method is used clinically for determination of the ratio of T4 to T8 cells in AIDS patients.

Recently, Brinchmann *et al.* (1989a) have developed a reliable method for isolation of human immunodeficiency virus (HIV) from cultures of naturally infected CD4$^+$ T cells. Brinchmann *et al.* (1990) also were able to demonstrate the inhibitory effect of activated CD8$^+$ T cells on the replication of HIV in naturally infected CD4$^+$ T cells. The results suggest that activated CD8$^+$ T cells secrete a soluble inhibitor of HIV replication. CD8$^+$ T cells from patients with AIDS showed reduced or no such inhibitory activity.

Beads coated with various monoclonal antibodies have been proven to be a powerful tool in immunophenotyping. Such beads will form rosettes with the corresponding cell types in a sample. This forms the basis of a method developed to provide for a rapid estimation of malignant cells in a blood sample (Skjønsberg *et al.*, 1990).

3.5. Functional Studies of Cell Subsets Isolated by Immunomagnetic Cell Separation

The possibility to directly isolate pure subpopulations of B and T cells has opened the way for functional studies of special cell types. In many cases the beads do not interfere with the functions to be studied and therefore need not be removed. Using a low bead-to-cell ratio, properly selected antibodies and cultivation of the isolated cells overnight often results in particle-free cell subsets for further studies. Early work on this application has been reviewed by Lea *et al.* (1988).

Most of the recent work has been done on T-cell subsets. Tumor infiltrating lymphocytes have been isolated from brain tumors (Bosnes and Hischberg, 1989) and from human renal cell carcinoma (Rayman *et al.*, 1989), and their cytotoxic effect on allogeneic or autologous tumor tissue has been studied *in vitro*. Hovdenes and co-workers (Hovdenes, 1989; Hovdenes *et al.*, 1989a–c) isolated CD4$^+$ and CD8$^+$ T cells from patients with inflammatory arthritides. Various functions of cells from synovial fluid, synovial tissue, and peripheral blood were compared, and the expression of activation markers was studied. Successful isolation of T-cell receptor (TCR)γ/δ^+ lymphocytes by means of magnetic beads was described in several papers (Brandtzaeg *et al.*, 1989; Bosnes *et al.*, 1989; Parker *et al.*, 1990). Isolation of CD45$^+$ cells with different CD45 isoforms made it possible to distinguish naive and memory T cells (Pilarski *et al.*, 1989). Based on the isolation of CD3$^+$ T cells, Leivestad *et al.* (1988) described a simple and sensitive bioassay for the detection of interleukin-2 (IL-2) activity. Activated T cells express special activation antigens on their surface (e.g., CD25). CD25$^+$ cells were isolated by Lundin *et al.* (1989) and studied in cloning experiments. T-cell subsets were isolated using magnetic beads in order to study their functions in self-tolerance (Frangoulis *et al.*, 1989) and their response to alloantigens (Leivestad *et al.*, 1989).

Using mAbs specific for HLA-DR, -DQ, or -DP molecules and highly purified T-cell subsets obtained by the use of antibody-coated magnetic beads, van Els *et al.* (1990) identified the major phenotype and major histocompatibility complex restriction determinant usage of the T-helper cells activated in the acute phase of GVDH.

Armitage and Goff (1988) isolated highly purified populations of MHM6$^+$ and MHM6$^-$ tonsil B cells using anti-CD23 mAbs.

Funderud *et al.* (1990) developed a fast and efficient method for direct isolation of resting B cells from peripheral blood by the use of Dynabeads with directly attached monoclonal antibodies. In an extension of this work, Rasmussen *et al.* (1992) detached the beads from the cells by a method similar to the one described above for isolation of stem cells, that is, by treatment with antibodies directed against the Fab part of the monoclonal antibodies. Packages containing Dynabeads with monoclonal antibodies specific for CD4-CD8-CD19 directly coupled to the beads and separate samples of detaching antibody are marketed as DETACHaBEAD systems.

3.6. Tissue Typing

Traditionally, HLA-I and HLA-II typing in connection with organ transplantation has been carried out by a microcytotoxic assay which necessitated isolation of mononuclear cells from peripheral blood by density-gradient centrifugation. This is a time-consuming and tedious process including washing steps with centrifugations in between. Vartdal *et al.* (1987a, 1988) have developed a microcytotoxic assay for HLA typing based on immuno-magnetic cell separation. An IgM mAb directed against CD8 T cells (HLA class I) and an IgM mAb specific for an HLA class II monomorphic epitope were attached directly to the beads. The isolation of the cells for typing can be carried out very rapidly from small samples of blood. The presence of beads did not interfere with the subsequent microcytotoxic typing. The method has especially been found to offer new possibilities in cadaveric transplant donor–recipient HLA matching.

Hansen and Hannestad (1989) recently applied the magnetic beads for a direct HLA typing. In this procedure, human monoclonal antibodies to polymorphic HLA determinants are coated onto magnetic beads, and HLA typing is performed by evaluation of rosette formation between cells and beads.

3.7. Various Types of Cell Separation with Magnetic Beads

Magnetic beads are finding use in steadily growing fields of cell separation processes. These include isolation of trophoblast cells (Douglas and King, 1989; Loke *et al.*, 1989;

Mueller *et al.*, 1990), eosinophils (Hansel *et al.*, 1990), and Langerhans cells (Hanau *et al.*, 1988).

Quite recently, Dynabeads have been applied for isolation of endothelial cells from whole blood. George *et al.* (1991) showed that endothelial cells even at low frequency could be isolated from blood using an IgG1 monoclonal antibody (S-Endo1) and Dynabeads and suggested that this procedure may represent a new approach to detect endothelial injury associated with vascular disorders such as thrombotic diseases and vasculities. Jackson (1991) described a simple, rapid technique for purifying endothelial cells derived from the microvasculature of human tissues. This technique is based on the selective binding of lectin to endothelial cells. Initially, the lectin is bound to the beads. The lectin-coated beads bound very rapidly to the endothelial cells, but not to human skin fibroblasts. The binding could be reversed by addition of fucose.

3.8. Isolation of Antigen-Specific B Cells

Egeland *et al.* (1988) have used immune complex-coated Dynabeads to enrich for rheumatoid factor-producing B lymphocytes from human peripheral blood. They obtained a 10^4–10^6-fold enrichment of specific B cells. After isolation of the B cells, they were directly transformed with Epstein–Barr virus for bulk expansion and later tested for antibody production. More than 90% of the growing lymphoblastoid cells were specific antibody producers. This technique may be used in human monoclonal antibody production. Starting with a homogeneous population of antigen-specific B lymphoblasts will increase the chances of successful cloning and will greatly increase the frequency of relevant hybrids after hybridization.

Ossendorp *et al.* (1989) applied thyroglobulin-coated Dynabeads to form rosettes with B-cell hybridoma cell lines. A 300-fold enrichment of thyroglobulin-specific cells was obtained.

4. IMMUNOMAGNETIC SEPARATION OF PROKARYOTIC CELLS

4.1. The Prokaryotic Cells

Bacteria belong to the prokaryotes. The major difference between prokaryotes and eukaryotes of importance for immunomagnetic separation is the presence of a rather rigid cell wall outside of the cell membrane in the former. Mycoplasma and the L-form of true bacteria lack this barrier to the surrounding environment, but most bacteria of medical, industrial, and environmental importance have this structure (Krieg and Holt, 1984). The cell wall is a complex, semirigid structure that helps to maintain the shape of the organism and to protect it against adverse environmental changes such as osmotic pressure. The basic component is a mixed polymer called peptidoglycan, consisting of *N*-acetylglucosamine and *N*-acetylmuramic acid. The thickness of the peptidoglycan layer has been used as one of the most important diagnostic criteria: bacteria with a thin peptidoglycan layer are designated gram-negative, and those with several peptidoglycan layers, often also with proteins and teichoic acids, gram-positive. The gram- negative bacteria do have a layer of lipopolysaccharides and lipoproteins outside the peptidoglycan, often called the outer membrane. Outside this structure, both gram-positive and gram-negative bacteria can possess different types of capsule materials, generally consisting of polysaccharides (Krieg and Holt, 1984).

4.2. Immunomagnetic Separation

Immunomagnetic separation of bacteria using Dynabeads coated with specific antibodies against surface antigens of cells has found several medical applications. Isolation of specific bacteria has generally been accomplished by inoculating heterogeneous samples into cultivation media offering growth preference to the target bacteria (Krieg and Holt, 1984). The term immunomagnetic enrichment, as many prefer to use, is based on the fact that bacteria immunologically bound to beads usually remain viable and can continue to multiply if nutritional requirements are provided (Olsvik et al., 1991c). The procedure involves mixed beads with the crude sample, and, after incubation for 10–30 min, the beads with the bound living bacteria are extracted using a magnet. The immunomagnetically isolated fraction can be washed before being inoculated on suitable growth media. Both polyclonal and monoclonal antibodies have been employed in this procedure with success (Olsvik et al., 1991c).

4.2.1. Bacterial Antigens Utilized for Immunomagnetic Separation

Several bacteria possess flagella that give the organisms some motility. The flagella are proteinaceous filaments about 15 nm in diameter but can be up to 2 μm in length. There can be a single flagellum or several flagella per cell. Fimbriae or phili are other surface structures on the bacterial surface suitable for immunological attachment. They consist of repeated protein units arranged in a helix to form a long filament, often used for bacterial attachment to eukaryotic cells. They can exist in several hundred copies per bacterial cell. Fimbrial antigens have been proposed to be designated F-antigens (Krieg and Holt, 1984). These structures have been used as targets for immunoseparation, and several authors have reported successful separation of specific bacteria from heterogeneous solutions using antibodies against fimbria coated on beads (Lund et al., 1988; Skjerve et al., 1990; Lund et al., 1991; Hornes et al., 1991).

Except for structural organelles such as flagella and fimbria, antigenic determinants on or in the cell wall are used for differentiating strains. The antigens used are the O-antigens, H-antigens, and K-antigens. The term O-antigens is used for the antigenic determinants from the repeated units of oligosaccharides which, in addition to lipid A and core polysaccharides, constitute lipopolysaccharides in bacteria. Using polyvalent sera, several hundred different O-antigens have been identified (Krieg and Holt, 1984). Polyclonal sera against specific O-groups have been utilized for immunomagnetic separation of pathogenic Escherichia coli (Olsvik et al., 1991c). Heterogeneous antibodies directed against the lipopolysaccharides of Salmonella species have also been utilized with success (Blackburn et al., 1991; Skjerve and Olsvik, 1991). Such organisms have been isolated from matrices such as different food stuffs. H-antigens are polysaccharides found in the capsule of some bacteria and are probably not well suited as targets for immunoseparations as the capsule is rather loosely associated with the cell, but interesting results have been described in immunoseparation of Staphylococcus aureus from clinical samples (Johne et al., 1989). In old cultures with a high percent of dead organisms and with large amounts of the target antigen found free in the solution, the efficiency of immunomagnetic separation might be reduced.

4.2.2. Advantages of Immunomagnetic Separation

The immunomagnetic separation technique has several advantages. The target bacteria is separated from the environment and is concentrated from a large volume to a volume suitable for cultivation on plates or in broths. Growth inhibitory reagents in the sample are also removed from the bacteria, which enhances cultivation. When pathogenic strains are present in samples that contain large numbers of nonpathogenic variants of the same species, selective growth media are normally less efficient in isolating the target organisms. However, immunomagnetic separation provides the potential for selective isolation of strains possessing specific surface epitopes such as fimbria that are associated with the ability to induce disease. The use of selective enrichment broths has been shown to favor *E. coli* strains of environmental origin compared to strains of human origin, and the majority of *E. coli* strains lost antimicrobial resistance genes and plasmids encoding heat-labile enterotoxin production (Hill and Carlise, 1981). These negative effects of selective media are not seen when immunomagnetic separation is used to achieve selective enrichment (Olsvik *et al.*, 1991c).

4.3. Identification of Bacteria after Immunomagnetic Separation

4.3.1. Traditional Methods

Bacteria attached to the beads can be stained with acridine orange to demonstrate viability and visualized under a fluorescence microscope (Lund *et al.*, 1988). This is a rapid and easy-to-perform approach for identification of a specific agent. Other workers have identified bacteria after electron-microscopic examination (see Fig. 3) (Johne *et al.*, 1989; Olsvik *et al.*, 1991c). Bacteria bound with antibodies to the surface of beads are generally viable and will grow when given appropriate conditions. The bacteria do not need to be detached from the beads, but will multiply directly, and apparently the attachment has no effect on growth (Skjerve *et al.*, 1990). Both plate and broth cultivation have been performed on several bacterial species bound to beads. However, quantitative enumeration of colony-forming units must take into consideration that each colony does not always reflect the product of one living cell (Skjerve *et al.*, 1990). Several cells might be attached to a cluster of beads to provide the origin of one colony.

4.3.2. Identification Using Nucleic Acid Technology after Immunomagnetic Separation

DNA technology provides new tools to clinical laboratories and improves both sensitivity and efficiency. The use of hybridization techniques for detection of either species-specific genes or genes encoding certain virulence factors and their locations on plasmids or on the genome is a valuable diagnostic tool (Olsvik *et al.*, 1991a). The development of the polymerase chain reaction (PCR) represents a new generation of DNA techniques available for diagnostic use (Olsvik *et al.*, 1991a,b). By employing PCR, it is

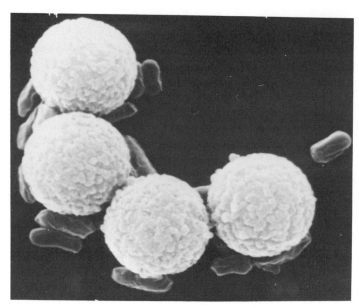

FIGURE 3. *Salmonella enteritidis* bound to M-280 particles using a direct linked polyvalent IgG against *Salmonella* species.

theoretically possible to identify genes encoding virulence factors directly in food or stool samples without precultivation of microorganisms. The extreme sensitivity of PCR can in many ways be compared with that of cultivation on nonselective media. One factor preventing diagnostic use of PCR directly on clinical samples has been the sensitivity of the Taq polymerase to inhibitor elements in samples such as feces and blood.

Immunomagnetic separation as a pre-PCR step has appeared to solve several of these problems. The bacteria in the sample are concentrated in a suitable volume and simultaneously removed from the specific Taq polymerase inhibitors. For some samples, precultivation can increase the number of target organisms in the sample to compensate for the low sensitivity in the presence of Taq inhibitors. Samples that have been frozen often contain nonviable cells. These bacterial cells can be extracted by immunomagnetic separation, and, although these cells are not suitable for cultivation, PCR has appeared to be a good detection system for such organisms (Hornes *et al.*, 1991).

Magnetic separation has also been applied to separation of DNA, and an interesting new assay designated DIANA (detection of immobilized amplified nucleic acids) has recently been applied on immunomagnetic-separated bacteria from clinical samples (Uhlén, 1989; Wahlberg *et al.*, 1990; Rimstad *et al.*, 1990; Hornes *et al.*, 1991; Olsvik *et al.*, 1991b).

REFERENCES

Armitage, R. J., and Goff, L. K., 1988, Functional interaction between B-cell subpopulations defined by CD23 expression, *Eur. J. Immunol.* **18:**1753–1760.

Berge, A., Ellingsen, T., Skjeltorp, A. T., and Ugelstad, J., 1990, Modelling of physical processes using monosized polymer particles, in: *Scientific Methods for the Study of Polymer Colloids and Their Applications*, (F. Candau and R. H. Ottewill, eds.), Kluwer Academic, Dordecht, The Netherlands, pp. 435–452.

Biéva, C. J., Vander Brugghen, F. J., and Stryckmans, P. A., 1989, Malignant leukemic cell separation by iron colloid immunomagnetic adsorption. *Exp. Hematol.* **17**:914–920.

Blackburn, C. de W., Patel, P. D., and Gibbs, P. A., 1991, Separation and detection of salmonellae using immunomagnetic particles, *Biofouling* **5**:143–156.

Bosnes, V., and Hirschberg, H., 1989, Immunomagnetic separation of infiltrating T lymphocytes from brain tumors, *J. Neurosurg.* **71**:218–223.

Bosnes, V., Halvorsen, R., Gaudernack, G., and Thorsby, E., 1989, Isolation of functionally active T cell receptor γδ-bearing lymphocytes from human peripheral blood, *J. Immunol. Methods* **118**:251–255.

Brandtzaeg, P., Bosnes, V., Halstensen, T. S., Scott, H., Sollid, L. M., and Valnes, K. N., 1989, T lymphocytes in human gut epithelium preferentially express the α/β antigen receptor and are often CD45/UCHL1-positive, *Scand. J. Immunol.* **30**:123–128.

Brinchmann, J. E., Vartdal, F., Gaudernack, G., Markussen, G., Funderud, S., Ugelstad, J., and Thorsby, E., 1988, Direct immunomagnetic quantification of lymphocyte subsets in blood, *Clin. Exp. Immunol.* **71**: 182–186.

Brinchmann, J. E., Gaudernack, G., Thorsby, E., Jonassen, T. Ø., and Vartdal, F., 1989a, Reliable isolation of human immunodeficiency virus from cultures of naturally infected CD4+ T cells, *J. Virol. Methods* **25**: 293–300.

Brinchmann, J. E., Leivestad, T., and Vartdal, F., 1989b, Quantification of lymphocyte subsets based on positive immunomagnetic selection of cells directly from blood, *J. Immunogenet.* **16**:177–183.

Brinchmann, J. E., Gaudernack, G., and Vartdal, F., 1990, CD8+ T cells inhibit HIV replication in naturally infected CD4+ T cells. Evidence for a soluble inhibitor. *J. Immunol.* **144**:2961–2966.

Champlin, R., Ho, W., Gajewski, J., Feig, S., Burnison, M., Holley, G., Greenberg, P., Lee, K., Schmid, I., and Giorgi, J., 1990, Selective depletion of CD8+ T lymphocytes for prevention of graft-versus-host disease after allogeneic bone marrow transplantation, *Blood* **76**:418–423.

Civin, C. I., Strauss, L. C., Fackler, M. J., Trischmann, T. M., Wiley, J. M., and Loken, M. R., 1990, Positive stem cell selection—basic science, in: *Bone Marrow Purging and Processing* (S. Gross, A. P. Gee, and D. A. Worthington-White, eds.), Wiley-Liss, New York, pp. 387–402.

Combaret, V., Favrot, M. C., Chauvin, F., Bouffet, E., Phillip, I., and Phillip, T., 1989, Immunomagnetic depletion of malignant cells from autologous bone marrow graft: From experimental models to clinical trials, *J. Immunogenet.* **16**:125–136.

Cottler-Fox, M., Bazar, L. S., and Deeg, H. J., 1990, Isolation of hemopoietic precursor cells from human marrow by negative selection using monoclonal antibodies and immunomagnetic beads, in: *Bone Marrow Purging and Processing* (S. Gross, A. P. Gee, and D. A. Worthington-White, eds.), Wiley-Liss, New York, pp. 277–284.

Douglas, G. C., and King, B. F., 1989, Isolation of pure villous cytotrophoblasts from term human placenta using immunomagnetic microspheres, *J. Immunol. Methods* **119**:259–268.

Egeland, T., Hovdenes, A., and Lea, T., 1988, Positive selection of antigen-specific B lymphocytes by means of immunomagnetic particles, *Scand. J. Immunol.* **27**:439–444.

Elias, A. D., Pap, S. A., and Bernal, S. D., 1990, Purging of small cell lung cancer-contaminated bone marrow by monoclonal antibodies and magnetic beads, in: *Bone Marrow Purging and Processing* (S. Gross, A. P. Gee, and D. A. Worthington-White, eds.), Wiley-Liss, New York, pp. 263–275.

Frame, J. N., Collins, N. H., Cartagena, T., Waldmann, H., O'Reilly, R. J., Dupong, B., and Kernan, N. A., 1989, T cell depletion of human bone marrow, *Transplantation* **47**:984–988.

Frangoulis, B., Pla, M., and Rammensee, H.-G., 1989, Alternative T cell receptor gene usage induced by self tolerance, *Eur. J. Immunol.* **19**:553–555.

Funderud, S., Nustad, K., Lea, T., Vartdal, F., Gaudernack, G., Stenstad, P., and Ugelstad, J., 1987, Fractionation of lymphocytes by immunomagnetic beads, in: *Lymphocytes—A Practical Approach*, (G. G. B. Klaus, ed.), IRL Press, Oxford, pp. 55–65.

Funderud, S., Erikstein, B., Åsheim, H. C., Nustad, K., Stokke, T., Blomhoff, H. K., Holte, H., and Smeland, E. B., 1990, Functional properties of CD19+ B lymphocytes positively selected from buffy coats by immunomagnetic separation, *Eur. J. Immunol.* **20**:201–206.

Gee, A. P., Mansour, V., and Weiler, M., 1989, T-cell depletion of human bone marrow, *J. Immunogenet.* **16**: 103–115.

Geisler, C., Møller, J., Plesner, T., Dickmeiss, E., Pallesen, G., Larsen, J. K., Jacobsen, N., and Svejgaard, A., 1989, Specific depletion of mature T lymphocytes from human bone marrow, *Scand. J. Immunol.* **29:** 617–625.

George, F., Poncelet, P., Brisson, C., Massot, O., Laurent, J. C., Charmasson, G., and Sampol, J., 1991, Isolation of endothelial cells from whole blood with S-Endol-coated magnetic beads, in: *Magnetic Separation Techniques Applied to Cellular and Molecular Biology* (J. Kemshead, ed.) (Wordsmiths' Conference Publications, Cedric Chivers Ltd., Bristol), pp. 223–234.

Hanau, D., Schmitt, D. A., Fabre, M., and Cazenave, J. P., 1988, A method for the rapid isolation of human epidermal Langerhans cells using immunomagnetic microspheres, *J. Invest. Dermatol.* **91:**274–279.

Hansel, T. T., Pound, J. D., and Thompson, R. A., 1990, Isolation of eosinophils from human blood, *J. Immunol. Methods* **127:**153–164.

Hansen, T., and Hannestad, K., 1989, Direct HLA typing by rosetting with immunomagnetic beads coated with specific antibodies, *J. Immunogenet.* **16:**137–139.

Hill, W. E., and Carlisle, C. L., 1981, Loss of plasmids during enrichment for *Escherichia coli*, *Appl. Environ. Microbiol.* **41:**1046–1048.

Hornes, E., Wasteson, Y., and Olsvik, Ø., 1991, Detection of *Escherichia coli* heat-stable enterotoxin genes in pig stool specimens by an immobilized, colorimetric, nested PCR, *J. Clin. Microbiol.* **29:**2375–2379.

Hovdenes, J., 1989, B-cell growth-promoting activity in supernatants from CD4$^+$ cells from synovial fluid and peripheral blood of patients with rheumatoid arthritis and other inflammatory arthritides, *Scand. J. Rheumatol.* **18:**385–392.

Hovdenes, J., Gaudernack, G., and Kvien, T. K., 1989a, Expression of activation markers on CD4$^+$ and CD8$^+$ cells from synovial fluid, synovial tissue and peripheral blood of patients with inflammatory arthritides, *Scand. J. Immunol.* **29:**631–639.

Hovdenes, J., Gaudernack, G., Kvien, T. K., Egeland, T., and Mellbye, O. J., 1989b, Functional study of purified CD4$^+$ and CD8$^+$ cells isolated from synovial fluid of patients with rheumatoid arthritis and other arthritides, *Scand. J. Immunol.* **29:**641–649.

Hovdenes, J., Gaudernack, G., Kvien, T. K., Hovdenes, A. B., and Egeland, T., 1989c, Mitogen-induced interleukin 2 and gamma interferon production by CD4$^+$ and CD8$^+$ cells of patients with inflammatory arthritides. A comparison between cells from synovial fluid and peripheral blood, *Scand. J. Immunol.* **30:** 597–603.

Jackson, C. J., 1991, Human endothelial cells bind to, and can be purified by, Ulex Europaeus 1 coated Dynabeads, in: *Magnetic Separation Techniques Applied to Cellular and Molecular Biology* (J. Kemshead, ed.) (Wordsmiths' Conference Publications, Cedric Chivers Ltd., Bristol), pp. 63–74.

Janssen, W. E., Lee, C., Gross, S., and Gee, A. P., 1989, Low antigen density leukemia cells: Selection and comparative resistance to antibody-mediated marrow purging, *Exp. Hematol.* **17:**252–257.

Janssen, W. E., Lee, C., Johnson, K. S., Spencer, C., Rios, A. M., Graham-Pole, J. R., and Gross, S., 1990, Immunomagnetic microsphere mediated purging of cALLa positive leukemic cells from bone marrow for autologous reinfusion, in: *Bone Marrow Purging and Processing* (S. Gross, A. P. Gee, and D. A. Worthington-White, eds.), Wiley-Liss, New York, pp. 285–292.

Johne, B., Jarp, J., and Haaheim, L. R., 1989, *Staphylococcus aureus* exopolysaccharide *in vivo* demonstrated by immunomagnetic separation and electron microscopy, *J. Clin. Microbiol.* **27:**1631–1635.

Kemshead, J. T., Elsom, G., and Patel, K., 1990, Immunomagnetic manipulation of bone marrow and tumor cells: An update, in: *Bone Marrow Purging and Processing* (S. Gross, A. P. Gee, and D. A. Worthington-White, eds.), Wiley-Liss, New York, pp. 235–251.

Knobloch, C., Spadinger, U., Rueber, E., and Friedrich, W., 1990, T cell depletion from human bone marrow using magnetic beads, *Bone Marrow Transpl.* **6:**21–24.

Krieg, N. R., and Holt, J. G., 1984, *Bergey's Manual of Systematic Bacteriology*, Williams and Wilkins, London.

Kvalheim, G., Funderud, S., Kvaløy, S., Gaudernack, G., Beiske, K., Jakobsen, E., Jacobsen, A. B., Pihl, A., and Fodstad, Ø., 1988a, Successful clinical use of an anti-HLA-DR monoclonal antibody for autologous bone marrow transplantation, *J. Natl. Cancer Inst.* **80:**1322–1325.

Kvalheim, G., Sørensen, O., Fodstad, Ø., Funderud, S., Kiesel, S., Dörken, B., Nustad, K., Jakobsen, E., Ugelstad, J., and Pihl, A., 1988b, Immunomagnetic removal of B-lymphoma cells from human bone marrow: A procedure for clinical use, *Bone Marrow Transpl.* **3:**31–41.

Kvalheim, G., Fjeld, J. G., Pihl, A., Funderud, S., Ugelstad, J., Fodstad, Ø., and Nustad, K., 1989, Immunomagnetic removal of B-lymphoma cells using a novel mono-sized magnetizable polymer bead, M-280, in conjunction with primary IgM and IgG antibodies, *Bone Marrow Transpl.* **4:**567–574.

Lea, T., Vartdal, F., Nustad, K., Funderud, S., Berge, A., Ellingsen, T., Schmid, R., Stenstad, P., and Ugelstad, J., 1988, Monosized magnetic polymer particles: Their use in separation of cells and subcellular components and in the study of lymphocyte function *in vitro*, *J. Mol. Recog.* **1:**9–18.

Leivestad, T., Gaudernack, G., Halvorsen, R., and Thorsby, E., 1988, Simple and sensitive bioassay for the detection of IL-2 activity, *J. Immunol. Methods* **114:**95–99.

Leivestad, T., Halvorsen, R., Gaudernack, G., and Thorsby, E., 1989, Ability of pure resting CD8[+] human T cells to respond to alloantigen, *Scand. J. Immunol.* **29:**543–553.

Loke, Y. W., Gardner, L., and Grabowska, A., 1989, Isolation of extravillous trophoblast cells by attachment to laminin-coated magnetic beads, *Placenta* **10:**407–415.

Lopez, M., Martinache, C., Bardinet, D., Mortel, O., Michon, J., and Combaret, V., 1989, Improved CFU-GM recoveries from marrows of neuroblastoma patients using a slightly modified immunomagnetic depletion technique, *Bone Marrow Transpl.* **4:**453–454.

Lund, A., Hellemann, A. L., and Vartdal, F., 1988, Rapid isolation of K88[+] *Escherichia coli* by using immunomagnetic particles, *J. Clin. Microbiol.* **26:**2572–2575.

Lund, A., Wasteson, Y., and Olsvik, Ø., 1991, Immunomagnetic separation and DNA hybridization for detection of enterotoxigenic *Escherichia coli* in a piglet model, *J. Clin. Microbiol.* **29:**1–12.

Lundin, K. E. A., Qvigstad, E., Sollid, L. M., Gjertsen, H. A., Gaudernack, G., and Thorsby, E., 1989, Positive selection of Tac-(CD25) positive cells following T-cell activation, *J. Immunogenet.* **16:**185–191.

Morgan, A., Eaves, J., Mockford, J. W., and Malkovska, V., 1989, Impaired mitogen responses of the non-leukaemic B cells from patients with chronic lymphocytic leukaemia, *Clin. Exp. Immunol.* **77:**239–244.

Mosbach, K., and Nilsson, K., 1982, New affinity techniques. Sulfonyl halides for the immobilization of affinity ligands and enzymes, Vol. 9 of *Affinity Chromatography and Related Techniques Anal. Chem. Symp. Ser.* (T. C. J. Gribnau, J. Visser, and R. J. F. Nivard, eds.) (Elsevier, Amsterdam) pp. 199–206.

Mueller, U. W., Hawes, C. S., Wright, A. E., Petropoulos, A., DeBoni, E., Firgaira, F. A., Morley, A. A., and Turner, D. R., 1990, Isolation of fetal trophoblast cells from peripheral blood of pregnant women, *Lancet* **336:**197–200.

Murakami, S., Shimazaki, C., Oku, N., Itoh, K., Takeda, N., Fujita, N., Ura, Y., and Nakagawa, M., 1990, Elimination of clonogenic leukemic cells from bone marrow using monoclonal antibodies and magnetic immunobeads, in: *Bone Marrow Purging and Processing* (S. Gross, A. P. Gee, and D. A. Worthington-White, eds.), Wiley-Liss, New York, pp. 253–261.

Olsvik, Ø., Hornes, E., Wasteson, Y., and Lund, A., 1991a, Detection of virulence determinants in enteric *Escherichia coli* using nucleic acid probes and polymerase chain reaction, in: *Molecular Pathogenesis of Gastro-intestinal Infections* (T. Wadstöm, P. H. Mäkelä, A.-M. Svennerholm, and H. Wolf-Watz, eds.), Plenum Press, New York (London), pp. 267–272.

Olsvik, Ø., Rimstad, E., Hornes, E., Strockbine, N., Wasteson, Y., Lund, A., and Wachsmuth, K., 1991b, A nested PCR followed by magnetic separation of amplified fragments for detection of *Escherichia coli* Shiga-like toxin genes, *Mol. Cell. Probes* **5:**429–435.

Olsvik, Ø., Wasteson, Y., Lund, A., and Hornes, E., 1991c, Pathogenic *Escherichia coli* found in food, *Int. J. Food Microbiol.* **12:**103–114.

Ossendorp, F. A., Bruning, P. F., Van den Brink, J. A. M., and De Boer, M., 1989, Efficient selection of high-affinity B cell hybridomas using antigen-coated magnetic beads, *J. Immunol. Methods* **120:**191–200.

Padmanabhan, R., Corsico, C., Holter, W., Howard, T., and Howard, B. H., 1989, Purification of transiently transfected cells by magnetic-affinity cell sorting, *J. Immunogenet.* **16:**91–102.

Parker, C. M., Groh, V., Band, H., Porcelli, S. A., Morita, C., Fabbi, M., Glass, D., and Strominger, J.L., 1990, Evidence for extrathymic changes in the T cell receptor γ/δ repertoire, *J. Exp. Med.* **171:**1597–1612.

Pilarski, L. M., Giilitzer, R., Zola, H., Shortman, K., and Scollay, R., 1989, Definition of the thymic generative lineage by selective expression of high molecular weight isoforms of CD45 (T200), *Eur. J. Immunol.* **19:**589–597.

Platsoucas, C. D., 1987, Biomedical applications of polymer particles with emphasis on cell separation, in: *Future Directions in Polymer Colloids*, NATO ASI Series E, No. 138 (M. S. El Asser and R. M. Fitch, eds.), M. Nijhoff, Dordrecht, The Netherlands, pp. 321–354.

Rasmussen, A.-M., Smeland, E. B., Erikstein, B. K., Caignault, L., and Funderud, S., 1992, A new method for detachment of Dynabeads from positively selected B lymphocytes, *J. Immunol. Methods* **146:**195–202.

Rayman, P., Finke, J. H., Alexander, J., Tubbs, R., Edinger, M., Pontes, E., Connely, B. and Bukowski, R., 1989, Isolation and functional analysis of CD4[+] and CD8[+] TIL subsets derived from human renal cell carcinoma (RCC), *Proc. Am. Assoc. Cancer Res.* **30:**1484.

Rimstad, E., Hornes, E., Olsvik, Ø., and Hyllseth, B., 1990, Identification of a double-stranded RNA virus by using polymerase chain reaction and magnetic separation of the synthesized DNA fragments, *J. Clin. Microbiol.* **28:**2275–2278.

Shimazaki, C., Inaba, T., Murakami, S., Fujita, N., Nakagawa, M., Gulati, S. C., Fried, J., Clarkson, B. D., Wisniewolski, R., and Wang, C. Y., 1990, Purging of myeloma cells from bone marrow using monoclonal antibodies and magnetic immunobeads in combination with 4-hydroperoxycyclophosphamide, in: *Bone Marrow Purging and Processing* (S. Gross, A. P. Gee, and D. A. Worthington-White, eds.), Wiley-Liss, New York, pp. 311–319.

Shpall, E. J., Anderson, I. C., Bast, R. C., Jr., Joines, W. T., Jones, R. B., Ross, M., Edwards, S., Eggleston, S., Johnston, C., Tepperberg, M., Affronti, M. L., Coniglio, D., Mathias, B., and Peters, W. P., 1990, Immunopharmacologic purging of breast cancer from bone marrow for autologous bone marrow transplantation, in: *Bone Marrow Purging and Processing* (S. Gross, A. P. Gee, and D. A. Worthington-White, eds.), Wiley-Liss, New York, pp. 321–336.

Skjerve, E., and Olsvik, Ø., 1991, Immunomagnetic separation of *Salmonella* from food, *Int. J. Food Microbiol.* **14:**11–18.

Skjerve, E., Rørvik, L. M., and Olsvik, Ø., 1990, Detection of *Listeria monocytogenes* in food by immunomagnetic separation, *Appl. Environ. Microbiol.* **56:**4378–4381.

Skjønsberg, C., Kiil Blomhoff, H., Gaudernack, G., Funderud, S., Beiske, K., and Smeland, E. B., 1990, Immunological typing of acute leukemias by rosetting with immunomagnetic beads in comparison with immunofluorescence staining, *Scand. J. Immunol.* **31:**567–575.

Smeland, E. B., Funderud, S., Kvalheim, G., and Egeland, T., 1991, Effective isolation of human progenitor cells; a new method for detachment of immunomagnetic beads from positively selected CD34+ cells, Abstract, 2nd International Symposium on Peripheral Blood Stem Cell Autografts, Mulhouse, France.

Treleaven, J. G., Gibson, F. M., Ugelstad, J., Rembaum, A., Philip, T., Caine, G. D., and Kemshead, J. T., 1984, The removal of neuroblastoma cells from bone marrow using monoclonal antibodies conjugated to magnetic microspheres, *Lancet* **14:**70–73.

Trickett, A. E., 1990, Tumour cell purging for autologous bone marrow transplantation, *Med. Lab. Sci.* **47:** 120–131.

Ugelstad, J., Mørk, P. C., Kaggerud, K. H., Ellingsen, T., and Berge, A., 1980, Swelling of oligomer–polymer particles. New methods of preparation of emulsions and polymer dispersions, *Adv. Colloid and Interface Sci.* **13:**101–140.

Ugelstad, J., Berge, A., Ellingsen, T., Aune, O., Kilaas, L., Nilsen, T. N., Schmid, R., Stenstad, P., Funderud, S., Kvalheim, G., Nustad, K., Lea, T., Vartdal, F., and Danielsen, H., 1988, Monosized magnetic particles and their use in selective cell separation, *Makromol. Chem., Macromol. Symp.* **17:**177–211.

Ugelstad, J., Kilaas, L., Stenstad, P., Ellingsen, T., Bjørgum, J., Aune, O., Nilsen, T. N., Schmid, R., and Berge, A., 1991, Preparation and application of monosized composite polymer particles, in: *Magnetic Separation Techniques Applied to Cellular and Molecular Biology* (J. Kemshead, ed.) (Wordsmiths' Conference Publications, Cedric Chivers Ltd., Bristol), pp. 235–254.

Uhlén, M., 1989, Magnetic separation of DNA, *Nature (London)* **340:**733–734.

van Els, C. A. C. M., Zantvoort, E., Jacobs, N., Bakker, A., van Rood, J. J., and Goulmy, E., 1990, Graft-versus-host disease associated T helper cell responses specific for minor histocompatibility antigens are mainly restricted by HLA-DR molecules, *Bone Marrow Transpl.* **5:**365–372.

Vartdal, F., Bratlie, A., Gaudernack, G., Funderud, S., Lea, T., and Thorsby, E., 1987a, Microcytotoxic HLA-typing of cells directly isolated from blood by means of antibody-coated microspheres, *Transplant. Proc.* **19:**655–657.

Vartdal, F., Kvalheim, G., Lea, T., Bosnes, V., Gaudernack, G., Ugelstad, J., and Albrechtsen, D., 1987b, Depletion of T lymphocytes from human bone marrow. Use of magnetic monosized polymer microspheres coated with T-lymphocyte-specific monoclonal antibodies, *Transplantation* **43:**366–371.

Vartdal, F., Bratlie, A., and Thorsby, E., 1988, Immunomagnetic tissue typing offers new possibilities in cadaveric transplant donor–recipient HLA matching, *Transplant. Proc.* **20:**384–385.

Wahlberg, J., Lundberg, J., Hultman, T., and Uhlén, M., 1990, General colorimetric method for DNA diagnostic allowing direct solid phase genomic sequencing of the positive sample, *Proc. Natl. Acad. Sci. U.S.A.* **87:**6569–6573.

IV

Affinity Chromatography with Biomimetic Ligands

17

Affinity Chromatography with Immobilized Dyes

Earle Stellwagen

1. INTRODUCTION

Affinity chromatography is a very attractive concept for protein purification in that it exploits the one property that distinguishes each protein, namely, its ability to complex one or a small number of molecules (bioligands) with high affinity. Unfortunately, this attractive concept has several practical limitations that restrict its usefulness. Firstly, a bioligand must be covalently attached (immobilized) to the chromatographic matrix in a manner which does not seriously diminish the affinity of the biofunctional site of the protein for the bioligand. Such attachment often requires execution of some adroit chemistry requiring either technical skills beyond that of a typical investigator or else adequate financial resources to purchase such a product. Secondly, a given immobilized bioligand will contribute to the purification of only a single or at best a small number of related proteins. Thus, a different affinity column must be available for each of the purified proteins required by an investigator. Thirdly, crude protein mixtures likely contain enzymes which can catalyze the hydrolysis of immobilized ligands, rendering them ineffective. Accordingly, affinity chromatography is commonly performed late in a purification scheme, where the concentrations of such hydrolytic enzymes are minimized by prior purification steps. Relegation of affinity chromatography to a late step in purification diminishes its selective potential.

During the 1960s, a cascade of events occurred which have significantly diminished these practical limitations on affinity chromatography. The cascade was initiated by the development of water-soluble reactive dyes by the commercial dye industry for the permanent coloration of cellulose fibers. These dyes form a covalent bond with cellulose and other polysaccharides under simple mild, conditions. Pharmacia used this chemistry to produce a colored macromolecule, blue dextran, which is commonly used to measure the void volume of size exclusion chromatographic columns. It was then observed that proteins which are normally included during size exclusion chromatography often coelute with the excluded blue dextran when a mixture of protein and blue dextran is applied to the column (Haeckel *et al.*, 1968). It was subsequently noted that the coelution of a protein and blue

Earle Stellwagen • Department of Biochemistry, University of Iowa, Iowa City, Iowa 52242.

Molecular Interactions in Bioseparations, edited by That T. Ngo. Plenum Press, New York, 1993.

dextran was eliminated when chromatography was done in solvents containing high concentrations of salt. Such strange behavior was noted for a variety of proteins, ranging from glutathionine reductase (Staal *et al.*, 1969) to blood coagulation factors (Swart and Hemker, 1970).

It was proposed that this strange behavior originated in the formation of a noncovalent, high-affinity complex between these proteins and the immobilized dye which extended from the surface of the blue dextran. This proposal was then tested by covalent attachment of either blue dextran or the blue dye to matrices appropriate to protein chromatography (Roschlau and Hess, 1971; Bohme *et al.*, 1972; Easterday and Easterday, 1974; Ryan and Vestling, 1974). As anticipated, a variety of proteins were found to be retained by the immobilized dye and to be quantitatively eluted by the addition of either salt or a bioligand to the chromatographic solvent. A significant advance was the subsequent demonstration that this chromatographic protocol could be used to purify proteins directly from crude cellular extracts with a good recovery, a significant increase in specific activity, and no damage to the immobilized dye. These results demonstrated that immobilized reactive dyes could achieve the potential purification power envisaged for affinity chromatography.

The experience accumulated over the past two decades indicates immobilized dye affinity chromatography to be an inexpensive, accessible, and versatile vehicle for the purification of hundreds of different proteins at a laboratory and a commercial scale (Dean and Watson, 1979; Kopperschlager *et al.*, 1982; Scawen and Atkinson, 1987). A variety of stable immobilized reactive dyes of various color can be purchased from supply houses at modest cost or can be synthesized from mobile reactive dyes very simply using common laboratory equipment. Protocols have been developed which reproducibly generate impressive fold purification values, sometimes to homogeneity in a single step, in good yield.

All that is required to assess whether immobilized dye chromatography can contribute to the purification of a protein of interest is a crude source of the protein, a qualitative assay for the functional of that protein, and a small sample of an immobilized dye. The latter could be supplied by a portion of colored cotton cloth since it represents a porous matrix likely colored by a reactive dye. If the protein functional disappears from the crude extract following exposure to the immobilized dye and reappears following exposure of the immobilized dye to $1M$ NaCl, then it is likely that immobilized dye chromatography can be used to advantage in the purification of the protein of interest. What follows is a more detailed discussion of the nature of the interaction between a model reactive dye and protein followed by a general discussion of the application of immobilized dye chromatography to protein purification. Additional details may be found in one or more of the recent reviews on immobilized dye chromatography since this topic is reviewed with great frequency (Lowe *et al.*, 1986; Scopes, 1986; Clonis *et al.*, 1987a; Clonis, 1988; Lowe *et al.*, 1990; Stellwagen, 1990).

2. INTERACTION OF A REACTIVE DYE WITH A MODEL PROTEIN

2.1. Structures of Reactive Dyes

The vast majority of studies of the interaction of reactive dyes with proteins involve the Cibacron dyes produced by Ciba-Geigy and the Procion dyes produced by Imperial Chemical Industries Ltd. (ICI). These dyes consist of a chromophore, such as an anthraquinone (blue), and aromatic azo (yellow, orange, or red) or a phthalocyanine-azo (green), covalently linked to a reactive chlorotriazine ring. The structure of the widely used blue dye reactive blue 2, called Cibacron blue 3G-A by Ciba-Geigy and Procion blue H-B by ICI, is illustrated in Fig. 1A. Reading from left to right, the dye consists of a sulfonated anthraquinone ring, a bridging diaminobenzene sulfonate, a monochlorotriazine, and a terminal aminobenzene sulfonate. It should be noted that this dye contains a mixture of apolar, polar, and ionic atoms and that its structure is rather flexible, consisting of a series of planar rings separated by amino groups which function as swivel points. The most well characterized interaction between a reactive dye and a protein involves the dye reactive blue 2 and the dimeric protein horse liver alcohol dehydrogenase. The interaction between these two molecules will be treated as a model system in the discussion which follows.

2.2. Mobile Dye–Protein Interaction

A crystallographic study of the reactive blue 2:alcohol dehydrogenase complex (Biellmann *et al.*, 1979) indicates that each protein monomer complexes a single dye molecule and that this complexation occurs in the NAD coenzyme binding cleft. The dye is bound in an L-shaped configuration oriented such that the anthraquinone ring occupies the hydrophobic pocket which binds the AMP portion of the coenzyme, the diaminobenzene and the triazine rings are bound to the cleft in a manner similar to that of the middle portion of the coenzyme, and the terminal aminobenzene sulfonate lies on the surface of the protein in an orientation somewhat distinct from that of the nicotinamide portion of the complexed coenzyme.

Solution measurements of the interaction of reactive blue 2 with alcohol dehydrogenase are consistent with the crystallographic results. Addition of the protein to solutions of the dye produces a marked changes in the visible absorption spectrum of the dye (Burton *et al.*, 1988). These spectral changes are consistent with the complexation of the dye in a hydrophobic pocket of the protein. The dissociation constant for the complex obtained from spectral measurements at 25°C in 0.1M tricine buffer, pH 8.5, ranged from 0.42 to 9.30 μM, dependent upon the position of the sulfonic acid group in the terminal aminobenzene sulfonate ring. Catalytic measurements (Liu and Stellwagen, 1986) indicated that the dye functions as a competitive inhibitor of the coenzyme NAD. An equimolar mixture of the meta and para isomers of the dye generated an inhibition constant for the dye:monomeric protein complex of 4.5 ± 0.8 μM. This value agrees with the average dissociation constant for the individual isomers obtained from spectral measurements.

The dye reactive blue 4 (Procion blue MX-R) is a truncated analog of reactive blue 2 containing two chloro groups on its triazine ring as shown in Fig. 1A. The presence of two chloro groups substantially enhances the reactivity of reactive blue 4 compared with reactive blue 2. Addition of reactive blue 4 to alcohol dehydrogenase led to a rapid inactivation of the enzyme, which was prevented by the presence of either NAD or NADH (Small *et al.*, 1982). The inactived enzyme contained one covalently bound dye molecule per protein monomer. Sequence analysis of this covalent complex demonstrated that reactive blue 4 was attached to cysteine residue 174, which lies adjacent to the coenzyme binding cleft in the crystallographic structure.

Taken together, these results clearly indicate that reactive blue 2 selectively binds to the coenzyme binding cleft of alcohol dehydrogenase. The dissociation constant for the enzyme:dye (ortho isomer) complex, 0.42 μM, is comparable with that of the enzyme: NADH complex, 0.35 μM, and significantly less than that of the enzyme:NAD complex,

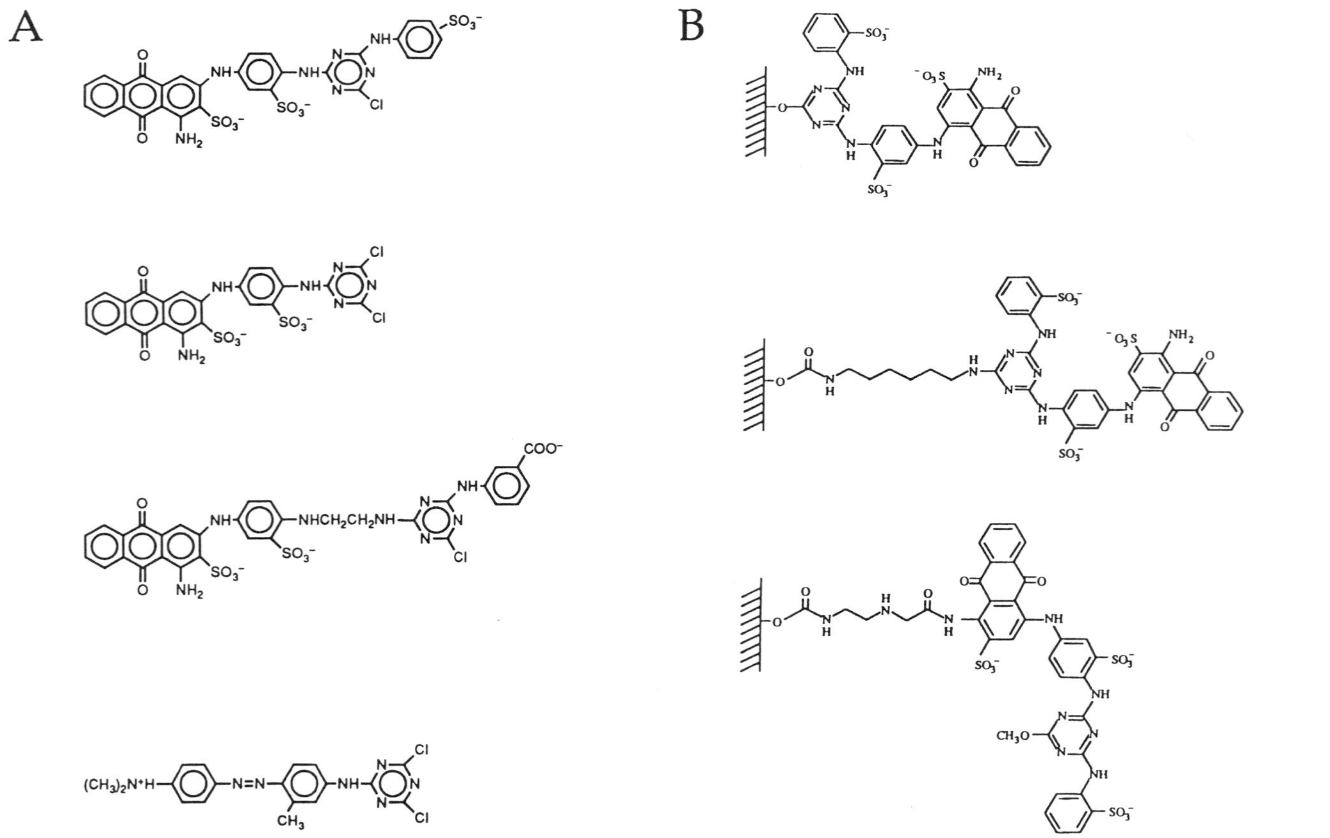

FIGURE 1. Structural formulas for mobile and immobile reactive dyes. (A) Structures of mobile dyes: from top to bottom, reactive blue 2 (Cibacron blue 3G-A, Procion blue H-B), reactive blue 4 (Procion blue MX-R), reactive blue 2 containing an internal aminoethyl spacer (Lowe et al., 1986), and a custom-designed cationic reactive dye (Clonis et al., 1987b). (B) Structures of immobilized dyes: from top to bottom, reactive blue 2 attached directly via its triazine ring, reactive blue 2 attached via its triazine ring using an aminohexyl spacer, and reactive blue 2 attached via its anthraquinone ring using a diaminoethylacetylamino spacer.

about 100 μM. These comparisons suggest that the dye can very effectively mimic the structure of the naturally occurring bioligand. Accordingly, reactive dyes which bind biofunctional sites on proteins can be termed biomimetic dyes (Lowe et al., 1986).

2.3.   Immobilized Dye–Protein Interaction

Sepharose has been covalently linked to reactive blue 2 in three different ways as illustrated in Fig. 1B: either directly to the triazine ring, to the triazine ring using a diaminohexyl spacer group, or to the anthraquinone ring using a diaminoethylacetylamino spacer group. Each of these immobilized dye preparations can retain alcohol dehydrogenase presented in solvents having an ionic strength less than 0.1M. The retained enzyme can be quantitatively eluted batchwise using solvents containing either 1M NaCl or 1mM NAD. Elution with NAD or NaCl gradients suggests that the affinity of the protein for the immobilized dye is approximately the same irrespective of whether the dye is linked to the triazine or to the anthraquinone ring with or without a spacer (Burton et al., 1988, 1990).

The affinity of alcohol dehydrogenase for reactive blue 2 immobilized directly via the triazine ring has been investigated by recycling partition chromatography (Hogg and Winzor, 1985) and by zonal chromatography (Liu and Stellwagen, 1986). If the density of the immobilized dye is sparse (i.e., a light blue column), most of the dimeric protein molecules can complex with only a single immobilized dye molecule with an association constant comparable to that of the mobile dye. However, if the density of the immobilized dye is large (i.e., a dark blue column), a significant fraction of the dimeric protein molecules can simultaneously complex with two immobilized dye molecules, one immobilized dye molecule in each NAD binding cleft. The affinity of the protein for the second immobilized dye molecule is about four times greater than that for the first immobilized dye molecule. This positive cooperativity likely results in part from the ease of formation of the second complex once the protein is anchored by the initial complexation.

2.4.   Protein Resolution

Immobilized reactive blue 2 has been employed to advantage in the purification of alcohol dehydrogenase from plant (Lamkin and King, 1976), yeast (Easterday and Easterday, 1974), and mammalian (Adinolfi and Hopkinson, 1978; Roy and Nishikawa, 1979) cellular extracts. The impure enzyme was applied to an immobilized dye column in a solution having an ionic strength below 0.1M. The enzyme was retained by the column and eluted by solvents containing either a fixed concentration of the competitive bioligand NAD or a linear gradient of NAD ranging in concentration from 0 to 5mM. The homodimeric isozymes A, B, and C of the mammalian enzyme can be resolved by a stepwise increase in the concentration of NAD (Adinolfi and Hopkinson, 1978). An analog of reactive blue 2 custom designed to more closely resemble the structure of NAD by placement of an aminoethyl spacer group between the interior diaminobenzene and the triazine ring (Fig. 1A) is reported to resolve a commercial preparation of the mammalian enzyme into two stable isozymes (Lowe et al., 1986).

Immobilized dye chromatography typically results in about a 50-fold enhancement in the specific activity of alcohol dehydrogenases. This enhancement was found sufficient to purify the enzyme from crude extracts of cottonseed to homogeneity in a single step (Lamkin and King, 1976). Immobilized dye chromatography of alcohol dehydrogenase has been scaled to a production level (Roy and Nishikawa, 1979) and has been also performed at high pressure by immobilization of the dye to a silica matrix (Lowe *et al.*, 1981). The results demonstrate that the interaction between reactive blue 2 and alcohol dehydrogenase can be used in a variety of chromatographic contexts to significantly enhance the fractional population of the enzyme with good recovery and laudable resolution.

3. PROTEIN PURIFICATION USING IMMOBILIZED DYE COLUMN CHROMATOGRAPHY

The principal application of immobilized dye column chromatography has been to purify proteins, a role which has enjoyed spectacular success largely due to the economy, stability, and capacity of immobilized reactive dye columns. While purification of some proteins from crude cellular extract to homogeneity has been achieved in a single step using immobilized dye chromatography, it is not a typical experience. Nonetheless, immobilized dye chromatography has significantly shortened the protocols required for purification of a great many proteins and has markedly improved their yield.

Most applications employ positive immobilized dye chromatography, in which the desired protein among others is retained by the column and then selectively eluted to enhance its fractional population. By contrast, negative immobilized dye chromatography has been used to selectively remove undesired proteins, such as albumin from serum or destructive catalysts such as proteases and nucleases, from protein solutions of interest. In some applications, negative and positive chromatography can be performed to advantage in tandem using two columns, each containing a different immobilized reactive dye (Scopes, 1986).

As in all affinity chromatography, selective enhancement of the concentration of a desired protein depends on several considerations: (1) the affinity of the desired protein for the immobile dye and for competitive mobile bioligands; (2) the concentration of the immobile dye accessible to the protein and the concentrations of the competitive mobile bioligands; and (3) the concentrations of other proteins and their affinities for the accessible immobile dye and for mobile bioligands. Accordingly, the chemical nature, accessible concentration, and density of the immobilized dye, the range of proteins and their relative concentrations in the sample injected, and the pH, temperature, ionic strength, and chemical composition of the chromatographic eluents all influence the retention of a protein. Each of these variables can be orchestrated to maximize the yield, the fold purification, and the fractional population of a desired protein.

3.1. Immobilized Dye Selection

It would appear that immobilized dyes can complex with high affinity to many protein crevices and pockets designed to complex bioligands. The flexible reactive dyes can apparently twist to complement both the topography and the polarity of the binding site for the bioligand, forming a multiplicity of interactions whose cumulative effects result in the observed high affinity of immobilized dyes for mobile proteins. While no one reactive dye can complement the biofunctional sites of all proteins, it appears that for most any protein a reactive dye can be identified which will complement its biofunctional site. Two commercial suppliers, Amicon Corporation and Sigma Chemical Company, market kits containing samples of a variety of reactive dyes immobilized to Sepharose for screening.

Additional mobile and immobilized reactive dyes are available from Affinity Chromatography Ltd., Aldrich Chemical Company, Gallard-Sclesinger Industries, ICN Biomedicals Inc., Polysciences Inc., Serva Fine Chemicals Inc., and Sigma Chemical Company. Reactive dyes that are not commercially produced for coloration can be chemically synthesized to extend the range of proteins which can be immobilized. A recent example is the synthesis of the anionic dye illustrated in Fig. 1 which, when immobilized, selectively retains proteases such as trypsin, thrombin, and carboxypeptides (Clonis *et al.*, 1987b). Such a column would be particularly useful for removal of protease contaminants by negative chromatography.

Mobile dyes are easily immobilized on a polysaccharide matrix (Clonis, 1988), the porous hydrophilic matrix Sepharose 6B-CL being the most popular. The matrix is suspended in an aqueous solution of reactive dye containing at least 6% NaCl and buffered at neutral pH to adsorb the dye to the matrix surface. The pH of the matrix suspension is then raised to about 10.5 to ionize some of the polysaccharide hydroxyl groups, which will then displace a chloride group from the triazine ring of the reactive dye, forming a covalent bond between the matrix and the dye.

Immobilized dye columns having polysaccharide matrices need to be washed periodically to remove the soluble oligosaccharide–dye complex which slowly accumulates as a result of matrix hydrolysis. Tightly bound proteins whose cumulative presence impedes retention and solvent flow can be eluted using a protein denaturant such as urea, guanidine hydrochloride, or an ionic detergent.

3.2. Protein Retention

Protein solutions should be introduced to immobilized dyes using solvent conditions which stabilize the biofunction of the desired protein. Whenever possible, the ionic strength of the protein solution should be less than $0.1M$, and the concentration of the bioligand specific for the protein of interest should be less than $1mM$. Minimization of the bioligand concentration is a concern because the immobilized dye and mobile bioligand compete for a common site on the desired protein. The greater the competition provided by the mobile bioligand, the weaker is the retention of the desired protein by the immobilized dye. High concentrations of nonspecific salts also weaken retention in several ways. Firstly, they weaken any electrostatic interaction between the immobilized polyanionic dye and the desired protein. Secondly, and more importantly, salts also function as chaotropic agents which weaken the hydrophobic interactions between the immobilized dye and the protein (Robinson *et al.*, 1981). Thirdly, salts decrease the numbers of immobilized dyes accessible to protein by enhancing the binding of immobilized dyes to the surface of the matrix (Liu and Stellwagen, 1987). As the number of available immobilized dyes is reduced, the more weakly retained proteins, perhaps including the desired protein, become mobile. Unwanted salts and bioligands can be removed most conveniently from protein solutions by selective precipitation of the proteins with polyethylene glycol 6000.

Since protein conformations are highly cooperative structures, complexation at one site can easily perturb the functionality of a distant site. Accordingly, it is not surprising to observe that the retention of a desired protein can be significantly modulated by complexation with divalent cations, allosteric effectors, and binding proteins at sites distant from that occupied by the immobilized dye (Clonis, 1988). Such modulation can be used to advantage in both the retention and the elution of a desired protein.

3.3. Protein Elution

The fold purification achieved by positive immobilized dye chromatography results principally from selective elution rather than selective retention. Since the affinity of a given protein for an immobilized ligand is sensitive in principal to a multiplicity of variables, such as pH, temperature, ionic strength, and ligand concentrations, elution protocols are developed to be workable rather than optimal. Clearly, the highest resolution among the retained proteins can be achieved using a shallow gradient of a competitive bioligand. However, the purity of the desired protein required, the cost of eluents, and the effort expended may dictate a lower resolution elution protocol involving batch elution with a fixed concentration of a competive bioligand, elution with a salt gradient, or even a batch elution with a fixed concentration of salt.

3.4. Large-Scale Protein Purification

Immobilized dye chromatography represents an attractive candidate for large-scale separations since the immobilized dye is cheap, stable, and versatile. While many large-scale applications of immobilized dye chromatography are likely proprietary, successful purifications at a pilot scale have been described (Scawen and Atkinson, 1987; Clonis, 1988). Economic viability at this and larger scales places a premium on screening for the optimal dye, a stable matrix, and a batch elution protocol using inexpensive components. These restrictions require considerably more experimentation at scale than is needed for a typical laboratory purification. The development of alternative matrices stable to a variety of harsh conditions including sterilization offers considerable promise for the use of immobilized dye chromatography in the production of diagnostic and therapeutic proteins (Lowe et al., 1990).

ACKNOWLEGMENTS

The author wishes to thank Professor Gerhart Kopperschlager for lively discussions of this topic. Financial support was provided by the U.S. Public Health Service research grant GM 22109 from the Institute of General Medical Sciences.

REFERENCES

Adinolfi, A., and Hopkinson, D. A., 1978, Blue Sepharose chromatography of human alcohol dehydrogenase: Evidence for interlocus and interallelic differences in affinity characteristics, Ann. Hum. Genet. **41**:399.

Biellmann, J. F., Samama, J. P., Branden, C. I., and Eklund, H., 1979, X-ray studies of the binding of Cibacron blue F3GA to liver alcohol dehydrogenase, Eur. J. Biochem. **102**:107.

Bohme, H., Kopperschlager, G., Schultz, J., and Hofmann, E., 1972, Affinity chromatography of phosphofructo-kinase using Cibacron blue F3G-A, J. Chromatogr. **69**:209.

Burton, S. J., Stead, C. V., and Lowe, C. R., 1988, Design and applications of biomimetic anthraquinone dyes: II. The interaction of C.I. reactive blue 2 analogues bearing terminal ring modifications with horse liver alcohol dehydrogenase, J. Chromatogr. **455**:201.

Bruton, S. J., Stead, C. V., and Lowe, C. R., 1990, Design and applications of biomimetic anthraquinone dyes: III. Anthraquinone-immobilized C.I. reactive blue 2 analogues and their interaction with horse liver alcohol dehydrogenase and other adenine nucleotide-binding proteins, J. Chromatogr. **508**:109.

Clonis, Y. D., 1988, The applications of reactive dyes in enzyme and protein downstream processing, in: *CRC Critical Reviews in Biotechnology*, Vol. 7 (G. G. Stewart, and I. Russell, eds.), CRC Press, Boca Raton, Florida, pp. 263–279.

Clonis, Y. D., Atkinson, T., Bruton, C. J., and Lowe, C. R., 1987a, *Reactive Dyes in Protein and Enzyme Technology*, Macmillan, Basingstoke, U.K.

Clonis, Y. D., Stead, C. V., and Lowe, C. R., 1987b, Novel cationic triazine dyes in protein purification, *Biotechnol. Bioeng.* **30**:621.

Dean, P. D. G., and Watson, D. H., 1979, Protein purification using immobilized triazine dyes, *J. Chromatogr.* **165**:301.

Easterday, R. L., and Easterday, I. M., 1974, Affinity chromatography of kinases and dehydrogenases on Sephadex and Sepharose dye derivative, in: *Immobilized Biochemicals and Affinity Chromatography* (R. B. Dunlop, ed.) (Plenum Press, New York), pp. 123–133.

Haeckel, R., Hess, B., Lauterborn, W., and Wuster, K., 1968, Purification and allosteric properties of yeast pyruvate kinase, *Hoppe-Seyler's Z. Physiol. Chem.* **349**:699.

Hogg, P. J., and Winzor, D. J., 1985, Effects of solute multivalency in quantitative affinity chromatography: Evidence for cooperative binding of horse liver alcohol dehydrogenase to blue Sepharose, *Arch. Biochem. Biophys.* **240**:70.

Kopperschlager, G., Bohme, H. J., and Hofmann, E., 1982, Cibacron blue F3G-A and related dyes as ligands in affinity chromatography, in: *Advances in Biochemical Engineering*, Vol. 25 (A. Fiechter, ed.), Springer-Verlag, Berlin, pp. 101–138.

Lamkin, G. E., and King, E. E., 1976, Blue Sepharose: A reusable affinity chromatography medium for purification of alcohol dehydrogenase, *Biochem. Biophys. Res. Commun.* **72**:560.

Liu, Y. C., and Stellwagen, E., 1986, Zonal chromatographic analysis of the interaction of alcohol dehydrogenase with blue-Sepharose, *J. Chromatogr.* **376**:149.

Liu, Y. C., and Stellwagen, E., 1987, Accessibility and multivalency of immobilized Cibacron blue F3GA, *J. Biol. Chem.* **262**:583.

Lowe, C. R., Glad, M., Larsson, P. O., Ohlson, S., Small, D. A. P., Atkinson, T., and Mosbach, K., 1981, High-performance liquid affinity chromatography of proteins on Cibacron blue F3G-A bonded silica, *J. Chromatogr.* **215**:303.

Lowe, C. R., Burton, S. J., Pearson, J. C., and Clonis, Y. D., 1986, Design and application of bio-mimetic dyes in biotechnology, *J. Chromatogr.* **376**:121.

Lowe, C. R., Burton, S. J., Burton, N., Stewart, D. J., Purvis, D. R., Pitfield, I., and Eapen, S., 1990, New developments in affinity chromatography, *J. Mol. Recog.* **3**:117.

Robinson, J. B., Jr., Strottmann, J. M., and Stellwagen, E., 1981, Prediction of neutral salt elution profiles for affinity chromatography, *Proc. Natl. Acad. Sci. U.S.A.* **78**:2287.

Roschlau, P., and Hess, B., 1971, Affinity chromatography of yeast pyruvate kinase with cibacronblau bound to Sephadex G-200, *Hoppe-Seyler's Z. Physiol. Chem.* **353**:441.

Roy, S. K., and Nishikawa, A. H., 1979, Large-scale isolation of equine liver alcohol dehydrogenase on a blue agarose gel, *Biotechnol. Bioeng.* **21**:775.

Ryan, L., and Vestling, C., 1974, Rapid purification of lactate dehydrogenase from rat liver and hepatoma: A new approach, *Arch. Biochem. Biophys.* **160**:279.

Scawen, M. D., and Atkinson, T., 1987, Large-scale dye-ligand chromatography, in: *Reactive Dyes in Protein and Enzyme Technology* (Y. D. Clonis, T. Atkinson, C. J. Bruton, and C. R. Lowe, eds.), Macmillan, Basingstoke, U.K., pp. 51–86.

Scopes, R. K., 1986, Strategies for enzyme isolation using dye-ligand and related adsorbents, *J. Chromatogr.* **376**:131.

Small, D. A. P., Lowe, C. R., Atkinson, T., and Bruton, C. J., 1982, Affinity labelling of enzymes with triazine dyes: Isolation of a peptide in the catalytic domain of horse-liver alcohol dehydrogenase using Procion blue MX-R as a structural probe, *Eur. J. Biochem.* **128**:119.

Staal, G., Vissar, J., and Veeger, C., 1969, Purification and properties of glutathione reductase of human erythrocytes, *Biochim. Biophys. Acta* **185**:39.

Stellwagen, E., 1990, Chromatography on immobilized reactive dyes, in: *Methods in Enzymology*, Vol. 182 (M. P. Deutscher, ed.), Academic Press, San Diego, California, pp. 343–357.

Swart, A. C. W., and Hemker, H. C., 1970, Separation of blood coagulation factors II, VII, IX, and X by gel filtration in the presence of dextrane blue, *Biochim. Biophys. Acta* **222**:692.

18

Pseudo-Biospecific Affinity Ligand Chromatography
The Case of Immobilized Histidine as a Universal Ligand

M. A. Vijayalakshmi

The term "pseudo-biospecific affinity ligands," introduced by Vijayalakshmi (1989), encompasses simple, small, and chemically defined molecules either of biological origin, such as single amino acids, or of nonbiological origin, such as triazine dyes from the textile industry or metal chelates.

The three classes of pseudo-biospecific affinity ligands—namely, amino acids, triazine dyes, and metal chelates—recognize protein molecules through somewhat similar mechanisms. These immobilized ligands all act as electron acceptors. The complementary donor groups in the proteins are invariably the NH-, SH-, or OH-containing lateral groups of amino acids such as histidine, cysteine, tryptophan, tyrosine, and perhaps serine. The feasibility of these interactions will depend on the solvent accessibility of these functional groups in the protein molecule.

Some proteins have been purified equally successfully by more than one of these three systems (Table 1). Often the optimal adsorption pH and ionic strength for purification are very similar with different ligands, especially in the case of dye-ligand and histidine-ligand affinity chromatography.

Among the three classes of pseudo-biospecific affinity ligands, the single amino acids, such as histidine and tryptophan have an inherent advantage over the other two classes, by virtue of their biological and hence nontoxic and biocompatible nature. In fact, the use of amino acids as specific ligands in some isolated specific case studies has been reported in the literature. For example, the purification of plasminogen and fibronectin using immobilized lysine and arginine, respectively, has been reported (Suamri *et al.*, 1976; Vuento and Vaheri, 1979).

The aromatic amino acids present a more versatile potential as immobilized ligands than the aliphatic amino acids. Tryptophan immobilized onto polysaccharide matrices was

M. A. Vijayalakshmi • Département Génie Biologique, Laboratoire d'Intéractions Moleculaires et de Technologie des Séparations, Université de Technologie de Compiègne, 60206 Compiègne Cédex, France.

Molecular Interactions in Bioseparations, edited by That T. Ngo. Plenum Press, New York, 1993.

TABLE 1
Some Proteins That Have Been Purified by More Than One
of the Pseudo-Biospecific Affinity Chromatographic (PBS-AC) Systems

| Protein | PBS-AC | Adsorption conditions | | | Reference(s) |
		pH	Ionic strength	Temperature	
IgG					
Monoclonal	IMAC-Zn	7.0	0.5M NaCl	20°C	Belew *et al.*, 1987
Polyclonal subclasses and monoclonal	HLAC	7.4	Low ionic strength	4°C	Kanoun *et al.*, 1986
Polyclonal	DLAC-CBF3GA	7.4	Low ionic strength	NS[b]	Byfield, 1988
Interferon-Hu-INF-β (fibroblast)	IMAC-Zn	7.2	PBS[c]	4°C	Edy *et al.*, 1977
	DLAC-CBF3GA	7.4	PBS	4°C	Jankowski *et al.*, 1976
	HLAC	7.4	PBS	4°C	E. Sulkowski, personal communication
Chymosin (aspartyl peptidase)	DLAC-CBF3GA	5.5	Low ionic strength	NS	Subramanian, 1987
	HLAC	5.5	Low ionic strength	4°C	Amourache and Vijayalakshmi, 1984
Collagenase (metalloprotease)	IMAC-Cu	7.5	1.0M NaCl	4°C	Berna and Vijayalakshmi, 1993
	DLAC-Reactive Red 120	7.5	Low ionic strength	4°C	Bond and VanWart, 1984
	HLAC	NS	NS	NS	Emöd *et al.*, 1977

[a] Abbreviations: IMAC, immobilized metal-ion affinity chromatography; HLAC, histidine-ligand affinity chromatography; DLAC, dye-ligand affinity chromatography; CBF3GA, Cibacron Blue F3G-A.
[b] NS, Not stated.
[c] PBS, Phosphate-buffered saline.

shown to be capable of separating interferon (Sulkowski *et al.*, 1976) and cellulases from *Trichoderma viridae* (Vijayalakshmi and Porath, 1979). The heterocyclic amino acid histidine, on the other hand, has proved to be a more versatile immobilized ligand, showing enormous potential in affinity chromatography purification of a plethora of proteins (Kanoun *et al.*, 1986).

Histidine has many properties which make it unique among the amino acids; these include its mild hydrophobicity, possibilities for weak charge-transfer interactions due to its imidazole ring, the wide range of its pK_a values, and its asymmetric carbon atom. Histidine residues also play a charge relay role in acid–base catalysis. Because of these properties, histidine can interact in many ways with proteins, depending upon conditions such as pH, temperature, and ionic strength. Moreover, when immobilized through the appropriate groups to polyhydroxylic matrices such as Sepharose or modified silica, specific dipole-induced interactions with proteins can occur.

Histidine's importance in biology is well elucidated. Histidine is often implicated in the catalytic mechanism of many enzymatic systems, mostly through proton transfer. The same residues are also involved in certain biorecognition events. Table 2 summarizes the biological systems in which histidine residues are involved and the mechanism through which they control the biological phenomenon.

TABLE 2
Examples Illustrating the Role of Histidine in Biological Phenomena

Biological system	Role of His	Reference
RNase A	His-12 proton donor	Nishikawa *et al.*, 1987;
RNase T	His-119 proton acceptor	Steyaert *et al.*, 1990
Serine protease	Acylation by the triad His-Asp-Ser	Blow *et al.*, 1969
Antigen, pH-dependent epitope	pH dependence of immunoadsorption equilibrium	Sada *et al.*, 1988
Collagenase	Inhibition	Harper, 1979
Phosphoglycerate keinase	Contribution to the mechanism of phosphorylation of phosphoglycerate to diphosphoglycerate	Fairbrother *et al.*, 1989
Human liver alcohol dehydrogenase	General catalysis	Torsten *et al.*, 1991
Diphtheria toxin	Binding of NAD^+	Papini *et al.*, 1989
Thymidylate synthase	Catalysis	Dev *et al.*, 1989
Mitochondrias F1F0 - ATPase (*E. coli*)	Nucleophilic attack between His-245 and Glu-219	Cain *et al.*, 1988

In this chapter, we try to show that, through the same type of interactions, immobilized histidine can recognize proteins and hence is useful for the purification of proteins.

1. PREPARATION OF SORBENTS WITH IMMOBILIZED HISTIDINE

Histidine can be immobilized onto OH-containing matrices with different orientations—through its α-NH_2 group or through its COOH function with or without the introduction of a spacer arm. Figure 1 shows the adsorbents that have been investigated for the purification of various proteins.

Moreover, different support matrices can be used for histidine coupling. Table 3 shows the various supports to which histidine has been chemically attached and examples of their use for the purification of proteins.

In addition to various insoluble particulate matrices, we have recently studied flat membranes of different chemical compositions, coupled with histidine (Manjini and Vijayalakshmi, 1993). The coupling chemistry was either through oxirane groups or through glutaraldehyde using the primary amine function (in the case of nylon and silica). The application of these membrane-based adsorbents for immunoglobulin G (IgG) purification was investigated. As shown in Table 4, their performance was comparable to that of a commercially available protein A-coupled membrane.

2. INTERACTION MECHANISM OF PROTEIN/PEPTIDE RECOGNITION BY IMMOBILIZED HISTIDINE

Histidine, with its mild hydrophobicity and possibilities for weak charge-transfer interactions due to its imidazole ring, wide range of pK_a values and the possibility for rapid changes in pK values depending on the environment (proton transfer), and the presence of an asymmetric carbon atom, can undergo multiple interactions with a protein molecule,

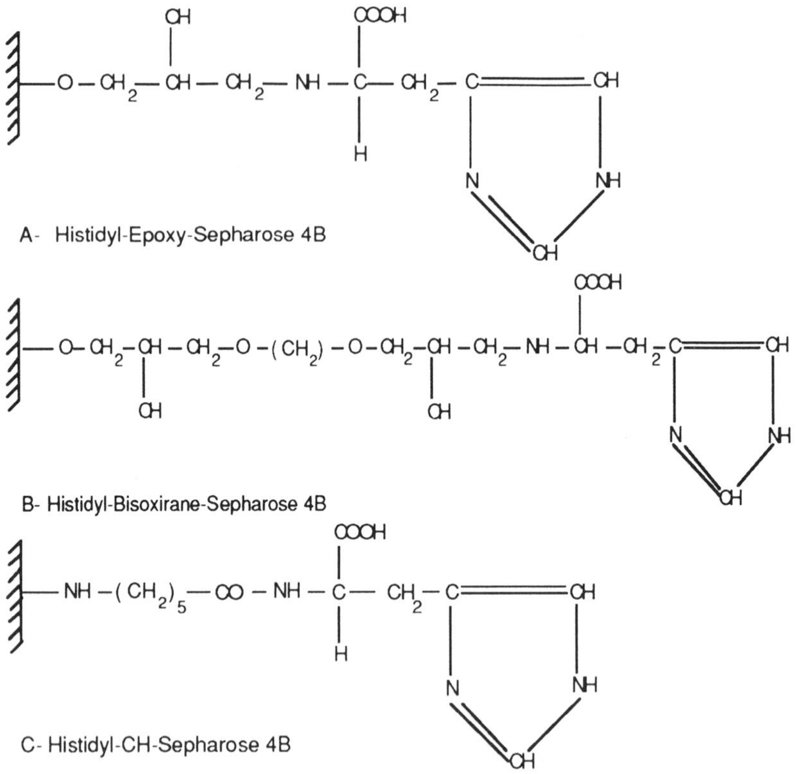

FIGURE 1. Schematic representation of the different imidazole-containing polysaccharide-based adsorbents.

depending on the pH, ionic strength, and temperature of the medium. Moreover, acyl histidine derivatives were reported to show unique detergent properties, which can be exploited for endotoxin removal. The interactions between phycocyanin chromopeptides and histidyl-Sepharose columns were shown to be mainly a weak charge transfer and proton exchange by the imidazole ring at pH values close to the pK values of both histidine (pH 6.5) and the chromopeptide (pH 5.8). Differences in the position of dipole groups (e.g., serine and the positively charged lysine and arginine residues) in the peptides contributed to the differential adsorption of the various chromopeptides (Rabier *et al.*, 1983). Several proteins and peptides have been purified using histidine-ligand affinity chromatography both in analytical high-performance liquid chromatography (HPLC) systems and on preparative scales. In all cases, the protein molecules were retained by the histidine ligand at or around the isoelectric point of the protein/peptide.

Histidine has a COOH group, an NH_2 group, and the imidazole side chain, all of which could contribute to charge-induced and hydrophobic interactions with a protein molecule. But what are their relative contributions? On the one hand, salt-promoted desorption suggests that electrostatic interactions with the —COOH groups predominate. However, when histidine was immobilized through its —COOH group, leaving the —NH_2 free, or when immobilized histamine was used, both systems showed the same optimal pH and ionic strength for the adsorption of human placental IgG1 (El-Kak and Vijayalakshmi, 1992). Moreover, imidazole-coupled gels have been shown to bind proteins through

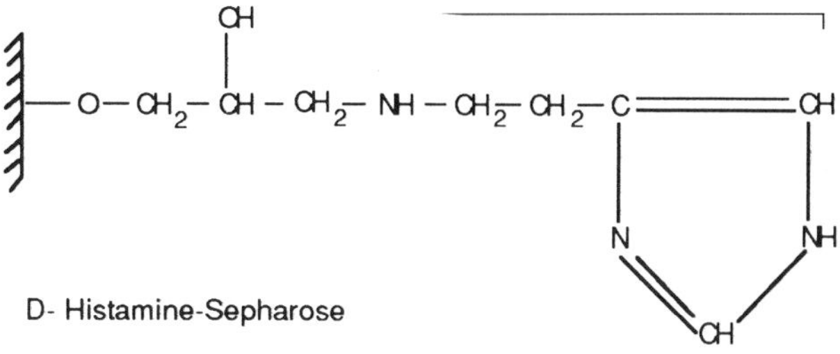

D- Histamine-Sepharose

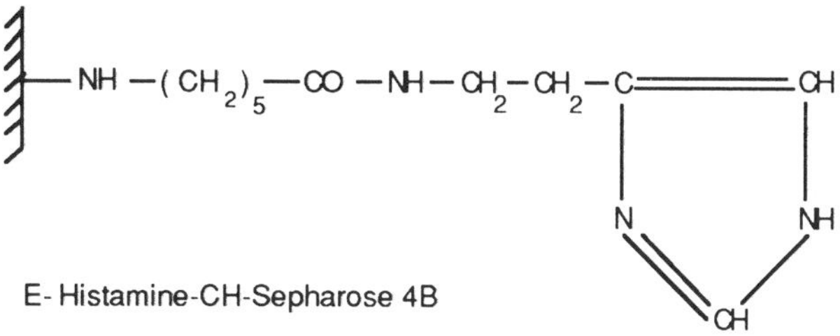

E- Histamine-CH-Sepharose 4B

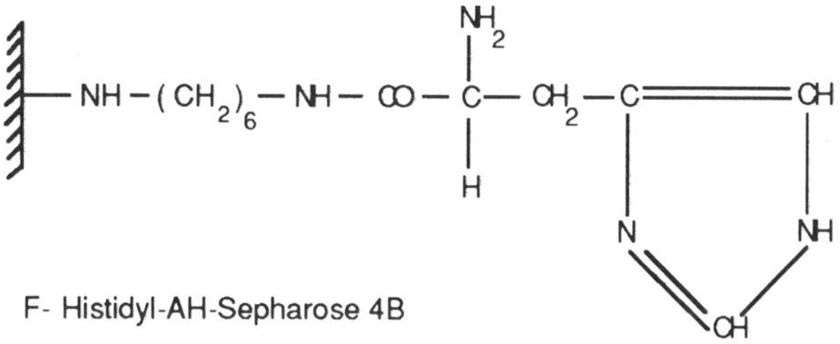

F- Histidyl-AH-Sepharose 4B

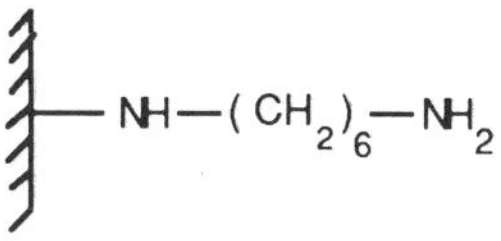

G- AH-Sepharose 4B

FIGURE 1 (Continued)

hydrophobic interactions. Furthermore, peptides and IgG which normally adsorb onto histidyl-Sepharose were not retained on that gel when imidazole or histidine was added to the mobile phase.

In addition, very recently it was shown that immobilized histidine could in some ways mimic protein A from *Staphylococcus aureus*, in terms of IgG binding. Both ligands bind IgG through its Fc fragment, and, moreover, the weak binding domain of protein A (N terminal) with a low binding strength has an essential histidine residue as its third amino acid (El-Kak *et al.*, 1992).

3. IMMOBILIZED HISTIDINE AS AN ALTERNATIVE TO IMMOBILIZED PROTEIN A FOR IMMUNOGLOBULIN (IgG) PURIFICATION

We have seen that adsorbents, both particulate supports and membrane-based supports, with immobilized histidine can selectively bind IgG from either plasma, placental

TABLE 3

Support Matrices Coupled with Histidine and Their Applications

Support to which His was coupled	Ligand concentration (μmol/g gel)	Applied to purification of	Capacity/ml gel
Sepharose 4B		Mouse mAbs (IgG1)	50 μg
Without spacer	37	IgG1 from human plasma	110 μg
		Factor VIII from human plasma	30 IU
		Depyrogenation of factor VIII from human plasma	ND[a]
		Myxaline (anticoagulant glycoprotein) from *Myxococcus xanthus*	3.7 IU
	Chymosin	ND	
With spacer	150	Factor VIII from human plasma	60 IU[b]
		IgG1 from human plasma	350 μg
Aminohexyl (AH)-Sepharose	52	IgG1 and IgG2 from human plasma	10.6 mg[c]
		Mouse mAbs (IgG1, IgG2b)	ND
		IgM from shark sera, mouse IgM	ND
Silica			
γ-Glycidoxypropyl (300 Å)	106	IgG1 from human plasma	57 mg[d]
Aminopropyl (100 Å)	166	IgG1 and IgG2 from human plasma	19 mg IgG1 and 6 mg IgG2
Sephacryl	350	Factor VIII from human plasma	75 IU
Fractogel TSK	24	Factor VIII from human plasma	8 IU
Nylon-methacrylate composite particles	ND	IgG1 human plasma	0.7 mg

[a]ND, Not determined.
[b]$K_d = 3.6 \times 10^{-5} M$.
[c]$K_d = 1.4 \times 10^{-6} M$.
[d]$K_d = 6.0 \times 10^{-6} M$.

TABLE 4

Comparison of Different Membranes

Membrane[a]	Specifications		Capacity for IgG (μg/cm^2)	Dissociation constant, K_d (M)
	Ligand	Pore size (μm)		
Nylon-methacrylate	Histidine	ND[b]	117	4.07×10^{-5}
PAL	Histidine	0.65	103	2.57×10^{-5}
Silica (FMC)	Histidine	1.00	145	9.67×10^{-5}
	Pore: 1.00μm,			
Nygene	Protein A	ND	173	5.69×10^{-5}

[a]Abbreviations: PAL, FMC and Nygene are trade names. PAL Biosupports, N.J., USA; FMC Bioproducts, ME., USA; Nygene Corporation N.Y., USA are the respective suppliers of PAL, FMC and Nygene Membranes.
[b]ND, Not Determined.

extract, or hybridoma cells in the monoclonal form. Moreover, their adsorption properties are similar to those of similar supports coupled with protein A from *Staphylococcus aureus* (Manjini and Vijayalakshmi, 1993; El-Kak *et al.*, 1992).

This mimicry of immobilized protein A by immobilized histidine has been further elucidated by the study of IgG from sources other than human sera.

As in the case of protein A-Sepharose 4B, histidyl-aminohexyl (AH)-Sepharose 4B shows different degrees of retention of IgG from various sources in the order pig > goat > sheep = monkey = horse > dog.

Furthermore, histidyl-AH-Sepharose showed good retention capacity for human, rat, and shark IgM, in contrast to protein A, which showed only a very low capacity for rat and shark antibodies.

Nevertheless, a systematic investigation of adsorption/desorption parameters for IgG retention onto histidyl-AH-Sepharose 4B such as pH, ionic strength, temperature, and the addition of chaotropic agents, such as urea and ethylene glycol, indicated very clearly a charge–charge mechanism of recognition, not one based on the net charge of the protein molecule and that of the histidine, but a more perfect charge balancing between the protonation of histidine due to α-NH$_2$ and imidazole together and the localized charges on the surface of the protein molecule. Moreover, a mechanism based on hydrogen bonding rather than a hydrophobic interaction mechanism was indicated due to the low energy requirement for the adsorption (El-Kak *et al.*, 1992).

In addition, an extensive study with model peptides and monoclonal antibody variants of a homologous series revealed a protonation mechanism based on H-bonding with Ser or Tyr residues of the peptides (Rabier *et al.*, 1984; El-Kak, 1991; El-Kak *et al.*, 1992).

4. SAMPLE PREPARATION

The upstream extraction methods used to prepare the crude extract of the protein to be purified play an important role in its retention behavior, as illustrated by the following examples.

Chymosin from calf or kid abomasum (sodium chloride extract) was selectively retained on a histidyl-Sepharose or histidyl-silica column; however, when the extraction was done with sodium chloride in the presence of sodium benzoate, no specific retention could be obtained (Amourache and Vijayalakshmi, 1984).

In the case of purification of IgG1 from human placental serum, a comparative study of ammonium sulfate precipitation and ethanol precipitation gave the results shown in Fig. 2.

In the case of an ammonium sulfate extract, all the proteins, including the IgG subclasses other than IgG1, were found in the breakthrough fractions (peak 1, Fig. 2a). In the case of ethanol-precipitated extracts, albumin was not retained (peak 1, Fig. 2b), and the IgG fractions other than IgG1 were slightly retarded (peak 2, Fig. 2b), whereas IgG1 was strongly retained and eluted with 200+mM sodium chloride (peak 2, Fig. 2a and peak 3, Fig. 2b) (Mbida *et al.*, 1989).

The purity of the extract used for the chromatographic step is another important factor. Table 5 shows a comparison of the purification of IgG1 from different placental extracts, S$_2$ and S$_4$, which had initial purities of 58 and 90%, respectively (Mbida *et al.*, 1989).

The moderate affinities (binding strength) observed with this kind of pseudo-

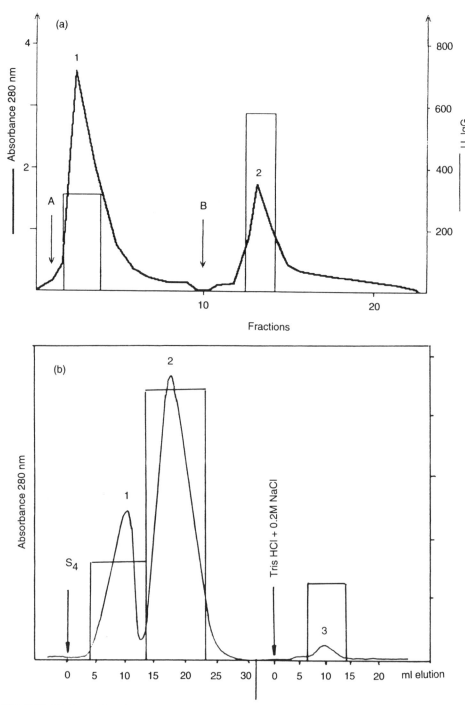

FIGURE 2. Elution profiles of placental serum extracts of varying degrees of initial purity (a) 0–50% ammonium sulfate extract; initial purity, 26%. (b) Ethanol fraction; initial purity, 90%. A: equilibrating buffer 25mM Tris-HCl pH 7.4; B: A + 0.2 M NaCl.

TABLE 5

Purification of IgG1 from Human Placental Serum:
Influence of the Purity of the Initial Extract on the Efficiency
of the Histidine-Ligand Affinity Chromatography (HLAC) Step

Precipitation method	Purity of initial extract (%)	Purity after HLAC (%)	Purification[a] (fold)
$(NH_4)_2SO_4$ (0–50%)	26	75	1.88
Ethanol			
Fraction S_2	58	95.6	4.16
Fraction S_4	90	98.9	6.45

[a]Purification fold is calculated with reference to a chloroform extract. Purity of chloroform extract, 14%; NaCl extract, 1.3%.

biospecific ligand, of the order of 10^{-4}–10^{-5} M, can explain this relation between the degree of purity of the initial extract and the efficiency of the chromatographic step. Various compounds may compete for the same sites, which in turn decreases the binding capacity for the protein of interest. A similar effect has been observed with metal-chelate adsorbents: the adsorption capacity of a purified enzyme was in the range of 50 mg of protein/ml of gel, whereas in the case of a crude extract it was only about 30 mg/ml, albeit with the specificity conserved (Krishnan *et al.*, 1987).

Similar observations were recorded during the purification of factor VIIIc on histidyl-Sepharose 4B columns. Table 6 shows the influence of the purity of the starting material on the final quality and yield of the product. While a starting material—a cryoprecipitate—with a specific activity of 1.59 U/mg yielded a final product with a specific activity of only 29.5 U/mg, another partially prepurified material with a specific activity of 159 U/mg yielded a final product with a specific activity of 875 U/mg (Ezzedine, 1990).

TABLE 6

Influence of the Purity of the Starting Material on the Separation
of Human Factor VIIIc on Histidyl-Bisoxirane-Sepharose 4B

Starting material[a]	Factor VIII (U/ml)		Specific activity (U/mg protein)		Total units		Yield (%)	
	Before HLAC	After HLAC	Before HLAC	After HLAC	Before HLAC	After HLAC	Before HLAC	After HLAC
Cryoprecipitate	4	16.8	1.59	29.5	400	395	100	99
CSD with pH adjusted to 6	1	3.8	ND[b]	ND	78	76	100	97.4
THP	30	35	150	875	3000	2800	100	93

[a]Abbreviations: CSD, Solvent Detergent treated Cryoprecipitate; THP, An internal code of National Blood Transfusion Centre, France, indicating very high purity extract.
[b]ND, Not determined.

5. CHROMATOGRAPHIC PARAMETERS

The chromatographic parameters for the adsorption and desorption steps can be used to fine-tune affinity-based recognition. The influence of these parameters is accentuated when biomimetic ligands with a very wide specificity and low binding strengths are employed (Vijayalakshmi, 1989; Ohlsson *et al.*, 1988). Moreover, a detailed knowledge of the retention mechanism of the desired molecule and the contaminants present along with it will enable one to adjust the buffer composition in a more precise manner to induce a high selectivity either at the adsorption step or at the desorption step.

We have seen above that the major contribution to the binding mechanism comes from a charge–charge mode of interaction. However, the optimal pH of adsorption of various proteins was found to be very close to their isoelectric point, where the net charge on the protein is zero (Vijayalakshmi, 1989). Moreover, at pHs below and above this optimum pH, the adsorption of a given protein was almost negligible. Nevertheless, the adsorption was favored by low-ionic-strength buffer. In fact, we had to adjust the resistivity of the injected sample to about 1100 ohms in order to adsorb factor VIIIc on histidyl-Sepharose 4B (Ezzedine, 1990).

6. ADSORPTION

Ideally, proteins should be adsorbed on the immobilized histidine adsorbents at low ionic strength and at pHs at or around the isoelectric point of the protein. Moreover, Tris buffer favored the adsorption of IgG and factor VIII as compared to a phosphate buffer at the same pH and molarity.

In the case of highly aromatic molecules, such as bioamines, however, a high salt concentration favored their adsorption.

7. DESORPTION

Desorption of proteins is invariably achieved by the addition of sodium chloride to the adsorption buffer. However, the optimal concentration of sodium chloride varies from $0.2M$ to $0.5M$ for different proteins. IgG1 and some proteases were successfully eluted by adding $0.2M$ sodium chloride to the initial buffer, whereas chymosin can be desorbed with $0.5M$ sodium chloride.

In fact, whereas IgG1 could be eluted from a histidyl-AH-Sepharose column by the addition of $0.2M$ NaCl to the adsorption buffer, IgG2 was eluted only at higher added NaCl concentrations.

Myxalin (a glycopeptide showing blood anticoagulant activity) from the culture medium of *Myxococcus xanthus* was eluted with $0.5M$ and $1.0M$ sodium chloride, using a stepwise gradient from 0 to $1.0M$ sodium chloride, while the contaminants were eluted at a lower sodium chloride concentration (Akoum *et al.*, 1989b).

In the case of factor VIIIc, an efficient desorption was achieved by both increasing the ionic strength and increasing the pH from 6.0 to 7.0. The ionic strength increase was achieved using a mixture of $0.1M$ Gly, $0.3M$ Lys, and $0.3M$ $CaCl_2$. This eluting buffer composition was chosen in order to avoid any activation of factor VIIIc and hence its denaturation (Ruttyn et al., 1989).

8. SCALING UP AND SCALING DOWN

Scaleup from 30 to 500 ml of adsorbent was easily achieved in the case of Myxalin recovery from the *Myxococcus xanthus* culture broth. The laboratory scale (30-ml bed volume) allowed the recovery of 15 mg of Myxalin from 1500 ml of microfiltered, heat-treated culture broth, whereas at a pilot scale we used a Sepragen radial flow column to recover 200 mg of Myxalin from 20 liters of the microfiltered, heat-treated culture broth. The efficiency of the operation in terms of both yield and purity of the product was unimpaired, or even improved, with scaleup of the operation (Table 7). Thus, this technique, developed on a laboratory scale, can be directly scaled up without difficulty (Akoum et al., 1989a).

The scaling down of this pseudoaffinity technique can be achieved by coupling histidine to an appropriate particle-size matrix after introducing the oxirane active groups on to the silica. As an example, we used a histidine-coupled Lichrosorb Si 60 matrix (0.8 × 10 cm) in a stainless steel column at 100-bar pressure and connected to a Waters HPLC system. The detection of residual chymosin activity in milk whey from cheese manufacture could be achieved by using the same chromatographic conditions (pH, ionic strength, and temperature) as in the open-column mode with a flow rate of 1.0 ml/min.

Other HPLC applications of histidine-ligand affinity chromatography include the analytical and preparative separation of IgG—subclasses IgG1 and IgG2—from serum extract or from hydridoma cell extracts, using silica particles of different size and porosity, with the histidine coupled through either its amino or its carboxyl group (Table 3).

TABLE 7
Purification of Myxalin by Two-Step Chromatography Using Linear-
and Radial-Flow Histidine-Ligand Affinity Chromatographic Columns

Sample	Activity (IU)		Purification (fold)		Activity yield	
	RF[a]	LF[b]	RF	LF	RF	LF
Crude extract	52	10.4	1	1	100	100
Chromatography on histidyl-Sepharose						
Fraction IV	31	4.2	13	9.2	59.6	40.4
Fraction V	41.8	5.6	33.5	21.5	80.4	53.8
Gel filtration						
Fraction IV	32.8	4.4	18.5	12.8	63	42.3
Fraction V	44.6	6	45.2	27.8	85.8	57.7

[a]RF, Radial Flow.
[b]LF, Linear flow.

9.   REGENERATION

The histidine-coupled adsorbents (all the types shown in Fig. 1) are easily regenerated by washing with 3 column volumes of 0.05M NaOH, followed by water and the starting buffer. However, this method of removing the nonspecifically and strongly adsorbed molecules is not suitable for silica-based sorbents. In the latter case, a longer wash with 1.0M NaCl followed by water was used. The efficiency of the regeneration was verified by the reproducibility of the purification efficiency and yield over 50 cycles.

10.   PREPARATION OF PYROGEN-FREE, PURIFIED SERUM PROTEINS: IgG AND FACTOR VIIIc

While histidine-coupled supports can be used for the purification of a wide range of proteins (from microbial sources, hydridoma cells, etc.), their usefulness for the purification of IgG and coagulation factor VIIIc has been particularly well demonstrated by extensive studies (Kanoun *et al.*, 1986; Mbida *et al.*, 1989; Ezzedine, 1990; El-Kak, 1991).

Moreover, it has been shown that histidine coupled Sepharose 4B could equally well be used for removing endotoxins of different origins and that the conditions for elution of endotoxins and those for elution of IgG and factor VIII were different. Thus, two columns of histidine-coupled Sepharose connected in series could be used to purify IgG and remove the pyrogens in a sequential operation.

11.   PURIFICATION OF IgG (SUBCLASSES) FROM HUMAN SERUM

Histidine-coupled aminohexyl-Sepharose 4B is the best suited sorbent for the purification of IgG (subclasses 1 and 2). In a typical procedure, Cohn fraction II from human plasma and histidyl-AH-Sepharose 4B equilibrated at pH 7.4 using 25mM Tris-HCl buffer at 4°C were employed. Elution was carried out using a step gradient of 0.2 to 1.0M NaCl added to the starting buffer. The IgG1 was specifically desorbed at 0.2M NaCl, while IgG2 was eluted at 1.0M NaCl (Fig. 3). The specificity and purity of the two subclasses were determined by immunoelectrophoresis using anti-IgG1 and anti-IgG2 antibodies, and the quantitative determination of IgG1 and IgG2 was done by the radial immunodiffusion technique. Table 8 summarizes the results obtained, and Fig. 4 shows sodium dodecyl sulfate-polyacrylamide gel electrophoresis (SDS-PAGE) analysis of the different fractions. Rechromatography on histidyl-AH-Sepharose 4B of the nonretained fraction and of the eluted IgG1 and IgG2 after the removal of the added NaCl confirmed the selectivity of the retention of IgG1 (El-Kak, 1991).

12.   PURIFICATION OF FACTOR VIIIc FROM HUMAN PLASMA

Histidine-coupled bisoxirane-Sepharose 4B was used for the recovery of factor VIIIc from either cryoprecipitate or a partially purified factor VIIIc. In the case of cryoprecipitate as starting material, 4 liters of plasma, corresponding to 4000 IU of factor VIIIc, were frozen and thawed at 4°C. This material was adsorbed onto an alumina gel, and the

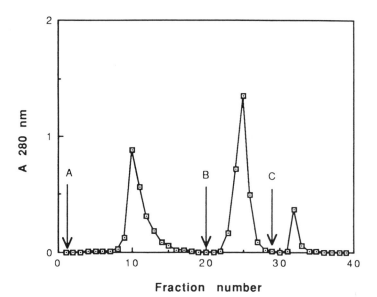

FIGURE 3. Typical elution profile of IgG subclasses from human placental extract/human serum extract on histidyl-aminohexyl-Sepharose 4B. A, 25mM Tris-HCl buffer, pH 7.4; B, 25mM Tris-HCl buffer, pH 7.4, a + 0.2M NaCl; C, 25mM Tris-HCl buffer, pH 7.4, a + 0.4M NaCl.

supernatant was subjected to diafiltration followed by precipitation at pH 6.0. The supernatant from this step, with a volume of 112 ml and containing 625 IU of factor VIIIc with a specific activity of 1.59 IU/mg of protein, was used for the chromatography on histidyl-bisoxirane-Sepharose 4B.

The partially purified factor VIIIc was obtained by the same procedure, except that an ion-exchange step was introduced prior to the histidyl-bisoxirane-Sepharose 4B chromatographic step. The product injected in this case was 100 ml at pH 6.0 containing 30 IU of factor VIIIc with a specific activity of 150 IU/mg of protein.

The chromatographic steps were:

1. Adsorption at pH 6.0 with 20mM Tris-HCl buffer at 4°C. In all the cases the injected material's ionic strength was adjusted to a resistivity of about 1100 ohms in order to

TABLE 8

Purification and Yield of IgG1 and IgG2 Subclasses from Human Plasma on Histidyl-AH-Sepharose 4B

Fraction	Protein (mg)	Total IgG (mg)*				Yield (%)		
		IgG1	IgG2	IgG3	IgG4	Protein	IgG1	IgG2
Fraction loaded	40	26.4	8.8	3.2	1.6	100	100	100
Peak 1 (nonretained)	17.1	4.9	7.1	3.2	1.6	43	18	81
Peak 2 (eluted with 0.2M NaCl)	21.7	21.3	0.43	0	0	54	80	5
Peak 3 (eluted with 1M NaCl)	1.13	0	1.13	0	0	3	0	13

*The quantities of IgG subclasses were determined by radial immunodiffusion method using anti IgG1; IgG2; IgG3 and anti IgG4 respectively from Janssen Biochimica Beerse, Belgium.

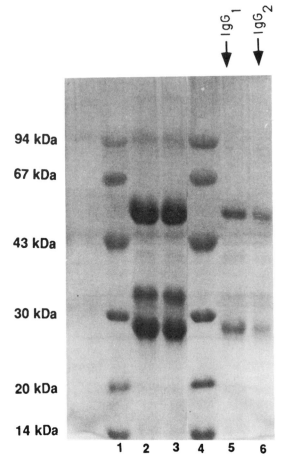

FIGURE 4. SDS-PAGE analysis of the different fractions described in caption to Fig. 3. Lanes 1 and 4, Molecular weight markers; lanes 2 and 3, samples before injection; lanes 5 and 6, pooled fractions eluted with buffer B and buffer C, respectively (see caption to Fig. 3).

ensure good adsorption. The adsorption was only partial if the pH of the starting material were below pH 5.8 or above pH 6.1 or even if NaCl concentrations as low as 20mM were employed to adjust the ionic strength.

2. Desorption was performed by using a solution composed of $0.1M$ Gly + $0.3M$ Lys + $0.3M$ CaCl$_2$ adjusted to pH 7.0. This eluent gave the best yield of nondenatured product, as lysine is known to stabilize factor VIIIc (Ruttyn *et al.*, 1989). However, when Tris-HCl buffers with different pHs and ionic strengths were used, the yields were very poor, perhaps owing to the low stability of factor VIIIc. Typical chromatographic profiles from the two starting materials are given in Figs. 5a and b. The compositions of the injected materials and the final purified preparations are given in Table 9.

13.   DEPYROGENATION (REMOVAL OF PYROGENS)

The pyrogen contamination of injectable therapeutics is a major concern in the pharmaceutical industry. Pyrogens are endotoxins of lipopolysaccharide nature, produced

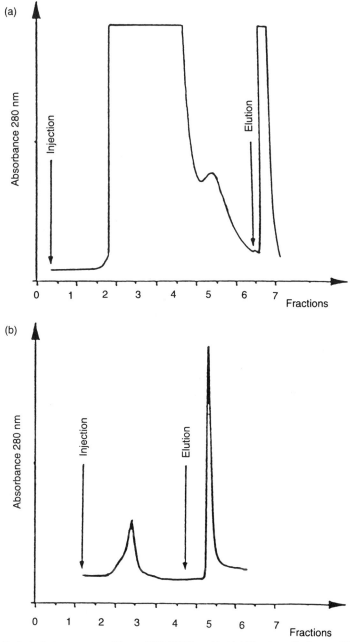

FIGURE 5. Typical elution patterns of factor VIIIc/VWF on histidyl-bisoxirane-Sepharose 4B. (a) Dialyzed cryoprecipitate with a specific activity of 1.59 U/mg; (b) prepurified cryoprecipitate with a specific activity of 150 U/mg (see text for the experimental conditions). The F VIII/VWF represent the complex Factor VIIIC coagulating + Von Willebrand Factor.

TABLE 9

Comparison of the Compositions of the Starting Material and
the Fraction Eluted from Histidyl-Bisoxirane-Sepharose 4B

Protein (quantity/ml)	Cryoprecipitate		Prepurified cryoprecipitate	
	Before HLAC	After HLAC	Before HLAC	After HLAC
Protein (mg)	2.51	0.568	0.2	0.04
Factor VIIIc (U)	4	16.8	30	30
Factor VIIIcAg (U)	5.2	18.2	65	35
vWf:Ag (U)	9	27.8	17	10
Fibrinogen (mg)	0.12	0.018	0.05	0.015
Fibronectin (mg)	0.165	0.295	0.027	0.01
IgG (mg)	0.4	0.02	0.12	—[a]
IgA (mg)	0.1	0.02	<0.025	—
IgM (mg)	0.03	0.02	—	—
Albumin (mg)	0.9	—	Trace	—

Factor VIIIc Ag: F VIII antigen units assayed by solid phase sandwich
technique using alloy antibodies from two multitransfused haemophilic
patients; VWF: Ag Von Willebrand factor antigen assayed by enzyme linked
immunosorbent assay.
[a]—, undetectable.

by gram-negative bacteria. Many approaches to removal of these pyrogens have been
attempted, ranging from heat denaturation to detergent treatment and including some
chromatographic methods (Pearson *et al.*, 1984). It is obvious that chromatographic
methods are preferable, owing to the minimum denaturation of the end product during the
process.

Minobe *et al.* (1982) used histamine-coupled Sepharose for the removal of pyrogens.
Histamine as a ligand is not very attractive in terms of regulatory restrictions. Histidine-
coupled Sepharose represents a better alternative, and we have studied pyrogen removal by
this sorbent. Table 10 shows the efficiency of pyrogen removal, using endotoxins of
different origins.

When endotoxin-spiked samples of factor VIIIc and IgG were tested, the retention of
endotoxins was optimum at pH 6.0–6.1, which is the same as the pH optimum for the
retention of factor VIIIc. However, the optimal conditions for desorption were different for
factor VIIIc and the endotoxins. The retained endotoxins were not coeluted with factor

TABLE 10

Adsorption of Endotoxins from Different Microorganisms at pH 6

Origin of endotoxin	Retention of endotoxin (%)	Concentration of endotoxin in the effluent (EU/ml)
Escherichia coli j5	95	7
Salmonella typhosa	90	10
Salmonella minnesota RE 595	88	13

TABLE 11

Purification of Factor VIIIc and Endotoxin Removal Using Bisoxirane-Sepharose 4B

Fraction	Factor VIIIc (U/ml)	Volume (ml)	Endotoxin (EU/ml)	Factor VIIIc (%)	Endotoxin (%)	Ratio Endotoxin: Factor VIIIc
Starting material	50	10	250	100	100	5
Unadsorbed	<0.05	30	0.19	<0.1	0.1	0
Fraction eluted	30	20	7	99	7	0.23

VIIIc when a $0.3M$ $CaCl_2$ + $0.1M$ Gly + $0.3M$ Lys buffer at pH 7.0 was used. Typical data for chromatography of endotoxin-spiked factor VIIIc are shown in Table 11.

14. CONCLUSION

The histidine-immobilized matrices are of enormous interest for the purification of different proteins, particularly serum proteins. In most of these cases, electrostatic forces and hydrogen bonds are the atomic forces implicated in protein recognition by the immobilized ligand. This is especially so when the histidine is coupled through its —COOH group, leaving the —NH$_2$ available for interaction.

Recent data generated by our group indicate that some enzymes, such as hydantoinase and catechol deoxygenase, seem to interact more selectively with histidyl-CH-Sepharose 4B (with the free —COOH group available for interaction; see Fig. 1) through a combination of electrostatic and hydrophobic interactions, facilitating their purification from the culture extracts of different *Pseudomonas* strains. It is to be noted that the substrates of these enzymes are invariably of an aromatic/hydrophobic nature.

It is thus obvious that histidine immobilized through the proper chemical function onto an insoluble support matrix can be a universal ligand that may be used for the successful and efficient purification of many proteins. Further, different charge-based and water-mediated interactions can be involved in the recognition of different proteins by this ligand immobilized through the appropriate chemical functions. This field, which is only at the beginning of its developments is thus extremely promising.

REFERENCES

Akoum, A., Vijayalakshmi, M. A., and Sigot, M., 1989a, Scale-up of "Myxalin" purification by pseudoaffinity method using a radial flow column, *Chromatographia* **28** (3/4) 157–160.

Akoum, A., Devichi, F., Kalyanpur, M., Neff, J. P., Vijayalakshmi, M. A., and Sigot, M., 1989b, Production and purification process for Myxalin: A new glycopeptide having anticoagulant property, *Process Biochem.* **1989** (April): 55–59.

Amourache, L., and Vijayalakshmi, M. A., 1984, Affinity chromatography of kid chymosin on histidyl-Sepharose, *J. Chromatogr.* **303**:385–390.

Belew, M., Yip, T. T., Andersson, L., and Ehrnström, R., 1987, High performance analytical applications of IMAC, *Anal. Biochem.* **164**:457–465.

Berna, P., and Vijayalakshmi, M. A., 1991, Immobilized metal-ion affinity chromatographic purification of collagenases from *Clostridium histolyticum* (in preparation).

Blow, D. M., Birktoft, J. J. and Hartley, B. S., 1969, Role of a buried acid group in the mechanism of action of chymotrypsin, *Nature* **221:**337–340.

Bond, M.D., and VanWart, H.E., 1984, Purification and separation of individual collagenases of *Clostridium histolyticum* using red dye ligand chromatography, *Biochemistry* **23:**3077–3085.

Byfield, P. G. H., 1988, Binding of prealbumin and immunoglobulin to Remazol yellow GGL, in: *Biotechnology of Plasma Proteins* (J. F. Stoltz and C. Rivat, eds.), Collaque INSERM, Vol. 175, pp. 185–189.

Cain, B. D. and Simoni, R. D., 1988, Interaction between Glu-219 and His-245 within the a subunit of F1Fo-ATPase, *Escherichia coli*, *J. Biol. Chem.* **263:**6606–6612.

Dev, I. K., Yates, B. B., Atashi, J., and Dallas W. S., 1989, Catalytic role of histidine 147 in *Escherichia coli* Thymidylate synthase, *J. Biol. Chem.* **264:**19132–19137.

Edy, V. G., Billiau, A., and De Somer, P., 1977, Purification of human fibroblast interferon by zinc chelate affinity chromatography, *J. Biol. Chem.* **252:**5934–5935.

El-Kak, A., 1991, Etude de l'interaction des immunoglobulines monoclonales et polyclonales avec un ligand de pseudobioaffinité l'acide aminé histidine. Aspects moléculaires, biochimiques et cinétiques, Ph.D. Thesis, Université de Technologie de Compiègne, France, pp. 134–135.

El-Kak, A., and Vijayalakshmi, M. A., 1991, Study of the separation of mouse monoclonal antibodies of pseudobioaffinity chromatography using matrix-linked histidine and histamine, *J. Chromatogr. Biomed. Appl.* **570:**29–41.

El-Kak, A., and Vijayalakshmi, M. A., 1992, Separation and purification of IgG1 and IgG2 from human plasma by pseudobioaffinity chromatography on histidyl-AH-Sepharose, BioSeparation (submitted).

El-Kak, A., Manjini, S., and Vijayalakshmi, M. A., 1991, Interaction of immunoglobulin G with immobilized histidine: Mechanistic and kinetic aspects, *J. Chromatogr.* **604:**29–37.

Emöd, I., Trocheris, I., and Keil, B., 1977, Separation of proteinases and in particular of clostripain and collagenase by affinity chromatography, in: *Affinity Chromatography* (O. Hoffmann-Ostenhof, M. Breitenbach, F. Koller, D. Kraft, and O. Scheiner, eds.), Pergamon Press, London, pp. 123–128.

Ezzedine, M., 1990, Utilisation d'un ligand general "Histidine" pour la purification et la depyrogenation du facteur anti-hémophilique a du plasma humain, Ph.D. Thesis. Université de Technologie de Compiègne, France, pp. 63–92.

Fairbrother, W. J., Walker, P. A., Minard, P., Littechild, J. A., Watson, H. C., and Williams, R. J. P., 1989, NMR analysis of site-specific mutants of yeast/Phosphoglycerate kinase: An investigation of the triose-binding site, *Eur. J. Biochem.* **183:**57–67.

Harper, E., 1979, Mechanism of action of collagenase. Irreversible inhibition by systeine; reversible inhibition by histidine or imidazole, *Federation proc.* **25:**790.

Jankowski, W. J., Muenchhausen, W. V., Sulkowski, E., and Carter, W. A., 1976, Binding of human interferons to immobilized CBF3GA: The nature of molecular interaction, *Biochemistry* **15:**5182–5187.

Kanoun, S., Amourache, L., Krishnan, S., and Vijayalakshmi, M. A., 1986, New support for the large-scale purification of proteins, *J. Chromatogr.* **376:**259–267.

Krishnan, S., Gaehel-Vignais, I., and Vijayalakshmi, M. A., 1987, A semi preparative isolation of carboxypeptidase isoenzymes from *Aspergillus niger* by a single immobilized metal ion affinity chromatography, *J. Chromatogr.* **397:**339–346.

Manjini, S., and Vijayalakshmi, M. A., 1993, Membrane based pseudobioaffinity chromatography of placental IgG using immobilized L-histidine, *J. Chromatogr.* (submitted).

Mbida, A., Kanoun, S., and Vijayalakshmi, M. A., 1989, Purification of IgG1-subclass from human placenta by pseudo affinity chromatography, in: *Biotechnology of Plasma Proteins* (J. F. Stoltz and C. Rivat, eds.), Colloque INSERM, Vol. 175, pp. 237–244.

Minobe, S., Watanabe, T., Sato, T., Tosa, T., and Chibata, I., 1982, Preparation of adsorbents for pyrogen adsorption, *J. Chromatogr.* **248:**401–408.

Muller-Schulte, D., Manjini, S., and Vijayalakshmi, M. A., 1991, Comparative affinity chromatographic studies using novel grafted polyamide and poly(vinyl alcohol) media, *J. Chromatogr.* **539:**307–314.

Nishikawa, S., Morioka, H., Kim, H. J., Fuchimura, K., Tanaka, T., Uesugi, S., Hakoshima, T., Tomita, K. I., Ohtsuka, E., and Ikehara, M., 1987, Two histidine residues are essential for Ribonuclease T1 activity as is the case for Ribonuclease A, *Biochemistry* **26:**8620–8624.

Ohlsson, S., Lundblad, A., and Zopf, D., 1988, Novel approach to affinity chromatography using "weak" monoclonal antibodies, *Anal. Biochem.* **169:**204–208.

Papini, E., Schiavo, G., Sandona, D., Rappuoli, R., and Montecucco, C., 1989, Histidine 21 is at the NAD+ binding site of Diphteria toxin, *J. Biol. Chem.* **264:**12385–12388.

Pearson, F. C., Dubczak, J., Nakashima, C., and Carpentier, D. F., 1984, *J. Parenter. Sci. Technol.* **38:**196–198.

Rabier, J., Vijayalakshmi, M. A., and Rüdiger, W., 1983, Affinity of phycocyanin chromopeptides to histidyl-Sepharose gels: A model for histidine–tetrapyrrol-interactions in biliproteins, *Z. Naturforsch.* **C38:**230–236.

Rabier, J., Vijayalakshmi, M. A., and Lambert, C., 1984, Affinity of phycoerythrin chromopeptides to histidyl, tyrosyl and tryptophyl sepharose gels, *J. Chromatogr.* **295:**215–219.

Ruttyn, Y., Brandin, M. P., and Vijayalakshmi, M. A., 1989, Chromatography of human plasma on aminohexyl Sepharose: Separation of factor VIII/v Wf and behaviour of factors II, VII, IX and X and antithrombin III. *J. Chromatogr.* **491:**299–308.

Sada, E., Katoh, S., Kiyokawa, A., and Kondo, A., 1988, Effect of histidine residues in antigenic sites on pH dependence of immuno-adsorption equilibrium, *Appl. Microb. Biotech.* **27:**528–532.

Steyaert, J., Hallenga, K., Wyns, L., and Stanssens, P., 1990, Histidine-40 of Ribonuclease T1 acts as base catalyst when the true catalytic base, glutamic acid-58, is replace by alanine, *Biochemistry* **29:**9064–9072.

Suamri, L., Spitz, F., and Arzadon L. J., 1976, Isolation and characterization of the affinity chromatography forms of human plasminogen and plasmins, *Biochem. Chem.* **251:**3693–3699.

Subramanian, S., 1987, Separation of chymosin and pepsin in calf rennet by dye ligand affinity chromatography, *Prep. Biochem.* **17**(3):297–306.

Sulkowski, E., Daley, W. M., and Carter, A. W., 1976, Interaction of human interferon with immobilized hydrophobic amino acids and dipeptides, *J. Biol. Chem.* **251:**5381–5385.

Torsten, E., Hurley, T. D., Edenberg, H. J., and Borson, W. F., 1991, General base catalysis in a glutamine for histidine mutant at position 51 of human liver Alcohol Dehydrogenase, *Biochemistry* **30:**1062–1068.

Vijayalakshmi, M. A., 1989, Pseudobiospecific ligand affinity chromatography, *Trends Biotechnol.* **7**(3):79–90.

Vijayalakshmi, M. A., and Porath, J., 1979, Charge-transfer and water-mediated adsorption, *J. Chromatogr.* **177:**201–208.

Vuento, M., and Vaheri, A., 1979, Purification of fibronectin from human plasma by affinity chromatography under non-denaturing condition, *Biochem. J.* **183:**331–337.

19

Synthetic Protein Surface Domains as Bioactive Stationary Phases

Metal Ion-Dependent Macromolecular Recognition and Biospecific Metal Ion Transfer

T. William Hutchens and Tai-Tung Yip

I. METAL ION-DEPENDENT MACROMOLECULAR INTERACTIONS

Significant developments toward the site- or domain-specific interaction of proteins with chemically defined stationary phases have been made during the last 10 to 20 years. Most of these efforts, however, were focused primarily on the need for improved protein purification strategies. One example was the introduction of new chemical methods for the immobilization of transition-metal ions (Porath *et al.*, 1975; Porath and Olin, 1983). Immobilized metal ion affinity chromatography (IMAC), although relatively slow to gain widespread acceptance, is clearly an important advancement in the field of protein purification. In our view, however, the use of immobilized metal ions can address questions that go beyond affinity chromatography and protein purification. Indeed, the interactions of peptides and proteins with surface-immobilized transition-metal ions presents an opportunity to investigate and ultimately *model* biospecific metal ion-dependent macromolecular recognition events. To help explore the development of this opportunity, we have evaluated (1) protein surface–metal ion interaction mechanisms (Hutchens *et al.*, 1988, 1989b; Hutchens and Li, 1988; Hutchens and Yip, 1990a,b, 1992a,b; Yip *et al.*, 1989; Yip and Hutchens, 1989), (2) metal ion-dependent macromolecular surface recognition (Hutchens *et al.*, 1981, 1989b), and (3) the use of immobilized metal ions to evaluate ligand-induced alterations in macro-

T. William Hutchens and Tai-Tung Yip • Protein Structure Laboratory, USDA/ARS Children's Nutrition Research Center, Department of Pediatrics, Baylor College of Medicine, Houston, Texas 77030. The work described in this chapter was supported in part by the U.S. Department of Agriculture, Agricultural Research Service Agreement No. 58-6250-1-003. The contents of this publication do not necessarily reflect the views or policies of the U.S. Department of Agriculture nor does mention of trade names, commercial products, or organizations imply endorsement by the U.S. Government.

Molecular Interactions in Bioseparations, edited by That T. Ngo. Plenum Press, New York, 1993.

molecular surface structure (Hutchens and Li, 1990; Hutchens and Yip, 1991b). We have shown both qualitative and quantitative examples of variability in affinity and selectivity in the biomolecular surface recognition of surface-immobilized metal ions. Until recently, our efforts in this area have been limited to the use of relatively simple, chemically defined (i.e., nonbiological) stationary-phase metal ion chelators of the type introduced by Porath and co-workers (Porath *et al.*, 1975; Porath and Olin, 1983).

Protein interactions with macromolecular surface-immobilized metal ions are biologically relevant. Metal ions *in vivo*, particularly within cells, are adsorbed (i.e., immobilized) on a variety of macromolecular surfaces. Such interactions can influence subsequent macromolecular recognition events of regulatory significance (Eisenstein *et al.*, 1989; Berg, 1990; Kaptein, 1991; Vallee *et al.*, 1991). If we are to exploit further our use of protein–metal ion interactions, we must understand better how *biospecific* macromolecular interactions of the type observed in nature are influenced by surface-bound metal ions.

We have shown that relatively minor differences in a given protein surface structure (ligand-induced) can result in a significant alteration in the affinity of that protein for the chemically defined types of immobilized chelator–metal ion complexes (e.g., Hutchens and Yip, 1991b, 1992b). This represented what we viewed as the inverse of protein surface recognition selectivity. It seems likely, therefore, that we may learn much more about protein surface recognition selectivity by using immobilized metal ion chelators that are biologically defined, namely, individual protein surface metal-binding domains.

The use of whole (intact) proteins for this purpose is generally very impractical above a certain size (e.g., 8–10 kDa) because it remains difficult to produce a large variety (and quantity) of defined structural mutants. Oriented immobilization of large proteins is also difficult. We chose, therefore, to begin with relatively small, well-defined metal-binding domains whose activities (i.e., metal binding and recognition) were determined largely by primary and/or secondary structure.

In light of this decision, our new objectives were (1) to produce adequate experimental models of protein surfaces, particularly those surface domains that are bioactive, (2) to understand the structural determinants of protein surface metal-binding reactions, especially those that affect subsequent macromolecular recognition events of regulatory significance, and (3) to identify means to affect the sequence- or surface-specific transfer of metal ions between proteins.

To accomplish these objectives, it was necessary to develop new and better methods for the rapid identification and characterization of individual protein surface metal-binding domains. We have shown that matrix-assisted UV laser desorption/ionization (MALDI) time-of-flight (TOF) mass spectrometry (Hutchens *et al.*, 1991, 1992b–d, 1993; Nelson and Hutchens, 1992; Yip and Hutchens, 1992b, 1993a, b; Yip *et al.*, 1993) can be used to evaluate specific peptide–metal ion complexes. We have now identified and characterized several specific metal-binding domains (from 5 to 71 residues in length) by this and/or a related procedure, electrospray ionization (ESI) mass spectrometry (Hutchens *et al.*, 1991, 1992a–d, 1993; Allen and Hutchens, 1992; Hutchens and Allen, 1992; Yip and Hutchens, 1992a, b, 1993a, b; Yip *et al.*, 1993). Once metal-binding protein surface domains have been identified, they can often be prepared synthetically, immobilized on a biocompatible stationary-phase surface, and shown to retain bioactivity (i.e., metal-binding capacity); selective and metal ion-dependent molecular recognition of these domains is demonstrated in this chapter.

2. IMMOBILIZED AND BIOACTIVE PROTEIN SURFACE METAL-BINDING DOMAINS: GENERAL EXPERIMENTAL DESIGN

The primary purpose of this chapter is to discuss strategies, outline methods that have been used with some degree of success, discuss problems, and present selected examples that may serve to increase general interest in the use of a biologically directed approach to the design and investigation of metal ion-dependent biomolecular interactions.

The strategy for construction and utilization of immobilized protein surface domains to investigate biologically relevant, sequence-specific metal ion recognition and transfer events is conceptually very simple; the complete experimental design, however, is cumbersome and detailed. Several different methodologies are involved. The preparation and successful use of stationary-phase reagents of the type described here generally requires completion of the following steps:

1. Demonstration of sequence-specific metal-binding sites exposed on the protein surface
2. Identification of an individual protein surface domain containing the metal-binding site(s)
3. Synthesis (or overexpression) of at least the essential amino acid sequence that defines the protein surface metal-binding domain in question
4. Purification (to homogeneity) of the synthetic (or recombinant) protein surface metal-binding domain
5. Verification of amino acid sequence and any posttranslational modifications that may affect metal-binding activity positively (e.g., phosphorylation) or negatively (e.g., intramolecular disulfide bond formation)
6. Confirmation of a defined structure in solution (e.g., composite secondary structure) and the dependence of observed structure on bound metal ions
7. Confirmation of metal-binding properties in solution (capacity, affinity, and specificity)
8. Immobilization of the bioactive peptide under conditions designed to preserve function [Note: Particularly for quantitative analyses, verification of ligand (i.e., immobilized protein surface domain) density is essential.]
9. Verification of metal-binding capacity and metal ion-dependent macromolecular recognition (i.e., bioactivity) of the immobilized protein surface domain.
10. Evaluation of immobilized protein domain stability.

3. MATERIALS AND METHODS

3.1. Materials

Fast-flow chelating (iminodiacetate) Sepharose was obtained from Pharmacia Fine Chemicals (Uppsala, Sweden). The model proteins (e.g., hen egg white lysozyme) and 2,5-dihydroxybenzoic acid (matrix for MALDI-TOF) were obtained from Sigma Chemical Company (St. Louis, Missouri). Human milk lactoferrin was purified in our laboratory by affinity chromatography on immobilized DNA, as described previously (Hutchens *et al.*, 1989a). Estrogen receptor proteins were prepared and radiolabeled, partially purified, and characterized as described previously (Hutchens *et al.*, 1990). Human plasma histidine-rich glycoprotein (HRG) was affinity-purified on immobilized tris(carboxymethyl)ethylene-diamine (TED)-Zn(II) ions (see below) exactly as described previously (Yip and Hutchens, 1991). Divinyl sulfone (DVS) was obtained from Fluka Chemie AG (Buchs, Switzerland). Imidazole (buffer grade), urea, copper sulfate, and zinc sulfate were obtained from E. Merck AG (Darmstadt, Germany). All other chemicals were of reagent grade.

3.2. Synthesis of Tris(carboxymethyl)ethylenediamine (TED)-Agarose Metal Chelate Gel

Immobilized tris(carboxymethyl)ethylenediamine (TED) groups on divinyl sulfone cross-linked agarose (6%) were synthesized and loaded with Zn(II) or Cu(II) ions as described previously (Porath and Olin, 1983; Yip and Hutchens, 1991).

3.3. Equilibration of Immobilized Iminodiacetate (IDA)-Sepharose with Metal Adsorbent

Except where indicated for temperature-controlled studies, all procedures were per-formed at room temperature (20–23°C). The chelating Sepharose was washed with several volumes of glass-distilled or Milli-Q water in a sintered glass funnel before being loaded with metal ions by equilibration with a 0.05–0.2M solution of copper sulfate. Excess (unadsorbed) metal ions were removed by washing the IDA-Cu(II) gel with 4 to 5 bed volumes of 0.1M sodium acetate buffer, pH 3.8–4.0. Finally, the loaded IDA-Me(II) gel was equilibrated into pH 7 column equilibration buffer consisting of 20mM sodium phosphate and 0.5M NaCl.

3.4. Equilibrium Protein Binding Analyses with Immobilized IDA-Cu(II) Ions

IDA-Sepharose gel was loaded with Cu(II) ions, equilibrated to pH 7, and allowed to settle in a 10-ml graduated cylinder to constant final bed volume (at least 30 min). A homogeneous gel suspension (50% v/v) was prepared, and 200-μl aliquots (100 μl of gel) were dispensed into duplicate or triplicate sets of small (3-ml) polystyrene incubation tubes containing 100 μl of protein solution at 6 to 10 different concentrations. After incubation for 30 min, with intermittent shaking, the tubes were centrifuged (30 s at 1200 rpm) to pellet the gel, and 100 μl of supernatant was removed for spectrophotometric or radio-metric determination of unbound protein concentrations. Equilibrium concentrations of unbound (free) protein (P) in the supernatant were subtracted from the total protein amounts added to determine ligand-bound protein (PL) at each concentration. These data were plotted according to the equation outlined by Scatchard (1949). The precise definition and use of variables to estimate parameters, such as total concentration of immobilized metal ion–protein interaction sites (L_t), were as described in detail by Hutchens et al. (1988).

3.5. Analysis of Protein Surface Structures

Protein crystal structures registered with the Brookhaven Protein Data Bank (Berstein et al., 1977) were evaluated using the program FRODO (Jones, 1982; version 6.6 by James W. Pflugrath, John S. Sack, and Mark A. Saper in the laboratory of Florante A. Quiocho at the Department of Biochemistry, Rice University, Houston, Texas) and the program ACCESS

(Lee and Richards, 1971; version 2.0 by B. Lee, F. M. Richards, T. J. Richmond, and M. D. Handschumacher, Yale University, New Haven, Connecticut) on a VAX 8810 computer equipped with an Evans and Sutherland PS390 molecular graphics terminal. The refined coordinates for human lactoferrin forms (Anderson *et al.*, 1989, 1990; Smith *et al.*, 1992) were generously provided by Dr. Edward N. Baker and colleagues at Massey University, New Zealand.

3.6. Isolation of Metal-Binding Protein Fragments by High-Performance Immobilized IDA–Metal Ion Interaction Chromatography

The TSK Chelate 5 PW (10-μm particle size, 750 mm × 7.5 mm i.d.; TosoHaas) was charged with Cu(II) ions as described above. Sample proteins were dissolved in column equilibration buffer (20m*M* sodium phosphate, 0.5*M* NaCl, pH 7.0) and applied (2 mg in 0.1 ml) to the column at a linear flow rate of 1.0 ml/min. Elution was initiated with a descending gradient of pH (pH 7 to 3.6) with programmed mixing of 50m*M* sodium phosphate (pH 3.8) with 0.5*M* NaCl into the column equilibration buffer. Columns were regenerated after each run by removing metal with 50m*M* EDTA (3 to 5 column volumes), after which they were washed with distilled water and recharged with Cu(II).

3.7. Identification of Protein Metal-Binding Domains by Mass Spectrometry

The sequential analysis of unfractionated mixtures of peptides in protein digests by mass spectrometry, performed before and after the addition of specific metal ions, greatly facilitates the identification of structurally defined protein metal-binding domains (Hutchens *et al.*, 1992d, 1993; Yip and Hutchens, 1993). Although our ability to characterize specific peptide–metal ion interactions has been improved significantly by these procedures, a variety of additional experimental procedures are often necessary to distinguish protein *surface* metal-binding domains from other metal-binding peptides exposed as a result of enzymatic digestion.

3.7.1. Proteolytic Digestion of Metal Ion-Binding Proteins

Purified proteins were dialyzed against 50m*M* EDTA, 20m*M* sodium phosphate, pH 7.0 and then against 20m*M* sodium phosphate, 0.15*M* NaCl, pH 7.0. Two micrograms of the dialyzed protein was incubated with 0.04 μg of trypsin or plasmin (Promega) in 30 μl of 5m*M* sodium phosphate (pH 7.5) at 37°C for 3.5 h; the protein samples digested were neither reduced nor denatured.

3.7.2. Matrix-Assisted UV Laser Desorption/Ionization (MALDI) Time-of-Flight (TOF) Mass Spectrometry

Aliquots of the peptide digests were diluted to ca. 10 nmol/ml and mixed (1:1) with a saturated aqueous solution of 2,5-dihydroxybenzoic acid (154.12 Da) in the presence and absence of copper sulfate (2 μL of a 20m*M* solution); 2μL of each mixture was applied to separate stainless steel probe tips (2-mm diameter) and air-dried at room temperature. Dried peptide-matrix deposits on the probe tips were washed in Milli-Q water to remove residual salts and redried. Mass spectrometry was performed on a Vestec model 2000 laser desorption linear time-of-flight mass spectrometer using the frequency-tripled output from a Q-switched neodymium–yttrium aluminum garnet (Nd–YAG) pulsed laser (355 nm, 5-ns pulse, Lumonics HY400). Ions desorbed by pulsed laser irradiation were accelerated to 30 keV and allowed to drift along a 2-m flight path (maintained at 30 μPa) to a 20-stage focused mesh electron multiplier. A LeCroy model TR8828D transient recorder (5-ns time resolution) and LeCroy 6010 MAGIC controller were used for the real-time signal averaging of multiple (100) laser shots. Data reduction was performed using PC-based software (LabCalc, Vestec).

3.7.3. Electrospray Ionization Mass Spectrometry (ESI)

Electrospray mass spectra were generated using a Vestec model 201 single-quadrupole mass spectrometer fitted with an electrospray ion source (Allen and Vestal, 1992) (Vestec Corporation, Houston, Texas) modified as described previously (Hutchens *et al.*, 1992b; Allen and Hutchens, 1992). The peptides were infused (ca. 2.5 nmol/ml) in water or 10mM ammonium acetate, pH 7.0, in the absence and presence of up to 100μM ZnSO$_4$ and/or CuSO$_4$. (Similar results were obtained with the use of sulfate and chloride metal salts.) Electrospray and data acquisition parameters, calibration, and data reduction based on the Fenn algorithm were performed exactly as described previously (Allen and Hutchens, 1992). The limited spectra presented here are displayed from mass/charge values (m/z) of 1020 to 1090 (ERDBD in the 8+ charge state).

3.7.4. Analysis of Peptide Mass and Assignment of Primary Sequence

The sequence-specific identification of individual peptide fragments was based, in part, on accurate (0.01%) molecular weight assignments obtained by MALDI-TOF mass spectrometry. The mass-dependent identification of both total and partial trypsin cleavage products (from proteins of known sequence) was performed on the program PROCOMP (Phillip C. Andrews, University of Michigan).

3.8. Synthesis of Specific Protein Surface Metal-Binding Domains

The four different protein *surface* metal-binding domains described here were se-lected, as indicated in each case, based upon a variety of published experimental evidence identifying the sequences chosen as being solvent-exposed and involved directly in either metal-binding and/or molecular recognition events known to be influenced by specific bound metal ions. The human HRG metal-binding domains (GHHPH)$_n$G (for n = 1, 2, 3, and 5), the human estrogen receptor dimerization domain (D473–L525) (ERDD) and DNA-binding domain (K180–M250) (ERDBD), and the N-terminal sequence of human β-casein (R1–K18) were each synthesized on an Applied Biosystems model 430A auto-mated peptide synthesizer using 2-(1H-benzotriazol-1-yl)-1,1,3,3 tetramethyluronium hexafluorophosphate as the coupling reagent in conjunction with 9-fluorenylmethyloxy-carbonyl/N-methylpyrrolidone (Fmoc/NMP) chemistry (FastMoc, Applied Biosystems, Foster City, California). The human β-casein R1–K18 phosphopeptide (five phosphoserine/phosphothreonine residues) was synthesized using the Fmoc; NMP/HOBt (N-hydroxy-benzotriazole) chemistry according to Applied Biosystems. All serine residues to be phosphorylated were coupled to the peptide chain without side-chain protection. The last amino acid, BOC-Arg, was manually coupled to the peptide chain at 60°C for 45 min. After removal of the Fmoc amino-terminus protecting group, the peptide was phosphorylated according to the procedure of Otvos *et al.* (1989). Finally, the peptide was deprotected and cleaved from the resin with 95% trifluoroacetic acid; thionanisole, 1,2-ethanedithiol, and phenol were used as scavengers. The phosphorylated and nonphosphorylated synthetic protein surface metal-binding domains were purified by a combination of reversed-phase and ion-exchange high-performance liquid chromatography (HPLC). The amino acid sequence was verified by sequential Edman degradation with an Applied Biosystems model 473A automated peptide sequence analyzer.

3.9. Amino Acid Analysis and Determination of Extinction Coefficient

The purified peptides were dialyzed into distilled water, hydrolyzed by acid vapor ($6N$ HCl at 150°C for 75 min), and derivatized with phenyl isothiocynate for quantitative amino acid analysis using an Applied Biosystems model 420A automated amino acid analyzer. The molar extinction coefficients of the apopeptides, the Cu-bound peptides, and the Zn-bound peptides were calculated at various wavelengths based on quantitative amino acid analyses of peptide solutions with known UV absorption properties.

3.10. Evaluation of Synthetic Protein Surface Metal-Binding Domains

3.10.1. Characterization of Intact Peptide–Metal Ion Complexes by MALDI-TOF and ESI Mass Spectrometry

A new and direct approach to the characterization of metal ion interactions with protein surface metal-binding domains is by mass spectrometry. As we have demonstrated recently, this may be accomplished using two new forms of soft-ionization mass spectrometry, namely, matrix-assisted laser desorption/ionization time-of-flight mass spectrometry and electrospray ionization mass spectrometry (Allen and Hutchens, 1992; Hutchens and Allen, 1992; Hutchens *et al.*, 1992a–c; Nelson and Hutchens, 1992).

3.10.2. Spectrophotometric Titration of Metal-Binding Domains with Metal Ions in Solution

Purified and lyophilized peptides were dissolved in either water alone or 10mM ammonium acetate (pH 7.0). For the cysteine-containing peptides (e.g., the "zinc-finger" DNA-binding domains), all solutions were prepared using Milli-Q water purged with nitrogen; the fully reduced status of the dissolved peptide was verified before each experiment by electrospray ionization mass spectrometry and by reaction with Ellman's reagent (5,5-dithionitrobenzoic acid) (Ellman, 1959). Solutions of peptide (5–20μM) were titrated with 2mM $CuSO_2$ or $ZnSO_4$ (or the chloride salts) at 24°C. Specific changes in UV absorbance were recorded from 300 to 200 nm. Incremental changes in metal ion-specific absorbance at several different wavelengths were plotted as a function of molar equivalents of added metal ion.

3.10.3. Estimation of Peptide-Bound Metal Ions by Atomic Absorption

Solutions of peptide were saturated with Zn or Cu, placed in a Spectrapor dialysis membrane (1000 molecular weight cutoff), and dialyzed against 5 changes (500 volumes each) of a nitrogen-purged solution of 10mM ammonium acetate (pH 7.0). Alternatively, samples were passed through a small column of Sephadex G-25 to remove free metal ions.

Aliquots of each of the peptide–metal ion complexes were analyzed against each of the *appropriate* control solutions (e.g., positive and negative controls including experimentally treated "blank" samples without protein or peptide) by atomic absorption on a Perkin-Elmer model 3030B atomic absorption spectrophotometer.

3.11. Structural Evaluation of Synthetic Protein Surface Metal-Binding Domains in Solution: Vacuum UV Circular Dichroism

Circular dichroism spectra into the far UV (170 nm) were collected on a Jasco J720 circular dichroism spectropolarimeter in a 0.05-mm quartz cell under nitrogen. Spectra were analyzed to predict secondary structure by the methods of Compton and Johnson (1986) and Manavalan and Johnson (1987).

3.12. Covalent Immobilization of Synthetic Protein Surface Metal-Binding Domains

Sepharose 6B (Pharmacia) was activated with divinyl sulfone (DVS) as described earlier by Porath and Axen (1976). Equivalent molar quantities of each of the peptides were coupled (10 μmol/g of gel) in 0.5M sodium bicarbonate at pH 8–9 for 20 h at 23–25°C. Remaining vinyl groups were blocked by incubation overnight with 10% glycine in 0.5M sodium carbonate at pH 9.4. Control preparations included DVS-activated agarose gel treated in an identical manner except for the addition of peptide. *Caution*: Divinyl sulfone is highly toxic and corrosive (see the Materials Safety Data Sheet). Wear gloves, eye protection, and a respirator. Perform all dispensing and washing steps in a chemical fume hood.

An alternative immobilization protocol involved the use of N-hydroxysuccinimide esters on a cross-linked BioGel (Bio-Rad Laboratories, Richmond, California) A-5m agarose gel (Affi-Gel 10, 75–300 mm). The Affi-Gel 10 agarose gel was washed with 5 volumes of isopropanol at −20°C and then mixed with the synthetic peptides (5 to 10 μmol/ml of gel) in isopropanol or dimethyl sulfoxide and allowed to incubate overnight at 4°C. Alternatively, the coupling can be carried out in the aqueous mode (see manufacturer's instructions). The Affi-Gel 10 must be quickly washed with 5 volumes of 50mM HEPES (pH 6.5–7) at 4°C, mixed with the peptide dissolved in the same buffer, and incubated overnight at 4°C. Remaining active groups were blocked by incubation with 10% glycine in 0.5M sodium carbonate at pH 9.4.

3.13. Metal-Binding Capacity of the Immobilized Protein Surface Metal-Binding Domains

Metal-binding capacities were confirmed by frontal analyses with copper sulfate (0.8mM) or [^{65}Zn]zinc sulfate (1.0mM) in 25mM HEPES buffer (pH 6.0–6.5). The void (dead) volumes were determined by pumping the same metal ion solutions into control columns of DVS-cross-linked agarose or Affi-Gel 10 agarose of the same dimension and bed volumes but prepared without immobilized peptides (i.e., glycine inactivated).

3.14. Evaluation of Immobilized Protein Surface Domain–DNA Interactions

Equal volumes of ERDBD immobilized on Affi-Gel 10 agarose were aliquoted into 1.5-ml capped tubes containing 200 μl of incubation buffer (10mM Tris-HCl, pH 8.0,

100mM NaCl, 10mM MgCl$_2$) $\pm$ 200μM ZnCl$_2$ $\pm$ 25 μg/ml of nonspecific calf thymus DNA. Synthetic oligonucleotides (30-mers) representing the human estrogen response element (ERE) DNA sequence (Klock *et al.*, 1987),

5$'$-CAAAGTCA**GGTCA**CAG**TGACC**TGATCAAAG-3$'$

3$'$-GTTTCAGT**CCAGT**GTC**ACTGG**ACTAGTTTC-5$'$

were prepared by Genosys Biotechnologies, Inc. (Houston, Texas). The oligomer probes were labeled with [γ^{32}P]-ATP (from ICN Pharmaceuticals Inc., Irvine, California) by incubation with T4 polynucleotide kinase before complementary strands were annealed (in 200mM NaCl) by heating to 95°C and cooling to room temperature (over 3h). Free [γ^{32}P]-ATP was removed by passing the reaction mixture through a Bio-Spin 6 chromatography column (Bio-Rad). Aliquots of 1 pmol of [^{32}P]-ERE were added to the immobilized ERDBD and mixed for 30 min at 23°C. Nonbound ERE probe DNA was removed by washing with 5 $\times$ 1 ml of incubation buffer. The bound ERE DNA was recovered by elution with 5 $\times$ 1 ml of incubation buffer containing 0.5M NaCl; further extraction with several column volumes of 3M urea and acid (0.1M citric acid, pH 2) did not reveal any additional radioactivity (recovery of the radiolabeled ERE was greater than 98%).

The metal ion-dependent interaction of soluble ERDBD with the soluble ERE DNA in solution was also monitored by nitrocellulose filter binding; similar results were obtained by both methods (Hutchens and Zhao, unpublished results).

4. WHAT HAS IMAC TAUGHT US ABOUT PROTEIN SURFACE METAL-BINDING DOMAINS?

Minor differences in the surface structures of some proteins can be initiated by their interaction with small ligand molecules. With respect to our understanding of protein surface interactions with immobilized metal ions, a revealing IMAC experiment involved the conformation-dependent retention of a *single* protein on IDA-Cu(II). Interpretation of these data in terms of variations in surface structure was possible because each of the different protein conformations was structurally defined by X-ray crystallography. Ligand-induced alterations in a protein conformation that result in significant alterations in the protein's affinity for another surface of immobilized metal ions illustrate the enormous potential for macromolecular recognition selectivity that may be operational *in vivo*. The following examples of surface interactions of human milk lactoferrin with stationary phases composed of simple, chemically defined (i.e., nonbiological) metal ion chelators are reviewed as evidence of the selectivity of natural protein surface metal-binding domains. This evidence was among the primary motivations to explore the development of new model systems based on immobilized protein surface metal-binding domains.

4.1. Biospecific (Ligand-Induced) Alterations in Protein Surface Recognition of Immobilized Metal Ions

The human milk protein lactoferrin is an 80-kDa protein composed of two similar lobes, each of which is constructed around a central metal ion binding site. The N-terminal lobe and the C-terminal lobe (each with two similar domains) have been shown by X-ray diffraction to exist in two different structural conformations (Anderson *et al.*, 1989, 1990).

The more compact or closed structure is favored when the protein is occupied with two bound metal ions (e.g., iron or copper). In the absence of a bound metal ion within each of its two lobes (i.e., apolactoferrin), however, an open conformation can exist (Anderson *et al.*, 1990). For example, metal binding at the N-lobe stabilizes a conformation in which nearly 50% of the N-lobe (25% of the entire protein) rotates 53° about the back end of the metal binding site. This ligand (metal ion)-induced conformational change cannot be detected by alterations in size or composite secondary structure (e.g., vacuum-UV circular dichroism). Although there is some evidence that the apo and holo forms are different in their electrophoretic mobilities, they cannot be separated by chromatography on ion-exchange or reversed-phase columns. We have shown, however, that this conformational change can be detected by a difference in the affinities of apo- and hololactoferrins for surface-immobilized Cu(II) ions (Hutchens and Yip, 1991b).

Figure 1 illustrates the sensitivity of surface-bound IDA-Cu(II) ions to ligand-induced alterations in lactoferrin surface properties. When purified apolactoferrin was applied to a column of immobilized Cu(II) ions (sample A), it eluted as two peaks. This was shown to be a result of the transfer of Cu(II) ions from the immobilized iminodiacetate to a portion of the added apolactoferrin sample by two simple tests. When the second peak (elution pH 4.30) derived from apolactoferrin sample A was isolated and reanalyzed under identical conditions (sample E), it eluted as a single peak with an elution pH exactly equal to its original elution pH; this elution pH was the same as that observed for lactoferrin deliberately saturated with Cu(II) ions prior to chromatography (sample B). In addition, the pH 4.30 peak and the lactoferrin saturated with Cu(II) both had UV/visible absorption spectra characteristic of lactoferrin with its specific metal binding sites occupied by Cu(II) ions. When the first peak (elution pH 4.86) derived from apolactoferrin sample A was isolated and reanalyzed under identical conditions (sample D), the elution profile again showed two peaks with elution pH values similar to those observed originally during the analysis of apolactoferrin (sample A). When the apolactoferrin was deliberately saturated with Fe(III) ions prior to analysis, only a single peak of hololactoferrin was eluted from the immobilized Cu(II) ions (sample C). Iron-saturated hololactoferrin, however, eluted from the column of immobilized Cu(II) ions at a pH (pH 4.47) which was intermediate between the elution pH of apolactoferrin (pH 4.86) and that of the Cu(II)-saturated hololactoferrin (pH 4.30); there was no loss of bound iron (Hutchens and Yip, 1991b). Thus, apolactoferrin, Fe(III)-saturated lactoferrin, and Cu(II)-saturated lactoferrin were all recognized differently by the same surface of immobilized Cu(II) ions and separated based upon these differences. Although modification of imidazole nitrogens by carboxyethylation with diethyl pyrocarbonate suggests that surface-exposed histidyl residues are most likely the primary residues involved in the interaction of lactoferrin with immobilized Cu(II) ions (Hutchens and Yip, 1991b), examination of the crystal structure of apolactoferrin and hololactoferrin did not reveal any differences in the degree to which the seven available histidyl residues in these two structures were exposed to solvent (Hutchens and Yip, 1991b). Note that the inclusion of $3M$ urea as a mobile phase-modifying reagent was necessary to resolve these three different conformational states of lactoferrin under relatively mild elution conditions. The presence of $3M$ urea may differentially affect the reactivity (i.e., pK_a) of individual His residues (e.g., Hutchens and Yip, 1991a). This possibility suggests differences in the structures of Fe(III)- and Cu(II)-saturated hololactoferrins *not yet apparent* from the high-resolution crystal structures of these three forms of the protein, which have recently been published (Anderson *et al.*, 1990; Smith *et al.*, 1992). To monitor

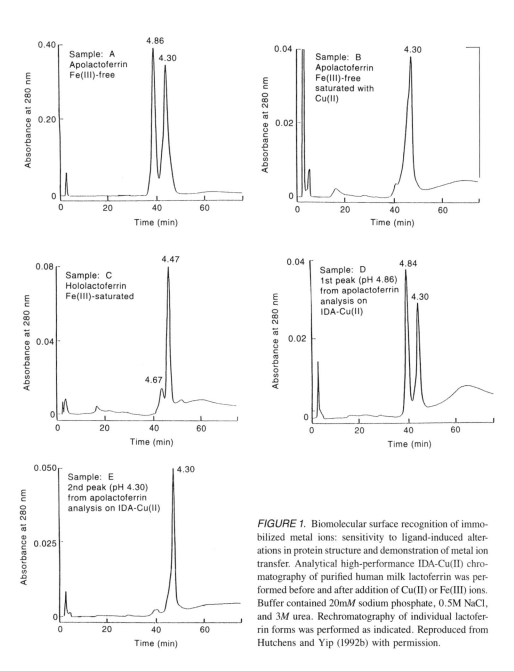

FIGURE 1. Biomolecular surface recognition of immobilized metal ions: sensitivity to ligand-induced alterations in protein structure and demonstration of metal ion transfer. Analytical high-performance IDA-Cu(II) chromatography of purified human milk lactoferrin was performed before and after addition of Cu(II) or Fe(III) ions. Buffer contained $20mM$ sodium phosphate, 0.5M NaCl, and $3M$ urea. Rechromatography of individual lactoferrin forms was performed as indicated. Reproduced from Hutchens and Yip (1992b) with permission.

and deconvolute the contribution of individual metal binding sites to the overall interaction heterogeneity observed, quantitative analyses are necessary.

4.2. Quantitative Analyses of Individual Factors Affecting Immobilized Metal Ion–Protein Interaction Heterogeneity

Quantitative evidence for the heterogeneity of lactoferrin interactions with the surface-immobilized IDA-Cu(II) ions is illustrated in Fig. 2A. The equilibrium binding of lactoferrin by agarose beads with immobilized IDA-Cu(II) revealed a curvilinear Scatchard plot and suggested the presence of multiple binding sites on the surface of lactoferrin with differing affinities for the immobilized Cu(II) ions. The inclusion of $3M$ urea decreased the overall affinity of lactoferrin for the immobilized Cu(II) ions but, in general, did not appear to eliminate the contribution of multiple types of interactions (affinities). It did seem to differentially affect the high-affinity class of sites relative to the others. The urea effects support other evidence that surface-bound water molecules are important determinants of protein–metal ion interactions (Hutchens and Yip, 1991a; Hutchens et al., 1992e).

Equilibrium binding assays are also useful for the evaluation of proteins that display only a single type of interaction affinity (thus possibly having a single protein surface metal-binding domain). The ability of urea to modify the *affinity* and, to a lesser extent, the interaction capacity of proteins for immobilized metal ions is demonstrated in Fig. 2B. In this case, hen egg white lysozyme (a relatively small protein with only one histidyl residue on its surface) was observed to vary markedly in its affinity for immobilized Cu(II) ions as a function of added urea (Fig. 2B). The effects of altered incubation temperature on the

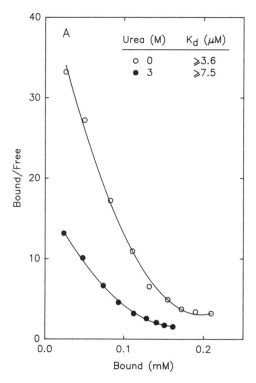

FIGURE 2. Proteins with single and multiple surface-exposed metal-binding domains: quantitative analyses of alterations in the protein surface recognition of surface-immobilized IDA-Cu(II) ligands. Scatchard plots show alterations in lactoferrin (A) and lysozyme (B, C) affinity for immobilized IDA-Cu(II) ligands as a function of the added mobile-phase modifier urea (A, B) or variation in temperature (C). The apparent equilibrium dissociation constants (K_d) are indicated for each of the profiles shown. In each case, 100 μl of IDA-Cu(II) gel was allowed to interact (30 min) with various protein concentrations in a 300-μl total reaction volume. The reaction buffer was $20mM$ sodium phosphate (pH 7.0) containing $0.5M$ NaCl and up to $3M$ urea as indicated. The equilibrium concentrations of unbound protein (P) were subtracted from total protein added to determine the immobilized ligand-bound protein concentrations (PL) (Hutchens et al., 1988). [Panel A reproduced from Hutchens and Yip (1992b) with permission.]

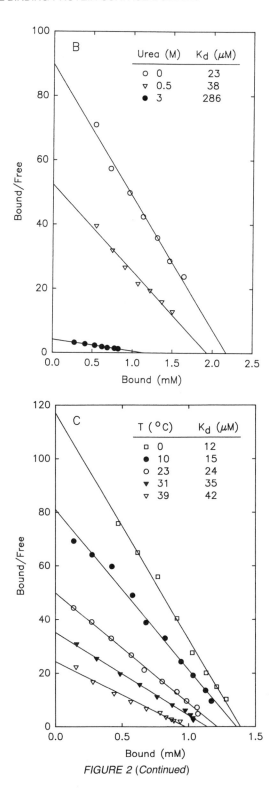

FIGURE 2 (Continued)

affinity, but not capacity, of lysozyme for immobilized Cu(II) ions were also readily apparent from the results of equilibrium binding assays (Fig. 2C). The use of equilibrium binding assays facilitates the evaluation of (1) protein–metal ion interaction heterogeneity and (2) experimental variables affecting protein–metal ion interaction specificity, capacity, and affinity.

To determine the contribution of local protein surface architecture to the interaction of these proteins with surface-immobilized metal ions, it is necessary to identify and/or isolate individual protein surface metal-binding domains. We have focused on those protein surface–metal ion interactions defined principally by contiguous residues of the primary sequence and/or secondary structure.

4.3. Individual Metal-Binding Sites and/or Domains on Protein Fragments: Retention of Affinity for Immobilized Metal Ions

In contrast to the spatial presentation of noncontiguous amino acid residues in a unique three-dimensional architecture to accomplish metal ion binding, the data presented in Fig. 3 and elsewhere (Yip *et al.*, 1989; Yip and Hutchens, 1989) suggest that the maintenance of secondary structure, or perhaps primary structure alone, may dictate metal binding affinity for surface-immobilized metal ions. The profiles presented in Fig. 3 resulted from the trypsin digestion of creatine phosphokinase (Fig. 3A) or bovine milk α-lactalbumin (Fig. 3B) followed by the resolution of the peptides in this mixture based upon their differing affinities for immobilized Cu(II) ions. Note the number of meal-binding fragments produced by the total and partial digestion of a single protein, particularly in the case of α-lactalbumin. There are only three His residues in this 123-residue protein; however, many different His-containing fragments can result from the total and partial digestion (only three result from total digestion) by trypsin. Metal-binding peptides from the plasmin digestion of purified bovine and porcine plasma HRG (see Yip and Hutchens, 1991) are being isolated by their affinity for TED-Zn(II) and used to sequence these proteins. Protein metal-binding domains from several different metal ion-binding and metal ion-transport proteins have been isolated in this manner (Yip and Hutchens, 1989) and identified (Yip and Hutchens, 1992b) as described below.

5. IMMOBILIZED PROTEIN SURFACE METAL-BINDING DOMAINS TO INVESTIGATE METAL ION-DEPENDENT MACROMOLECULAR INTERACTIONS

Four different examples of selective and metal ion-dependent molecular recognition of immobilized protein domains are presented. Each example illustrates the use of these model protein surface domains for different purposes. First, we illustrate how the immobilized HRG metal-binding domain (GHHPH)$_n$G was used to investigate the affinity and specificity of this unique sequence for different transition-metal ions and to isolate plasma proteins that may interact with this domain as a result of metal ion occupancy. Second, in a case where the native protein has yet to be isolated in quantities sufficient for such characterizations, we show how a synthetic dimerization domain of the human estrogen receptor has been used to investigate the role of metal ions in the interference of steroid-induced allosteric transitions between hormone-binding subunits. Third, we show that "zinc-finger" DNA-binding

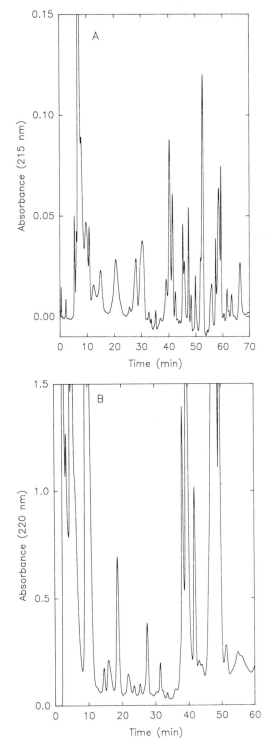

FIGURE 3. Retention of metal-binding properties by protein fragments: interaction with surface-immobilized Cu(II) ions. Separation of the trypsin digest products of creatine phosphokinase (A) or bovine milk α-lactalbumin (B) on a high-performance column of immobilized IDA-Cu(II) with a descending gradient of pH is shown.

domains can be used to investigate the metal ion-dependent recognition of specific DNA response element sequences. Finally, we show how an immobilized iron-binding domain from human β-casein was used to investigate macromolecular metal ion transfer.

6. IMMOBILIZED METAL-BINDING DOMAINS FROM HISTIDINE-RICH GLYCOPROTEIN

The human plasma Cu(II)- and Zn(II)-transport protein known as histidine-rich glycoprotein (HRG) (Koide *et al.*, 1986) was digested with either trypsin or plasmin and analyzed by mass spectrometry before and after the addition of metal ions (Hutchens *et al.*, 1992d,e). Because the sequence of this protein is known, it was possible to identify sequence-specific peptide–metal ion interactions based upon discrete shifts in peptide mass corresponding to one or more bound metal ions (Hutchens *et al.*, 1992d). These data, and data reported previously by other investigators (Morgan, 1985; Mulhoberac *et al.*, 1985), have been used to establish the domain of surface-exposed residues necessary for HRG interaction with metal ions, heme, and other organic ligands (Morgan, 1978).

6.1. Synthesis, Verification of Metal-Binding Properties in Solution, and Immobilization of HRG Surface Metal Binding Domains

A unique sequence of 26 residues on the surface of human HRG (Fig. 4A) defines the metal-binding domain of interest. This domain, composed of only three different types of amino acid residues (60% His), represents a tandem repeat (five units) of only five residues (GHHPH). Because the amino acid sequence $GHHPH_n$ exists elsewhere in HRG, in the form of one repeat ($n = 1$) up to five tandem repeats ($n = 5$), several individual peptides were synthesized ($n = 1, 2, 3,$ and 5), purified to homogeneity, and evaluated individually for their metal-binding properties in solution. Each of the synthetic peptides was found experimentally to bind transition-metal ions such as Zn(II) and Cu(II) to an extent that can be predicted by their primary sequence (composition and relative location of histidyl residues) (Table 1). An example of the MALDI-TOF mass spectra demonstrating the direct verification of metal-binding activity for the intact 26-residue $(GHHPH)_5G$ metal-binding domain is illustrated in Fig. 4B. Each of the different HRG peptides was then immobilized on Affi-Gel 10 and evaluated for *retention* of metal-binding capacity and affinity.

6.2. Specific Metal-Binding Properties of the Immobilized HRG Surface Metal-Binding Domains

Figure 5 presents the analysis of Cu(II)-binding capacity and affinity for columns of the immobilized HRG peptide where the $(GHHPH)_n$ sequence existed either as a single repeat ($n = 1$) or as a peptide with as many as five internal repeats ($n = 5$). Figure 5A demonstrates that, within experimental variation, the metal-binding capacities of the immobilized metal-binding domains $(GHHPG)_nG$ for $n = 1, 2, 3,$ and 5 were similar to those expected on the basis of analyses by mass spectrometry and titration of these peptides with metal ions in solution. Thus, the metal-binding capacity of the immobilized HRG peptides appeared to have been preserved without alteration. It is important to note that the type of immobilization chemistry will determine whether or not this observation can be made. In Fig. 5B (see also Hutchens and Yip, 1992a), we demonstrated that the use of

A

HUMAN PLASMA HISTIDINE-RICH GLYCOPROTEIN (HRG)

```
          10         20         30         40         50         60
    VSPTDCSAVE PEAEKALDLI NKRRRDGYLF QLLRIADAHL DRVENTTVYY LVLDVQESDC

          70         80         90        100        110        120
    SVLSRKYWND CEPPDSRRPS EIVIGQCKVI ATRHSHESQD LRVIDFNCTT SSVSSALANT

         130        140        150        160        170        180
    KDSPVLIDFF EDTERYRKQA NKALEKYKEE NDDFASFRVD RIERVARVRG GEGTGYFVDF

         190        200        210        220        230        240
    SVRNCPRHHF PRHPNVFGFC RADLFYDVEA LDLESPKNLV INCEVFDPQE HENINGVPPH

         250        260        270        280        290        300
    LGHPFHWGGH ERSSTTKPPF KPHGSRDHHH PHKPHEHGPP PPPDERDHSH GPPLPQGPPP

         310        320        330        340        350        360
    LLPMSCSSCQ HATFGTNGAQ RHSHNNNSSD LHPHKHHSHE QHPHGHHPHA HHPHEHDTHR

         370        380        390        400        410        420
    QHPH***GHHPHG   HHPHGHHPHG   HHPHGHHPH***C HDFQDYGPCD PPPHNQGHCC HGHGPPPGHL

         430        440        450        460        470        480
    RRRGPGKGPR PFHCRQIGSV YRLPPLRKGE VLPLPEANFP SFPLPHHKHP LKPDNQPFPQ

         490        500        507
    SVSESCPGKF KSGFPQVSMF FTHTFPK
```

B

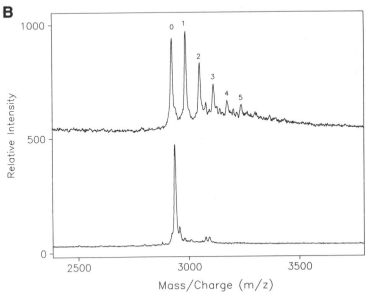

FIGURE 4. Synthetic protein surface metal-binding domains: verification of metal-binding capacity by matrix-assisted UV laser desorption/ionization time-of-flight mass spectrometry (MALDI-TOF). (A) The 26-residue peptide $(GHHPH)_5G$ from the surface of histidine-rich glycoprotein (HRG). (B) MALDI-TOF analyses of the synthetic HRG peptide $(GHHPH)_nG$ for $n = 5$ performed before (bottom profile) and after (top profile) addition of metal ions to the peptide in solution. Peaks labeled 0 through 5 represent HRG peptides with 0 to 5 bound Cu(II) ions.

TABLE 1
Summary of Synthetic HRG Peptide Metal-Binding Properties[a]

Analytical method	Concentration of HRG peptide (GHHPH)$_5$G	Concentration of bound Cu(II) (observed)	Cu(II):(GHHPH)$_5$G ratio
Dialysis and atomic absorption	16.7 μM	83 μM	5:1
Spectrophotometric titration (234 nm)	16.7 μM	85.5 μM	5:1
ESI mass spectrometry	8 μM	—[b]	5:1
MALDI-TOF mass spectrometry	<1 pmol	—[c]	5:1[d]

[a]Reproduced from Hutchens and Yip (1992b) with permission.
[b]ESI mass spectrometry was performed in the presence of excess CuCl$_2$ (separation of bound and free Cu is not necessary before analysis) [see Hutchens et al., 1992b].
[c]MALDI-TOF mass spectrometry was performed with the sample in the solid state (in the presence of excess CuCl$_2$)
[d]Values reported represent the maximum number of bound metal ions observed under the conditions used [see Hutchens et al. (1991, 1992b–e) and Nelson and Hutchens (1992)].

divinyl sulfone as a chemical cross-linking reagent yielded columns of immobilized HRG peptides almost indistinguishable to their metal-binding capacities despite a variation in the value of n from 1 to 5. The immobilized HRG peptides were next evaluated for metal ion-dependent recognition of specific peptides and proteins.

6.3. Specific and Metal Ion-Dependent Macromolecular Recognition of the Immobilized HRG Surface Metal-Binding Domain

Human plasma was collected into tubes containing EDTA and protease inhibitors (soybean trypsin inhibitor, benzamidine, D-phenylalanyl-L-arginine chloromethyl ketone, Na-tosyl-L-lysine chloromethyl ketone, and phenylmethylsulfonyl fluoride). All plasma was dialyzed at 4°C against 50mM EDTA, 20mM sodium phosphate, pH 7.0, and then against 20mM sodium phosphate with 0.5M sodium chloride, pH 7.0. The immobilized HRG peptide column was loaded with Cu(II) ions at pH 6.5 and equilibrated with 20mM sodium phosphate containing 0.5M NaCl, pH 7.0. Dialyzed plasma was applied; non-adsorbed and low-affinity binding proteins were eluted with column equilibration buffer. Adsorbed proteins were eluted with a descending pH gradient (pH 7 to 4) and, finally, 50mM EDTA.

As shown in Fig. 6A, in the absence of metal ions, the immobilized HRG peptides did not bind any plasma proteins under the conditions employed. In the presence of added Cu(II) ions, however, several specific plasma proteins were capable of recognizing the HRG peptides (Fig. 6B). Comparative separations of the same plasma samples were performed in parallel using columns of immobilized IDA-Cu(II) and TED-Cu(II) under identical experimental conditions. Most of the proteins were tightly bound by IDA-Cu(II) and required EDTA to elute (not shown). In contrast, the elution patterns of adsorbed plasma proteins from the columns of immobilized HRG metal-binding domain peptides and TED-Cu(II) (Fig. 6C) were similar. However, the sodium dodecyl sulfate (SDS)-polyacrylamide gel electrophoresis profiles of the plasma proteins that interacted with the HRG peptides in a metal ion-dependent manner were quite distinct from those of the plasma proteins retained by TED-Cu(II) (Fig. 6D). Thus, the sequence-specific presentation of metal ions by protein

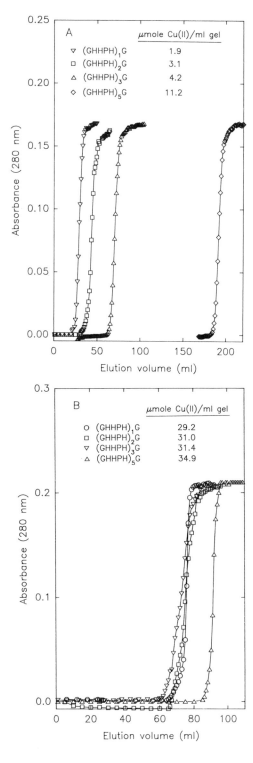

FIGURE 5. Immobilized protein surface metal-binding domains: quantitative evaluation of Cu(II)-binding capacity by frontal analysis chromatography. The individual HRG peptides $(GHHPH)_nG$ with n = 1, 2, 3, and 5 were immobilized on Affi-Gel 10 (A) or DVS-activated agarose (B) and verified to bind metal ions by frontal analyses. The calculated metal-binding capacities are as indicated. [Panel A reproduced from Hutchens and Yip (1992b) with permission.]

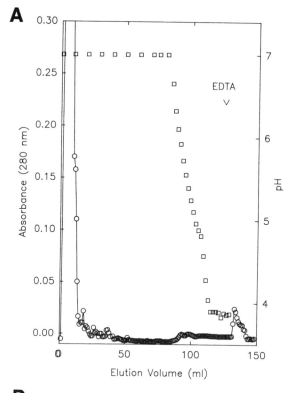

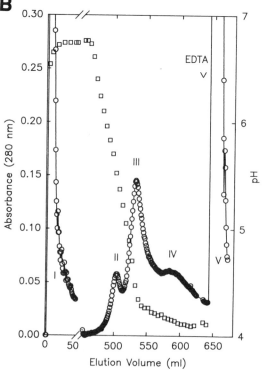

FIGURE 6. Bioselective and metal ion-dependent biomolecular recognition of immobilized protein surface metal-binding domains. Plasma proteins were allowed to interact with the immobilized 26-residue HRG peptide $(GHHPH)_5G$ in the absence (A) and presence (B) of bound Cu(II) ions and with immobilized TED-Cu(II) ions under identical conditions (C). The protein compositions of eluted plasma proteins in peaks labeled I through IV and V (EDTA eluted) were determined by silver staining after reduction, denaturation, and separation by sodium dodecyl sulfate polyacrylamide gradient gel electrophoresis (D). [Panels A and B reproduced from Hutchens and Yip (1992b) with permission.]

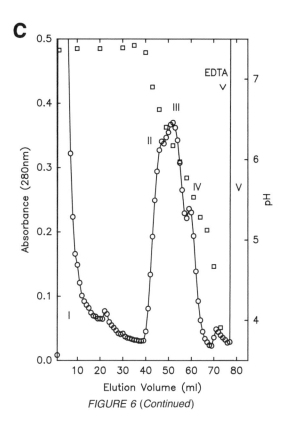

FIGURE 6 (*Continued*)

surface metal-binding domains is clearly different from that of chemically defined (non-biological) metal ion chelators.

7. IMMOBILIZED AND FUNCTIONAL DIMERIZATION DOMAINS: INTERACTIONS AFFECTED BY SPECIFIC TRANSITION METAL IONS

The use of immobilized protein surface metal-binding domains to investigate the contribution of specific metal ions to specific protein–protein recognition events is perhaps best illustrated thus far by the example of the intranuclear estrogen receptor protein (Green *et al.*, 1986). The human estrogen receptor is proposed to consist of two 595-residue (66-kDa) hormone- and DNA-binding subunits (along with other receptor-associated components) that are believed to undergo dimerization, a process that facilitates interaction with its palindromic DNA response element. We have tentatively defined a portion of the human estrogen receptor dimerization domain as a 53-residue helical sequence of amino acids (D473–L525) containing seven His residues (no Cys residues). This assignment was based primarily upon previous investigations by Fawell and co-workers (e.g., Fawell *et al.*, 1990) of receptor cDNA deletion mutations. We have demonstrated previously that Cu(II) and Zn(II) differentially affect the allosteric modulation of receptor affinity for its steroid hormone; Cu(II) stabilizes the receptor dimer in a noncooperative, high-affinity hormone

D

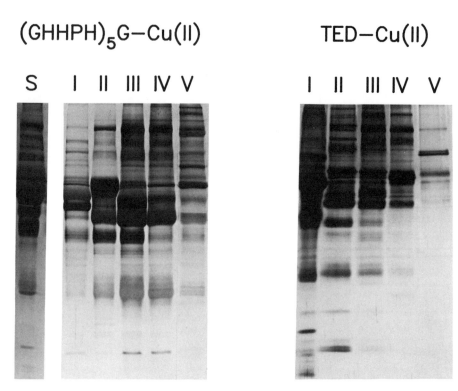

FIGURE 6 (Continued)

binding state (Hutchens *et al.*, unpublished results). Because suitable quantities of the purified estrogen receptor protein are not yet available, to investigate receptor dimerization further, the estrogen receptor dimerization domain (ERDD) peptide was synthesized, purified, evaluated for composite secondary structure by vacuum-UV circular dichroism (predominantly helical), and characterized extensively for its ability to bind Zn(II) and Cu(II) ions in solution (see Table 2); the metal-binding properties of the synthetic dimerization domain were based on analyses which included atomic absorption, two different forms of soft-ionization mass spectrometry, and spectrophotometric titrations. The synthetic ERDD peptide (DHIHRVLDKI TDTLIHLMAK AGLTLQQQHQ RLA-QLLLILS HIRHMSNKGM EHL) was immobilized using DVS-activated agarose 6B and Affi-Gel 10 agarose. The interaction of the soluble dimerization domain with the immobilized dimerization domain was investigated in the presence and absence of added Zn(II) ions (not shown) or Cu(II) ions. The profile shown in Fig. 7A demonstrates that the immobilized ERDD was able to bind the soluble ERDD more effectively (i.e., higher affinity) in the presence of added Cu(II) ions. Figure 7B demonstrates that the immobilized ERDD was also capable of interaction with the intact estrogen receptor subunit, which had

TABLE 2

Summary of Synthetic Estrogen Receptor Dimerization Domain (ERDD) Metal-Binding Properties

Analytical method	Concentration of ERDD (D473–L525)	Concentration of bound Cu(II) (observed)	Cu(II):ERDD ratio
Dialysis and atomic absorption	13.6 μM	40.7 μM	3:1
Spectrophotometric titration (225 nm)	21.3 μM	68.2 μM	3.2:1
ESI mass spectrometry	8.1 μM	—[a]	2:1–3:1
MALDI-TOF mass spectrometry	<1 pmol	—[b]	3:1[c]

[a]ESI mass spectrometry was performed in the presence of excess $CuCl_2$ (separation of bound and free Cu is not necessary before analysis) [see Hutchens et al. (1992b)].
[b]MALDI-TOF mass spectrometry was performed with sample in the solid state (in the presence of excess $CuCl_2$).
[c]Values reported represent the maximum number of bound metal ions observed under the conditions used [see Hutchens et al. (1991, 1992b–e) and Nelson and Hutchens (1992)].

been partially purified by DNA-affinity chromatography from preparations of calf uterine cytosol. The Cu(II)-dependent interaction of the intact receptor with the immobilized dimerization domain was quite strong, perhaps due to the localized orientation of multiple His residues on the dimerization domain surface (7/21 total His residues). This pseudo dimerization reaction was not easily disrupted by low pH or by the inclusion of EDTA in the column elution buffer. In fact, as described previously in the case of the estrogen receptor interaction with immobilized IDA-Cu(II) ions, the inclusion of $3M$ urea was required to elute the receptor. Considerable purification resulted from the high affinity interaction of the intact receptor with the immobilized dimerization domain (Hutchens et al., unpublished results).

It is significant, we believe, that in contrast to the estrogen receptor interaction with immobilized IDA-Cu(II) ions (Hutchens and Li, 1988; Hutchens et al., 1989b), the interaction of the intact 67-kDa receptor with the dimerization domain–Cu(II) complex was not eliminated upon removal of the DNA-binding domain from the receptor by proteolysis with trypsin. We believe that this is another example to demonstrate the importance of biospecific metal ion presentation for immobilized metal ion-dependent macromolecular recognition phenomena. It is also clear that the metal-binding sites at the dimerization/hormone-binding domain are separate and distinct from the metal-binding sites at the DNA-binding domain.

8. IMMOBILIZED DNA-BINDING DOMAINS WITH RETENTION OF METAL ION-DEPENDENT, SEQUENCE-SPECIFIC DNA BINDING PROPERTIES

The ability to immobilize protein surface metal-binding domains involved in DNA recognition can also be illustrated. A 71-residue peptide representing the DNA-binding domain of the human estrogen receptor (Kumar et al., 1986) was synthesized (Fig. 8A), characterized extensively with respect to its transition-metal ion-binding properties (Fig. 8B), and immobilized in the presence of bound Zn(II) ions. The metal-binding properties of the soluble and immobilized estrogen receptor DNA-binding domain (ERDBD) are summarized in Table 3 and in Fig. 8B. A 30-base pair DNA sequence

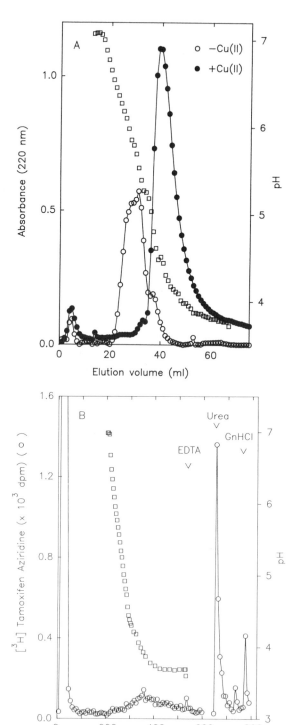

FIGURE 7. Selective and metal ion-dependent biomolecular recognition of the immobilized 53-residue estrogen receptor dimerization domain (ERDD). Panel A shows the interaction of soluble ERDD with the immobilized ERDD in the absence (○) and presence (●) of bound Cu(II) ions. Panels B and C show the interaction of covalently labeled estrogen receptor proteins with the immobilized ERDD before (B) and after removal of the DNA-binding domain by treatment with trypsin (C). Intact estrogen receptor protein subunits (67 kDa) present in calf uterine cytosol were affinity labeled with [³H]tamoxifen aziridine and partially purified by transformation to the DNA-binding form followed by affinity chromatography on columns of immobilized DNA. The elution of bound protein from the column of immobilized ERDD was initiated by introduction of a descending gradient of pH (□). Arrows mark the sequential introduction of 50mM EDTA, 3M urea, and 8M guanidine hydrochloride (GnHCl) upon completion of the pH gradient. Elution of radiolabeled estrogen receptor proteins (○) upon introduction of 3M urea was verified by immunoblot analyses with monoclonal antibodies (Hutchens et al., unpublished results).

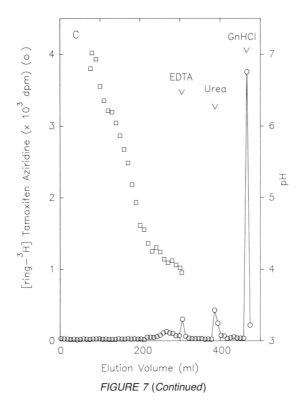

FIGURE 7 (Continued)

containing the estrogen response element (ERE) was labeled with ^{32}P and incubated with the immobilized ERE in the presence and absence of bound Zn(II) and/or excess non-specific calf thymus DNA. The ability of the immobilized ERDBD to bind the ERE DNA in the presence and absence of bound Zn(II) ions is summarized in Table 4. We believe that this approach may lead to the construction of immobilized quaternary structures necessary to reconstitute transcriptionally active machinery.

9. MODEL PROTEIN SURFACES FOR THE INVESTIGATION OF BIOSPECIFIC METAL ION TRANSFER BETWEEN MACROMOLECULAR SURFACES

Immobilized protein surface domains are potentially useful models for the evaluation of sequence- or domain-specific metal ion transfer between macromolecular surfaces. Figure 9A shows a portion of the MALDI-TOF mass spectra obtained for the trypsin digest products of purified human β-casein. Obtained in the absence and the presence of added Fe(II) ions, these spectra offer the first direct evidence of the phosphorylation-dependent Fe(II)-binding properties of the 18-residue phosphopeptide from the amino terminus of human β-casein. The iron-binding properties of the intact β-casein and the R1–K18 peptide were verified in separate experiments by spectrophotometric titration, atomic absorption,

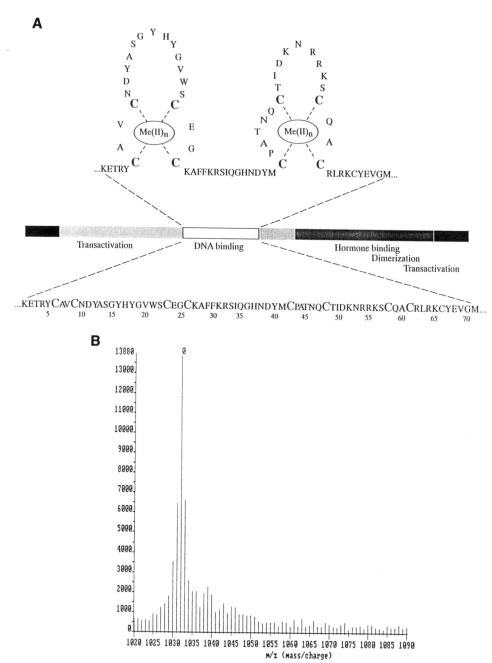

FIGURE 8. Synthesis, characterization, and immobilization of the 71-residue human estrogen receptor DNA-binding domain (ERDBD) with retention of function. The ERDBD (K180–M250) peptide (A) was synthesized and purified to homogeneity; the fully reduced status of the nine Cys residues was confirmed by reaction with Ellman's reagent (Ellman, 1959) and by electrospray ionization (ESI) mass spectrometry. The fully reduced 8248.5-Da ERDBD without bound metal ions (apopeptide) (B, top profile, peak labeled 0) was verified to be capable of binding up to two Zn(II) ions (B, middle profile, peaks labeled 1 and 2) or up to four Cu(II) ions (B, bottom profile, peaks labeled 3 and 4) by ESI mass spectrometry; these data were confirmed by spectrophotometric titration and

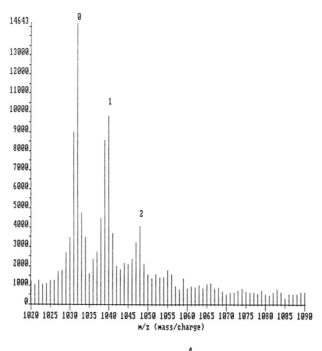

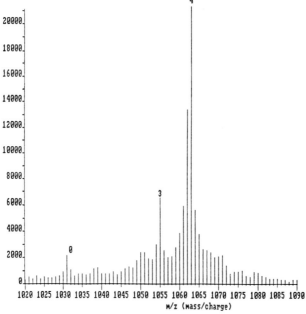

FIGURE 8 (*Continued*). atomic absorption (Table 3). The ERDBD apopeptide was occupied with Zn(II) ions and immobilized to agarose. The metal ion-dependent interaction of the immobilized ERDBD with specific ^{32}P-labeled DNA (ERE) in the presence and absence of excess nonspecific DNA is summarized in Table 4.

TABLE 3

Metal Binding Properties of the Synthetic Estrogen Receptor DNA-Binding Domain (ERDBD)

Analytical method	Concentration of ERDBD (K180–M250)	Concentration of bound Zn(II) (observed)	Concentration of bound Cu(II) (observed)	Zn(II):ERDBD ratio	Cu(II):ERDD ratio
Dialysis and atomic absorption	10.9 μM	25.1 μM	44.1 μM	2:1	4:1
Spectrophotometric titration (228 or 235 nm)	9.5 μM	18.1 μM	36.9 μM	2:1	4:1
ESI mass spectrometry	8.1 μM	—[a]	—[a]	2:1	4:1
MALDI-TOF mass spectrometry	<1 pmol	—[b]	—[b]	2:1[c]	4:1[c]

[a] ESI mass spectrometry performed in the presence of excess $CuCl_2$ or $ZnCl_2$ (separation of bound and free metal ions is not necessary before analysis) (see Allen and Hutchens (1992), Hutchens and Allen (1992), and Hutchens et al., 1992a,b).
[b] MALDI-TOF mass spectrometry performed with sample in the solid state (in the presence of $CuCl_2$ or $ZnCl_2$).
[c] Values reported represent the maximum number of bound metal ions observed under the conditions used.

TABLE 4

Effects of Bound Zn(II) Ions on the Specific DNA
(ERE)-Binding Properties of the Immobilized Human
Estrogen Receptor DNA-Binding Domain (ERDBD)

Peptide	Metal ion present	Nonspecific DNA present	Percent ^{32}P-labeled ERE bound
−ERDBD	+Zn(II)	No	1.34
−ERDBD	+Zn(II)	Yes	0.32
+ERDBD	−Zn(II)	Yes	0.9
+ERDBD	+Zn(II)	No	32.4[a]
+ERDBD	+Zn(II)	Yes	13.7[a]

[a]Net percent bound, i.e., after correction for binding to immobilized
ERDBD in the absence of Zn(II).

and mass spectrometry both *before* and *after* enzymatic dephosphorylation. Each of these other analytical methods verified that specific iron binding was phosphorylation-dependent (Yip and Hutchens, 1993b; Yip *et al.*, 1993). To better evaluate the specificity of Fe(II) transfer to and from the R1–K18 peptide in the presence of one or more other factors (e.g., calcium) or in its native environment (e.g., in the presence of added milk), the intact casein and the R1–K18 phosphopeptide were immobilized on Affi-Gel 10 agarose. The R1–K18 phosphopeptide to be immobilized was isolated from native casein in this case, but the R1– K18 domain has also been synthesized and phosphorylated chemically in our laboratory (Yip and Hutchens, 1992a, 1993; Otvos *et al.*, 1989). Immobilization was performed in the presence of added metal ions to help ensure the preservation of metal binding sites. The ^{59}Fe(II)-binding and ^{59}Fe(II)-transfer properties of this immobilized protein surface metal-binding domain were then evaluated in the presence of sequestered ^{59}Fe (i.e., ^{59}Fe-labeled hololactoferrin) and apolactoferrin, respectively. The presumed competitive effects of added Ca(II) ions (Hallberg *et al.*, 1992) on these activities were also monitored. The results are summarized in Table 5. Similar attempts to evaluate these types of interactive relationships with the use of peptides and/or proteins in free solution have been unsuccessful.

10. OTHER IMMOBILIZED PROTEIN SURFACE METAL-BINDING DOMAINS TO BE INVESTIGATED

We are interested in (1) the identification of protein surface metal-binding domains, (2) alterations to domain structure and metal-binding activity resulting from protein surface modifications (e.g., ligand-induced) or degradation, and (3) the generation (i.e., exposure) of new metal-binding protein domains and fragments by posttranslational modifications, including proteolysis. Table 6 summarizes the protein surface metal-binding domains identified to date and those used for the investigation and modeling of metal ion-dependent recognition and metal ion transfer events.

11. DISCUSSION

We believe that the identification and immobilization of protein surface domains as a *biospecific* approach to the investigation of metal ion interactions with the surface of

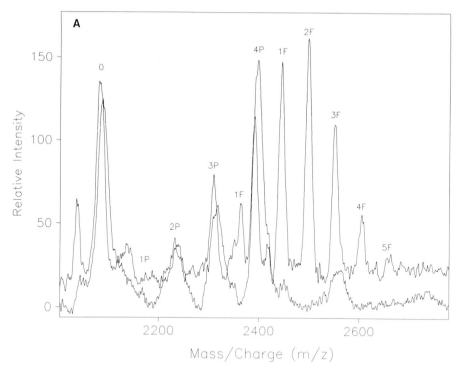

B

Human β-Casein (R1-K18)

Arg-Glu-Thr-Ile-Glu-Ser-Leu-Ser-Ser-Ser-Glu-Glu-Ser-Ile-Thr-Glu-Tyr-Lys

Ⓟ   Ⓟ   ⓅⓅⓅ

Molecular mass = 2088.23 Da
plus 79.98 Da per phosphate group

FIGURE 9. Identification by MALDI-TOF of the 18-residue Fe(II)-binding phosphopeptide produced by trypsin digestion of native human β-casein. MALDI-TOF analyses of the unfractionated digest mixture shown here (A) were performed before *in situ* dephosphorylation (Yip and Hutchens, 1992a, 1993a) either in the absence (bottom profile) or presence (top profile) of added Fe(II) ions. Peaks labeled 0, 1P, 2P, 3P, and 4P represent the R1–K18 N-terminal domain with 0 to 4 phosphorylated serine/threonine residues. Peaks labeled 1F to 5F represent either the 3P or 4P form of the R1–K18 phosphopeptide with up to 5 bound Fe(II) ions. The mass, amino acid sequence, and sites of phosphorylation are shown in panel B. A summary of the Fe(II)- binding and Fe(II)-transfer properties of this peptide (R1–K18) relative to intact β-casein is presented in Table 5. Reproduced from Hutchens and Yip (1992b) with permission.

macromolecules is important from the perspective of both molecular biology and physiology. One example of the physiological importance, and possible practical use, of these investigations may be illustrated by the requirement of preterm infants for bioavailable trace-metal ions and minerals. Maturation of the gastrointestinal tract in preterm and newborn infants determines the rate and extent of dietary protein digestion. This, in turn, determines the type and concentration of proteins, protein fragments, and small peptides

TABLE 5

Biospecific Metal Ion Transfer between Casein and Lactoferrin Evaluated
with Immobilized Protein-Surface Metal-Binding Domains:
A Comparison with Metal Ion Transfer between Proteins in Free Solution[a]

Metal ion transfer	Competitive metal ion present	Immobilized protein surface metal-binding domain	
		Casein	(R1–K18)-4P
%Fe(II) transferred *to* apolactoferrin	—	13.3	11.4
	Ca(II)	11.7	9.6
%Fe(II) transferred *from* hololactoferrin	—	17.7	23.9
	Ca(II)	1.7	8.6

[a]Reproduced from Hutchens and Yip (1992b) with permission.

available to interact with essential metal ions (e.g., zinc and iron). Because all factors affecting dietary transition-metal ion availability in these infants are important in practice (all preterm infants are born in a metal ion-deficient state), we need to understand the structural determinants of peptide interactions with metal ions. It may then be possible to identify and/or design specific carriers of metal ions to improve the bioavailability of essential dietary elements.

We have used several peptides and proteins with well-defined sequences and surface structures to explore differences between the predicted and observed interactions with nonbiological surface-immobilized transition-metal ions under a variety of experimental conditions (e.g., Hutchens and Li, 1988; Hutchens et al., 1988, 1989b; Hutchens and Yip,

TABLE 6

Immobilized Protein Surface Metal-Binding Domains Identified and/or Evaluated to Date
for Evidence of Metal Ion-Dependent Molecular Recognition and/or Metal Ion Transfer

Native protein model	Sequence of immobilized domain	No. of residues	Metal ions bound	Recognition targets
Histidine-rich glyco-protein, $(GHHPH)_nG$ (for $n = 1$ to 5)	G365–H389	6–26	Cu(II)/Zn(II)	Plasma proteins, milk proteins, and lymphocyte receptor proteins
Estrogen receptor dimerization domain	D473–L525	53	Cu(II)/Zn(II)	Estrogen receptor monomers, heat shock proteins, and transcription factors
Estrogen receptor DNA-binding domain	K180–M250	71	Zn(II)/Cu(II)	Sequence-specific DNA (response elements), transcription factors
β-Casein	R1–K18 ± 4P	18	Fe(II)/Ca(II)	Milk proteins, cell receptors
	E2–K18 ± 4P	17	Fe(II)/Ca(II)	Milk proteins, cell receptors
Hemopexin	D203–R216	14	Cu(II)/Zn(II)	Plasma proteins, heme
Albumin	H535–K565	31	Cu(II)/Zn(II)	Plasma proteins
Ceruloplasmin	E2–K23	22	Cu(II)	Plasma proteins, cells

1990a,b, 1991a,b; Yip *et al.*, 1989; Yip and Hutchens, 1989). Some of these interactions were evaluated as a function of protein-surface reaction site geometry to quantify variables such as interaction selectivity, affinity, and capacity. To date, however, despite considerable efforts, the interactions of peptides and protein surface domains with either free or surface-immobilized metal ions are not understood sufficiently to predict those surface patterns used successfully by nature to achieve a specified degree of interaction specificity and affinity.

If we are to understand these relationships, we believe that simple biopolymer model surfaces are required to build upon what we have learned from the construction and use of simple chemical (i.e., nonbiological) surfaces. The first steps in this process—identification and characterization of native protein-surface metal-binding domains—are perhaps the most important and the most difficult. Few techniques have been developed to identify specific peptide–metal ion interactions in complex mixtures, *particularly* in the case of relatively *low-affinity* interactions involving *low-molecular-weight* peptides. The high-resolution separation of peptides typically requires experimental conditions incompatible with the retention of bound metal ions (e.g., high-performance reserved-phase liquid chromatography). The techniques available presently to investigate protein–metal ion interactions in complex mixtures include chromatography (e.g., size-exclusion, ion-exchange) or SDS-polyacrylamide gel electrophoresis followed by transfer of the proteins to a membrane for blotting (i.e., probing) with radioactive metal ions; these procedures are not appropriate for the resolution of low-molecular-weight peptides. Furthermore, these procedures depend on separate means of resolution and detection that often result in a low recovery and disruption of the peptide- and/or protein–metal ion interaction. Our evaluation of synthetic metal-binding peptides by MALDI-TOF and ESI mass spectrometry has demonstrated that peptide-bound metal ions can remain stable during the desorption and/or ionization processes (Hutchens *et al.*, 1991, 1992a–d, 1993; Nelson and Hutchens, 1992; Allen and Hutchens, 1992; Hutchens and Allen, 1992; Yip and Hutchens, 1992a, b, 1993a, b; Yip *et al.*, 1993). While these techniques have certainly improved our ability to characterize soluble peptide–metal ion complexes, they do not help us understand these complexes after immobilization.

Indeed, a major problem with the investigation of any surface, particularly surfaces with heterogeneous collections of reactive groups, is the absence of available technologies for the detailed investigation of immobilized ligand structure. Thus, we can only infer retention of structure based on a complete characterization of the bioreactive domain before and after immobilization. Evidence for the retention of metal-binding capacity, affinity, and recognition selectivity after immobilization can be taken as good but not definitive evidence.

In many other ways, the results presented here are preliminary, and numerous questions remain. Regarding immobilization, we have not yet evaluated fully the contribution of bound metal ions to the oriented versus disoriented immobilization of domains during the covalent attachment process. Certainly, we have observed that occupancy of the specific metal binding sites can reduce the involvement of or protect these residues during the attachment process, but the conditions to optimize this protection are not yet known. We also need to investigate more completely the role of water in selective metal ion immobilization and during subsequent metal ion-dependent molecular interactions. Protein surfaces have regions with bound water and regions without any evidence of bound water. Reagents that affect the structure of water, for example, chaotropic salts and urea, are known to affect the affinity and capacity of protein interactions with surface-immobilized metal ions (Hutchens and Yip 1990a, 1991a). Furthermore, water is a good ligand for most transition-metal ions. We have recently reported what may be direct evidence for the contribution of bound water to the metal-binding capacity of a peptide (Hutchens and Yip 1990a, 1991a; Hutchens *et al.*, 1992a).

In the case of metal ion-dependent macromolecular interactions on model protein surfaces, it is difficult at this time to distinguish two possible mechanisms of metal ion participation, that is, the direct participation of metal ion coordination sites with both interacting surfaces versus an indirect effect of the bound metal ion on one surface to induce and/or stabilize the structural conformation required for such an interaction. We believe that we have experienced examples of both mechanisms.

In summary, our work demonstrates that investigations of metal ion-dependent molecular recognition events can at least be assisted by the use of model surfaces of stationary metal ions comprised of synthetic bioactive peptides that represent naturally occurring metal-binding domains identified from the surface of known metal-transport proteins. It will be important to compare the results obtained by this approach to results from the quantitative analyses of these interactions by equilibrium sedimentation.

Nevertheless, based on these trials, we can now begin to investigate the *de novo* design of artificial protein surface structures modeled after biologically active protein surface domains. We can design these artificial protein surfaces so that their activity is effectively modulated by specific bound metal ions. Ultimately, the construction of such model protein surfaces on a stationary phase may facilitate the reconstruction of biologically active quaternary structures.

ACKNOWLEDGMENTS

We thank Chee Ming Li and Gary L. Cook in our lab for their help with those portions of these investigations which included synthesis and sequence analyses of both natural and synthetic peptides. We thank Dr. Xiao-Yen Zhao in our lab for her help in the analyses of the metal ion-dependent ERDBD–ERE interactions in solution. We also thank Dr. Kenji Nakahara and Dr. Naohito Kaneko for their extensive efforts to characterize various forms of the immobilized ERDD.

REFERENCES

Allen, M. H., and Hutchens, T. W., 1992, Electrospray ionization mass spectrometry for the detection of discrete peptide/metal ion complexes involving multiple cysteine (sulfur) ligands, *Rapid Commun. Mass Spectrom.* **6:** 308–312.

Allen, M. H., and Vestal, M. L., 1992, Design and performance of a novel electrospray interface, *J. Am. Soc. Mass Spectrom.* **13:**18–32.

Anderson, B. F., Baker, H. M., Rice, D. W., and Baker, E. N., 1989, Structure of human lactoferrin: Crystallographic structure analysis and refinement at 2.8 A resolution, *J. Mol. Biol.* **209:**711–734.

Anderson, B. F., Baker, H. M., Norris, G. E., Rumball, S. V., and Baker, E. N., 1990, Apolactoferrin structure demonstrates ligand-induced conformational change in transferrins, *Nature (London)* **344:**784–787.

Berg, J. M., 1990, Zinc finger domains: Hypotheses and current knowledge, *Annu. Rev. Biophys. Biophys. Chem.* **19:**405–421.

Bernstein, F. C., Koetzle, T. F., Williams, G. J. B., Meyer, E. F., Jr., Brice, M. D., Rodgers, J. R., Kennard, O.,

Shimanouchi, T., and Tasumi, M., 1977, The protein data bank: A computer-based archival file for macromolecular structures, *J. Mol. Biol.* **112:**535–542.

Compton, L. A., and Johnson, W. C., Jr., 1986, Analysis of protein circular dichroism spectra for secondary structure using a simple matrix multiplication, *Anal. Biochem.* **155:**155–167.

DeLa Cadena, R. A., and Colman, R. W., 1992, The sequence HGLGHGHEQQHGLGHGH in the light chain of high molecular weight kininogen serves as a primary structural feature for zinc-dependent binding to an ionic surface, *Protein Sci.* **1:**151–160.

Edman, P., and Henschen, A., 1975, Sequence determination, in: *Protein Sequence Determination*, (S. B. Needleman, ed.), Springer-Verlag, New York, pp. 232–279.

Eisenstein, E., Markby, D. W., and Schachman, H. K., 1989, Changes in stability and allosteric properties of aspartate transcarbamoylase resulting from amino acid substitutions in the zinc-binding domain of the regulatory chains, *Proc. Natl. Acad. Sci. U.S.A.* **86:**3094–3098.

Ellman, G. L., 1959, Tissue sulfhydryl groups, *Arch. Biochem. Biophys.* **82:**70–77.

Fawell, S. E., Lees, J. A., White, R., and Parker, M. G., 1990, Characterization and colocalization of steroid binding and dimerization activities in the mouse estrogen receptor, *Cell* **60:**953–962.

Green, S., Walter, P., Kumar, V., Krust, A., Bornert, J. M., Argos, P., and Chambon, P., 1986, Human estrogen receptor cDNA: Sequence, expression and homology to v-*erb*-A, *Nature (London)* **320:**134–141.

Hallberg, L., Rossander-Hulten, L., Brune, M., and Gleerup, A., 1992, Bioavailability in man of iron in human milk and cow's milk in relation to their calcium contents, *Pediatr. Res.* **31:**524–527.

Hutchens, T. W., and Allen, M. H., 1992, Differences in the conformational state of a zinc-finger DNA-binding protein domain occupied by zinc and copper revealed by electrospray ionization mass spectrometry, *Rapid Commun. Mass Spectrom.* **6:**469–473.

Hutchens, T. W., and Li, C. M., 1988, Estrogen receptor interaction with immobilized metals: differential molecular recognition of Zn^{2+}, Cu^{2+}, and Ni^{2+} and separation of receptor isoforms, *J. Mol. Recog.* **1:**80–92.

Hutchens, T. W., and Li, C. M., 1990, Ligand-binding properties of estrogen receptor proteins after interaction with surface-immobilized Zn(II) ions: Evidence for localized surface interactions and minimal conformational changes, *J. Mol. Recog.* **3:**174–179.

Hutchens, T. W., and Yip, T.-T., 1990a, Protein interaction with surface-immobilized metal ions: Quantitative ecaluation of variations in affinity and binding capacity, *Anal. Biochem.* **191:**160–168.

Hutchens, T. W., and Yip, T.-T., 1990b, The differential interaction of peptides and protein surface structures with free metal ions and surface-immobilized metal ions, *J. Chromatogr.* **500:**531–542.

Hutchens, T. W., and Yip, T.-T., 1991a, Protein interactions with macromolecular surface-immobilized metal ions: Structure-dependent variations in affinity and binding capacity with temperature and urea concentration, *J. Inorg. Biochem.* **42:**105–118.

Hutchens, T. W., and Yip, T.-T., 1991b, Metal ligand-induced alterations in the surface structures of lactoferrin and transferrin probed by interaction with immobilized Cu(II) ions, *J. Chromatogr.* **536:**1–15.

Hutchens, T. W., and Yip, T.-T., 1992a, Synthetic metal-binding protein surface domains for metal ion-dependent interaction chromatography. II. Immobilization of synthetic metal-binding peptides from metal ion transport proteins as model bioactive protein surface domains, *J. Chromatogr.* **604:**133–141.

Hutchens, T. W., and Yip, T.-T., 1992b, Model protein surface domains for the investigation of metal ion-dependent macromolecular interactions and biospecific metal ion transfer, *Methods: A Companion to Methods in Enzymology* **4:**79–96.

Hutchens, T. W., Markland, F. S., and Hawkins, E. F., 1981, Physiochemical analysis of reversible molybdate effects on different molecular forms of glucocorticoid receptors, *Biochem. Biophys. Res. Commun.* **103:**60–67.

Hutchens, T. W., Yip, T.-T., and Porath, J., 1988, Protein interaction with immobilized ligands. Quantitative analysis of equilibrium partition data and comparison with analytical affinity chromatographic data using immobilized metal ion adsorbents, *Anal. Biochem.* **170:**168–182.

Hutchens, T. W., Henry, J. F., and Yip, T.-T., 1989a, Purification and characterization of intact lactoferrin found in the urine of human milk-fed preterm infants, *Clin. Chem.* **35:**1928–1933.

Hutchens, T. W., Li, C. M., Sato, Y., and Yip, T.-T., 1989b, Multiple DNA-binding estrogen receptor forms resolved by interaction with immobilized metal ions. Identification of a metal-binding domain, *J. Biol. Chem.* **264:**17206–17212.

Hutchens, T. W., McNaught, R. W., Yip, T.-T., Li, C.M., Suzuki, T., and Besch, P. K., 1990, Unique molecular properties of a urea- and salt-stable DNA-binding estrogen receptor dimer covalently-labeled with the antiestrogen [^{3}H]desmethylnafoxidine aziridine, *J. Mol. Endocrinol.* **4:**255–267.

Hutchens, T. W., Nelson, R. W., and Yip, T.-T., 1991, The evaluation of peptide/metal ion interactions by UV laser desorption time-of-flight mass spectrometry, *J. Mol. Recog.* **4**:151–153.

Hutchens, T. W., Allen, M. H., Li, C. M., and Yip, T.-T., 1992a, Occupancy of a C_2-C_2 "zinc-finger" protein domain by copper: Direct observation by electrospray ionization mass spectrometry, *FEBS Lett.* **309**: 170–174.

Hutchens, T. W., Nelson, R. W., Allen, M. H., Li, C. M., and Yip, T.-T., 1992b, Peptide–metal ion interactions in solution: Detection by laser desorption time-of-flight mass spectrometry and electrospray ionization mass spectrometry, *Biol. Mass Spectrom.* **21**:151–159.

Hutchens, T. W., Nelson, R. W., Li, C. M., and Yip, T.-T., 1992c, Synthetic metal-binding protein surface domains for metal ion-dependent interaction chromatography, I. Analysis of bound metal ions by matrix-assisted UV laser desorption time-of-flight mass spectrometry, *J. Chromatogr.* **604**:125–132.

Hutchens, T. W., Nelson, R. W., and Yip, T.-T., 1992d, Recognition of transition metal ions by peptides. Identification of specific metal-binding peptides in proteolytic digest maps by UV laser desorption time-of-flight mass spectrometry, *FEBS Lett.* **296**:99–102.

Hutchens, T. W., Nelson, R. W., and Yip, T.-T., 1992e, Effects of bound water on peptide metal-binding stoichiometry: An evaluation by matrix-assisted UV laser desorption time-of-flight mass spectrometry, 6th Annual Symposium of the Protein Society, San Diego, California, July 25–29, 1992.

Hutchens, T. W., Nelson, R. W., and Yip, T.-T., 1993, Identification of conserved protein surface metal-binding sites in related proteins by mass spectrometry, in: *Techniques in Protein Chemistry IV* (R. H. Angeletti, ed.), Academic Press, New York, pp. 33–40.

Jones, T. A., 1982, FRODO: A graphics fitting program for macromolecules, in: *Computational Crystallography* (D. Sayre, ed.) Clarendon Press, Oxford, pp. 303–317.

Kaptein, R., 1991, Zinc-finger structures, *Curr. Opin. Struct. Biol.* **2**:109–115.

Klock, G., Strahle, U., and Schutz, G., 1987, Oestrogen and glucocorticoid responsive elements are closely related but distinct, *Nature (London)* **329**:734–736.

Koide, T., Foster, D., Yoshitake, S., and Davie, E. W., 1986, Amino acid sequence of human histidine-rich glycoprotein derived from the nucleotide sequence of its cDNA, *Biochemistry* **25**:2220–2225.

Kumar, V., Green, S., Staub, A., and Chambon, P., 1986, Localisation of the oestradiol-binding and putative DNA-binding domains of the human oestrogen receptor, *EMBO J.* **5**:2231–2236.

Lee, B., and Richards, F. M., 1971, The interpretation of protein structures: Estimation of static accessibility, *J. Mol. Biol.* **55**:379–400.

Manavalan, P., and Johnson, W. C., Jr., 1987, Variable selection method improves the prediction of protein secondary structure from circular dichroism spectra, *Anal. Biochem.* **167**:76–95.

Morgan, W. T., 1978, Human serum histidine-rich glycoprotein. I. Interactions with heme, metal ions and organic ligands, *Biochim. Biophys. Acta* **533**:319–333.

Morgan, W. T., 1985, The histidine-rich glycoprotein of serum has a domain rich in histidine, proline, and glycine that binds metal ions, *Biochemistry* **24**:1496–1501.

Muhoberac, B. B., Burch, M. K., and Morgan, W. T., 1988, Paramagnetic probes of the domain structure of histidine-rich glycoprotein, *Biochemistry* **27**:746–752.

Nelson, R. W., and Hutchens, T. W., 1992, Mass spectrometric analysis of a transition-metal-binding peptide using matrix-assisted laser desorption time-of-flight mass spectrometry. A demonstration of probe tip chemistry, *Rapid Commun. Mass Spectrom.* **6**:4–8.

Otvos, L., Jr., Elekes, I., and Lee, V. M.-Y., 1989, Solid-phase synthesis of phosphopeptides, *Int. J. Pept. Protein Res.* **34**:129–133.

Porath, J., and Axen, R., 1976, Immobilization of enzymes to agar, agarose, and Sephadex supports, *Methods Enzymol.* **44**:19–38.

Porath, J., and Olin, B., 1983, Immobilized metal ion affinity adsorption and immobilized metal ion affinity chromatography of biomaterials. Serum protein affinities for gel-immobilized iron and nickel ions, *Biochemistry* **22**:1621–1630.

Porath, J., Carlsson, J., Olsson, I., and Belfrage, G., 1975, Metal chelate affinity chromatography, a new approach to protein fractionation, *Nature (London)* **258**:598–599.

Scatchard, G., (1949), The attractions of proteins for small molecules and ions, *Ann. N. Y. Acad. Sci.* **51**:660–672.

Smith, C. A., Anderson, B. F., Baker, H. M., and Baker, E. N., 1992, Metal substitution in transferrins: The crystal structure of human copper-lactoferrin at 2.1-A resolution, *Biochemistry* **31**:4527–4533.

Vallee, B. L., Coleman, J. E., and Auld, D. S., 1991, Zinc fingers, zinc clusters, and zinc twists in DNA-binding protein domains, *Proc. Natl. Acad. Sci. U.S.A.* **88**:999–1003.

Yip, T.-T., and Hutchens, T. W., 1989, Development of high-performance immobilized metal affinity chroma-
      tography for the separation of synthetic peptides and proteolytic digestion products, in: *Protein Recognition of
      Immobilized Ligands* (T. W. Hutchens, ed.), Alan R. Liss, New York, pp. 45–56.
Yip, T.-T., and Hutchens, T. W., 1991, Metal ion affinity adsorption of a Zn(II)-transport protein in maternal
      plasma during lactation: Structural characterization and identification as histidine-rich glycoprotein, *Protein
      Expression and Purification* **2:**355–362.
Yip, T.-T., and Hutchens, T. W., 1992a, Mapping and sequence-specific identification of phosphopeptides in
      unfractionated protein digest mixtures by matrix-assisted laser desorption/ionization time-of-flight mass
      spectrometry, *FEBS Lett.* **308:**149–153.
Yip, T.-T., and Hutchens, T. W., 1992b, Protein surface metal binding domains identified in unfractionated
      proteolytic digests of metal transport and storage proteins by matrix-assisted UV laser desorption time-of-
      flight mass spectrometry, 40th ASMS Conference on Mass Spectrometry and Allied Topics, Washington
      D.C., May 31–June 5, 1992, pp. 1915–1916.
Yip, T.-T., and Hutchens, T. W., 1993a, Protein phosphorylation: Sequence-specific identification of *in vivo*
      phosphorylation sites by MALDI-TOF mass spectrometry, in: *Techniques in Protein Chemistry IV* (R. H.
      Angeletti, ed.), Academic Press, New York, pp. 201–210.
Yip, T.-T., and Hutchens, T. W., 1993b, Evaluation of phosphorylation-dependent protein bioactivity by matrix-
      assisted laser desorption time-of-flight mass spectrometry, 41st ASMS Conference on Mass Spectrometry
      and Allied Topics, San Francisco, May 30–June 4, 1993, in press.
Yip, T. T., Nakagawa, Y., and Porath, J., 1989, Evaluation of the interaction of peptides with Cu(II), Ni(II), and
      Zn(II) by high-performance immobilized metal ion affinity chromatography, *Anal. Biochem.* **183:**159–171.
Yip, T.-T., Li, C. M., Yip, C., and Hutchens, T. W. 1993, Effects of phosphorylation site density on metal binding
      specificity: Direct visualization by MALDI-TOF and ESI mass spectrometry, *Protein Sci.* **2**(1):92.

20

Affinity Chromatography with Immobilized Benzeneboronates

Milan J. Beneš, Alexandra Štambergova, and William H. Scouten

1. INTRODUCTION

Boronate affinity chromatography is based on the unique ability of ionized boronic acids to reversibly form cyclic (five- or six-membered ring) esters with diol groups under mild conditions. An aqueous medium and a 1,2-diol in a *cis*-coplanar geometry (Fig. 1) or a 1,3-diol of proper geometry are required for this interaction. Such diol functions are present in many biologically significant structures, particularly saccharides, and, therefore, boronate chromatography occupies an important position in analytical and preparative chromatography of these substances. Among the advantages of boronate chromatography, in comparison with many other affinity methods, are the availability and stability of the ligands. Boronate chromatography has already been subject of several reviews (Bergold and Scouten, 1983; Carlsohn and Hartmann, 1979; Dean *et al.*, 1983; Fulton, 1981; Mazzeo and Krull, 1989). We have, therefore, restricted this chapter to a discussion of the basic principles of the process and a description of developments over the past few years (chiefly since 1983).

2. BASIC PRINCIPLES

2.1. Adsorption

Formation of cyclic anionic boronates in aqueous media depends on the properties of each component: the diol, the matrix, and the application buffer, particularly its pH and ionic strength.

Boronic acids, since they are Lewis acids, do not ionize directly by deprotonation, but, rather, hydration is first required, followed by deprotonation of the intermediate formed (Fig. 2). Alternately, boronate ions can be formed by the direct addition of hydroxyl ions (Fig. 2). In this way, the planar trigonal arrangement is changed to the more stable

Milan J. Beneš • Institute of Macromolecular Chemistry, Academy of Sciences of the Czech Republic, Prague, Czech Republic. *Alexandra Štambergova* • Institute of Toxicology and Forensic Chemistry, Faculty of General Medicine, Charles University, Prague, Czech Republic. *William H. Scouten* • Chemistry Department, Biotechnology Center, Utah State University, Logan, Utah 84321.

Molecular Interactions in Bioseparations, edited by That T. Ngo. Plenum Press, New York, 1993.

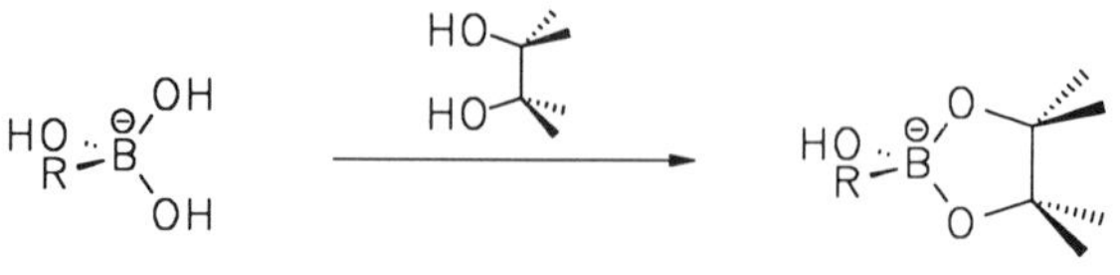

FIGURE 1. Complex formation between vicinal diol and boronate.

tetrahedral one (Lorand and Edwards, 1959). In such a tetrahedral configuration, favorable conditions exist for the formation and stability of cyclic esters and, therefore, for binding of appropriate analytes on immobilized boronates. Boronic acids are weak acids and ionize significantly only at pH values near their pK_as. Adsorption of an analyte should therefore be carried out at this or higher pH. A recent report (Singhal *et al.*, 1991) suggested another mechanism for the formation of anionic cyclic boronates, namely, the direct reaction of boronic groups with diols. This mechanism is only worth considering at pH values lower than the pK_a. Also at lower pH values, anionic cyclic boronates revert to the more strained planar (neutral) structures, which decompose to the starting components. This process can be used for desorption of diols from boronate matrices. Alternatively, elution can be effected with competing diols or by the addition of boric acid.

2.2. Synthesis of Immobilized Boronates

The immobilization of the boronate ligand on an appropriate activated matrix is the most widely used procedure for the preparation of boronate adsorbents for biochemical purposes. For the majority of adsorbents, 3-aminobenzeneboronic acid (previously called 3-aminophenylboronic acid) is the most commonly immobilized boronate. This is because it is readily available and easy to couple to most matrices. Other ligands used are derivatives of benzeneboronic acid, including the 4-aminomethyl (Yurkevich *et al.*, 1975), 3-nitro-4-carboxy, 2-nitro-4-carboxy (Soundararajan *et al.*, 1989), 3-amino-2 (or 4)-nitro (Johnson, 1981), and 3-nitro-4-(6-aminohexanoylamino) derivatives (Myoehaenen *et al.*, 1981). Introduction of electron-withdrawing nitro and carboxyl substituents on position 3 or 4 lowers the pK_a from 8.86 for benzeneboronic acid to 7.0 for 3-nitro-4-carboxybenzene-boronic acid and, thereby, also lowers the pH for the adsorption process. Surprisingly, a nitro group in position 2 does not act in the same way but rather appears to form a complex with the neighboring boron.

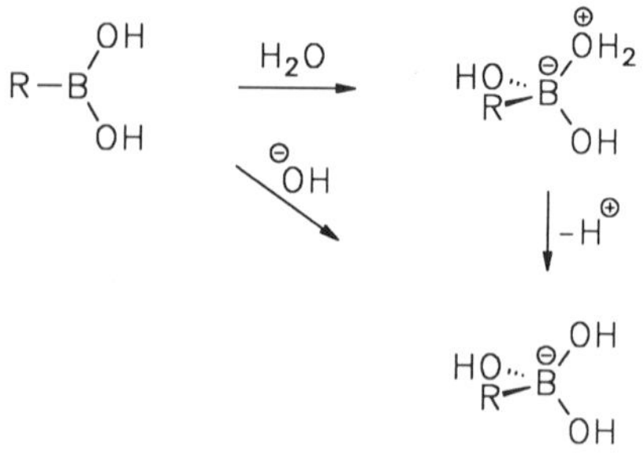

FIGURE 2. Formation of boronate anion.

Substituents on the aromatic ring not only influence the stability of the ester formed, but also the kinetics of its formation. A tertiary amino group accelerates the ester formation if its position allows coordination with the boronate. The same effect also occurs when an appropriate amine is present in aqueous solution with the diol.

Aliphatic boronic acids have not been used as ligands. Two reasons for this are that they have a higher pK_a than aromatic boronic acids and that they are not sufficiently stable, since they are easily oxidized by atmospheric oxygen.

Boronate adsorbents can be prepared by polymerization or polycondensation of monomers containing boronate groups (Carlsohn and Hartman, 1979; Wulff and Kirstein, 1990). These materials are commonly based on vinylbenzeneboronic acid and are used chiefly for the separation of carbohydrates and their derivatives. They are not suitable, due to their hydrophobic character, for protein chromatography. Use of these materials is discussed by Wulff in Chapter 23 of this volume.

2.3. Diols as Adsorbates

The formation of cyclic boronate esters is dependent upon whether a 1,2- or 1,3-diol is utilized, and it is further influenced by the diol conformation and the substituents in the vicinity of the hydroxyls, which may sterically or electronically inhibit or promote the coordination of the hydroxyls. Stable esters are formed from 1,2-diols only when the hydroxyl groups are in a synperiplanar conformation (Fig. 3). This conformation can be fixed by the participation of the carbon atoms bearing the hydroxyl groups in an aromatic or alicyclic ring. Appropriate conformations can also be preferred if there are bulky substituents in the vicinity of the boronate ligand. The capacity of many diols for the formation of boronate esters can be judged from the available chromatographic data utilizing immobilized boronates and the formation constants of complexes between diols and nonimmobilized boronic or boric acids (Shvarts, 1990) that have been collected.

Carbohydrates represent the first group of substances yielding boronates with sufficient stability for chromatographic separations. Among these are the furanose or pyranose forms of fructose, mannose, glucose, ribose, and sucrose and acyclic alcoholic saccharides such as sorbitol and mannitol (both groups listed in approximate order of decreasing stability). Other polyhydroxy compounds (e.g., erythritol) can also be used for this purpose. In these cases, the participation of a third hydroxyl group is sometimes suspected to enhance the stability of the boronate esters formed. Substances containing accessible covalently bound saccharides, including some glycoproteins and ribose-containing nucleosides, nucleotides, and nucleic acids, can also be separated by boronate chromatography. Conversely, low binding is shown by many common 1,2-diols such as ethylene glycol and glycerol. The same is also probably true for 1,2- and 1,3-aminoalcohols. If they are contained in buffer solutions, however, interactions of such compounds with boronates have to be considered. Separation of diol analytes can sometimes be inhibited, for example, by

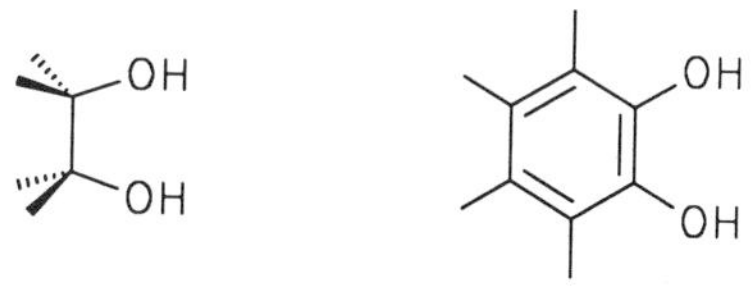

FIGURE 3. Examples of 1,2-diols in synperiplanar conformation.

Tris, triethanolamine, or serine, or, conversely, the formation of diol esters can be improved, for example, by HEPES buffer. The interactions of boronic acids with hydroxy-carboxylic acids (salicylic, tartaric, ascorbic, etc.) and their amides have not been employed as yet for chromatographic purposes.

2.4. Matrices

Beaded hydrophilic materials with low nonspecific adsorption are chiefly used as supports for boronate ligands. They are based on natural polysaccharides (agarose, cellulose), synthetic polymers (acrylamide or methacrylate type), and modified silica and glass. Boronate ligands containing an amino function are immobilized either by classical procedures, yielding amide-type linkages (activation of the carrier with cyanogen bromide or using carbodiimides), or by formation of more hydrolytically stable amine bonds (using carriers containing epoxy or tosylate groups).

The concentration of boronate ligands on the adsorbent is very important. For chromatography of low-molecular-weight substances (saccharides or catechol derivatives), the concentration of boronate can be rather high (100–300 μmol/ml). When high-molecular-weight polymeric analytes are used, the concentration of boronate is usually 10 times higher (10–100 μmol/ml) than the concentration of ligand employed in typical affinity chromatography (<30 μmol/ml).

There are many commercially available boronate adsorbents for different modes of affinity chromatography, including high-performance liquid chromatography (HPLC), as listed in Table 1.

2.5. Secondary Interactions

The chromatographic behavior of analytes on boronic adsorbents is often determined by factors other than the formation of diol esters. Additional interactions can occur due to boronate groups (ionic, coordination, and hydrogen bonding) and to other groups that are used for immobilization such as the spacer arms (hydrophobic and ionic bonding). It is also

TABLE 1
Commercially Available Boronate Adsorbents

Name (designation)	Concentration of ligand (μmol/ml)	Matrix	Distributor
Matrix Gel PBA 60	60–100	Agarose	Amicon
Matrix Gel PBA 30	30–50	Agarose	Amicon
Matrix Gel PBA 10	10–15	Agarose	Amicon
Affi-Gel 601	130	Polyacrylamide	Bio-Rad
20244-H	100	Polyacrylamide	Pierce
Glyco-Gel B		Agarose	Pierce
—		Silica	Serva
A4046	300–600	Acrylic	Sigma
A8530	5–15	Agarose	Sigma
Separon HEMA BIO 1000 PBA	100–150	Acrylic	Tessek Ltd.

necessary to take into account interactions with the matrix alone or the action of remaining groups that may have been introduced onto the matrix for the immobilization of the boronate ligand, for example, carboxylic acids and amines. Some of the complicating effects can be successfully eliminated by choosing appropriate chromatographic conditions. Normally, this can be accomplished by regulation of the ionic strength (hydrophobic interactions are more pronounced at higher values) and pH. Additionally, the presence of certain bivalent ions (especially Mg^{2+}) sometimes increases the specificity of the boronate–diol interactions. This is not only true in the case of negatively charged nucleosides but also in the case of glycosylated hemoglobin. The presence of organic solvents is sometimes desirable and useful (e.g., methanol and dimethyl sulfoxide) to improve the solubility of some analytes and to decrease hydrophobic interactions. The effect of temperature on binding to boronate adsorbents has not been systematically studied. For carbohydrate chromatography, it has been reported that an increase in temperature results in an increase of adsorption.

3. CHROMATOGRAPHIC APPLICATIONS

Boronic adsorbents may be used either for direct affinity chromatography or for the extraction of an analyte followed by elution and further analysis by another chromatographic method.

3.1. Carbohydrates

The first chromatographic utilization of boron-containing cyclic complexes was not typical boronate chromatography. Carbohydrates containing appropriate diols form anionic complexes with boric acid (molar ratio 2:1 or 1:1). These complexes are negatively charged and were separated by early investigators on ion-exchange columns. This approach facilitated the purification and analysis of sugars.

The use of boronate adsorbents for the isolation and separation of mono- and oligosaccharides was described some time ago (Glad et al., 1980; Weith et al., 1970) but has not gained favor as an analytical tool, since HPLC is a much simpler technique (Ben-Bassat and Grushka, 1991). Isomeric pentose phosphates have been separated on boronate cellulose (Gascon et al., 1981).

Probably most applications of boronate affinity chromatography are concerned with the purification of carbohydrate-containing compounds such as glycoproteins and ribonucleic acids, ribonucleotides, and ribonucleosides.

Glucose, and other aldoses, glycosylate the amino-terminal valine of the β-chain of hemoglobin, as well as its ε-lysyl amines (Mayer and Freedman, 1983). In this process a Schiff's base is formed with the sugar. This is subsequently irreversibly converted by Amadori rearrangement to the ketoamine, yielding hemoglobin A_{Ic}. The concentration of hemoglobin A_{Ic} reflects the blood glucose concentration over the past one to two months. Together with the determination of glucose in blood or urine, this is an important indicator for the clinical analysis of diabetes. Boronate affinity chromatography is a rapid, simple, precise, and accurate alternative to other methods used for the determination of diabetic hemoglobin such as ion-exchange chromatography and colorimetric methods (Flueckiger et al., 1984; Gould et al., 1982; Herold et al., 1983; Hjerten and Li Jin-Ping, 1990; Klenk et al., 1982; Mallia et al., 1981; Middle et al., 1983; Reid and Gisch, 1989; Yue et al., 1982).

Boronate affinity chromatography has also been used for the determination of glycosylated albumin (Ducrocq *et al.*, 1987; Rendell *et al.*, 1985; Woo *et al.*, 1987), which is formed by essentially the same mechanism as hemoglobin A_{Ic}. A good correlation is found between the concentration of glycosylated albumin in the blood and the mean blood glucose value for the preceding 20 days. This determination constitutes a good short-term glycemic index and is an alternative to determination of glycosylated hemoglobin in certain specific instances.

3.2. Nucleosides, Nucleotides, and RNAs

An extraordinary interest exists in chromatography of nucleic acids, and especially their low-molecular-weight components. DNA and deoxyribonucleotides, as well as $3'$-phosphorylated ribonucleotides, bind poorly to boronate, since they do not contain free $2',3'$-dihydroxyribose groups. Separation of RNA from DNA and from $3'$-phosphorylated ribonucleotides can be easily carried out by boronate affinity chromatography. In this case, the nature of the base plays a large role in the nucleoside binding. The strongest binding is caused by the purine bases, adenine and guanine. Divalent cations are added to buffer solutions to improve the binding of nucleotides, since the negative charge of the nucleotide phosphate groups, and therefore repulsive forces with the boronate group, is thus partly neutralized.

Nucleotides and nucleosides contained in body fluids yield important information about the metabolism of tRNA, and thus determination of the excretion levels and patterns of unmodified, modified, and hypermodified ribonucleosides in urine or plasma is a powerful, noninvasive screening method to investigate disorders in ribonucleoside and RNA metabolism. High-performance liquid affinity chromatography with boronate ligands (Kuo *et al.*, 1978) for the analysis of nucleosides has been refined to include on-line multidimensional high-performance liquid affinity chromatography and reverse-phase HPLC (Schlimme *et al.*, 1986). The separation of nucleotides and nucleosides has been described by Okayama (1980), and the nucleotide modifying group from the inactive iron protein of nitrogenase has been isolated utilizing boronate affinity columns (Pope *et al.*, 1985).

In nucleic acid chromatography, the porosity and character of the boronate matrix is of considerable importance. For example, boronate polyacrylamide and poly(2-hydroxyethyl methacrylate) specifically adsorb diols containing mononucleotides or short oligonucleo-tides (Štambergova and Vinš, 1991), but boronate agarose is the most suitable material for the fractionation of macromolecular RNA (Pace and Pace, 1980). Boronate agarose is also useful for the isolation of capped small nuclear RNA and mRNA (Wilk *et al.*, 1982). Boronate cellulose has been applied for the purification of tRNA. Cellulose and poly-acrylamide boronate derivatives are also useful for the isolation and determination of terminal polynucleotide fragments (Rosenberg, 1974).

Boronates with bound nucleotides are also useful as group-specific affinity adsorbents for the isolation of certain nucleotide-dependent enzymes. For example, yeast glucose-6-phosphate dehydrogenase is tightly bound to immobilized boronate–NADP complex (Bouriotis *et al.*, 1981) and on NAD–boronate complex (Maestas *et al.*, 1980).

3.3. Polyphenols

Boronate adsorbents may be used for the separation of simple catechols (Sugumaran and Lipke, 1982), mono-, di-, and trihydroxy aromatic carboxylic acids, and flavonoids (plant phenolic dyestuffs with flavone skeletons) with hydroxyl groups on adjacent carbons (Elliger and Rabin, 1981). Likewise, maysin, responsible for corn silk insect resistance, was isolated from a mixture of flavonoids from Zapolate Chico corn silk. To minimize nonspecific binding of the flavonoids to the polyacrylamide matrix, a wash buffer containing 20% methanol was used.

Another group of diphenolic substances widely studied by boronate chromatography are the important hormones the catecholamines (aminoalkyl derivatives of catechol). In this case, the boronate adsorbents are usually used for the specific isolation of the hormones from body fluids. After desorption, the catecholamines are further analyzed by other methods (Allenmark, 1982; Kemper *et al.*, 1984a,b). For example, adrenaline, noradrenaline, and dopamine have been isolated in this fashion from urine or plasma and further quantified by ion-exchange chromatography (Bauersfeld *et al.*, 1984), by normal phase HPLC (Gelijkens and De Leenheer, 1980; Speek *et al.*, 1983), or by reversed-phase HPLC (Glad *et al.*, 1983). Similarly, proteins containing 3,4-dihydroxyphenylalanine (DOPA) were isolated on boronate-agarose columns (Hawkins *et al.*, 1986).

3.4. Other Ligates

Several enzymes are inhibited by boric acid and both aliphatic and aromatic boronic acids. These enzymes can be successfully chromatographed on boronate adsorbents (Akparov and Stepanov, 1990). Since these enzymes do not contain typical 1,2-diols, it has been assumed that the binding of the boronate occurs via an active-site serine hydroxyl and the backbone peptide bond, catalyzed by a neighboring histidine imidazole. The optimal pH value for enzyme adsorption in such chromatographic separations is approximately 7.5, which is considerably lower than the usual pH for boronate affinity chromatography. Among the active-site serine enzymes that can be chromatographed in this fashion are the serine proteinases (chymotrypsin, trypsin, subtilisin, etc.), β-lactamases (Cartwright and Waley, 1984), and esterases (cholinesterase).

The mechanism of the binding of α-glucosidase from yeast on boronate-agarose in the presence of sugars is also not completely clear. The pH optimum of this separation is 7.4, similar to that seen in the case of active-site serine enzymes. The mechanism of fractionation of membrane proteins on phenylboronic acid-agarose (Williams *et al.*, 1982).

4. CONCLUSIONS

What is the future for boronate affinity chromatography? Above all, for further progress to occur, the problem of obtaining a boronate matrix with an appropriate pK_a (<7.5) must be solved, since its restricted operating pH range is the chief limitation of boronate chromatography. This problem might be circumvented by binding an appropriate immobilized ligand giving a sufficiently stable complex with boric acid (Fig. 4). The

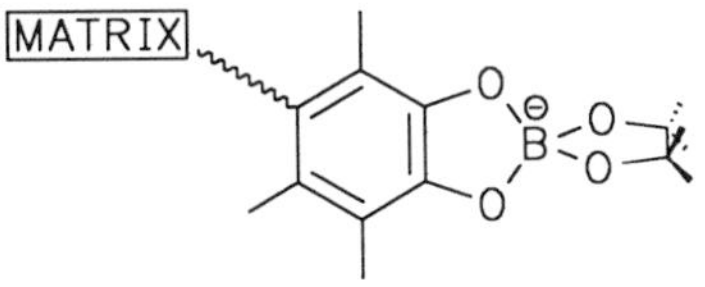

FIGURE 4. Potential borate gel for improved complexation of diols.

remaining free valences could subsequently be utilized for adsorption of the analyte from solution. Once the problem of lowering the operating pH is solved, many new applications of boronate chromatography will quickly follow.

REFERENCES

Akparov, V. Kh., and Stepanov, V. M., 1990, Substituted boric acids as ligands in chromatographic systems with specificity for the catalytic center of enzymes, *Usp. Biol. Khim.* **31**:97–114 [in Russian].

Allenmark, S., 1982, Analysis of catecholamines by HPLC, *J. Liq. Chromatogr.* **5**(S1):1–44.

Bauersfeld, W., Ratge, D., Knoll, E., and Wisser, H., 1984, Determination of catecholamines in plasma by HPLC and amperometric detection, *Fresenius' Z. Anal. Chem.* **317**:679–680.

Ben-Bassat, A. A., and Grushka, E., 1991, High performance liquid chromatography of mono and oligosaccharide, *J. Liq. Chromatogr.* **14**:1051–1112.

Bergold, A., and Scouten, W. H., 1983, Boronate chromatography, in: *Solid Phase Biochemistry: Analytical and Synthetic Aspects* (W. H. Scouten, ed.), John Wiley & Sons, pp. 149–187.

Bouriotis, V., Galpin, I., and Dean, P. D. G., 1981, Application of immobilized phenylboronic acids as supports for group-specific ligands in the affinity chromatography of enzymes, *J. Chromatogr.* **210**:267–278.

Carlsohn, H., and Hartmann, M., 1979, Synthesis and application of polymers modified by boronic acid, *Acta Polym.* **30**:420–426 [in German].

Cartwright, S. L., and Waley, S. G., 1984, Purification of beta-lactamases by affinity chromatography on phenyl-boronic acid agarose, *Biochem. J.* **221**:505–512.

Dean, P. D. G., Middle, F. A., Longstaff, C., Bannister, A., and Dembinski, J. J., 1983, Application of immobilized boronic acids, in: *Affinity Chromatography and Biological Recognition* (I. M. Chaiken, M. Wilchek, and I. Parikh, eds.), Academic Press, Orlando, Florida, p. 431.

Ducrocq, R., Le Bonniec, B., Carlier, O., and Assan, R., 1987, Measurement of glycated albumin in diabetic patient by biospecific affinity chromatography, *J. Chromatogr.* **419**:75–83.

Elliger, C. A., and Rabin, L. B., 1981, Separation of plant polyphenolics by chromatography on a boronate resin, *J. Chromatogr.* **216**:261–268.

Flueckiger, R., Woodtli, T., and Berger, W., 1984, Quantitation of glycosylated hemoglobin by boronate affinity chromatography, *Diabetes* **33**:73–76.

Fulton, S., 1981, Boronate Ligands in Biochemical Separation, Amicon Co., Danvers, Massachusetts.

Gascon, A., Wood, T., and Chitemerese, L., 1981, The separation of isomeric pentose phosphates from each other and the preparation of D-xylulose 5-phosphate and D-ribulose 5-phosphate by column chromatography, *Anal. Biochem.* **118**:4–9.

Gelijkens, C. F., and De Leenheer, A. P., 1980, Simple method for isolation of free urinary catecholamines by boric acid gel chromatography, *J. Chromatogr.* **183**:78–82.

Glad, M., Ohlson, S., Hansson, L., Maansson, M. O., and Mosbach, K., 1980, High-performance liquid affinity chromatography of nucleosides, nucleotides, and carbohydrates with boronic acid-substituted microparticulate silica, *J. Chromatogr.* **200**:254–260.

Glad, M., Hansson, L., and Hansson, C., 1983, Recent application of high-performance liquid affinity chromatography (HPLAC) with boronic acid silica, in: *Affinity Chromatography and Biological Recognition* (I. M. Chaiken, M. Wilchek, and I. Parikh, eds.), Academic Press, Orlando, Florida, pp. 253–254.

Gould, B. J., Hall, P. M., and Cook, J. G. H., 1982, Measurement of glycosylated hemoglobins using an affinity chromatography method, *Clin. Chim. Acta* **125**:41.

Hawkins, C. J., Lavin, M. F., Parry, D. L., and Ross, I. L., 1986, Isolation of 3,4-dihydroxyphenylalanine-containing proteins using boronate affinity chromatography, *Anal. Biochem.* **159**:187–190.

Herold, D. A., Boyd, J. C., Bruns, D. E., Emerson, J. C., Burns, K. G., Bray, R. E., Vandenhoff, G. E., Freedlende, A. E., and Fertier, G. A., 1983, Measurements of glycosylated hemoglobin using boronate chromatography, *Ann. Clin. Lab. Sci.* **13**:482–488.

Hjerten, S., and Li Jin-Ping, 1990, High-performance liquid chromatography of proteins on deformed non-porous agarose beads. Fast boronate affinity chromatography of hemoglobin at neutral pH, *J. Chromatogr.* **500**: 543–553.

Johnson, D. J. B., 1981, Synthesis of a nitrobenzeneboronic acid substituted polyacrylamide and its use in purification isoaccepting transfer ribonucleic acids, *Biochemistry* **20**:6103–6108.

Kemper, K., Hagemeier, E., Boos, K. S., and Schlimme, E., 1984a, Direct clean-up and analysis of urinary catecholamines, *J. Chromatogr.* **336:**374–379.

Kemper, K., Hagemeier, E., Ahrens, D., Boos, K. S., and Schlimme, E., 1984b, Group-selective prefractionation and analysis of urinary catecholamines by on-line HPLAC-HPLC chromatography, *Chromatographia* **19:**288–291.

Klenk, D. C., Hermanson, G. T., Krohn, R. I., Fujimoto, E. K., Mallia, A. K., Smith, P. K., England, J. D., Wiedmayer, H. M., Little, R. R., and Goldstein, D. E., 1982, Determination of glycosylated hemoglobin by affinity chromatography: Comparison with colorimetric and ion-exchange methods and effects of common interferences, *Clin. Chem.* **28:**2088–2094.

Kuo, K. C., Gehrke, C. W., McCune, R. A., Waalkes, T. P., and Borek, E., 1978, Rapid quantitative high-performance liquid column chromatography of pseudouridine, *J. Chromatogr.* **145:**383–392.

Lorand, J. P., and Edwards, J. O., 1959, Polyol complexes and structure of the benzeneboronate ion, *J. Org. Chem.* **24:**769–774.

Maestas, R. R., Prieto, J. R., Kuehn, G. D., and Hageman, J.-H., 1980, Polyacrylamide-boronate beads saturated with biomolecules: A new general support for affinity chromatography of enzymes, *J. Chromatogr.* **189:**225–231.

Mallia, A. K., Hermanson, G. T., Krohn, R. L., Fujimoto, E. K., and Smith, P. K., 1981, Preparation and use of boronic acid affinity support for separation and quantitation of glycosylated hemoglobins, *Anal. Lett.* **14:**649–661.

Mayer, T. K., and Freedman, Z. R., 1983, Protein glycosylation in diabetes mellitus: A review of laboratory measurements and of their clinical utility, *Clin. Chim. Acta* **127:**147–184.

Mazzeo, J. R., and Krull, I. S., 1989, Immobilized boronates for the isolation and separation of bioanalytes, *BioChromatography* **4**(3):124–130.

Middle, F. A., Bannister, A., Bellingham, A. J., and Dean, P. D. G., 1983, Separation of glycosylated hemoglobins using immobilized phenylboronic acid. Effect of ligand concentration, column operating conditions and comparison with ion exchange and isoelectric focusing, *Biochem. J.* **209:**771–779.

Myoehaenen, T. A., Bouriotis, V., and Dean, P. D. G., 1981, Affinity chromatography on yeast α-glucosidase using ligand-mediated chromatography on immobilized phenylboronic acid, *Biochem. J.* **197:**683–688.

Okayama, H., 1980, Application of boronic acid derivatives to affinity chromatography, *Tanpakushitsu Kakusan Koso, Bessatsu* **22:**72–75.

Pace, B., and Pace, N. R., 1980, The chromatography of RNA and oligoribonucleotides on boronate-substituted agarose and polyacrylamide, *Anal. Biochem.* **107:**128–135.

Pope, M. R., Murrell, S. A., and Ludden, P. W., 1985, Purification and properties of the heat-released nucleotide-modifying group from the inactive iron protein of nitrogenase from *Rhodospirillum rubrum*, *Biochemistry* **24:**2374–2380.

Reid, T. S., and Gisch, D. J., 1989, Determination of glycosylated and nonglycosylated hemoglobin, using a fast affinity boronate HPLC column, *J. High Resolut. Chromatogr.* **12**(4):249–250.

Rendell, M., Kao, G., Mecherikunnel, P., Petersen, B., Dubaney, R., Nierenberg, J., Rasbold, K., Klenk, D., and Smith, P. K., 1985, Aminophenylboronic acid affinity chromatography and thiobarbituric acid colorimetry compared for measuring glycated albumin, *Clin. Chem.* **31:**229–234.

Rosenberg, M., Wiebers, J. L., and Gilham, P. T., 1972, Studies on interactions of nucleotides, polynucleotides and nucleic acids with dihydroboryl-substituted cellulose, *Biochemistry* **11:**3623–3628.

Schlimme, E., Boos, K.-S., Hagemeier, E., Kemper, K., Meyer, V., Hobler, H., Schnelle, T., and Weise, M., 1986, Direct clean-up and analysis of ribonucleosides in physiological fluids, *J. Chromatogr.* **378:**349–360.

Shvarts, E. M., 1990, *Interaction of Boric Acid with Alcohols and Hydroxy Acids*, Zinatne, Riga, Latvian SSR, [in Russian].

Singhal, R. P., Ramamurthy, B., Govindraj, N., and Sarvar, Y., 1991, New ligands for boronate affinity chromatography. Synthesis and properties, *J. Chromatogr.* **543:**17–38.

Soundararajan, S., Badawi, M., Kohlrust, C. M., and Hageman, J. H., 1989, Boronic acids for affinity chromatography: Spectral methods for determination of ionization and diol-binding constants, *Anal. Biochem.* **178:**125–134.

Speek, A. J., Odink, J., Schrijver, J., and Schreurs, W. H. P., 1983, HPLC determination of urinary free catecholamines with electrochemical detection after prepurification on immobilized boric acid, *Clin. Chim. Acta* **128:**103–113.

Štambergova, A., and Vinš, I., 1991, unpublished results.

Sugumaran, M., and Lipke, H., 1982, A procedure for isolation and concentration of catechols by chromatography on dihydroxyboryl cellulose, *Anal. Biochem.* **121:**251–256.

Weith, H. L., Wibers, J. L., and Gilham, P. T., 1970, Synthesis of cellulose derivatives containing the dihydroboryl group and a study of their capacity to form specific complexes with sugars and nucleic acid component, *Biochemistry* **9**:4396–4401.

Wilk, H.-E., Kecskemethy, N., and Schaefer, K. P., 1982, *M*-Amino-phenylboronate agarose specifically binds capped snRNA and mRNA, *Nucleic Acids Res.* **10**:7621.

Williams, G. T., Johnstone, A. P., and Dean, P. D. G., 1982, Fractionation of membrane proteins on phenylboronic acid—agarose, *Biochem. J.* **205**:167–171.

Woo, J., Weinstock, R. S., Ozark, C., and Sunderji, S., 1987, Glycated albumin by affinity chromatography and radioimmunoassay in the management of diabetes mellitus, *J. Clin. Lab. Anal.* **1**:163–169.

Wulff, G., and Kirstein, G., 1990, Measuring the optical activity of chiral imprints in insoluble highly cross-linked polymers, *Angew. Chem. Int. Ed. Engl.* **29**:684–686.

Yue, D. K., McLennan, S., Church, D. B., and Turtle, J. R., 1982, The measurement of glycosylated hemoglobin in man and animals by aminophenylboronic acid affinity chromatography, *Diabetes* **31**:701–705.

Yurkevich, A. M., Kolodkina, I. I., Ivanova, E. A., and Pichuzchkina, E. I., 1975, Interaction of polyols with polymers containing N-substituted [(4-boronylphenyl)methyl]ammonio groups, *Carbohydr. Res.* **43**: 215–224.

21

Affinity Chromatographic Removal of Pyrogens

Tetsuya Tosa, Tadashi Sato, Taizo Watanabe, and Satoshi Minobe

1. INTRODUCTION

Pyrogens are classified into two groups, exogenous and endogenous. Exogenous pyrogens originating in the cell wall of gram-negative bacteria have the strongest pyrogenicity (Nowotny, 1969), and for them synonyms such as endotoxins, lipopolysaccharides (LPS), and O-antigens are used. If a pharmacologically active substance is contaminated with pyrogens, its intravenous administration produces transient fever in homoiothermic animals. Therefore, it is necessary to remove any pyrogen present in medicines or drugs for intravenous administration.

To date, basically, three different approaches have been used to remove pyrogens from aqueous solutions. These are (1) ultrafiltration, (2) adsorption to activated charcoal, and (3) ion-exchange chromatography.

The ultrafiltration approach (Sweadner *et al.*, 1977) is useful for pyrogen removal from a solution containing low-molecular-weight substances. However, it would be unsatisfactory for removing pyrogens from a solution containing other, desirable high-molecular-weight substances, because some pyrogens are also high-molecular-weight substances and this technique is dependent on a difference between the physical size of the pyrogen molecules and that of the product being purified (Sharma, 1986). Traditional adsorption techniques such as adsorption to activated charcoal or ion-exchange chromatography (Nolan *et al.*, 1975) would also be unsatisfactory, because there is a high risk of adsorption of the desired product to the adsorbent (Sharma, 1986).

Thus, the traditional methods for removal of pyrogens have several shortcomings. The most unfavorable defect is that these methods cannot be applied to the removal of pyrogens from relatively unstable high-molecular-weight substances such as enzymes, hormones, and so on, because biologically active high-molecular-weight substances such as insulin, interferon, and interleukin have recently been produced by microorganisms through the use

Tetsuya Tosa • Research and Development, Tanabe Seiyaku Co., Ltd., Osaka 532, Japan.   *Tadashi Sato* • Analytical Chemistry Research Laboratory, Tanabe Seiyaku Co., Ltd., Osaka 532, Japan.   *Taizo Watanabe and Satoshi Minobe* • Research Laboratory of Applied Biochemistry, Tanabe Seiyaku Co., Ltd., Osaka 532, Japan.

Molecular Interactions in Bioseparations, edited by That T. Ngo. Plenum Press, New York, 1993.

of DNA recombinant technology and developed as medicines. Moreover, as government regulations concerning pyrogen content in medicines have become strict, the need for efficient methods for removal of pyrogens has become apparent.

As a result, a number of attempts have been made to use affinity chromatographic approaches to remove pyrogens. These are based on the specific adsorption of pyrogens and therefore might be useful for removal of pyrogens from various solutions containing high-molecular-weight substances. Two approaches have been reported. One approach is based on the use of affinity adsorbents that contain polymyxin B as a ligand (Issekutz, 1983; Kodama et al., 1990), and the other uses immobilized protamine (Helander and Vaara, 1987) as the adsorbent. However, polymyxin B is strongly toxic to the central nervous system and the kidneys (Srinivasa and Ramachandran, 1979) and thus is not suitable as a ligand for removing pyrogens from a solution for intravenous injection. Immobilized protamine is also not suitable for practical use, because protamine is a protein and relatively unstable. Therefore, we considered the development of new adsorbents in order to overcome the defects of these other methods for removal of pyrogens. In this work, we developed immobilized histamine (Minobe et al., 1983) and immobilized histidine (Minobe et al., 1988). However, from the practical point of view, we chose immobilized histidine as the most favorable adsorbent for removal of pyrogens. Thus, in this chapter, we mainly describe the preparation, characteristics, and applications of immobilized histidine.

2. PREPARATION OF ADSORBENTS SUITABLE FOR PYROGEN ADSORPTION

Since the beginning of the 1960s, we have been studying techniques for the immobilization of enzymes and microbial cells and have successfully implemented industrial applications of immobilized enzymes and immobilized microbial cells. On the basis of this background, we have extended the utilization of the immobilization techniques to the preparation of immobilized biomaterials as adsorbents for general-ligand affinity chromatography and developed adsorbents for pyrogen adsorption.

2.1. Selection of Ligands

Kanoh et al. (1968) reported that ribonucleic acid has a high affinity for pyrogens and that it is very difficult to remove pyrogens from nucleic acids. Also, we have found that the removal of pyrogens from histidine is very difficult. Therefore, we postulated that if a nucleic acid-related compound or histidine or a related compound is immobilized on a water-insoluble matrix, the immobilized preparation may be used as a specific adsorbent for pyrogens. Thus, various heterocyclic compounds containing nitrogen were immobilized on aminohexyl cellulose with glutaraldehyde and their affinities for pyrogens were compared using the column method (Minobe et al., 1982). The results are shown in Table 1. All compounds tested showed high affinity for pyrogens, especially adenine, cytosine, histamine, and histidine. Although the ligand contents of the adsorbents differed, no effect of the content of histamine in the adsorbent in the range 0.5–15.5 μmol/ml of adsorbent on the affinity for pyrogens was observed (Minobe et al., 1982).

TABLE 1
TABLE 1
Selection of Ligands Possessing Affinity for Pyrogens[a]

Ligand	Ligand content of adsorbent (μmol/ml of adsorbent)	Concentration of pyrogen in effluent (ng/ml)
Adenine	0.4	0.2
2-Amino-4,6-dimethylpyrimidine	7.8	1.6
2-Amino-4-hydroxy-6-methylpyrimidine	8.0	1.3
Cytosine	7.6	0.7
5-Methylcytosine	9.4	1.4
Ethacridine	2.1	3.7
Histamine	9.4	0.1
Histidine	1.8	0.2

[a]Aminohexyl cellulose was activated with glutaraldehyde at pH 7.2, and each ligand was immobilized on this activated matrix. Then, the adsorbents were reduced with sodium borohydride. A 400-ml volume of pyrogen solution (*E. coli* 0128: B12, LPS 1000 ng/ml) was passed through the column packed with 8 ml of the adsorbent, and the concentration of pyrogen in the effluent was measured by the *Limulus* amebocyte lysate (LAL) test.

2.2. Selection of Matrices

In order to select matrices suitable for the preparation of adsorbents for pyrogens, histamine was covalently bound to various matrices possessing amino groups (Minobe *et al.*, 1982). As shown in Table 2, cellulose and agarose were suitable matrices for the preparation of adsorbents having high affinity for pyrogens.

2.3. Effect of Chain Length of Spacer on Adsorption of Pyrogens

In affinity chromatography, it is well known that the accessibility of macromolecules toward a ligand increases with extension of the spacer arm. Therefore, the effect of chain

TABLE 2
Selection of Matrices Suitable for Pyrogen Adsorbents[a]

Matrix	Content of histamine (μmol/ml of adsorbent)	Concentration of pyrogen in effluent (ng/ml)
AH-cellulose	9.4	0.1
AH-Sepharose 4B[b]	3.1	1.4
AH-Sepharose CL-4B[c]	4.9	0.7
AH-Toyopearl HW-55	16.4	470
AH-Toyopearl HW-65	21.5	390
AH-polystyrene	36.8	460
AH-Diaion CR-10	32.5	850
Diaion WA-21	52.5	710
AH-Diaion WK-11	5.5	1100
AH-Diaion WK-20	11.8	650

[a]Immobilization of histamine to each matrix and measurement of the affinity of each adsorbent for pyrogens were carried out by the same method as described in footnote *a* to Table 1.
[b]Hexamethylenediamine was bound to cyanogen bromide-activated Sepharose 4B.
[c]Hexamethylenediamine was bound to epichlorohydrin-activated Sepharose CL-4B.

length of the spacer on the affinity of the adsorbent for pyrogens was investigated (Minobe *et al.*, 1982). When the chain length of the spacer was 20–29 Å, the adsorbent showed the highest affinity for pyrogens. In Figure 1, our results concerning adsorbents suitable for the adsorption of pyrogens are summarized.

2.4. Preparation of Immobilized Histidine

Among the adsorbents that we considered, we chose immobilized histidine as a pyrogen adsorbent on the basis of its cost, safety, and ease of use. The immobilized histidine adsorbent was prepared by immobilization of histidine on aminohexyl (AH)-Sepharose CL-4B using epichlorohydrin (Minobe *et al.*, 1988). The histidine content was 12 μmol/ml of adsorbent.

3. CHARACTERISTICS OF IMMOBILIZED HISTIDINE

3.1. Adsorption of Pyrogens

Immobilized histidine showed a high affinity for pyrogens at low ionic strength (μ) and over a wide pH range and temperature (Minobe *et al.*, 1988; Takenaga *et al.*, 1991). Examples of the effect of ionic strength and pH on adsorption of pyrogens are shown in Fig. 2. The adsorption capacity was 0.53 mg of LPS (*Escherichia coli* 0128: B12) per milliliter of the adsorbent (Minobe *et al.*, 1988) or 0.31 mg of LPS (*E. coli* UKT-B) per milliliter of the adsorbent (Matsumae *et al.*, 1990). The apparent dissociation constant (K_d) of immobilized histidine was $1.57 \times 10^{-9}\,M$ for LPS (*E. coli* 0128: B12; Minobe *et al.*, 1988) and $7.3 \times 10^{-13}\,M$ for LPS (*E. coli* UKT-B; Matsumae *et al.*, 1990) when the molecular weight of the LPS was taken as 10^6. Umeda *et al.* (1983) reported that the K_d of polymyxin-Sepharose for LPS (*Salmonella minnesota* R595) was $2.46 \times 10^{-9}\,M$. Although different kinds of LPS were used in these various studies, the results show that the affinity of immobilized histidine for LPS is comparable to or higher than that of polymyxin-Sepharose.

The effect of temperature on adsorption of LPS was investigated at varying ionic strengths. At each ionic strength, the adsorption of LPS on immobilized histidine increased with increasing temperature (Matsumae *et al.*, 1990).

Pyrogens originating from various microorganisms are different from each other.

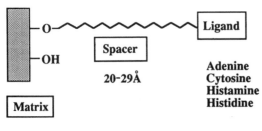

FIGURE 1. Adsorbent suitable for adsorption of pyrogens.

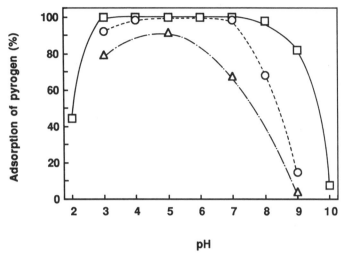

FIGURE 2. Effects of ionic strength and pH on adsorption of pyrogen to immobilized histidine. Adsorption of pyrogen was carried out by the column method using 8 ml of adsorbent and 40 ml of pyrogen solution (LPS, *E. coli* 0128: B12, 10 ng/ml) at a flow rate of 24 ml/h. □, $\mu = 0.02$; ○, $\mu = 0.05$; △, $\mu = 0.1$.

Therefore, the adsorption of pyrogens originating from various microorganisms was investigated. The various kinds of endotoxins tested were adsorbed well on immobilized histidine. Immobilized histidine was able to decrease the concentration of endotoxins in water from 1000 ng/ml to less than 0.01 ng/ml (Matsumae *et al.*, 1990).

3.2. Adsorption of Various Substances Other than Endotoxins

The adsorption of various substances on immobilized histidine was investigated under the conditions under which endotoxins were well adsorbed (Matsumae *et al.*, 1990). All substances tested, except for the acidic high-molecular-weight substances, were less well adsorbed under the conditions of pH 4.0–9.5 and $\mu = 0.02$. However, acidic high-molecular-weight substances were adsorbed at a pH more basic than the isoelectric point (pI) of the substances under the condition $\mu = 0.02$. The adsorption of these substances decreased with increasing ionic strength. When the ionic strength was increased ($\mu = 0.05, 0.1$), most acidic high-molecular-weight substances, except for DNA, RNA, heparin, and pepsin, were scarcely adsorbed. Therefore, it is possible to remove pyrogens from most substances by this approach.

3.3. Mechanism of the Selectivity

In clarifying the mechanism of adsorption of pyrogens to immobilized histidine, we must consider the characteristics of pyrogens. LPS consist of a hydrophilic polysaccharide and a hydrophobic lipid A and are, therefore, amphiphilic. The lipid A component is responsible for many biological activities (Galanos *et al.*, 1977). It is well known that LPS are chemically heterogeneous. Although the basic chemical structures (lipid A) of LPS originating from various gram-negative bacteria are similar, there are variations in the glycosyl and fatty acyl residues. The heterogeneity of LPS originating from the same strain has been reported (Rosner *et al.*, 1979). Moreover, LPS show unusual molecular weight characteristics. For example, LPS range in size from subunits of molecular weight 10,000–20,000 to aggregates of 0.1-μm diameter (Hannecart-Pokorni *et al.*, 1973). In aqueous solutions, the hydrophobic regions of LPS molecules tend to associate and form large aggregates. Divalent cations, chelating reagents, and detergents affect the extent of aggregation (Nelsen, 1978; Sweadner *et al.*, 1977).

In addition to polymyxin B, substances known to have a high affinity for LPS include anti-LPS factor (Aketagawa *et al.*, 1986), tachyplesin (Nakamura *et al.*, 1988), poly-phemusins (Miyata *et al.*, 1989), LPS binding protein (LBP; Tobias *et al.*, 1989), and bactericidal/permeability-increasing protein (BPI; Ooi *et al.*, 1987). The mechanism of binding of LPS to these substances at the molecular level is not clear yet. However, these substances are amphipathic and have positively charged groups, and Aketagawa *et al.* (1986) suggested that the cationic regions of anti-LPS factor interacted with the anionic regions of LPS, especially with phosphate groups in the lipid A moiety, and that anti-LPS factor bound to the lipid A moiety of LPS through the NH_2-terminal hydrophobic region.

L-Histidine is bound to the matrix through its α-amino group, and the free imid-azole and the carboxyl group play an important role in the binding mechanism. However, because no neutral or positively charged substances were adsorbed on immobilized histidine under any conditions (Matsumae *et al.*, 1990), it is presumed that the anionic nature of the carboxyl group was neutralized by an amino group or/and an imidazole group; thus, the charge of the adsorbent was cationic below pH 9 because of secondary amino groups originating from AH-Sepharose CL-4B. This suggests that ionic interactions are involved in the binding between the adsorbent and the pyrogens or negatively charged substances. However, Pommerening *et al.* (1979) have already suggested that the imidazole group is a good hydrophobic ligand, and the spacer is also hydrophobic. Accordingly, because pyrogens also have a hydrophobic nature, it is reasonable to assume that both ionic and hydrophobic interactions contribute to the binding between pyrogens and the adsorbent. Participation of hydrophobic interactions is supported by the fact that pyrogens are adsorbed to the adsorbent under conditions of high ionic strength ($\mu > 3$) and that the adsorption of pyrogens increases at high temperature (Matsumae *et al.*, 1990). The adsorption capacity of immobilized histidine for pyrogens is 0.3–0.5 nmol of LPS/ml of adsorbent when the molecular weight of the pyrogen is 10^6. On the other hand, the histi-dine content of the adsorbent is 12 μmol/ml of adsorbent. Thus, the adsorption capacity for pyrogens is very low compared with the histidine content of the adsorbent. This indicates that one large pyrogen molecule is adsorbed in the ionic and hydrophobic re-gion comprising several molecules of histidine. This suggests that the binding force between pyrogens and the adsorbent is stronger than that between other molecules and the adsorbent.

In considering the adsorption of a pyrogen on immobilized histidine in a solution containing other substances, the interactions between the adsorbent and the pyrogen, between the adsorbent and the other substances, and between the pyrogen and the other substances should be taken into account. The adsorption of a pyrogen on immobilized histidine in bovine serum albumin (BSA) solution has been investigated (Minobe *et al.*, 1988). The pyrogen was adsorbed less from a solution containing BSA and the pyrogen than from a solution containing the pyrogen only. This phenomenon is considered to be due to the interaction between BSA and the pyrogen. Especially at pH values lower than the pI of BSA, the pyrogen may be bound to BSA via ionic interaction. For endotoxin–protein dissociation, Karplus *et al.* (1987) reported the use of a dialyzable surfactant, octyl-β-D-glucopyranoside.

4. APPLICATIONS OF IMMOBILIZED HISTIDINE

4.1. Removal of Pyrogens from Various Substances

The removal of pyrogens from various substances has been investigated (Minobe *et al.*, 1988; Tosa *et al.*, 1987). Table 3 shows data for the removal of pyrogens from various low-molecular-weight substances. The substances were dissolved in pyrogen-free water, and LPS (*E. coli* 0128: B12) was added at a concentration of 100 ng/ml. The sample solutions were then passed through a column packed with immobilized histidine. After treatment, the pyrogen concentration in the effluent was below 0.1 ng/ml in all cases. On the other hand, the recoveries of the substances, except for FAD, were 100%. FAD was adsorbed slightly at low ionic strength, but at $\mu = 0.05$, no FAD was adsorbed.

Table 4 shows data for the removal of pyrogens from various high-molecular-weight substances. The sample solutions containing environmental pyrogens were passed through a column packed with immobilized histidine. The pyrogen concentration in the effluent was below 0.1 ng/ml in all cases. On the other hand, the recoveries of the substances were more than 91%.

In the removal of pyrogens from any given substance, the conditions of treatment should be optimized. After such optimization, the removal of pyrogens from tumor necrosis factor and lysozyme was accomplished (Matsumae *et al.*, 1990).

Thus, immobilized histidine is useful as a tool for the selective removal of pyrogens, not only from low-molecular-weight substances, but also from high-molecular-weight substances such as enzymes and hormones.

TABLE 3
Removal of Pyrogens in Various Low-Molecular-Weight Substances by Immobilized Histidine[a]

	Sample solution		
Substance	Concentration (mg/ml)	Concentration of pyrogen in effluent (ng/ml)	Recovery of substance (%)
L-Alanine	20	<0.1	100
L-Proline	20	<0.1	100
Glucose	50	<0.1	100
6-Aminopenicillanic acid	0.8	<0.1	100
Cytosine	5	<0.1	100
FAD	2.5	<0.1	92

[a]Each substance was dissolved in pyrogen-free water, and LPS (*E. coli* 0128: B12) was added at a concentration of 100 ng/ml. Each sample solution was passed through the column packed with 8 ml of immobilized histidine at a flow rate of 24 ml/h.

TABLE 4
Removal of Pyrogens in Various High-Molecular-Weight Substances by Immobilized Histidine[a]

Sample solution			Conditions for treatment	Concentration of pyrogen (ng/ml)		Recovery of substance (%)
Substance	Concentration (mg/ml)	pH	Salt concentration (M)	Before	After	
Asparaginase	5.8	6.0	0.05	15.4	<0.1	94
Bovine serum albumin	5.0	7.0	0.05	17.0[b]	<0.1	95
Gelatin	50	6.9	0.05	0.7	<0.1	91
Human serum albumin	250	6.9	—	0.5	<0.1	99
Myoglobin	1.0	7.0	0.05	4.7	<0.1	99

[a]Each substance was dissolved in buffer solution at low ionic strength. Each sample solution was passed through the column packed with 8 ml of immobilized histidine at a flow rate of 24 ml/h.
[b]LPS (*E. coli* 0128: B12) was added.

4.2. Assay of Pyrogens

The *Limulus* amebocyte lysate (LAL) test has been used for sensitive detection of pyrogens. However, the LAL test response is inhibited or enhanced by many substances such as amino acids and antibiotics. In order to overcome these problems, we have investigated an improvement of the LAL test employing immobilized histidine. Pyrogens in the sample solution were selectively adsorbed on immobilized histidine, and the adsorbed pyrogens were separated and determined with the LAL test (Watanabe *et al.*, 1990). In the gel-clot technique, many substances showed false negatives or false positives. On the other hand, the concentration of pyrogens could be accurately assayed by this method. This method could be used widely as the means for assaying pyrogens in solutions containing LAL-inhibiting or LAL-enhancing substances—not only low-molecular-weight substances but also high-molecular-weight substances such as proteins, serum, and plasma (Minobe *et al.*, 1991; Nawata *et al.*, 1992; Watanabe *et al.*, 1992).

5. CONCLUSION

As discussed in this chapter, immobilized histidine has high affinity and selectivity for pyrogens and can be used for the removal of pyrogens from various substances and the assay of pyrogens in various solutions containing LAL-inhibiting or LAL-enhancing substances. Immobilized histidine was commercialized in Japan in 1988 as a pyrogen removal adsorbent and is sold under the trade name PyroSep. We hope that this adsorbent will be used widely as a tool for the selective removal of pyrogens from various substances, especially biologically active high-molecular-weight substances, and for pyrogen assay.

REFERENCES

Aketagawa, J., Miyata, T., Ohtsubo, S., Nakamura, T., Morita, T., Hayashida, H., Miyata, T., and Iwanaga, S., 1986, Primary structure of limulus anticoagulant anti-lipopolysaccharide factor, *J. Biol. Chem.* **261**:7357–7365.

Galanos, C., Lüderitz, O., Rietschel, E. T., and Westphal, O., 1977, Newer aspects of the chemistry and biology of bacterial lipopolysaccharides, with special reference to their lipid A component, in: *International Review of Biochemistry, Biochemistry of Lipid II*, Vol. 14 (T. W. Goodwin, ed.), University Press, Baltimore, pp. 239–335.

Hannecart-Pokorni, E., Dekegel, D., and Depuydt, F., 1973, Macromolecular structure of lipopolysaccharides from gram-negative bacteria, *Eur. J. Biochem.* **38:**6–13.

Helander, I. M., and Vaara, M., 1987, Reversible binding of *Salmonella typhimurium* lipopolysaccharides by immobilized protamine, *Eur. J. Biochem.* **163:**51–55.

Issekutz, A. C., 1983, Removal of gram-negative endotoxin from solutions by affinity chromatography, *J. Immunol. Methods* **61:**275–281.

Kanoh, S., Kohlhage, H., and Siegert, R., 1968, Pyrogenic principle of the ribonucleic acid from the yeast *Candida utilis*, *J. Bacteriol.* **96:**738–743.

Karplus, T. E., Ulevitch, R. J., and Wilson, C. B., 1987, A new method for reduction of endotoxin concentration from protein solutions, *J. Immunol. Methods* **105:**211–220.

Kodama, M., Hanasawa, K., and Tani, T., 1990, New therapeutic method against septic shock—removal of endotoxin using extracorporeal circulation, in: *Advances in Experimental Medicine and Biology*, Vol. 256 (Endotoxin) (H. Friedman, T. W. Klein, M. Nakano, and A. Nowotny, eds.), Plenum Press, New York, pp. 653–664.

Matsumae, H., Minobe, S., Kindan, K., Watanabe, T., Sato, T., and Tosa, T., 1990, Specific removal of endotoxin from protein solutions by immobilized histidine, *Biotechnol. Appl. Biochem.* **12:**129–140.

Minobe, S., Nawata, M., Watanabe, T., Sato, T., and Tosa, T., 1991, Specific assay for endotoxin using immobilized histidine and *Limulus* amebocyte lysate, *Anal. Biochem.* **198:**292–297.

Minobe, S., Watanabe, T., Sato, T., Tosa, T., and Chibata, I., 1982, Preparation of adsorbents for pyrogen adsorption, *J. Chromatogr.* **248:**401–408.

Minobe, S., Sato, T., Tosa, T., and Chibata, I., 1983, Characteristics of immobilized histamine for pyrogen adsorption, *J. Chromatogr.* **262:**193–198.

Minobe, S., Watanabe, T., Sato, T., and Tosa, T., 1988, Characteristics and applications of adsorbents for pyrogen removal, *Biotechnol. Appl. Biochem.* **10:**143–153.

Miyata, T., Tokunaga, F., Yoneya, T., Yoshikawa, K., Iwanaga, S., Niwa, M., Takao, T., and Shimonishi, Y., 1989, Antimicrobial peptides, isolated from horseshoe crab hemocytes, Tachyplesin II, and Polyphemusins I and II: Chemical structures and biological activity, *J. Biochem.* **106:**663–668.

Nakamura, T., Furunaka, H., Miyata, I., Tokunaga, F., Muta, T., and Iwanaga, S., 1988, Tachyplesin, a class of antimicrobial peptide from hemocytes of the horseshoe crab (*Tachypleus tridentatus*), *J. Biol. Chem.* **263:**16709–16713.

Nawata, M., Minobe, S., Hase, M., Watanabe, T., Sato, T., and Tosa, T., 1992, Specific assay for endotoxin using immobilized histidine, *Limulus* amoebocyte lysate and a chromogenic substrate, *J. Chromatogr.* **597:**415–424.

Nelsen, L. L., 1978, Removal of pyrogens from parenteral solutions by ultrafiltration, *Pharmacol. Technol.* **48:**46–49.

Nolan, J. P., McDevitt, J. J., and Goldmann, G. S., 1975, Endotoxin binding by charged and uncharged resins, *Proc. Soc. Exp. Biol. Med.* **149:**766–770.

Nowotny, A., 1969, Molecular aspects of endotoxic reaction, *Bacteriol. Rev.* **33:**72–98.

Ooi, C. E., Weiss, J., Elsbach, P., Frangione, B., and Mannion, B., 1987, A 25-kDa NH2-terminal fragment carries all the antibacterial activities of the human neutrophil 60-kDa bactericidal/permeability-increasing protein, *J. Biol. Chem.* **262:**14891–14897.

Pommerening, K., Kühn, M., Jung, H., Buttergereit, K., Mohr, P., Stanberg, J., and Benes, M., 1979, Affinity chromatography of haemoproteins: 1. Synthesis of various imidazole containing matrices and their interaction with haemoglobin, *Int. J. Biol. Macromol.* **1:**79–88.

Rosner, M. R., Tang, J. Y., Barzilay, I., and Khorana, H. G., 1979, Structure of the lipopolysaccharide from an *Escherichia coli* heptose-less mutant: 1. Chemical degradations and identification of products, *J. Biol. Chem.* **254:**5906–5917.

Sharma, S. K., 1986, Endotoxin detection and elimination in biotechnology, *Biotechnol. Appl. Biochem.* **8:**5–22.

Srinivasa, B. R., and Ramachandran, L. K., 1979, The polymyxins, *J. Sci. Ind. Res.* **38:**695–709.

Sweadner, K. J., Forte, M., and Nelsen, L. L., 1977, Filtration removal of endotoxin (pyrogens) in solution in different states of aggregation, *Appl. Environ. Microbiol.* **34:**382–385.

Takenaga, Y., Nawata, M., Sakata, N., Senuma, M., Watanabe, T., Sato, T., and Tosa, T., 1991, Removal of pyrogen using immobilized histidine by batchwise method, *Kagaku Kogaku Ronbunshu* **17:**204–206.

Tobias, P. S., Soldau, K., and Ulevitch, J., 1989, Identification of lipid A binding site in the acute phase reactant lipopolysaccharide binding protein, *J. Biol. Chem.* **264:**10867–10871.

Tosa, T., Sato, T., Watanabe, T., Minobe, S., and Chibata, I., 1987, Adsorbents for removal of pyrogen, *Ann. N. Y. Acad. Sci.* **501:**395–402.

Umeda, M., Asakawa, M., Terada, T., Tatsumi, Y., Kashiwara, N., Kawakita, J., Yamagami, S., Kishimoto, T., Maekawa, M., and Niwa, M., 1983, Adsorption and removal of endotoxin by polymyxin-Sepharose, its properties and applications, *Jpn. J. Artif. Organs* **12:**258–278.

Watanabe, T., Minobe, S., Sato, T., and Tosa, T., 1990, Specific assay of endotoxin using immobilized histidine, Abstracts of the First Congress of the International Endotoxin Society, May 10–12, p. 36.

Watanabe, T., Minobe, S., Nawata, M., Shibatani, T., and Tosa, T., 1992, Assay of endotoxin in plasma using immobilized histidine and *Limulus* amebocyte lysate—Basic experiments—, Abstracts of the Second Conference of the International Endotoxin Society (August 17–20, 1992), p. 147.

22

Molecular Interactions in Hydrophobic Chromatography

Patrick Hubert and Edith Dellacherie

1. INTRODUCTION

The hydrophobic effect is perhaps one of the most important factors in the organization of complex structural entities such as cell membranes and organelles from simple constituent molecules. In biology, the folding of globular proteins, the binding of small molecules by proteins to ensure their transport in the circulating blood, the recognition of steroid or peptide hormones by their receptors, the binding of substrates to enzymes, and the self-association of (phospho)lipids to form a biological membrane bilayer are among the many events where hydrophobic interactions play a major role.

On the other hand, the understanding of the mode of action of biomolecules, whatever their role in the biological assembly, implies knowledge of their structure and therefore requires that they be selectively isolated from their natural environment, preferably in a form as close as possible to their native functional state. Considering the presence of thousands of proteins, hundreds of RNAs, and multiple DNA and polysaccharide components in a single cell, it is not surprising that the advancement of our comprehension of biological phenomena is intimately correlated to the development of separation techniques.

In this regards, there is little doubt that the emergence, 20 years ago, of affinity chromatography is one of the most outstanding milestones in the development of biochemistry. This technique has been the subject of numerous theoretical studies and reports of successful purifications, such that it is nowadays almost compulsorily included in purification protocols, as routinely as are classical nonspecific techniques such as ammonium sulfate precipitation and gel permeation chromatography.

The fortuitous observation that, during affinity chromatography experiments, retention was dependent on the length of the spacer arm intercalated between the ligand and the matrix gave rise to a new type of chromatography, named by Hjertén (1973) "hydrophobic interaction chromatography" (HIC). It was first developed on soft gels, before significant advances in the preparation of mechanically resistant, macroporous, and biocompatible microparticles suitable for the chromatography of biological macro-

Patrick Hubert and Edith Dellacherie • Laboratoire de Chimie-Physique Macromoléculaire, URA CNRS 494, ENSIC, 54001 Nancy Cédex, France.

Molecular Interactions in Bioseparations, edited by That T. Ngo. Plenum Press, New York, 1993.

molecules allowed the experiments to be performed with both high resolution and medium to high flow rates.

A different approach, based on principles that are the "reverse" of those involved in what was considered as normal partition liquid chromatography, that is, a polar stationary phase equilibrated with a nonpolar mobile phase, rapidly found applications in the purification of small molecules. However, as in the case of HIC, reversed-phase (RP) chromatography did not make a crucial breakthrough as a technique for the purification of biological macromolecules until high-performance liquid chromatography (HPLC) supports combining hydrophilic, soft surfaces with immobilized hydrophobic ligands became available.

Although both techniques are governed by hydrophobic interactions, the fact that they have evolved differently and that they do not exhibit a lot of similarities, insofar as the mobile phases employed, the operating conditions, and the scope of applications are concerned, has led to these similar approaches being considered as completely different.

Anyhow, the considerable number of articles on the use of these techniques is convincing evidence in favor of considering them as a major tool, comparably with affinity chromatography, in the arsenal of separation techniques available to the biochemist.

2. THE HYDROPHOBIC INTERACTION

The hydrophobic interaction is a very complex phenomenon comprising various physicochemical and thermodynamic aspects and has been the subject of controversy among experts, which is beyond the scope of the present chapter (Tanford, 1980; Dill, 1990; Herzfeld, 1991; Murphy et al., 1990; Israelachvili and Pashley, 1982).

From an etymological point of view, the terminology "hydrophobic" (i.e., "hating water") is a somewhat misleading description of the mutual interaction between nonpolar solutes or chains, in the presence of water. As a matter of fact, it gives the impression that the association between these nonpolar solute arises mainly from their inherent repulsion, or "phobia," for water, accordingly resulting in a "like-to-like" association.

Actually, nonpolar solutes dissolved in an organic solvent exhibit no or only a very limited tendency for self-association. The situation is quite different when the organic solvent is exchanged for water. The fact that nonpolar solutes will effectively, in this unique case, tend to adhere to each other demonstrates that water must play the major role in the hydrophobic interaction process.

A direct way to demonstrate this dominant role of water in a nonpolar solute–water system, is to consider the free energy of association between the two substances involved, that is, their tendency for mutual attraction (these energies are related to the surface tensions of the pure liquids and to the interfacial tension at the nonpolar solute–water interface).

The free energy of attraction between water and, for example, hexane or octane is about -40 erg/cm^2 of contact area. The free energy of attraction of these hydrocarbons for themselves is also about -40 erg/cm^2. Obviously, neither a clear-cut affinity of a hydrocarbon chain for itself nor its marked repulsion for water is evidenced, and these data cannot afford satisfactory explanations for the phenomena observed during hydrophobic interaction. The actual reason, as already pointed out, must be sought in the unique properties of water.

In fact, the free energy of attraction of water molecules for themselves is around -140

erg/cm^2. The practical consequence of this enormous self-attraction of water is that when nonpolar solutes are dissolved in water, the system tends to optimize the water–water interactions or conversely minimize nonpolar solute–water contacts. In other words, nonpolar solutes are forced to cluster so that the water–water associations can occur as much as possible.

The hydrophobic interaction may therefore be understood, on the macroscopic scale, as a consequence of the immoderate like of water molecules for themselves, rather than, as currently believed and suggested by the term "hydrophobic," of dislike of nonpolar chains for water.

Another way to approach a better understanding of the phenomenon and obtain a somewhat less rough view consists in considering the system in terms of thermodynamics.

Water molecules do not move around as free, separate molecules, but in complexes in which the molecules are kept together by strong hydrogen bonds, which form an isotropic network that must be disrupted or distorted when any solute is dissolved (Tanford, 1980; Lewin, 1974; Kavanau, 1964).

If the solute is ionic or strongly polar, it can form strong bonds with water molecules which, in terms of energy, more than compensate for the disruption or distortion of the bonds existing in pure water. Consequently, these substances tend to be readily soluble in water.

No such compensation occurs with nonpolar groups, which are unable to form hydrogen bonds with water, and their dissolution in water is accordingly resisted. The system must therefore reorganize so that the perturbation created by the intrusion of foreign molecules is minimized. To do so, water molecules at the surface of the cavity created by a nonpolar solute must rearrange themselves, in order to regenerate as many broken hydrogen bonds as possible. This process involves two effects:

First, the initial disruption of hydrogen bonds is followed by the formation of new ones. The energy lost in breaking hydrogen bonds is partly compensated for by the energy gained during the formation of new ones (Frank and Evans, 1945). This globally results in some enthalpy change close to zero or only slightly positive ($\Delta H \cong 0$).

Second, these rearrangements of hydrogen bonds result in a higher degree of local order than exists in pure liquid water, thereby producing a decrease in entropy ($\Delta S < 0$).

Together, these two effects contribute to an energetically unfavorable or only slightly favorable state ($\Delta G = \Delta H - T\Delta S \cong 0$, responsible for the very low solubility of hydrocarbons in water.

When two nonpolar solutes approach each other in aqueous solution (Fig. 1), that is, when hydrophobic interaction occurs, some of the ordered water molecules surrounding the individual nonpolar molecules will be expelled and adopt the less ordered bulk water structure (Lewin, 1974). This is accompanied by a large entropy increase ($\Delta S > 0$). Since the associated enthalpy change is again close to zero (as during the solubilization process, the energy lost during bond breaking is more or less compensated for by the energy gained during hydrogen bond re-formation), ΔG can be approximated as $\Delta G \cong -T\Delta S$, and the reaction can occur spontaneously since ΔG becomes negative. The hydrophobic interaction may therefore be regarded as an entropy-driven phenomenon.

Both macroscopic and thermodynamic analyses provide evidence that no forces exist between nonpolar solutes. As stated by Van Oss and Good (1988), the hydrophobic interaction can be quantitatively accounted for mainly by the hydrogen-bonding energy of cohesion of water plus, to a negligible extent, a van der Waals attraction between the

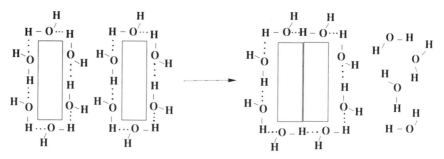

FIGURE 1. Schematic representation of the hydrophobic interaction between two nonpolar solutes dissolved in water.

dissolved hydrophobic solutes. The terminology "hydrophobic forces" frequently encountered in the literature is thus largely erroneous and misleading.

Many of the practical rules that are associated with the hydrophobic interaction concept—for example, the phenomenon is nonspecific, the interaction is enhanced by a temperature increase or by the presence of high concentrations of certain salts—are regarded somewhat as axiomatic principles and are more or less taken for granted, owing to their common use. However, a better understanding of the thermodynamic basis underlying the hydrophobic interaction would unquestionably be of considerable help in the handling of these rules.

For instance, the effect of temperature on the strength of the hydrophobic interaction simply lies in the formulation of the energy of the system. Since the positive enthalpy term is small and does not influence the interaction to a great extent, ΔG can be approximated as $\Delta G \cong -T\Delta S$, and the evolution toward an energetically more favorable state ($\Delta G < 0$) is obviously enhanced by a temperature increase, as long as the condition $\Delta H \ll T\Delta S$ is fulfilled (Jennissen and Botzet, 1979).

As pointed out previously, the hydrophobic interaction is a spontaneous process, simply because the entropy gained during it allows the system to pass from an energetically not very favorable state ($\Delta G \cong 0$ for dissolution of the individual apolar solutes) into an energetically stabilized one ($\Delta G < 0$ after the hydrophobic association takes place).

It follows that, if any external parameter is able by itself to bring about the entropy increase necessary for the system to attain an energetically favorable state, then the hydrophobic interaction becomes groundless and does not occur.

Conversely, any external parameter leading to some entropy decrease, that is, creating some extra order in the structure of water, would reinforce the need for some compensating effect and accordingly would result in a stronger hydrophobic interaction.

2.1. The Effect of Salts and Other Additives

The dissolution of any solute in water proceeds by some initial disruption or distortion of the ordering of H_2O molecules. This first event is followed by a reorganization, which, in the case of ions, frequently results in a large negative entropy, more negative than that in pure water, indicative of the ordering of water molecules around the ions by ion–dipole forces (Tanford, 1980).

The introduction of a nonpolar solute in such a system is therefore even more resisted than in pure water. The entropy increase associated with the hydrophobic interaction process is accordingly even more indispensable than in pure water, to provide the necessary compensation toward a thermodynamically stabilized state. The introduction of a high salt concentration in a nonpolar solute–water system therefore favors the establishment of hydrophobic interactions or reinforces the effect if it preexists in pure water. Conversely, decreasing the ionic strength loosens the association and may lead to its complete disruption.

This effect depends not only on the salt concentration but also on the nature of the salt (Hatefi and Hanstein, 1969; Nishikawa and Bailon, 1975; Påhlman et $al.$, 1977). The ability of ions to increase hydrophobic interactions is in the order

$$SCN^- < ClO_4^- < NO_3^- < I^- < Br^- < SO_4^{2-}, PO_4^{3-}, CH_3COO^-, F^-$$
$$Ba^{2-} < Ca^{2+} < Mg^{2-} < Li^+ < Na^+ < K^+ < NH_4^+$$

essentially overlapping that established as early as 1888 and known as the Hofmeister series (Hofmeister, 1887–1888).

According to Geng et $al.$ (1990), the ability of salts to increase the hydrophobic interaction decreases in the following order:

$$Na_2SO_4 > (NH_4)_2SO_4 > NaCl > KCl > NaBr > NH_4Cl$$

These salts are frequently referred to as "salting-out" agents. They are well known to decrease the solubility of most nonelectrolytes in water. Also, precipitation by concentrated salt solutions (e.g., ammonium sulfate) has long proved one of the most satisfactory methods for the separation of proteins.

As pointed out by Melander and Horvath (1977), the effect of salts on hydrophobic interactions roughly parallels their ability to increase the surface tension of water (the solvophobic theory). However, certain salts (e.g., $MgCl_2$) do not follow this trend, since, although they increase the surface tension of water, they do not, for instance, enhance the binding of a protein to hydrophobic columns as much as expected from the surface tension increment (Raymond et $al.$, 1981) or may even weaken the binding (Miller and Karger, 1985). Certain other salts (e.g., $BaCl_2$, $CaCl_2$, NaSCN) may enter into specific interactions with proteins and produce quite irregular effects (Påhlman et $al.$, 1977; Arakawa, 1986; Arakawa and Timasheff, 1984; Szepesy and Horvath, 1988).

A very important category of anions must be considered apart from the above classification. Those located on its left side (from SCN^- to I^-) have so little ability to increase hydrophobic interactions that they turn out to actually decrease them (Hatefi and Hanstein, 1969; Lee et $al.$, 1979). The dissolution of these anions is accompanied by a large positive entropy change, regarded by various authors as a possible structure-breaking effect of these ions on water. According to Frank and Evans (1945), a possible explanation is that around these ions, beyond the first saturated region of water molecules, there is a belt in which the water structure is broken down, or melted or depolymerized, as compared to that in ordinary water.

It is thus clear that such chaotropic ions are able to bring about the disorder, that is, the entropy increase, necessary for the stabilization of the system such that the hydrophobic association is no longer indispensable. The immediate consequence is that the hydrophobic interaction does not occur in the presence of such chaotropic salts or is disrupted upon their addition to the system.

This ability to bring disorder to the system is not restricted to the above-mentioned chaotropic anions. Hammes and co-workers (Hammes and Swann, 1967; Hammes and Roberts, 1968), for instance, concluded that urea and guanidine chloride also have solvent structure-breaking properties. In this respect, these substances may also be regarded as potential hydrophobic interaction-disrupting agents.

Similarly, as the hydrophobic interaction process mainly depends on the unique properties of water to form strongly hydrogen-bonded networks, it is thus clear that the hydrophobic interaction is virtually exclusive to aqueous solutions. In other words, the interaction decreases when the water is "diluted" by the addition of other solvents such as glycerol or ethylene glycol (Hofstee, 1973) whose molecules do not self-associate to the same extent as do H_2O molecules. This effect is observed with other organic solvents, to an extent depending on their concentration and their ability to alter the physicochemical properties of water. This ability, essentially paralleling their ability to decrease the surface tension of water or lower its polarity, leads to the classification reported in Table 1. These organic modifiers play an important role in reversed-phase chromatography and will be discussed in more detail in Section 4, which deals with this topic.

All these chaotropic salts or lyotropic solvents able to bring some disorder to the system and thus prevent the hydrophobic interaction from taking place or lead to its disruption are frequently referred to as "salting-in" agents. In contrast to the "salting-out" substances, these "salting-in" agents favor the transfer of apolar groups to water and accordingly lead to enhanced solubility.

2.2. Hydrophobicity Determinations

The hydrophobic interaction plays a primordial role in many aspects of physicochemical or biological events (Tanford, 1980; Kauzmann, 1959; Lee and Richards, 1971; Nemethy and Scheraga, 1962; Richmond, 1984). Its ability to act as an organizing force, for example, in the formation of surface films, micelles, and lipid bilayers, and its importance for the stability, conformation, and function of biological macromolecules and cells have naturally prompted some investigators to establish a hydrophobicity scale within a given class of biological substances.

In the study of protein organization, theoretical developments have led to a picture of the globular protein molecule as one having most of its apolar residues inside the

TABLE 1
Eluotropic Series of Organic Modifiers
(Decreasing Solvent Strength from Top to Bottom)[a]

Dioxane
n-Propanol
Acetone
Isopropanol
Acetonitrile
Ethanol
Acetic acid
Methanol

[a]From Colin and Guiochon, 1977.

molecule (to avoid contact with the surrounding water) and most of its charged groups outside (to promote contact with water). On the basis of this simple statement, Waugh, as early as 1954, estimated the hydrophobicity of a protein simply by determining the fraction of residues that were nonpolar. He defined tryptophan, isoleucine, tyrosine, phenylalanine, proline, leucine, and valine as nonpolar residues and found that proteins had nonpolar side-chains (NPS) frequencies between 0.21 and 0.47 (Waugh, 1954). Fisher (1964) developed this early model by defining what he called a polarity ration, $p = V_e/V_i$, where V_e and V_i are the external (shell) and internal (core) volumes, respectively.

However, obviously both approaches have a serious drawback, namely, the arbitrary definition of amino acid side chains as polar or nonpolar.

In 1962, Tanford proposed the introduction of a new parameter, $H\phi_{ave}$, to describe the average hydrophobicity of a protein. From the free energy of transfer from water to ethanol (as estimated from solubility measurements) of individual amino acids, $H\phi_{ave}$ was calculated as the sum of the side-chain hydrophobicities of the constituent amino acids. A comparison of the three different parameters (NPS, p, and $H\phi_{ave}$), allowed Bigelow (1967) to propose a classification of over 150 proteins and to conclude that NPS and $H\phi_{ave}$, which were closely correlated to each other, had more general validity than Fisher's p values (Table 2).

The pioneering work by Tanford (Tanford, 1962; Nozaki and Tanford, 1971) has been followed by various studies which have gradually provided refined values for hydrophobicities of the amino acid residues of proteins (Janin, 1979; Wolfenden et al., 1981). Some of these theoretical approaches are based on the empirical examination of known protein structures (Wertz and Scheraga, 1978; Rose et al., 1985). Various hydrophobicity scales based on, for example, the environment of the different amino acids or the fraction of amino acids buried in the protein have thus been proposed, and these have been reviewed by Cornette et al. (1987). Another recent theory based on hydrophobic moments of segments of protein secondary structure has been developed by Eisenberg et al. (1986).

The free energy of transfer of a solute from water to an organic solvent, that is, its hydrophobicity, can be directly estimated from measurements in biphasic systems, and a continuous scale for hydrophobicity can thus be derived from the determination of the

TABLE 2
Hydrophobicity Parameters Calculated for
a Few Proteins from Their Amino Acid Composition[a]

Protein	NPS	p	$H\Phi_{ave}$
Fibroin (*Bombyx mori*)	0.02	0.45	480
α-Hydroxysteroid dehydrogenase	0.30	1.11	1010
β-Hydroxysteroid dehydrogenase	0.32	0.79	1040
$\Delta_{5\to4}$ 3-Ketosteroid isomerase	0.33	0.85	1040
Hemoglobin γ-chain (human)	0.35	0.93	1080
Hemoglobin α-chain (human)	0.35	0.87	1110
Hemoglobin β-chain (human)	0.38	0.81	1130
Serum albumin (bovine)	0.36	1.09	1120
Lysozyme (papaya)	0.37	1.10	1150
Gramicidin S	0.80	0.29	2020

[a]From Bigelow, 1967.

logarithm of the partition coefficient. The n-octanol–water partitioning system is widely used for this purpose and is considered as the common reference (Masimov *et al.*, 1984), although the use of other organic solvents, for example, N-cyclohexyl-2-pyrrolidinone, in place of n-octanol has been sometimes recommended (Lawson *et al.*, 1984).

Obviously, such systems cannot be used with biomolecules, and another method that also involves biphasic systems has been proposed. This method is derived from the "aqueous polymer two-phase system" technique widely described by Albertsson (1971) for the analysis and separation of biological macromolecules and cell particles. Generally, proteins are placed in a dextran–poly(ethylene glycol) (PEG) system where some of the polyether is esterified with fatty acids. Proteins that interact with hydrocarbon groups will partition more in favor of the upper PEG-rich phase, and the difference $\Delta \log K$ between the logarithm of the partition coefficient of the protein in a system with and without fatty acid ester is taken as a measure of the hydrophobicity. Numerous examples of the use of such two-phase systems have been reported in the literature (Shanbhag and Axelsson, 1975).

Certain drawbacks inherent to the partition techniques, for example, the fact that the conventional octanol–water shake-flask method is tedious, time-consuming, and of limited applicability to ionic or volatile compounds or in the presence of impurities, have led to the development of chromatographic characterizations of hydrophobicity (Keshavarz and Nakai, 1979) and to the application of chromatographically determined partitioning data in quantitative structure–activity relationships (QSAR).

Much effort was initially directed toward the development of high-performance liquid chromatography (HPLC systems) to mimic the conventional n-octanol–water partitioning system. Several procedures based on coating a stationary phase with octanol and using n-octanol-saturated aqueous eluents were reported. Later on, the stable reversed-phase HPLC systems most often used for hydrophobicity assessment employed alkyl (usually octyl or octadecyl) ligands chemically bonded onto a silica support surface, a styrene–divinylbenzene copolymer stationary phase, or other polymer-coated reversed-phase materials (Heinemann *et al.*, 1987) such as alkylpolysiloxane-coated silica or polybutadiene-coated alumina (Kaliszan *et al.*, 1988).

Finally, various fluorescence techniques have been proposed for the evaluation of protein hydrophobicity. They can be classified into the fluorescence quenching methods (Eftink *et al.*, 1977) and the fluorescence probe methods (Sklar *et al.*, 1977; Kato and Nakai, 1980).

2.3. Hydrophobic Interaction at the Chromatographic Level

The study of the hydrophobic interaction between two solutes in the presence of water is naturally hampered by the inherently low solubility of such compounds in water.

However, highly hydrophobic and insoluble compounds may be molecularly dispersed in an aqueous milieu by covalent attachment to an insoluble, but hydrophilic matrix. The systematic study of hydrophobic phenomena can thus be carried out and can consequently lead to applications in separation procedures.

As early as 1948, Tiselius noticed that proteins that could be precipitated at high concentrations of salts were adsorbed (at much lower salt concentration) to supports that, in the absence of salts, exhibited no tendency to bind these proteins. Similarly, in 1950, the concept that substances could be separated through the participation of weak, van der

Waals-type interactions between the solute and a nonpolar stationary phase was established by Boscott (1947), Boldingh (1948), and Howard and Martin (1950).

However, this type of chromatography did not develop significantly until the mid-seventies, when crucial advances in the preparation and chemical substitution of polymers led to the availability of hydrophilic supports carrying various hydrophobic groups.

Historically, the chromatographic separation of polypeptides and proteins via hydrophobic mechanisms has evolved out of studies with chemically modified polymeric gels. At an early stage, matrices were adsorptively coated with nonpolar stationary-phase components such as long-chain alkanes. However, these systems lacked long-term stability, and only the introduction of chemically bonded n-alkyl agaroses and similar materials permitted most of the difficulties to be circumvented. Later on, the emergence of HPLC supports—meso- and macroporous, chemically bonded, hydrocarbonaceous supports of small particle diameters, typically in the range of 5–10 μm with narrow particle and pore size distribution—afforded a rapid, high-resolution method for the analysis and purification of biological macromolecules.

Two classes of chromatographic techniques are designed for the separation of substances according to their hydrophobic behavior. These are reversed-phase (RP) and hydrophobic interaction chromatography (HIC). In both cases, hydrophobic groups are immobilized and used as the functioning stationary phase. The most outstanding physical difference between RP and HIC matrices is the density of exposed hydrophobic ligands bound to the chromatography support. The density of hydrophobic groups in RP packings is commonly 10 to 100 times higher than that in HIC packings. In addition, HIC is generally restricted to rather short alkyl chains (C_2–C_6) whereas upper members of the alkyl series (C_{12}–C_{18}) are often encountered in RPLC.

Although both types of chromatography are governed by hydrophobic interactions, this large difference in stationary-phase hydrophobicity has progressively led to these similar approaches being considered as completely different. As a matter of fact, in contrast with HIC, where the interaction takes place with ligands more or less dispersed on the support, the RP technique is often considered as a form of liquid–liquid chromatography, where the retention is governed by partition between two liquid phases (Brandts et al., 1986).

As both approaches have evolved differently and do not exhibit a great deal of similarities insofar as the stationary and mobile phases employed, the operating conditions, and the scope of applications are concerned, they will be discussed separately in the following sections.

3. HYDROPHOBIC INTERACTION CHROMATOGRAPHY (HIC)

3.1. HIC Stationary Phases: Matrices, Ligands, and Immobilization Reagents

3.1.1. Matrices

Studies on HIC of proteins originated from the fortuitous observation that during affinity chromatography experiments, retention times were dependent on the length of the spacer arm intercalated between the affinity ligand and the matrix (e.g., Lee et al., 1979; Shaltiel and Er-el, 1973; Er-el et al., 1972; Hofstee, 1973). Since at that time most affinity chromatography experiments were conducted on cyanogen bromide-activated agarose

(Axén *et al.*, 1967), it is not surprising that this type of matrix was initially selected for HIC experiments as well and still remains in common use nowadays.

In the early 1980s great advances have been made in chemical modification of surfaces, leading to siliceous and macroporous rigid polymeric stationary phases having "soft" surfaces, and thus eligible for HPLC of proteins or other biopolymers. This development has engendered a family of composite stationary phases which combine the mechanical stability of rigid macroporous microparticles and the highly hydrated nature of the polysaccharide gels traditionally used in chromatography of biopolymers. The introduction between 1982 and 1985 of stationary phases comprising macroporous siliceous particles with a biocompatible surface containing hydrophobic binding sites filled the gap between gels and HPLC supports in the field of HIC.

Most HPHIC supports are macroporous (pore size in the range 300–1000 Å), have bead sizes around 5–15 μm, and are based either on silica, organic polymer resins, or highly cross-linked agarose. Recently, nonporous pellicular supports of small particle diameter (~2–5 μm) have been studied, and a commercially available support of this type, containing chemically bound butyl groups (TSK gel butyl NPR), has been developed (Kato *et al.*, 1989).

The main commercially available HIC supports—soft gels as well as HPLC matrices—are listed in Tables 3 and 4.

3.1.2. Ligands

Most ligands used in HIC are alkyl chains or phenyl groups. After the initial article by Er-el *et al.* (1972), reporting on the preparation of a homologous series of alkyl chain-derivatized Sepharose supports (Sepharose-C_n, *n* from 1 to 12), papers dealing with HIC were published for almost a decade with this polysaccharide support and this type of hydrophobic ligands. The present commercially available supports, both gels and rigid HPHIC matrices, are generally restricted to the lower homologs (from propyl to octyl) (Tables 3 and 4).

On the other hand, many articles have been published on the purification of certain proteins by chromatography at very high salt concentration [e.g., 3–4M ($NH_4)_2SO_4$] on unsubstituted Sepharose (Ashton and Anderson, 1981a; Sawatzki *et al.*, 1981; Lascu *et al.*, 1986). This type of chromatography, called "salting-out chromatography," takes place at a

TABLE 3
Commercially Available HIC Gels (A Nonexhaustive List)

Supplier	Product name	Matrix
Pharmacia LKB (Uppsala, Sweden)	Phenyl-Sepharose CL-4B	Agarose
	Octyl-Sepharose CL-4B	Agarose
	Phenyl-Sepharose FF	Agarose
IBF-Sepracor (Villeneuve-la-Garenne, France)	Spherosil M,LS	Silica coated with a hydrophobic, ionic polymer (40–100 μm)
J. T. Baker France (Paris, France)	Bakerbond WP HI-Propyl, Prepscale	Silica (40 μm)
Tosoh Corp. (Tokyo, Japan)	Fractogel-TSK-Butyl	Polyvinyl gel (25–150 μm)

TABLE 4
Commercially Available HPHIC Packings (A Nonexhaustive List)

Supplier	Product name	Matrix
Pharmacia LKB (Uppsala, Sweden)	Phenyl Superose	Cross-linked agarose (10 μm)
	Alkyl Superose (C_5)	Cross-linked agarose (10 μm)
J. T. Baker France (Paris, France)	WP HI-Propyl	Silica (5 and 15 μm)
Tosoh Corp. (Tokyo, Japan)	TSK-Gel Phenyl-5 PW	Hydroxylated polymer (10 μm)
SynChrom, Inc. (Linden, New Jersey)	SynChropak HIC (propyl, hydroxy-propyl, benzyl, methyl, pentyl)	Silica coated with polyamide (6.5 μm)
Alltech Associates Inc. (Deerfield, Illinois)	HEMA-HIC	Hydroxyethyl methacrylate/ dimethacrylate polymer (10 μm)
Beckman (Gagny, France)	Spherogel CAA-HIC	Methyl polyether linked to silica (5 μm)
	Ultrapore C8	Silica (5 μm)
Brownlee Labs (Santa Clara, California)	Aquapore HIC	Silica (7 μm)
	Polybore Phenyl HIC	Polymeric resin (10 μm)

salt concentration slightly lower than that required for precipitation. However, according to Ashton and Anderson (1981b), the phenomenon cannot be due to precipitation brought about by competition between the gel and the protein for water of solvation, since adsorption can also occur in batch conditions where competition for available water is insignificant. These various articles relate salting-out chromatography to HIC, the role of hydrophobic ligands being assumed to be played by the anhydrogalactosyl residues of Sepharose or the cross-linking reagents introduced in the polysaccharide structure.

Rather recently, mild hydrophobic interaction chromatography was introduced by Ling and Mattiasson (1983). Ligands frequently used in this approach are polymers, such as poly(vinyl alcohol) or polyethers. Both gels (Ling and Mattiasson, 1983; Shibusawa et al., 1987) and silica-based rigid matrices (J. P. Chang et al., 1985; S. H. Chang et al., 1976; Miller et al., 1984) have been used in this version of HIC.

3.1.3. Immobilization Reagents

Earliest stationary phases were prepared by reacting aryl- or alkylamines with agarose activated by the cyanogen bromide method (Er-el et al., 1972; Axén et al., 1967). This coupling technique mainly affords N-substituted isoureas, which are expected to contain positively charged nitrogen under certain pH conditions (Svensson, 1973). Furthermore, N-carbamates and N-substituted imidocarbonates may also be present in non-negligible amounts. These matrices may therefore exhibit mixed electrostatic and hydrophobic interactions with proteins (Hofstee, 1973; Hjertén, 1973). Wilchek and Miron (1976) acetylated CNBr-coupled amines, and this resulted in a significant reduction of the charges on the gel, but unfortunately also in an enhanced tendency of the ligands to leakage. The introduction by Porath et al. (1973) of another activation procedure for agarose, based on the intermediary formation of an epoxide, opened the way for the preparation of non-charged, purely hydrophobic stationary phases. Alternatively, agarose hydroxyl groups may be directly reacted with desired alkylglycidyl ethers (Hjertén et al., 1974), prepared according to the method described by Ulbrich *et al.* (1964). Another possibility for synthesizing neutral gels was introduced by Jost *et al.* (1974). It consisted in the reaction of alkylhydrazides on agarose and afforded neutral gels above pH 4.

Since silanol groups at the surface of silica bear a weak negative charge, the surface must be modified before it becomes suitable for chromatography experiments. This is accomplished by derivatizing surface silanols with hydrophilic organosilanes such as *N*-acetylaminopropylsilane (Englehardt and Mathes, 1977) or γ-glycidoxypropyltrimeth-oxysilane (Regnier and Noel, 1976), to both neutralize the surface and make it hydrophilic.

Epoxy activation can be performed as well on silica supports using γ-glycidoxypropyl-trimethoxysilane as the activation reagent (J. P. Chang *et al.*, 1985; S. H. Chang *et al.*, 1976). The same compound was used by Hjertén *et al.* (1986) to immobilize hydrophobic ligands on highly cross-linked agarose gels used as medium-pressure liquid chromatography packings.

A different approach for silica, reported by Miller *et al.* (1984) to produce charge-free supports, consisted in a two-step procedure: conversion of the desired hydrophobic ligand from its OH form to an allyl ether (Williamson's synthesis) followed by hydrosilylation to the triethoxysilane by an adaptation of the method of Speier *et al.* (1957).

The density of immobilized hydrophobic ligands is an important parameter, with respect to the binding and elution conditions. This density can be accurately determined by using radioactive ligands or estimated by various other procedures, including nuclear magnetic resonance (NMR) spectroscopy, elementary C analysis, and gas chromatography (Rosengren *et al.*, 1975; Johansson and Drevin, 1985). Typically, octyl- and phenyl-Sepharose CL-4B gels contain approximately 40 μmol of hydrophobic substituent/ml of gel, corresponding to a substitution of 0.2 mol of ligand/mol of galactose.

3.2. Adsorption and Elution

Generally, alkyl chain-substituted adsorbents with C_n in the range C_2–C_{10} and substitution ratios from 10 to 100 μmol/ml of gel are suitable for most purification problems.

As was stressed in section 2, hydrophobic interactions are favored by a high ionic strength or an increase in temperature. However, when biomaterials are concerned, the effect of temperature is seldom exploited, owing to the risk of denaturation that would result.

On the other hand, pH also plays an important role during HIC of proteins. As a matter of fact, the presence in these biological substances of high amounts of ionizable (i.e., highly hydrophilic), groups (NH_3^+, COO^-) greatly influences their retention. However, there does not seem to be any obvious relationship between retention and isoelectric point of individual proteins, although empirical observations suggest that retention increases as the pH decreases from neutrality (Fausnaugh *et al.*, 1984).

For proteins with poor solubility in buffers of high salt concentration, for example, membrane proteins, HIC adsorbents with long alkyl chains are recommended, since they allow a suitable adsorption at lower ionic strength. On the other hand, a phenyl group has about the same hydrophobicity as a pentyl group, although it can have a quite different selectivity owing to additional π–π interaction.

Elution, whether done stepwise or with a gradient, can usually be achieved in three different ways:

1. by decreasing the ionic strength of the mobile phase. This results in the elution of solutes in the order of their increasing hydrophobic character.
2. by increasing the entropy of the system. As already mentioned, water structure-breaking additives disrupt hydrophobic interactions. Organic solvents, usually ethylene glycol up to a concentration of 80% or propanol at lower concentrations, are commonly used to obtain this effect. The addition of these polarity-decreasing agents can be carried out after or concomitantly with the decrease of salt concentration. Other previously mentioned entropy-increasing substances, such as chaotropic salts (NaSCN, KI), may also afford elution. However, these substances are generally avoided when biological substances are concerned, owing to the structure-breaking properties of these ions. Similarly, the use of urea or guanidine chloride is restricted to regeneration purposes, in order to remove strongly adsorbed species from the stationary phase before it can be used for subsequent purifications.
3. by addition of detergents. Detergents are strong hydrophobic compounds; they act as displacers and have been mainly used for the purification of membrane proteins (Buckley and Wetlaufer, 1990). Their drawback is that they remain strongly adsorbed on the stationary phase, thus requiring extensive washings with various alcohols before subsequent runs.

The retention/elution mechanism of proteins on HIC columns has been widely studied. Besides the model describing the favorable entropy increase which results when hydrophobic groups on both the matrix and the protein interact and are thus excluded from their polar environment in the mobile phase (Jennissen and Botzet, 1979; Fausnaugh et al., 1984), a number of other approaches have been proposed.

In their solvophobic theory, Melander and Horvath (1977) demonstrated that, in the absence of special binding effects, an increase in salt molality in the mobile phase or a change to a salt of greater molal surface tension increment resulted in increased retention of proteins. They proposed the following equation:

$$\log k' = \log k'_0 + mC_s$$

relating the logarithm of the capacity factor k' of the solute to the salt concentration in the mobile phase (C_s), $\log k'_0$ being the extrapolated value of $\log k'$ at zero salt concentration.

This equation is derived from the linear relationship between the logarithm of the capacity factor and the surface tension of the mobile phase (Horvath et al., 1976), together with the linear relationship between surface tension and salt concentration (Heydweiller, 1910). Melander and Horvath proposed that the coefficient m was a linear function of the hydrophobic area of contact between the protein and the stationary phase.

Geng et al. (1990) recently proposed a different analysis of the phenomenon. They elaborated a model in which the retention mechanism of proteins is a stoichiometric displacement process, with water as the displacing agent. The adsorption of a protein should be accompanied by the release of a stoichiometric number, Z, of water molecules to the solution, from the interface between the protein and the hydrophobic ligands bound to the stationary phase. The capacity factor k' is related to Z and to the concentration of water in the mobile phase, according to the following relationship:

$$\log k' = \log I - Z \log [H_2O]$$

where the intercept of the equation, $\log I$, contains a number of constants related to the affinity of the protein for the hydrophobic ligands.

3.3. Mild Hydrophobic Interaction Chromatography

The terminology "mild hydrophobic interaction chromatography" was first introduced by Ling and Mattiasson (1983), who immobilized poly(vinyl alcohol) and poly-(ethylene glycol) (PEG) on Sepharose. The idea was that mildly hydrophobic polymers, such as those generally used in the technique based on partitioning between aqueous two-phase systems described by Albertsson (1971), could be immobilized onto inert supports to afford stationary phases less denaturing toward biomolecules than those involved in HIC with traditional alkyl chain ligands.

A similar concept was exploited very early by Morris (1963) and Anker (1971), but their aqueous polymer phase was merely soaked up by the support beads. The idea was revived in the early 1980s when different groups started to covalently immobilize these polymers on the support. The designation of these polymers as having a mild hydrophobic character is justifiable when comparing their surface tension to that of other hydrophobic ligands (Fig. 2). The surface tension of PEGs, in the range $\gamma_{sv} = 42\text{--}46$ mJ/m^2 (Körösi and Kovats, 1981), is situated midway between that of water ($\gamma_{sv} = 70$mJ/m^2) and that of low-energy materials such as C_6–C_{16} alkanes ($\gamma_{sv} = 18\text{--}28$ mJ/m^2; Jarvis *et al.*, 1979).

The same concept was applied with rigid supports, and the first report on the immobilization of PEGs of increasing molecular weight on silica for HPHIC of proteins was made by S. H. Chang *et al.*, as early as 1976. This was followed by articles by J. P. Chang *et al.* (1985), who used PEG 400 as the ligand, and Miller *et al.* (1984), who prepared silica supports having the structure

$$\equiv Si - (CH_2)_3 - O - (CH_2 - CH_2 - O)n - R$$

with $n = 1\text{--}3$ and R being an alkyl chain, from methyl to butyl.

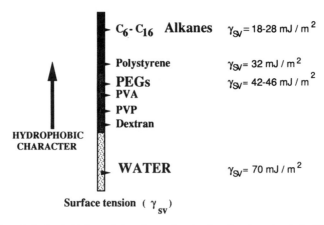

FIGURE 2. Schematic hydrophobicity scale based on surface tension of various chemicals. PEG, Poly(ethylene glycol); PVA, poly (vinyl alcohol); PVP, polyvinylpyrrolidone.

In our group, we attempted the purification of enzymes acting on hydrophobic substrates taken in the lipid and steroid fields (for a review, see Hubert *et al.*, 1991). Experiments were conducted on actual fractionation problems, with complex crude extracts. Stationary phases were prepared by covalent immobilization of various polymers—PEG, Pluronic [a copolymer of PEG and poly(propylene glycol)], poly(propylene glycol) (PPG), and poly(tetramethylene glycol) (PTMG)—on Sepharose 6B, after its preliminary activation by carbonyldiimidazole.

Figure 3 presents the chromatographic profile of the stepwise elution of three enzymes—$\Delta_{5\rightarrow4}$3-ketosteroid isomerase (isomerase), 3α-hydroxysteroid dehydrogenase (α-HSD), and 3β,17β-hydroxysteroid dehydrogenase (β-HSD)—obtained from a crude extract of *Pseudomonas testosteroni*. The chromatography was carried out with Sepharose-Pluronic as the stationary phase and led to a total separation of the three enzymes from the bulk of contaminants and from one another, with good recoveries of enzymatic activities (α-HSD 60%; β-HSD 40%; isomerase 65%) and almost quantitative removal of the contaminants in each fraction (Mathis *et al.*, 1989) (Table 5). Somewhat inferior results, in terms of resolution and recovery of enzymatic activities, were obtained with this extract when the chromatography was carried out with Sepharose-PEG as the stationary phase. This shows that small modifications of the hydrophobicity of the stationary phase may lead to significantly different purification qualities. A careful selection of chromatographic conditions may consequently enable complicated challenges to be met—in the present case, good purification ratios and high recoveries for three closely related enzymes by stepwise elution and starting from the same crude extract—with this mild version of HIC.

Similar purification results (Fig. 4) were obtained for an esterase from *Bacillus*

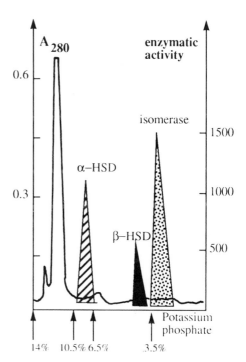

FIGURE 3. Stepwise elution of the *Pseudomonas testosteroni* crude extract on Sepharose-Pluronic with decreasing concentrations of potassium phosphate, pH 7.0. The filled triangles correspond to the various enzymatic activities. (From Mathis *et al.*, 1989.)

TABLE 5

Percentages of Contaminants in the Fractions
Containing the Three Enzymes Purified
on Two Different Mild Hydrophobic Supports[a]

Stationary phase	Percentage[b] of contaminants in the fraction containing:		
	α-HSD	β-HSD	Isomerase
Sepharose-PEG	5.7	2.8	0.5
Sepharose-Pluronic	1.5	2.0	ND[c]

[a]From Mathis et al., 1989.
[b]Results are % of the total proteins in the crude extract.
[c]ND, Not detectable.

pumilus (86% yield, 140-fold purification) on a Sepharose-Pluronic stationary phase (Mathis *et al.*, 1988), without the necessity, often encountered with lipolytic enzymes, to perform elution in the presence of detergents.

4. REVERSED-PHASE CHROMATOGRAPHY (RPLC)

The terminology "reversed-phase chromatography" (RPLC) may sound rather weird to certain readers not fully accustomed to the various chromatographic separation procedures. Simply, this name was proposed to describe the opposite of the until then normal practice of partition liquid chromatography, where the ligands are polar groups (e.g., amino, nitrile, diol) and the elution is performed with mobile phases consisting of apolar

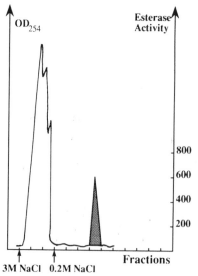

FIGURE 4. Stepwise elution of the *Bacillus pumilus* crude extract on Sepharose-Pluronic with NaCl aqueous solutions. The cross-hatched triangle corresponds to the enzymatic activity. (From Mathis *et al.*, 1988.)

solvents (e.g., hexane, octane) or of mixtures of apolar solvents and polar modifiers (e.g., chloroform, tetrahydrofuran, acetonitrile). In the case of RPLC, this situation is "reversed," since the stationary phase is prepared by immobilization of apolar ligands and the elution is carried out with an aqueous, polar mobile phase.

As pointed out previously in this chapter, the main difference between HIC- and RP-type chromatographies is the density of hydrophobic ligand immobilized on the chromatographic support and, accordingly, the overall hydrophobic character of the chromatographic packing. In contrast to HIC, where, for example, alkyl chains may be considered as disperse interacting sites, the distribution of hydrophobic chains grafted on the RPLC support is so dense (10 to 100 times as dense as for HIC-type phases) that these molecules form a continuous liquidlike layer almost completely covering the surface of the support.

Whether the mechanism of retention of small molecules should be related to some liquid–liquid partitioning between the stationary and the mobile phase or to some interaction between this apolar layer and the hydrophobic parts of the solute is not clearly established, and a number of models have been proposed to explain the chromatographic mechanism of RPLC and the contribution of mobile and stationary phases to retention and selectivity (Horvath *et al.*, 1976; Karger *et al.*, 1976; Tijssen *et al.*, 1976; Jandera *et al.*, 1982). However, when macromolecules are concerned, the maximum size of an extended alkyl chain (e.g., 21 Å for an octadecyl) makes it very unlikely that a protein whose size is greater than 100 Å would penetrate into the liquidlike layer covering the surface of the support. As in HIC, an adsorption–desorption mechanism based on hydrophobic interaction can be proposed, but, in this case, the strength of the interaction is so high that the use of quite different mobile phase conditions from those employed in HIC becomes necessary.

In fact, the main practical consequence of this high density of hydrophobic ligands is that the adsorption is far stronger than that involved in HIC-type chromatography. Accordingly, the relatively mild elution conditions—for example, descending salt gradients—which have led to HIC sometimes being referred to as "reversed-phase without denaturation" are no longer effective. One has then to turn to more severe elution procedures, which must be selected with great care, bearing in mind the multiple and sometimes contradictory goals of selectivity, speed, mass recovery, biological activity retention, and so on. As a matter of fact, although it is commonly believed that the propensity of these hydrophobic supports to cause denaturation is a rather general phenomenon, the accumulated data on polypeptides and proteins in the molecular weight range 5000–7000, indicates, according to Hearn (1989), that denaturation is often merely the consequence of inappropriate selection of operating conditions.

4.1. The Stationary Phase

Irrespective of considerations concerning the ligand itself, the criteria for selecting a suitable RPLC support are not different from those followed in all kinds of high-performance liquid chromatography, namely, small particle size to afford high resolution at high flow rate and mechanical stability to withstand high pressures. Although HPLC techniques have been exploited for a long time in the separation of low-molecular-weight compounds, the advent of high-performance supports specifically designed for biopolymer separations was not until the early 1980s.

Typically, 5–10-μm particles having a nominal pore diameter in the range 250–500 Å and specific surfaces as low as 50 m^2/g are the most commonly used materials although the question about the optimal pore size is still debated. In a recent article, Fulton *et al.* (1991) suggested that the limitation in liquid chromatography caused by stagnant mobile-phase mass transfer can be solved by using either nonporous 1–2-μm particles (Unger *et al.*, 1986; Kalghatgi and Horvath, 1988) or large particles (20 μm) having transecting pores (6000–8000 Å) that allow convective rather than diffusive transport within the particles (Afeyan *et al.*, 1990, 1991).

The inert, hydrophilic surface character that polysaccharide or polyacrylamide supports possess and which is widely exploited for the purification of biopolymers by means of all kinds of low-/medium-pressure chromatographic techniques has not yet been extrapolated to HPLC systems, owing to the present lack of knowledge about how to prepare small, high-pressure-stable particles from these gels. Accordingly, two types of supports have been developed, namely, surface-modified silica matrices and totally organic gels.

4.1.1. Silica-Based Packings

Although the first commercially available silica-based RPLC columns were introduced more than ten years ago, this type of stationary phase is still the main choice for polypeptide analyses, and more than 65% of HPLC analyses are performed on such columns. This is due to their excellent mechanical resistance, their reasonable chemical stability under a rather wide variety of elution conditions, and their permanent porosity, in contrast with swelling carbohydrate gels such as cross-linked dextrans.

The introduction of hydrophobic ligands is usually carried out, as for HIC supports, by chemical bonding via reaction with the silanol groups, particularly by formation of Si—O—Si—C bonds, which are reasonably stable under the conditions of RPLC. This silanization reaction may be performed using mono-, bi-, or trifunctional organosilanes.

It is generally accepted that only 25 to 50% of silanol groups can be derivatized. The silica surface therefore still contains large amounts of free silanols, as well as other OH groups resulting from the hydrolysis of residual chloro or alkoxy groups when the silanization reaction is carried out using di- or trifunctional organosilanes. A subsequent treatment, "end-capping" by, for example, trimethylchlorosilane [ClSi(CH$_3$)$_3$] is generally recommended.

Another way of modifying silica to make it a suitable support for the separation of biomolecules consists in coating its surface with polymers. The deposited polymers are either cross-linked or chemically bonded to precapped (trimethylsilylated) silica materials. Silanol groups are not needed for the immobilization and may either be partly masked (~75%) by precapping or remain unchanged on the silica surface underneath the coating layer. Polymers of the polysiloxane or polybutadiene type have been used for this purpose (Schomburg *et al.*, 1983; Figge *et al.*, 1986), and the grafting of styrene–methylvinyl-dichlorosilane copolymers onto the silica surface has been recently reported by Kurganov *et al.* (1983).

A large variety of nonpolar functional groups have been immobilized onto silica, both as monolayers and as polymeric layers. However, most commercially available RP silica supports are based on the use of octyl, octadecyl, or diphenyl chains.

According to many experimental findings, the length of the alkyl chains on the surface of brush-type RP packings influences the total quality of resolution but exerts no significant effect on the retention of proteins, unlike that of small organic molecules. In the latter case, the strength of the interaction increases with the length of the alkyl chain and its density on the support. More generally, it was shown (Rosset *et al.*, 1982) that the logarithm of the capacity factor increased linearly with the surface area of the immobilized hydrocarbon ligand (product $N \times S$, where N is the number of immobilized ligands/nm^2 of support and S is the surface area of the hydrocarbon chain). A relation established by Bondi (1964) allows the calculation of S versus the length of the ligand chain.

4.1.2. Organic Gels

Several polymer-based RPLC columns have been available for some time, but although they theoretically offer some advantages over silica-based supports, they are not yet of widespread use for polypeptide or protein analysis and purification. Nevertheless, the problems associated with silica-based stationary phases—limited chemical stability especially at high pH, potential interference of residual silanol groups, and lack of column-to-column reproducibility even with different batches from the same manufacturer—undoubtedly are factors that may have prompted investigators in the 1980s to try to develop new, purely organic HPLC supports.

Purely organic supports are generally prepared by the polymerization of some monomer(s) in the presence of a cross-linking agent. The porosity of the particle is determined by the percentage of cross-linker used; the lower this percentage is, the greater the porosity but also the softer the resulting matrix. The preparation of supports, both macroporous and rigid, did not develop quickly because of these opposite effects of the cross-linking reagent until Moore (1964) synthesized macroporous poly(styrene–divinylbenzene) materials, with a high percentage of divinylbenzene. The polymerization was carried out in the presence of a porogen—a compound soluble in the monomer but not in the polymer—and afforded rigid spherical particles containing large voids/pores. Similar materials were subsequently used in microparticulate form for RP chromatography (Tweeten and Tweeten, 1986; Lee, 1988; Lloyd, 1991).

4.2. The Mobile Phase

Although stationary-phase characteristics, including those of, for example, the silica support (chemical treatment, particle size, pore structure, etc.) and of the immobilized hydrophobic ligand (alkyl or aryl nature, density, surface coverage, etc.) are important, chromatographic performance is predominantly influenced by the organic and ionic makeup of the mobile phase.

As in HIC, entropy drives the interaction of the protein with the hydrophobic surface, but the interaction is so strong on RPLC columns that proteins may not be eluted with buffers. Desorption is accomplished through the introduction of an organic additive in the mobile phase. The organic solvent may interact with both the hydrophobic matrix and the protein, thereby displacing the protein, or it may disorder the water structure so that the protein may be eluted from the column (Fausnaugh *et al.*, 1984).

Organic modifiers such as acetonitrile, methanol, ethanol, and propanol have been the most popular mobile-phase additives, but the use of butanol, tetrahydrofuran, and acetone has also been reported. Organic solvents can be ranked according to decreasing eluotropic strength as compared to water: propanol > acetonitrile > methanol > water.

More generally, the accumulation of empirical observations from the considerable number of reports on RPLC purification of proteins has led to typical conditions such as low pH, organic solvent combination (usually in the range 0–50% v/v organic solvent), almost systematic use of low concentrations of trifluoracetic acid, low ionic strength (10–100mM), flow rate between 0.5 and 2 ml/min, and ambient temperature. Elution with an acetic acid gradient in water, without organic modifiers, was recently reported to afford excellent results in the case of poly(styrene–divinylbenzene) stationary phases (Melinder, 1991).

Within reasonable retention times commonly practiced in RPLC, that is, for values of the capacity factor in the range $1 \leq k \leq 20$, the logarithm of the capacity factor k of a solute can generally be approximated as

$$\log k = \log k_w - S\varphi$$

where k_w is the capacity factor in pure water, φ is the volume fraction of the organic modifier in the mobile phase, and S is the slope of the plot of $\log k$ versus φ (Snyder et al., 1983; Stadalius et al., 1984).

The value of the slope S depends on the solute and the conditions of the chromatography. It is frequently supposed to reflect the magnitude of the non-polar contact area between the solute and the hydrophobic ligands of the stationary phase. Typical S values for small peptides are in the range 2–10, somewhat lower than those observed for polypeptides or proteins (Aguilar et al., 1985), which can exhibit steep nonlinear relationships between $\log k$ and φ for S values greater than 20.

This relationship between $\log k$ and the volume fraction φ of the organic modifier in the mobile phase parallels the expression relating the capacity factor of the solute with the salt concentration (C_s) in the mobile phase in the solvophobic theory of HIC, presented by Melander and Horvath (1977).

Ion-Pairing Reagents in the Mobile Phase (Hetaeric Chromatography)

Another important domain of HPLC analysis of biomolecules related to the use of RP packings must be mentioned although its principles locate it at the boundary between hydrophobic and ion-exchange chromatographies. The addition to the mobile phase of ion-pairing agents with a hydrophobic moiety which can adsorb onto the RP stationary phase and an ionic group which can pair with the charged groups on the biomolecule results in enhanced retention and gave birth to a new concept designated as ion-pairing or hetaeric chromatography (Horvath et al., 1977; Hearn, 1984a).

Typical additives used include alkylsulfonic acids (RSO_3^-), alkyl sulfonates, fluorinated alkylcarboxylic acids, amine salts, tetraalkylammonium salts (R_4N^+), ClO_4^-, and PO_4^{3-} (for a review, see Hearn, 1984a).

Whether the secondary equilibrium phenomena involved in this type of chromatography correspond to ion-pair formation or to ion-exchange selectivity is still controversial. In the former the interaction is viewed as, first, formation of an ion-pair between the biomolecule and the additive and then retention of the ion pair by the hydrophobic stationary phase. In the latter, the additive is first retained by hydrophobic interaction and then the biomolecule exchanges more or less selectively, as a charged species, with the co-ion that accompanies the charged additive.

5. DISCUSSION

At the chromatographic level, the cohabitation of a solute protein with external substances such as a stationary phase and its different components (matrix backbone, ligands, silanol groups, etc.) and a mobile phase with various additives (salts, organic solvents, chaotropic agents, etc.) obviously results in some perturbation of the subtle equilibrium of the forces maintaining the structure of the protein.

In fact, the energy of folding of a protein from a disorganized coil to the native well-structured conformation necessary for its functional activity involves various intertwined covalent and noncovalent forces—including the hydrophobic interaction—and it is now well established that when a polypeptide or a protein comes in contact with a surface, structural and conformational changes can occur as the species responds to the environment. Various physicochemical or spectroscopic techniques, including ellipsometry (Morrisey *et al.*, 1976), Raman spectroscopy (Aurengo *et al.*, 1982), circular dichroism (McMillin and Walton, 1974), Fourier transform infrared (FTIR) spectroscopy (Gendreau *et al.*, 1981), and fluorescence spectroscopy (Oroszlan *et al.*, 1990), have been used to explore these conformational modifications.

The extent of the perturbation is obviously dependent on the physicochemical nature of both the stationary and the mobile phase, but is also dependent on other parameters such as the temperature, the contact time, and above all the inherent lability of the biomolecule. In addition to the large differences that may exist between proteins in a complex mixture and which result in separation, slight, though significant, differences may as well exist for different conformers of the same solute and lead to distorted chromatographic peak shape or multiple peaks, depending on whether the conformational changes are reversible, with a half-life comparable to the chromatographic elution time. Ultimately, the alteration can be so extensive that it results in denaturation, thus precluding the recovery of certain proteins in their native functional form or of certain multisubunit proteins.

The subtle and fragile equilibrium of the forces responsible for the cohesion of tertiary/quaternary structures of the protein edifice and their potential perturbation by the components of both mobile and stationary phases can clearly be expected to have different implications for HIC and RPLC, given the large differences in the experimental conditions employed in these two techniques, which have been frequently pointed out throughout this chapter. It is obvious from examination of the considerable amount of literature reporting on the use of either HIC or RPLC for purification purposes that HIC is generally considered as a soft, nondenaturing technique in contrast to RPLC. Whether the denaturation reported during RPLC experiments is merely due to some inappropriate selection of experimental conditions, as sometimes stated, or to the severe conditions involved will probably be debated for quite some time.

Several separations of polypeptide mixtures using RPLC suggest that, when the peptide is small enough so that conformational considerations may be disregarded, the elution position of a peptide appears to be determined by the sum of the hydrophobicities of its constituent amino acids. O'Hare and Nice (1979) were able to predict with reasonable accuracy the elution behavior of peptides containing up to 15 residues by summing the fragmental hydrophobic constants of the five most hydrophobic residues. In his approach, Meek (1980) reached a similar conclusion with polypeptides containing up to 20 amino acid residues.

In a very recent article, Champney (1990) used C_4 reversed-phase chromatography to separate 52 ribosomal proteins. A very strong correlation was found between observed retention times and those predicted on the basis of the total hydrophobicity and chain length of the protein (Champney, 1990). Considering that, with the elution conditions used (0.1% trifluoroacetic acid, pH 2), these proteins should be fully denatured, thus allowing maximal interaction between the amino acid sequences and the hydrophobic column matrix, this result is additional support for the hypothesis that retention can be directly related to the global hydrophobicity, provided that conformational considerations may be disregarded.

The situation is quite different for longer polypeptides or proteins where spatial arrangement plays a major role. Using lysozyme variants, Fausnaugh and Regnier (1986) established the existence of a chromatographic contact surface area between the proteins and the hydrophobic sorbent. This notion of contact surface area, along with the stoichiometric displacement model (Geng et al., 1990), were used by Kutikani et al. (1986) to explain the differences in retention for 30 variants of interleukin-2 on an RP support. They concluded that RPLC retention data corresponded to changes in the relative contact area between the solute and the stationary phase for the different variants, resulting from a higher degree of unfolding due to the elution conditions.

Regnier and co-workers have been very active in this field of comparative study of HIC and RPLC. As early as 1984 (Fausnaugh et al., 1984), they reported on the retention times of a series of proteins, on both HIC and RPLC supports and concluded that they could be explained by examining the structures of these proteins during the chromatographic process. In HIC, the proteins retain their native structure and most of the hydrophobic residues are buried in the interior of the molecule, whereas in RPLC the acids and organic solvents partially unfold a protein's active conformation, thereby exposing the more hydrophobic interiors. In a very recent article, Chicz and Regnier (1990) showed that conditions could be found where a HIC support could separate proteins differing in only one amino acid (genetically engineered variants of subtilisin) and concluded that RPLC was of much less utility for the resolution of these variants.

Similarly, comparative studies of HIC and RPLC of polypeptides differing from each other by single-residue replacement and exhibiting tertiary structures were recently reported (Heinitz et al., 1988; Alpert, 1988). They led to the conclusion that tertiary structure plays a dominant role in the HIC retention mechanism and that differences in solute selectivity between HIC and RPLC are due to alterations in the three-dimensional structure of the polypeptides, resulting from the mobile-phase denaturing conditions used in RPLC.

Although a long list of articles reporting on difficulties encountered during RPLC of proteins could be provided, an even longer list of achievements could be presented. As stated by Hearn (1984a), in numerous studies low mass recovery, reduced separation capacities, and generally unsatisfactory resolution have been obtained with a variety of proteins with n-alkyl silica RP supports (Cohen et al., 1984). However, related studies have demonstrated that excellent resolution and almost quantitative mass recovery could be obtained with the same polypeptides and proteins under slightly different chromatographic conditions (Hearn, 1984a). This apparently contradictory behavior is typical of many globular proteins chromatographed on alkyl silicas, where a small change in eluent or stationary-phase conditions can have a major effect on the chromatographic performance and can ultimately be associated with a dramatic loss of recovered mass and biological function. Under certain chromatographic conditions and especially in RP chromatography, slow dynamic interconversion may result in broad asymmetric peaks or even multiple peaks corresponding to native, unfolded, or denatured forms of a protein, when the overall time constants for the equilibration between these forms are of comparable magnitude to the time of separation.

Besides the fact that RPLC can be readily applied to the separation of proteins which are resistant to denaturation or easily renature (Berchtold *et al.*, 1983) or in cases when the purification is performed in the context of primary structure determination, this technique may certainly lead as well to spectacular achievements when retention of the native form is required, provided that some care is taken to select the optimal mobile- and stationary-phase conditions. With respect to the arsenal of separation techniques available to the biochemist, HIC and RPLC would thus be better considered as complementary rather than adversary techniques.

REFERENCES

Afeyan, N. B., Gordon, N. F., Mazsaroff, I., Varady, L., Fulton, S. P., Yang, Y. B., and Regnier, F. E., 1990, Flow-through particles for the high-performance liquid chromatographic separation of biomolecules: Perfusion chromatography, *J. Chromatogr.* **519**:1–29.

Afeyan, N. B., Fulton, S. P., and Regnier, F. E., 1991, Perfusion chromatography packing materials for proteins and peptides, *J. Chromatogr.* **544**:267–279.

Aguilar, M. I., Hodder, A. N., and Hearn, M. T. W., 1985, High-performance liquid chromatography of amino acids, peptides and proteins, *J. Chromatogr.* **327**:115–138.

Albertsson, P. Å., 1971, *Partition of Cell Particles and Macromolecules*, 2nd ed., Almqvist and Wiksell, Stockholm, and Wiley-Interscience, New York.

Alpert, A. J., 1988, Hydrophobic interaction chromatography of peptides as an alternative to reversed-phase chromatography, *J. Chromatogr.* **444**:269–274.

Anker, H. S., 1971, On partition chromatography of proteins, *Biochim. Biophys. Acta* **269**:290–291.

Arakawa, T., 1986, Thermodynamic analysis of the effect of concentrated salts on protein interaction with hydrophobic and polysaccharide columns, *Arch. Biochem. Biophys.* **248**:101–105.

Arakawa, T., and Timasheff, S. N., 1984, Mechanism of proteins salting-in and salting-out by divalent cation salts: Balance between hydration and salt binding, *Biochemistry* **23**:5912–5923.

Ashton, A. R., and Anderson, L. E., 1981a, A novel procedure for the rapid purification of plastocyanin from (*Pisum sativum*) leaves, *Biochem. J.* **193**:375–378.

Ashton, A. R., and Anderson, L. E., 1981b, Purification of multiple pea ferredoxins by chromatography at high ionic strength on unsubstituted Sepharose 4B, *Biochim. Biophys. Acta* **667**:452–456.

Aurengo, A., Masson, M., and Dupeyrat, E., 1982, Optimizing of ultra-thin non resonant Raman spectra provided by combining a light-trapping device, a high N.A. objective and a spectrometer, *Appl. Opt.* **22**:602–608.

Axén, R., Porath, J., and Ernback, S., 1967, Chemical coupling of peptides and proteins to polysaccharides by means of cyanogen halides, *Nature (London)* **214**:1302–1304.

Berchtold, M. W., Heizmann, C. W., and Wilson, K. J., 1983, Calcium-binding proteins: A comparative study of their behavior during high-performance liquid chromatography using gradient elution on reverse-phase supports, *Anal. Biochem.* **129**:120–131.

Bigelow, C. C., 1967, On the average hydrophobicity of proteins and the relation between it and protein structure, *J. Theor. Biol.* **16**:187–211.

Boldingh, J., 1948, Application of partition chromatography to mixtures insoluble in water, *Experimenta* **4**: 270–271.

Bondi, A., 1964, Van der Waals volumes and radii, *J. Phys. Chem.* **68**:441–451.

Boscott, R. J., 1947, Solvent-treated cellulose acetate as the stationary phase in partition chromatography, *Nature (London)* **159**:342.

Brandts, P. M., Middelkoop, C. M., Gelsema, W. J., and Deligny, C. L., 1986, Hydrophobic interaction chromatography of simple compounds on alkyl-agaroses with different alkyl chain lengths and chain densities. Mechanism and thermodynamics, *J. Chromatogr.* **356**:247–259.

Buckley, J. J., and Wetlaufer, D. B., 1990, Surfactant-mediated hydrophobic interaction chromatography of proteins: Gradient elution, *J. Chromatogr.* **518:**99–110.

Champney, W. S., 1990, Reversed-phase chromatography of *Escherichia coli* ribosomal proteins. Correlation of retention time with chain length and hydrophobicity, *J. Chromatogr.* **522:**163–170.

Chang, J. P., El Rassi Z., and Horvath, C., 1985, Silica-bound polyethylene glycol as stationary phase for separation of proteins by high-performance liquid chromatography, *J. Chromatogr.* **319:**396–399.

Chang, S. H., Gooding, K. M., and Regnier, F. E., 1976, Use of oxiranes in the preparation of bonded phase supports, *J. Chromatogr.* **120:**321–333.

Chicz, R. M., and Regnier, F. E., 1990, Microenvironmental contributions to the chromatographic behavior of subtilisin in hydrophobic-interaction and reversed-phase chromatography, *J. Chromatogr.* **500:**503–518.

Cohen, K. A., Schellenberg, K., Benedek, K., Karger, B. L., Grego, B., and Hearn, M. T. W., 1984, Mobile phase and temperature effects in the reversed-phase chromatographic separation of proteins, *Anal. Biochem.* **140:**223–235.

Colin, H., and Guiochon, G., 1977, Introduction to reversed-phase high-performance liquid chromatography, *J. Chromatogr.* **141:**289–312.

Cornette, J. L., Cease, K. B., Margalit, H., Spouge, J. L., Berzofsky, J. A., and Delisi, C., 1987, Hydrophobicity scales and computational techniques for detecting amphipatic structures in proteins, *J. Mol. Biol.* **195:** 659–685.

Dill, K. A., 1990, The meaning of hydrophobicity, *Science* **250:**297.

Eftink, M. R., Zajicek, J. L., and Ghiron, C. A., 1977, A hydrophobic quencher of protein fluorescence: 2,2,2-Trichloroethanol, *Biochim. Biophys. Acta* **491:**473–481.

Eisenberg, D., Wilcox, W., and McLachlan, A. D., 1986, Hydrophobicity and amphiphilicity in protein structure, *J. Cell. Biochem.* **31:**155–161.

Engelhardt, H., and Mathes, D., 1977, Chemically-bonded stationary phases for aqueous high-performance exclusion chromatography, *J. Chromatogr.* **142:**311–320.

Er-el, Z., Zaidenzaig, Y., and Shaltiel, S., 1972, Hydrocarbon-coated Sepharoses. Use in the purification of glycogen phosphorylase, *Biochem. Biophys. Res. Commun.* **49:**383–390.

Fausnaugh, J. L., and Regnier, F. E., 1986, Solute and mobile phase contributions to retention in hydrophobic interaction chromatography of proteins, *J. Chromatogr.* **359:**131–146.

Fausnaugh, J. L., Kennedy, L. A., and Regnier, F. E., 1984, Comparison of hydrophobic interaction and reversed-phase chromatography of proteins, *J. Chromatogr.* **317:**141–155.

Figge, H., Deege, A., Köhler, J., and Schomburg, G., 1986, Stationary phases for reversed-phase liquid chromatography. Coating of silica by polymers of various polarities, *J. Chromatogr.* **351:**393–408.

Fisher, H. F., 1964, A limiting law relating the size and shape of protein molecules to their composition, *Proc. Natl. Acad. Sci. U.S.A.* **51:**1285–1291.

Frank, H. S., and Evans, M. W., 1945, Entropy in binary liquid mixtures; partial molal entropy in dilute solutions; structure and thermodynamics in aqueous electrolytes, *J. Chem. Phys.* **13:**507–532.

Fulton, S. P., Afeyan, N. B., Gordon, N. F., and Regnier, F. E., 1991, Very high speed separation of proteins with a 20-μm reversed-phase sorbent, *J. Chromatogr.* **547:**452–456.

Gendreau, R. M., Winters, S., Leininger, R. I., Fink, D., and Hassler, C. R., 1981, Fourier transform infrared spectroscopy of protein adsorption from whole blood: *Ex vivo* dog studies, *Appl. Spectrosc.* **35:**353–357.

Geng, X., Guo, L., and Chang, J., 1990, Study of the retention mechanism of proteins in hydrophobic interaction chromatography, *J. Chromatogr.* **507:**1–23.

Hammes, G. G., and Roberts, P. B., 1968, Cooperativity and solvent–macromolecule interactions in aqueous solutions of polyethylene glycol and polyethylene glycol–urea, *J. Am. Chem. Soc.* **90:**7119–7122.

Hammes, G. G., and Swann, J. C., 1967, Influence of denaturing agents on solvent structure, *Biochemistry* **6:**1591–1596.

Hatefi, Y., and Hanstein, W. G., 1969, Solubilization of particulate proteins and non electrolytes by chaotropic agents, *Proc. Natl. Acad. Sci. U.S.A.* **62:**1129–1136.

Hearn, M. T. W., 1984a, in: *Ion-Pair Chromatography*, Marcel Dekker, New York, pp. 1–296.

Hearn, M. T. W., 1984b, Reversed-phase high-performance liquid chromatography, *Methods Enzymol.* **104:**190–212.

Hearn, M. T. W., 1989, High resolution reversed phase chromatography, in: *Protein Purification. Principles, High Resolution Methods, and Applications* (J. C. Janson and L. Rydén, eds.), VCH Publishers, New York, pp. 175–206.

Heinemann, G., Köhler, J., and Schomburg, G., 1987, New polymer coated anion-exchange HPLC-phases: Immobilization of poly(2-hydroxy-3*N*-ethylenediamino) butadiene on silica and alumina, *Chromatographia* **23:**435–441.

Heinitz, M. L., Flanigan, E., Orlowski, R. C., and Regnier, F. E., 1988, Correlation of calcitonin structure with chromatographic retention in high-performance liquid chromatography, *J. Chromatogr.* **443**:229–245.

Herzfeld, J., 1991, Understanding hydrophobic behavior, *Science* **253**:88.

Heydweiller, G., 1910, Über physikalische eigenscnasten von lösungen in ihren zusammenhang, *Ann. Phys.* **33**:145.

Hjertén, S., 1973, Some general aspects of hydrophobic interaction chromatography, *J. Chromatogr.* **87**:325–331.

Hjertén, S., Rosengren, J., and Påhlman, S., 1974, Hydrophobic interaction chromatography. The synthesis and the use of some alkyl derivatives of agarose, *J. Chromatogr.* **101**:281–288.

Hjertén, S., Yao, K., Eriksson, K. O., and Johansson, B., 1986, Gradient and isocratic high-performance hydrophobic interaction chromatography of proteins on agarose columns, *J. Chromatogr.* **359**:99–109.

Hofmeister, T., 1887–88, Zur lehre von der wirkung der salze, *Arch. Exp. Path. U. Pharmakol.* **24**:247.

Hofstee, B. H. J., 1973, Hydrophobic affinity chromatography of proteins, *Anal. Biochem.* **52**:430–448.

Hofstee, B. H. J., and Otillio, N. F., 1973, Immobilization of enzymes through noncovalent binding to substituted agaroses, *Biochem. Biophys. Res. Commun.* **53**:1137–1144.

Horvath, C., Melander, W. R., and Molnar, I., 1976, Solvophobic interactions in liquid chromatography with non-polar stationary phases, *J. Chromatogr.* **125**:129–156.

Horvath, C., Melander, W., Molnar, I., and Molnar, P., 1977, Enhancement of retention by ion-pair formation in liquid chromatography with non polar stationary phases, *Anal. Chem.* **49**:2295–2305.

Howard, G. A., and Martin, A. J. P., 1950, Separation of the C_{12}–C_{18} fatty acids by reversed-phase partition chromatography, *Biochem. J.* **46**:532–538.

Hubert, P., Mathis, R., and Dellacherie, E., 1991, Polymer ligands for mild hydrophobic interaction chromatography. Principles, achievements and future trends, *J. Chromatogr.* **539**:297–306.

Israelachvili, J., and Pashley, R., 1982, The hydrophobic interaction is long-range, decaying exponentially with distance, *Nature (London)* **300**:341–342.

Jandera, P., Colin, H., and Guiochon, G., 1982, Interaction indexes for prediction of retention in reversed-phase liquid chromatography, *Anal. Chem.* **54**:435–441.

Janin, J., 1979, Surface and inside volumes in globular proteins, *Nature (London)* **277**:491–492.

Jarvis, N., Fox, R. B., and Zesnian, W. A., 1979, Contact angle, wettability and adhesion, in: ACS Advances in Chemistry Series, American Chemical Society, Washington D.C., p. 223.

Jennissen, H. P., and Botzet, G., 1979, Protein binding to two-dimensional hydrophobic binding-site lattices: Adsorption hysteresis on immobilized butyl residues, *Int. J. Biol. Macromol.* **1**:171–179.

Johansson, B. L., and Drevin, I., 1985, Three independent methods for quantitative determination of octyl covalently coupled to Sepharose CL-4B, *J. Chromatogr.* **346**:255–263.

Jost, R., Miron, T., and Wilchek, M., 1974, Mode of adsorption of proteins to aliphatic and aromatic amines coupled to cyanogen bromide-activated agarose, *Biochim. Biophys. Acta* **362**:75–82.

Kalghatgi, K., and Horvath, C., 1988, Rapid peptide mapping by high-performance liquid chromatography, *J. Chromatogr.* **443**:343–354.

Kaliszan, R., Blain, R. W., and Hartwick, R. A., 1988, A new HPLC method of hydrophobicity evaluation employing poly(butadiene)-coated alumina columns, *Chromatographia* **25**:5–7.

Karger, B. L., Gant, J. R., Hartkopf, A., and Weiner, P. H., 1976, Hydrophobic effects in reversed-phase liquid chromatography, *J. Chromatogr.* **128**:65–78.

Kato, A., and Nakai, S., 1980, Hydrophobicity determined by a fluorescence probe method and its correlation with surface properties of proteins, *Biochim. Biophys. Acta* **624**:13–20.

Kato, Y., Kitamura, T., Nakatani, S., and Hashimoto, T., 1989, High-performance hydrophobic interaction chromatography of proteins on a pellicular support based on hydrophilic resin, *J. Chromatogr.* **483**:401–405.

Kauzmann, W., 1959, Some factors in the interpretation of protein denaturation, *Adv. Protein Chem.* **14**:1–63.

Kavanau, J. L., 1964, *Water and Solute–Water Interactions*, Holden-Day, San Francisco.

Keshavarz, E., and Nakai, S., 1979, The relationship between hydrophobicity and interfacial tension of proteins, *Biochim. Biophys. Acta* **576**:269–279.

Körösi, G., and Kovats, E., 1981, Density and surface tension of 83 organic liquids, *J. Chem. Eng. Data* **26**:323–332.

Kurganov, A., Tevlin, A., and Davankov, V., 1983, High-performance ligand-exchange chromatography of enantiomers, *J. Chromatogr.* **261**:223–233.

Kutikani, M., Johnson, D., and Snyder, L. R., 1986, Model of protein conformation in the reverse-phase separation of interleukin-2 muteins, *J. Chromatogr.* **371**:313–333.

Lascu, I., Abrudan, I., Muresan, L., Presecan, E., Vonica, A., and Proinov, I., 1986, Salting-out chromatography on unsubstituted Sepharose CL-6B as a convenient method for purifying proteins from dilute crude extracts, *J. Chromatogr.* **357**:436–439.

Lawson, E. Q., Sadler, A. J., Harmatz, D., Brandau, D. T., Micanovic, R., McElroy, R. D., and Middaugh, C. R., 1984, A simple experimental model for hydrophobic interactions in proteins, *J. Biol. Chem.* **259:**2910–2912.

Lee, B., and Richards, F. M., 1971, Interpretation of protein structures: Estimation of static accessibility, *J. Mol. Biol.* **55:**379–400.

Lee, D. P., 1988, Chromatographic evaluation of large-pore and non-porous polymeric reversed phases, *J. Chromatogr.* **443:**143–153.

Lee, J. C., Gekko, K., and Timasheff, S. N., 1979, *Methods Enzymol.* **61:**26–49.

Lewin, S., 1974, *Displacement of Water and Its Control of Biochemical Reactions*, Academic Press, London.

Ling, T. G. I., and Mattiasson, B., 1983, Poly(ethyleneglycol)- and poly(vinyl alcohol)-substituted carbohydrate gels for "mild" hydrophobic chromatography, *J. Chromatogr.* **254:**83–89.

Lloyd, L. L., 1991, Rigid macroporous copolymers as stationary phases in high- performance liquid chromatography, *J. Chromatogr.* **544:**201–217.

Masimov, A. A., Zaslavsky, B. Y., Gasanov, A. A., Davidovich, Y. A., and Rogozhin, S. V., 1984, Thermodynamic properties of aqueous solutions of macromolecular compounds. Effect of degree of acetylation of poly(vinyl alcohol) on the properties of its aqueous solutions, *J. Chromatogr.* **284:**349–355.

Mathis, R., Mourey, A., and Hubert, P., 1988, One-step purification of a lipolytic enzyme from *Bacillus pumilus* by mild hydrophobic interaction chromatography on polyoxyalkylene glycol-bound Sepharose 6B, *Appl. Environ. Microbiol.* **54:**1307–1308.

Mathis, R., Hubert, P., and Dellacherie, E., 1989, Polyoxyalkyleneglycols immobilized for the sequential extraction of three enzymes from a crude extract of *Pseudomonas testosteroni, J. Chromatogr.* **474:**396–399.

McMillin, G. R., and Walton, A. G., 1974, Circular dichroism technique for the study of adsorbed protein structure, *J. Colloid Interface Sci.* **48:**345–349.

Meek, J. L., 1980, Prediction of peptide retention times in high-pressure liquid chromatography on the basis of amino acid composition, *Proc. Natl. Acad. Sci. U.S.A.* **77:**1632–1636.

Melander, W., and Horvath, C., 1977, Salt effects on hydrophobic interactions in precipitation and chromatography of proteins: An interpretation of the lyotropic series, *Arch. Biochem. Biophys.* **183:**200–215.

Melinder, B. S., 1991, Use of polymeric reversed-phase columns for the characterization of polypeptides extracted from human pancreata, *J. Chromatogr.* **542:**83–99.

Miller, N. T., and Karger, B. L., 1985, High-performance hydrophobic interaction chromatography on ether-bonded phases. Chromatographic characteristics and gradient optimization, *J. Chromatogr.* **326:**45–61.

Miller, N. T., Feibush, B., and Karger, B. L., 1984, Wide-pore silica-based ether-bonded phases for separation of proteins by high-performance hydrophobic-interaction and size-exclusion chromatography, *J. Chromatogr.* **316:**519–536.

Moore, J. C., 1964, Gel permeation chromatography. I. New method for molecular weight distribution of high polymers, *J. Polym. Sci., Part B* **2:**835–843.

Morris, C. J. O. R., 1963, A new method of protein chromatography, *Protides Biol. Fluids. Proc. Colloq.* **10:**325–328.

Morrisey, B. M., Smith, L. E., Stromberg, R. R., and Fenstermaker, C. A., 1976, Ellipsometric investigation of the effect of potential on blood protein conformation and adsorbance, *J. Colloid Interface Sci.* **56:**557–563.

Murphy, K. P., Privalov, P. L., and Gill, S. J., 1990, Common features of protein unfolding and dissolution of hydrophobic compounds, *Science* **247:**559–561.

Nemethy, G., and Scheraga, H. A., 1962, The structure of water and hydrophobic bonding in proteins. III. The thermodynamic properties of hydrophobic bonds in proteins, *J. Phys. Chem.* **66:**1773–1789.

Nishikawa, A. H., and Bailon, P., 1975, Affinity purification methods. Lyotropic salt effects in hydrophobic chromatography, *Anal. Biochem.* **68:**274–280.

Nozaki, Y., and Tanford, C., 1971, Solubility of amino acids and two glycine peptides in aqueous ethanol and dioxane solutions. Establishment of a hydrophobicity scale, *J. Biol. Chem.* **246:**2211–2217.

O'Hare, M. J., and Nice, E. C., 1979, Hydrophobic high-performance liquid chromatography of hormonal polypeptides and proteins on alkylsilane-bonded silica, *J. Chromatogr.* **171:**209–226.

Oroszlan, P., Blanco, R., Lu, X. M., Yarmursh, D., and Karger, B. L., 1990, Intrinsic fluorescence studies of the kinetic mechanism of unfolding of α-lactalbumin on weakly hydrophobic chromatographic surfaces, *J. Chromatogr.* **500:**481–502.

Påhlman, S., Rosengren, J., and Hjertén, S., 1977, Hydrophobic interaction chromatography on uncharged Sepharose derivatives. Effects of neutral salts on the adsorption of proteins, *J. Chromatogr.* **131:**99–108.

Porath, J., Sundberg, L., Fornstedt, N., and Olsson, I., 1973, Salting-out in amphiphilic gels as a new approach to hydrophobic adsorption, *Nature* **245:**465–466.

Raymond, J., Azanza, J. L., and Fotso, M., 1981, Hydrophobic interaction chromatography: A new method for sunflower protein purification, *J. Chromatogr.* **212:**199–209.

Regnier, F. E., and Noel, R., 1976, Glycerol propyl silane bonded phases for aqueous high-performance exclusion chromatography, *J. Chromatogr. Sci.* **14:**316–320.

Richmond, T. J., 1984, Solvent accessible surface area and excluded volume in proteins. Analytical equations for overlapping spheres and implications for the hydrophobic effect, *J. Mol. Biol.* **178:**63–89.

Rose, G. D., Geselowitz, A. R., Lesser, G. J., Lee, R. H., and Zehfus, M. H., 1985, Hydrophobicity of amino acid residues in globular proteins, *Science* **229:**834–838.

Rosengren, J., Påhlman, S., Glad, M., and Hjertén, S., 1975, Hydrophobic interaction chromatography on non-charged Sepharose derivatives. Binding of a model protein, related to ionic strength, hydrophobicity of the substituent, and degree of substitution (determined by NMR), *Biochim. Biophys. Acta* **412:**51–61.

Rosset, R., Caude, M., and Jardy, A., 1982, *Manuel Pratique de Chromatographie en Phase Liquide*, Masson, Paris.

Sawatzki, G., Anselstetter, V., and Kubanek, B., 1981, Isolation of mouse transferrin using salting-out chromatography on Sepharose CL-6B, *Biochim. Biophys. Acta* **667:**132–138.

Schomburg, G., Deege, A., Köhler, J., and Bien-Vogelsang, U., 1983, Immobilization of stationary liquids in reversed- and normal-phase liquid chromatography, *J. Chromatogr.* **282:**27–39.

Shaltiel, S., and Er-el, Z., 1973, Hydrophobic chromatography: Use for purification of glycogen synthetase, *Proc. Natl. Acad. Sci. U.S.A.* **70:**778–781.

Shanbhag, V. P., and Axelsson, C. G., 1975, Hydrophobic interaction determined by partition in aqueous two-phase systems, *Eur. J. Biochem.* **60:**17–22.

Shibusawa, Y., Matsumoto, U., and Takatori, M., 1987, Surface affinity chromatography separation of blood cells, *J. Chromatogr.* **398:**153–164.

Sklar, L. A., Hudson, B. S., and Simoni, R. D., 1977, Conjugated polyene fatty acids as fluorescence probes: Binding to serum albumin, *Biochemistry* **16:**5100–5108.

Snyder, L. R., Stadalius, M. A., and Quarry, M. A., 1983, Gradient elution in reversed-phase HPLC-separation of macromolecules, *Anal. Chem.* **55:**1412A.

Speier, J. L., Webster, J. A., and Barnes, G. H., 1957, The addition of silicon hydrides to olefinic double bonds, *J. Am. Chem. Soc.* **79:**974–979.

Stadalius, M. A., Gold, H. S., and Snyder, L. R., 1984, Optimization model for the gradient elution separation of peptide mixtures by reversed-phase high-performance liquid chromatography, *J. Chromatogr.* **296:**31–59.

Svensson, B., 1973, Use of isoelectric focusing to characterize the bonds established during coupling of cyanogen bromide-activated amylodextrin to subtilisin type novo, *FEBS Lett.* **29:**167–169.

Szepesy, L., and Horvath, C., 1988, Specific salt effects in hydrophobic interaction chromatography of proteins, *Chromatographia* **26:**13–18.

Tanford, C., 1962, Contribution of hydrophobic interactions to the stability of the globular conformation of proteins, *J. Am. Chem. Soc.* **84:**4240–4247.

Tanford, C., 1980, *The Hydrophobic Effect*, 2nd ed., John Wiley & Sons, New York.

Tijssen, R., Billiet, H. A. H., and Schoenmakers, P. J., 1976, Use of the solubility parameter for predicting selectivity and retention in chromatography, *J. Chromatogr.* **122:**185–203.

Tiselius, A., 1948, Adsorption separation by salting-out, *Ark. Kemi. Mineral Geol.* **26B:**1.

Tweeten, K. A., and Tweeten, T. N., 1986, Reversed-phase chromatography of proteins on resin-based wide-pore supports, *J. Chromatogr.* **359:**111–119.

Ulbrich, V., Makes, J., and Juracek, M., 1964, Identification of glycidyl ethers, *Collect. Czech. Chem. Commun.* **29:**1466–1475.

Unger, K. K., Jilge, G., Kinkel, J. N., and Hearn, M. T. W., 1986, Evaluation of advanced silica packings for the separation of biopolymers by high-performance liquid chromatography, *J. Chromatogr.* **359:**61–72.

Van Oss, C. J., and Good, R. J., 1988, On the mechanism of "hydrophobic" interactions, *J. Dispersion Sci. Technol.* **9:**355–362.

Waugh, D. F., 1954, Protein–protein interaction, *Adv. Protein Chem.* **9:**325–437.

Wertz, D. H., and Scheraga, H. A., 1978, Influence of water on protein structure. Analysis of the preferences of amino acid residues for the inside or outside and for specific conformations in a protein molecule, *Macromolecules* **11:**9–15.

Wilchek, M., and Miron, T., 1976, On the mode of adsorption of proteins to "hydrophobic columns", *Biochem. Biophys. Res. Commun.* **72:**108–113.

Wolfenden, R., Anderson, L., Cullis, P. M., and Southgate, C. C. B., 1981, Affinities of amino acid side chains for solvent water, *Biochemistry* **20:**849–855.

V

Novel Concepts and Applications

23

Biorecognition in Molecularly Imprinted Polymers
Concept, Chemistry, and Application

Günter Wulff

1. INTRODUCTION

It seems appropriate to subdivide the field of affinity chromatography into a bioselective and a chemoselective one. Bioselective affinity chromatography is based upon defined biological-type recognition systems between immobilized ligands and biomolecules (proteins, nucleic acids, low-molecular-weight biomolecules, etc.). The interaction between the ligand and the biomolecule is of a rather complex nature involving different types of noncovalent interactions (such as van der Waals forces, hydrophobic interactions, hydrogen bonding, dipole–dipole interactions, charge-transfer interactions, and Coulombic interactions). For a highly selective interaction, the orientation in space of the functional groups acting as binding sites is an important factor. This orientation facilitates a highly cooperative combination of interactions. In addition to the orientation of the functional groups, an exact steric fit of the two complementary compounds considerably improves selectivity (shape selectivity).

Chemoselective affinity chromatography is characterized by more or less selective nonbiological-type interactions between all kinds of binding sites and substrates. It includes, for example, hydrophobic affinity chromatography, covalent affinity chromatography, and affinity chromatography with immobilized dye ligands, with immobilized phenyl boronate, and with immobilized cyclodextrins.

It was the purpose of our work to develop a chemoselective affinity chromatography acting in a similar mode as bioselective affinity chromatography. This means that the orientation of the binding sites has to be controlled in order to obtain certain cooperativities in binding and interactions and an additional shape selectivity.

Günter Wulff • Institute of Organic Chemistry and Macromolecular Chemistry, Heinrich-Heine-University Düsseldorf, 40225 Düsseldorf, Germany.

Molecular Interactions in Bioseparations, edited by That T. Ngo. Plenum Press, New York, 1993.

2. CONCEPT

Usually, functional groups acting as binding sites for chemoselective affinity chromatography are attached to a suitable support without any control of their relative orientation. In nearly all known cases, just one type of binding site is introduced. Much higher selectivity can be expected if two, three, or more different types of binding sites are used and if they are introduced in a certain orientation with respect to each other.

Different functional groups can be introduced into a synthetic polymer by copolymerization of the appropriate monomers bearing the desired functionalities. By this method one obtains a polymer with randomly distributed functional groups (Fig. 1a). Another possibility involves the grafting of side chains containing the desired arrangement of binding sites onto the parent polymer (Fig. 1b). A third possibility is the polymerization of monomers with the desired arrangement of binding sites already fixed. In this case the groups are localized in the main chain one after another. This type of orientation of binding sites exists in certain hormone receptors for interaction with hormones (Fig. 1c). This arrangement has been called "continuate word" by Schwyzer (1970).

In antibodies and natural enzymes the functional groups responsible for the specificity are located at quite distant points from each other along the peptide chain. They are brought

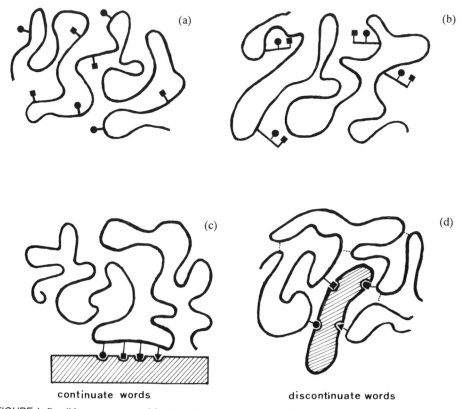

continuate words   discontinuate words

FIGURE 1. Possible arrangements of functional groups in synthetic and natural polymers (Wulff et al., 1973).

into a fixed spatial relationship by a specific folding of the chain. In this case, both the functional-group sequence in the chain and the peptide's tertiary structure, that is, its topochemistry, are decisive (Fig. 1d). This type of arrangement has been termed "discontinuate word." Only in this case can complex, three-dimensional steric arrangements of the functional groups be obtained. In this chapter we report on our investigations to obtain such "discontinuate word" arrangements in cross-linked polymers.

One of the most selective interactions in bioselective affinity chromatography is that between antigens and antibodies. Antibodies can be generated in animals against substances acting as antigens. A procedure introduced quite some time ago by our group (Wulff and Sarhan, 1972; Wulff *et al.*, 1973, 1977; Wulff, 1986) was similar in principle to the generation of an antibody. In this case, polymerizable binding-site groups are bound by covalent or noncovalent interactions to a suitable template molecule (see Fig. 2). This template monomer is copolymerized in the presence of a high amount of cross-linking agent. After splitting off of the template molecule from the polymer, microcavities are obtained having a shape and an arrangement of functional groups corresponding to an imprint of the template. If there is sufficient cross-linking, the polymer chains are in a fixed arrangement and the cavities retain their structure. The binding sites in this polymer are located at quite different points along the polymer chain, and they are held in a fixed spatial relationship with respect to one another by the cross-linking; that is, the desired "discontinuate word" arrangement is achieved.

This approach to preparing a cavity differs from those of Cram (1988), Lehn (1988), Breslow *et al.* (1988), Murakami (1983), and others, who used crown-type compounds, cyclodextrins, cyclophanes, and other ring systems to provide a low-molecular-weight moiety bearing the desired stereochemical information. On the other hand, our method has some similarity with that of Dickey (1949), who precipitated silicic acid in the presence of certain templates to produce silicas with an affinity for the molecule used as template.

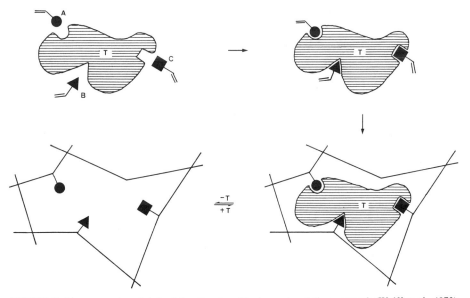

FIGURE 2. The preparation of defined functional cavities by an imprinting approach (Wulff *et al.*, 1973).

3. CHIRAL CAVITIES IN POLYMERS PREPARED BY IMPRINTING WITH TEMPLATES

As an example of the preparation of cavities with a defined arrangement of binding-site groups and with a predetermined shape, the polymerization of **1** is considered. Phenyl-α-D-

mannopyranoside (**2**) acts as the template. Two molecules of 4-vinylphenylboronic acid are bound to the template molecule by esterification with each pair of hydroxyl groups. The boronic acid was chosen as the binding-site group because it undergoes an easily reversible interaction with diol groupings. The template molecule **2** is chiral and optically active, and for this reason the cavities produced should be chiral as well. Therefore, after cleavage of the original template, the accuracy of the steric arrangement of the binding sites in the cavity could be tested by the ability of the polymer to resolve the racemate of the template. The monomer **1** has been extensively used for the optimization of the imprinting method (Wulff *et al.*, 1977, 1982, 1987c; Wulff, 1986).

The monomer **1** was copolymerized by free-radical initiation in the presence of an inert solvent (porogenic agent) with a large amount of a bifunctional cross-linking agent. Under these conditions, macroporous polymers possessing a permanent pore structure and a high inner surface area were obtained. These polymers therefore exhibited good accessibility and low swelling characteristics and hence a limited mobility of the polymer chains.

The template can be split off by water or methanol to an extent of up to 95% (see Fig. 3). When this polymer is treated with the racemate of the template in a batch procedure under equilibrium conditions, the enantiomer that has been used as the template for the preparation of the polymer is taken up preferentially. If the specificity is expressed by the separation factor α, which is the ratio of the distribution coefficients of D- and L-forms between solution and polymer, α values ranging from 1.20 to 5.11 were obtained, depending on the equilibrium conditions and the polymer structure. Values of α as high as 5.0 result from the simple batch procedure with a maximal enrichment of the D-form at the polymer to the extent of 70–80%.

The enantiomer selectivity of these polymers is strongly dependent on the type and amount of cross-linking agent used during the polymerization (Wulff *et al.*, 1982, 1987c; Wulff, 1986). With ethylene glycol dimethacrylate as the cross-linking agent, it was observed that for polymers containing <10% cross-linking (Fig. 4), virtually no specificity was observed. Up to 50% cross-linking, the α value increases linearly to 1.50. From 50 to 66.7% cross-linking, a dramatic increase of the α value from 1.50 to 3.04 was

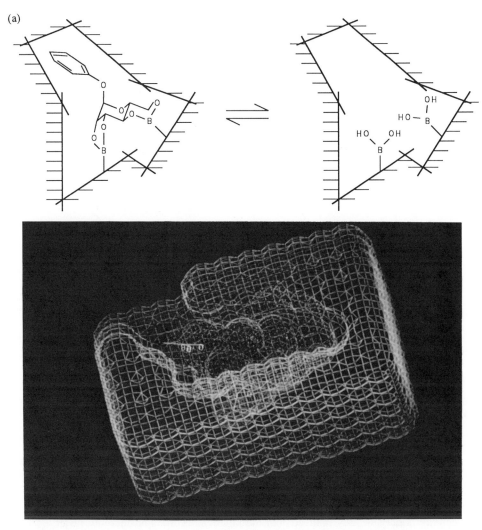

FIGURE 3. (a) Schematic representation of the removal and uptake of the template **2** from an imprinted polymer (Wulff *et al.*, 1977). (b) Computer graphical representation of the imprinted cavity carrying **2** as the template.

observed, thus implying a fourfold increase in selectivity over this range. With further increase in cross-linking up to 95%, the specificity rises to $\alpha = 3.66$. On the other hand, with the use of butanediol dimethacrylate and especially *p*-divinylbenzene as cross-linking agents, a much lower specificity is observed as a function of cross-linking percentage.

In addition to the requirement of rigidity for cavity stability, the polymers should, at the same time, possess some degree of flexibility. This is necessary to enable a fast reversible binding of the substrates within the cavities. Cavities of accurate shape but without any flexibility present kinetic hindrance to reversible binding.

In order to obtain a fast attainment of equilibrium during the batch procedure or a chromatographic operation, the consequences of variations in the polymer structure were investigated (G. Wulff, K. Jacoby, and J. Steinke, unpublished results). These investiga-

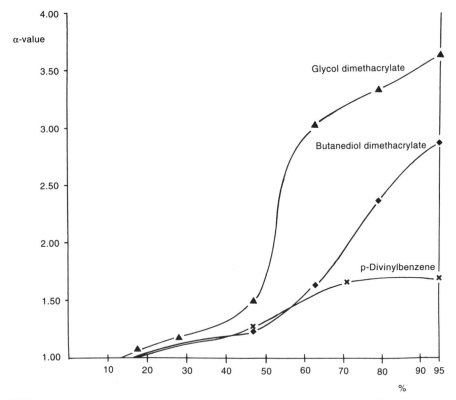

FIGURE 4. Dependence of the specificity of the polymers for racemate resolution upon the kind and amount of cross-linking agent (Wulff *et al.*, 1982, 1987c).

tions showed (see Tables 1 and 2) that polymers prepared at lower temperatures are less selective, but the attainment of equilibrium becomes faster. It is more advantageous to use only 70% of cross-linking agent and 25% of methyl methacrylate. In this case selectivity is only slightly reduced, but the kinetics become considerably faster.

Polymers obtained with ethylene glycol dimethacrylate as cross-linker retained their specificity for a long period. Even under high pressure in a high-performance liquid chromatography (HPLC) column, the activity remained for months. This was true even when the column was used at 70–80°C. On the other hand, polymers cross-linked with divinylbenzene gradually lost their specificity at higher temperatures. Interestingly, at 60°C the α value for racemic resolution was further increased to 5.11, as is shown in Table 2. Higher selectivity at increased temperature had already been observed during earlier chromatographic studies (Wulff and Vesper, 1978; Wulff and Minárik, 1986).

4. CHROMATOGRAPHIC SEPARATIONS

If these polymers are to be used as stationary phases in HPLC for the separation of racemates, the chromatographic conditions have to be carefully optimized (Wulff *et al.*,

TABLE 1
Optimization of Polymer Structure[a,b]

Polymer	Cross-linker, EGDMA (%)	MMA (%)	Initiator	Polymerization conditions
1	95	—	AIBN	65°C, 2 days
2	95	—	ADVN	20°C, 12 days
3	70	25	AIBN	65°C, 2 days
4	70	25	ADVN	20°C, 12 days
5	70	25	HPPN	3°C, 5 days, 65°C, 2 days

[a]G. Wulff, K. Jakoby, and J. Steinke, 1991, unpublished results.
[b]Five percent of 1 together with ethylene glycol dimethacrylate (EGDMA) and in some cases methyl methacrylate (MMA) was polymerized after addition of 1 ml of THF as the porogen and 8 mg of 2,2′-azobis(isobutyronitrile) (AIBN), 12 mg of 2,2′-azobis(2,4-dimethylvaleronitrile) (ADVN), or 17 mg of hexylene glycol peroxyneodecanoate (HPPN), respectively, as the initiator per gram of monomer mixture.

1986c; Wulff and Minárik, 1988). Due to the specific separation mechanism, mass transfer during chromatography is somewhat slow and substances tend to show tailing. Tailing during chromatography can be reduced by addition of certain bases which strongly accelerate the attainment of equilibrium between the free diol groupings and the polymer-bound boronic esters. By addition of these bases, the covalent binding reaction becomes as fast as noncovalent binding-site interactions (Wulff et al., 1984; Lauer et al., 1985). Mass transfer in chromatography can be further increased if the temperature is increased. Figure 6 shows the results from the chromatography of the racemate of phenyl-α-mannoside at different temperatures (Wulff and Minárik, 1986). It is striking that the first peak (L-phenyl-L-mannopyranoside) is relatively sharp whereas the second shows considerable spreading. In this case the separation is complete, but at a separation factor of $\alpha = 3$ the peaks should show much better resolution. Detailed investigations have shown that the

TABLE 2
Properties of Polymers Described in Table 1[a]

Polymer	Inner surface area (m²/g)	Splitting percentage[b]	Swelling ability in methanol	Separation factor[c] $\alpha(20°C)$ (methanol)	Separation factor[c] $\alpha(60°C)$ (methanol)	$t_{1/2}$ to reach equilibrium[d] (20°C, min)
1	556	81.9	1.37	4.45	5.11	193
2	240	96.1	1.95	4.13	4.25	174
3	247	87.3	1.59	4.30	4.46	158
4	198	95.9	1.94	3.54	3.73	152
5	332	86.1	1.79	3.87	4.27	152

[a]G. Wulff, K. Jakoby, and J. Steinke, 1991, unpublished results.
[b]Splitting percentage: Percentage of templates that could be removed from the polymer.
[c]Separation factor α was determined in the batch procedure by equilibration of the racemate of 2 with the polymers at 20°C or 60°C. α is the ratio of the distribution coefficient between polymer and solution of the D-form to that of the L-form.
[d]The time $t_{1/2}$ to reach half of the possible enantiomeric enrichment was measured at 20°C.

enantiomer not acting as the template is only retarded by a one-point binding in the cavity. The template molecule, on the other hand, can undergo a two-point binding in the cavity by formation of two boronic diester bonds (Wulff *et al.*, 1986b; Wulff and Kirstein, 1990). This two-point binding leads to a retardation of the mass transfer. The spreading of the second peak can be strongly reduced by gradient elution (Wulff and Minárik, 1990). Figure 5 shows such a separation with a resolution of $r_s = 4.3$ (calculated at half peak width). Columns of this kind can also be used in preparative separations (Wulff *et al.*, 1986c; Wulff and Minárik, 1988).

Besides sugar derivatives, amino acid derivatives, hydroxycarboxylic acids, dialde-hydes, and several other classes of compounds have been successfully used as templates by our group (Wulff and Sarhan, 1982; Wulff, 1986). Other research groups have utilized the principle of molecular imprinting and have widened the scope of this approach by using a variety of templates (see, e.g., Shea *et al.*, 1980; Damen and Neckers, 1980; Fujii *et al.*, 1985; Sarhan and El-Zahab, 1987; Sellergren *et al.*, 1988; Morihara *et al.*, 1988; Braco *et al.*, 1990; Anderson and Mosbach, 1990). The group of Mosbach has successfully concentrated on the use of noncovalent interactions for fixation of the template during polymerization (for a review, see Chapter 24 of this volume).

In most cases, for racemic resolution of racemates on this type of chromatographic support (as with other chromatographic racemic separations), the racemates need to be transformed to specific derivatives. In more recent investigations it was possible to use free sugars as templates for the imprinting procedure. In these cases the separation of the racemate of the free sugar was possible. During these investigations, some results leading to a better understanding of the separation mechanism were obtained (Wulff and Schauhoff, 1991; Wulff and Haarer, 1991).

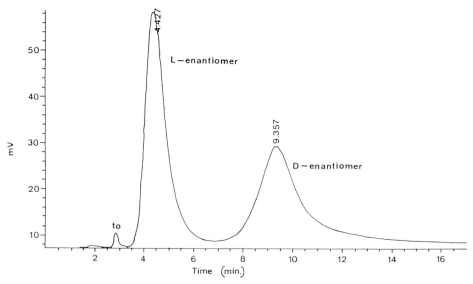

FIGURE 5. Chromatographic racemic resolution of phenyl-α-D,L-mannopyranoside on imprinted polymers (see Fig. 3). Chromatography at 70°C; flow rate, 1 ml/min; gradient elution with acetonitrile containing 5% aqueous NH₃ (25%) and water containing 5% NH₃ (25%) ranging from 9:1 to 5:5 (Wulff and Minárik, 1990).

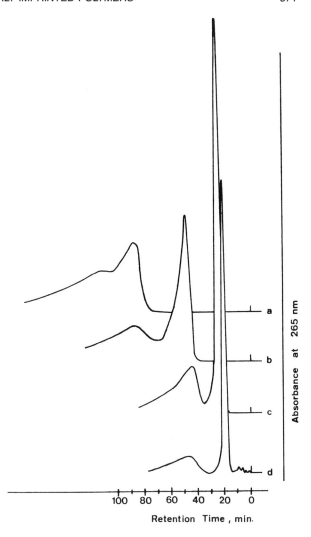

FIGURE 6. Chromatographic separation at different temperatures of
racemic phenyl-α-mannopyranoside
with a support prepared from **1**: (a)
25°C; (b) 40°C; (c) 65°C; (d) 80°C
(Wulff and Minárik, 1986).

5. THE MECHANISM OF MOLECULAR RECOGNITION

The mechanism of molecular recognition in the molecular imprinting procedure is still
under debate. The high selectivity observed is believed, on the one hand, to be mainly
due to the shape of the cavity, whereas the binding sites within the cavity are primarily
responsible for the driving force to bring the substrate inside the cavity. On the other hand,
molecular recognition could also be due to the spatial arrangement of the functional groups
(binding sites) within the cavity. Certainly, the shape of the cavity plays an important role
since even templates with only a one-point binding generate selective cavities (Sarhan and
El-Zahab, 1987; Wulff and Lohmar, 1979; Wulff *et al.*, 1980; Wulff and Gimpel, 1982).
In a recent paper, Shea and Sasaki (1989) showed that shape selectivity may be the most
important factor for molecular recognition. Our own results, however, lead to the conclusion

that the orientation of the functional groups in the recognition site is the predominating factor.

This conclusion could be drawn from attempts to prepare polymers for the resolution of free sugar racemates, in order to avoid the need for derivatization. For this purpose, β-D-fructopyranose-2,3;4,5-bis-O-(4-vinylphenyl)boronate (**3a**) and α-D-galactopyranose-

4a R=H   3a R=H
4b R=Benzyl   3b R=Benzyl
4c R=Methyl   3c R=Methyl

1,2;3,4-bis-O-(4-vinylphenyl)boronate (**4a**), easily prepared from their parent free sugars in a single step, were polymerized as described for **1** (Wulff and Schauhoff, 1991). After splitting off of the respective templates, these polymers were used for the racemic resolution of the racemates of the templates, and we obtained relatively high separation factors α in the batch procedure. Thus, the racemates of fructose and galactose can be separated on polymers prepared from **3a** and **4a** (see Table 3).

Using 6-O-benzyl-D-galactose as the template for the preparation of polymers instead of D-galactose, we achieved a higher selectivity for the racemic resolution of free D,L-galactose. In this case the monomer **4b** is better defined than **4a**, and transesterification is inhibited (Wulff and Haarer, 1991). The benzyl ether **4b** is more advantageous than the

TABLE 3
Selectivity in Racemic Resolution in the Batch Procedure

Polymer prepared from template monomer	Racemate	Separation factor, α
1	Phenyl-α-D,L-mannoside	5.0
1	D,L-Mannose	1.6
1	D,L-Fructose	1.34
3a	D,L-Fructose	1.63
3c	1-O-Methyl-D,L-fructose	1.76
3b	1-O-Benzyl-D,L-fructose	1.64
4a	D,L-Galactose	1.38
4c	D,L-Galactose	1.39
4b	D,L-Galactose	1.58
3a	D,L-Galactose	0.85 (1.17)[a]
4a	D,L-Fructose	0.80 (1.25)
4b	D,L-Fructose	0.71 (1.41)

[a]The reciprocal values, i.e., K_L/K_D, are given in parenthesis. This facilitates the comparison of selectivities.

methyl ether **4c**, because the benzyl group enlarges the cavities and facilitates the uptake of the substrate.

On the other hand, the selectivity for the separation of racemates having larger substituents such as benzyl is not enhanced on polymers prepared from the benzyl ether monomer **4b** or **3b**. Higher selectivity is observed for racemates with smaller substituents such as methyl or hydrogen. This is in contrast to the separation of phenyl-α-D,L-mannopyranoside, which is much easier than that of D,L-mannose. The reasons for the difference here may be twofold. In phenylmannopyranoside there is no mutarotational equilibrium possible, and all the substance is present as the correct isomer. Secondly, the benzyl ether is certainly much more flexible than the phenyl glycoside and therefore hinders the uniform imprinting of the cavities and also the embedding of the substrates in a certain conformation.

Surprisingly, polymers prepared from **3a** adsorb D-fructose from D,L-fructose but L-galactose from D,L-galactose. Similarly, polymers prepared from **4a** adsorb D-galactose but L-fructose from the corresponding racemates. In this context it was also investigated whether other sugar racemates can be separated on a polymer prepared with phenyl-α-D-mannopyranoside as the template. This polymer is able to separate the racemate of the free sugar D,L-mannose with an α value of 1.60. It was not possible to separate the racemates of arabinose, glucose, galactose, xylose, lyxose, or ribose on this polymer. Only D,L-fructose could be separated, with α = 1.34, D-fructose being the preferentially adsorbed enantiomer (see Table 3).

The inverse selectivity observed on polymers prepared from D-fructose and from D-galactose can be explained by inspecting molecular models of the preferred conformations of the free sugars. Figure 7 shows ball-and-stick models of β-D- and β-L-fructopyranose and of α-D- and α-L-galactopyranose. Since binding occurs under thermodynamic control, the most stable diboronates will be formed. In the case of galactose this is from the α-D- and α-L-galactopyranose. Even if this form is only present as 29% of the mutarotational equilibrium, the equilibrium will be shifted during binding, similar to the case of the formation of the monomer *4a* in solution. It can be seen that the spatial arrangements of the hydroxyl functions responsible for the interaction with the boronic acid binding sites at the polymer (oxygens marked black) are identical for D-galactose and L-fructose as well as for D-fructose and L-galactose. On the contrary, for both the enantiomers of mannose, the hydroxyl groups possess a different steric arrangement.

On the other hand, the polymer prepared from **1** with phenyl α-D-mannopyranoside as template has the appropriate orientation of the boronic acids in the cavity to also accommodate the free D-mannose. By comparison of the molecular models of β-D-fructopyranose (in Fig. 7) and phenyl-α-D-mannopyranoside (Fig. 8), it is evident that the hydroxyl groups at positions 2,3 and 4,6 in the mannoside possess the same spatial arrangement as those at positions 1,3 and 4,5 of the β-D-fructopyranose. D-Fructose can therefore be bound, and thus a polymer from **1** shows selectivity in racemic resolution of D,L-fructose.

These results have a number of important consequences for the concept of molecular imprinting with templates (Wulff and Schauhoff, 1991):

(a) They provide a deeper insight into the question pertaining to the origin of the selectivity for racemic resolution. The chiral construction of the cavities is stabilized by means of the cross-linking points in the polymer chains. This type of chirality can arise from the configuration of the cross-linking points as well as from asymmetric conformations

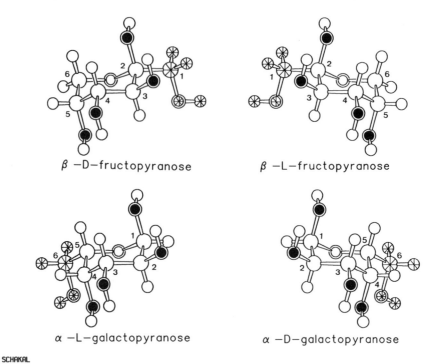

β −D−fructopyranose   β −L−fructopyranose

α −L−galactopyranose   α −D−galactopyranose

SCHAKAL

FIGURE 7. Ball-and-stick models of the preferred conformations of β-D-fructopyranose, α-L-galactopyranose, β-L-fructopyranose, and α-D-galactopyranose (Wulff and Schauhoff, 1991).

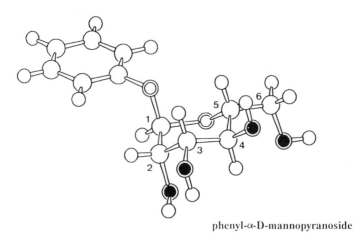

phenyl-α-D-mannopyranoside

FIGURE 8. Ball-and-stick model of 2.

of the polymer chains that are stabilized by cross-linking (Wulff, 1989). The chiral configurations of the linear portions of the chains are not expected to contribute to the asymmetry of the cavity since no asymmetric cyclopolymerization (Wulff *et al.*, 1987b) is possible for the examples presented in this chapter. The asymmetry of the empty cavities can also be analyzed by measuring the optical activity (Wulff and Kirstein, 1990). This is measured by suspending the polymer in a solvent that has the same refractive index as the polymer. In this case the optical rotation is not caused by individual chiral centers, as is usually the case, but by the empty imprints as a whole.

The results from the equilibration with different racemates clearly suggest that for selectivity, at least for the present cases, the orientation of the functional groups inside the cavity is primarily responsible for molecular recognition. Shape selectivity is of secondary importance. A comparison of the structures of β-D-fructopyranose and α-L-galacto-pyranose, which are both preferentially taken up by the polymer prepared from **3**, shows that they do not differ (in their preferred conformation) in the orientation of the interacting hydroxyl groups (marked black in Fig. 7), but the hydroxymethyl groups (marked with spokes in Fig. 7) are at different places in the molecule. This gives these compounds considerably different shapes, but, because of the coincidence of the other hydroxyls, the selectivity did not change but was only reduced to some extent. The same is true when phenyl-α-D-mannopyranoside is used as the template. Both D-mannose and D-fructose can be incorporated into the polymer. The large phenyl substituent in the mannoside considerably alters the shape (compare β-D-fructopyranose in Fig. 7 with Fig. 8). Still, only the extent of selectivity is reduced on equilibration with D,L-mannose or D,L-fructose on the polymer prepared from **1**.

(b) If the orientation of the interacting groups inside the cavity is the principal factor governing the selectivity, other molecules besides the templates with a similar orientation of their functional groups can be separated. In this case it is possible to prepare a well-defined polymer, as in the case of **1** from phenyl-α-D-mannopyranoside, and use it to separate D,L-mannose or D,L-fructose. Furthermore, the concept of inverse selectivity might be of importance in a situation where pure enantiomers of the racemate in question are not easily accessible.

(c) In contrast to imprinting with noncovalent interactions (Sellergren *et al.*, 1988; Andersson *et al.*, this volume, Chapter 24), in our systems the functional groups are present in a defined number and orientation per active center. With noncovalent interactions, an excess (usually fourfold) of binding sites needs to be incorporated to obtain reasonable selectivity. This implies that only one-quarter of the functional groups are arranged in an oriented manner, the rest being arranged irregularly all over the polymer. With covalent interactions, certain functionalities can be introduced in a defined number and orientation into the active site, and these groups can be transformed into other functionalities. This is especially true for the boronic acid moiety, which can be easily replaced by a variety of other functional groups (Wulff and Dhal, 1987).

(d) It is possible not only to prepare active centers with a defined orientation of functional groups and with a particular shape to selectively bind certain substances or catalyze certain reactions, but also to introduce additional pockets or niches at the active centers. This can be achieved by attaching bulky substituents to the template molecule. Thus, in the case of **1**, the phenyl ring provides additional space in the active center.

6. EXACT PLACEMENT OF FUNCTIONAL GROUPS ON THE SURFACES OF RIGID MATRICES

The selectivity observed in the type of racemic resolution discussed above is a result of the combination of an accurate cavity-shape fitting and the exactness of the arrangement of the functional groups. Since the arrangement of functional groups is the predominating factor, it was of prime interest to examine whether the arrangement of the functional groups alone could give rise to selectivity by using a distinct distance between two functional groups. In other words, the question is whether a two-dimensional instead of a three-dimensional information transfer will be adequate to bring about selectivity.

We therefore attempted to introduce two amino groups onto a more or less planar surface of silica arranged at a distinct distance with the aid of a template (see Fig. 9a). With this method it was possible to locate two functional groups on a silica surface via siloxane bond formation (see Fig. 9b), thus enabling an investigation of the selectivity due to distance accuracy alone (Wulff *et al.*, 1986a, 1987a).

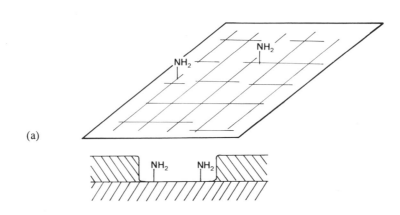

FIGURE 9. Introduction of two amino groups on the surface of a rigid matrix by a template molecule such as **5** or **6** after removal of the template. (a) Schematic representation of the two amino groups on a planar surface or on a surface where the remaining groups around have been capped. (b) Two amino groups on the surface of silica attached via a siloxane grouping. The remaining silanol groups have been blocked by trimethyl silyl groups.

Two amino groups were attached to the surface of silica at a distance of 0.72 and 1.05 nm from one another using the template monomers **5** and **6**. The attachment to the surface is

$$\text{Monomer 5} \quad H_3CO-\underset{\underset{CH_3}{|}}{\overset{\overset{CH_3}{|}}{Si}}-\!\!\bigcirc\!\!-N\!\!=\!\!\overset{H}{C}-\!\!\bigcirc\!\!-\overset{H}{C}\!\!=\!\!N-\!\!\bigcirc\!\!-\underset{\underset{CH_3}{|}}{\overset{\overset{CH_3}{|}}{Si}}-OCH_3$$

$$\text{Monomer 6} \quad H_3CO-\underset{\underset{CH_3}{|}}{\overset{\overset{CH_3}{|}}{Si}}-\!\!\bigcirc\!\!-N\!\!=\!\!\overset{H}{C}-\!\!\bigcirc\!\!-CH_2-\!\!\bigcirc\!\!-\overset{H}{C}\!\!=\!\!N-\!\!\bigcirc\!\!-\underset{\underset{CH_3}{|}}{\overset{\overset{CH_3}{|}}{Si}}-OCH_3$$

$$\text{Monomer 7} \quad H_2N-(CH_2)_3-Si(OCH_3)_3$$

achieved through the formation of siloxane bonds by condensation between the methoxy-silane group of **5** and **6** and the silanol groups on the surface of the silica. Most of the remaining silanol groups were afterwards capped by reaction with hexamethyldisilazane to avoid nonspecific adsorption (see Fig. 9b). Over 95% of the templates could be split off. Unlike the situation with polymers, in this case the position of the two amino groups should not be changed as a result of chain mobility, swelling, or shrinking. The distance can only be altered by conformational changes within the functional group part. For comparison purposes, a silica with randomly distributed amino groups was prepared from **7**. In order to elucidate the role of distance accuracy, the selectivity was determined by equilibration with an equimolar mixture of the two template dialdehydes **8** and **9** (see Table 4). Both the silicas showed a significant difference in binding, preferring their own templates, with α values of 1.74 and 1.67. This clearly suggests that by using distance selectivity alone and with differences of only 0.33 nm (between **8** and **9**), substrate selectivity can be observed.

Chromatography with silicas of this type could also be used to separate dicarboxylic acids that differ in the distance between their carboxyl groups. Good separations were observed compared with those obtained under identical conditions on silica with randomly distributed amino groups.

Imprinting on the surface of silica is thus a further extension of the original imprinting method.

7. CONCLUSION

By designing new types of adsorbents for affinity chromatography, it was possible to use the principles of biorecognition for nonbiological systems. Synthetic polymers and silicas with high selectivity were prepared with the aid of template molecules by an imprinting procedure. In this way, binding sites are prepared having a specific shape and an arrangement of binding-site groups corresponding to the chemical nature of the template. These adsorbents show highly selective binding for the templates used. Thus, it was possible to separate the racemate of the template with separation factors α of about 5. The adsorbents can be used in a batch procedure or in an HPLC mode.

In contrast to the typical bioselective affinity chromatography adsorbents with high-molecular-weight ligands, these adsorbents show good chemical and thermal stability. Columns of this type can be used at 80°C at high pressure for a long time without loss of selectivity. Furthermore, it is possible to use the principles of biorecognition also in

TABLE 4

Selectivity of Modified Silicas With Each of the Two Amino Groups in a Defined Distance
(Wulff et al., 1986a)

	Splitting percentage	Distance r of groups (nm)	Apparent binding constants of:		Selectivity α'
			OHC—⬡—CHO **8**	OHC—⬡—CH₂—⬡—CHO **9**	
Silica modified with **5**	<95%	0.72	4.91	2.58	1.74
Silica modified with **6**	>95%	1.05	9.07	13.77	1.67
Silica modified with **7** (at random)	—	—	2.26	2.05	—

nonaqueous solvents and to use binding-site-type interactions not known in biological systems.

In contrast to chemoselective affinity chromatography adsorbents and in accordance with their bioselective counterparts, these adsorbents show selectivity only for the template or for a substance of very similar structure. For substrates of different chemical structure, therefore, a new adsorbent must be prepared.

A further extension of this work is toward the construction of specific reagents and catalysts. In this respect, it was possible to perform an asymmetric synthesis of α-amino acids within a chiral cavity templated with the end product of the reaction (Wulff and Vietmeier, 1989). An important application of these polymers in the future will probably be their use as novel asymmetric catalysts. With molecular imprinting there is a possibility to prepare analogs of the active sites of enzymes. By incorporation of suitable binding and catalytic sites in the correct orientation, it might be possible to prepare enzyme-analog-built polymers. Work in this direction is under way in different laboratories around the world.

ACKNOWLEDGMENTS

Thanks are due to Fonds der Chemischen Industrie, to the Minister für Wissenschaft und Forschung des Landes Nordrhein-Westfalen, and to Deutsche Forschungsgemeinschaft for financial support.

REFERENCES

Anderson, L. I., and Mosbach, K., 1990, Enantiomeric resolution on molecularly imprinted polymers prepared with only non-covalent and non-ionic interactions, *J. Chromatogr.* **516**:313–322.

Braco, L., Dabulis, K., and Klibanov, A. M., 1990, Production of abiotic receptors by molecular imprinting of proteins, *Proc. Natl. Acad. Sci. U.S.A.* **87**:274–277.

Breslow, R., Chmielewski, J., Foley, D., and Johnson, B., 1988, Optically active amino acid synthesis by artificial transaminase enzymes, *Tetrahedron* **44**:5515–5524.

Cram, D. J., 1988, Molecular hosts and guests, and their complexes, *Angew. Chem. Int. Ed. Engl.* **27**:1009–1014.

Damen, J., and Neckers, D. C., 1980, Stereoselective syntheses via a photochemical template effect, *J. Am. Chem. Soc.* **102**:3265–3267.

Dickey, F. H., 1949, Preparation of specific adsorbents, *Proc. Natl. Acad. Sci. U.S.A.* **35**:227–229.

Fujii, Y., Matsutani, K., and Kikuchi, K., 1985, Formation of a specific coordination cavity for a chiral amino acid by template synthesis of a polymer Schiff base cobalt(III) complex, *J. Chem. Soc. Chem. Commun.* **1985**: 415–417.

Lauer, M., Böhnke, H., Grotstollen, R., Salehnia, M., and Wulff, G., 1985, Zur Chemie von Haftgruppen IV. Über eine ausserordentliche Erhöhung der Reaktivität von Arylboronsäuren durch Nachbargruppen, *Chem. Ber.* **118**:246–260.

Lehn, J. M., 1988, Supramolecular chemistry—molecules, supermolecules, and molecular functional units, *Angew. Chem. Int. Ed. Engl.* **27**:89–114.

Morihara, K., Kurihara, S., and Suzuki, J., 1988, Footprint catalysis. I. A new method for designing "tailor-made" catalysts with substrate specificity: Silica (alumina) catalysts for butanolysis of benzoic anhydride, *Bull. Chem. Soc. Jpn.* **61**:3991–3998.

Murakami, Y., 1983, Functionalized cyclophanes as catalysts and enzyme models, *Top. Curr. Chem.* **115**:107–155.

Sarhan, A., and El-Zahab, M. A., 1987, Racemic resolution of mandelic acid on polymers with chiral cavities. Enzyme-analogue stereospecific conversion of configuration, *Makromol. Chem. Rapid Commun.* **8**:555–561.

Schwyzer, R., 1970, Organization and read-out of biological information in polypeptides, *Proceedings of the Fourth International Congress on Pharmacology*, Vol. 5, pp. 196–209: *Chem. Abstracts* 1971, **74**: 38232.

Sellergren, B., Lepistö, M., and Mosbach, K., 1988, Highly enantioselective and substrate-selective polymers obtained by molecular imprinting utilizing noncovalent interactions. NMR and chromatographic studies on the nature of recognition, *J. Am. Chem. Soc.* **110**:5853–5860.

Shea, K. J., and Sasaki, D. Y., 1989, On the control of microenvironment shape of functionalized network polymers prepared by template polymerization, *J. Am. Chem. Soc.* **111**:3442–3444.

Shea, K. J., Thompson, E. A., Pandey, S. D., and Beauchamps, P. S., 1980, Template synthesis of macromolecules. Synthesis and chemistry of functionalized macroporous polydivinylbenzene, *J. Am. Chem. Soc.* **102**:3149–3155.

Wulff, G., 1986, Molecular recognition in polymers prepared by imprinting with templates, in: *Polymeric Reagents and Catalysts* (W. T. Ford, ed.), ACS Symposium Series 308, American Chemical Society, Washington, D.C., pp. 186–230.

Wulff, G., 1989, Main-chain chirality and optical activity in polymers consisting of C—C-chains, *Angew. Chem. Int. Ed. Engl.* **28**:21–37.

Wulff, G., and Dhal, P. K., 1987, Design of vinyl functional copolymers with main chain chirality through chemical modification, *Makromol. Chem.* **188**:2847–2856.

Wulff, G., and Gimpel, J., 1982, Über den Einfluss der Flexibilität der Haftgruppen auf die Racemattrennungsfähigkeit, *Makromol. Chem.* **183**:2469–2477.

Wulff, G., and Haarer, J., 1991, The preparation of defined chiral cavities for the racemic resolution of free sugars, *Makromol. Chem.* **192**:1329–1338.

Wulff, G., and Kirstein, G., 1990, Measuring the optical activity of chiral imprints in insoluble highly cross-linked polymers, *Angew. Chem. Int. Ed. Engl.* **29**:684–686.

Wulff, G., and Lohmar, E., 1979, Specific binding effects in chiral microcavities of crosslinked polymers, *Isr. J. Chem.* **18**:279–284.

Wulff, G., and Minárik, M., 1986, Enzyme-analogue built polymers XX. Pronounced effect of temperature on racemic resolution using template-imprinted polymeric sorbents, *J. High Resolut. Chromatogr. Commun.* **9**:607–608.

Wulff, G., and Minárik, M., 1988, Tailor-made sorbents. A modular approach to chiral separation, in: *Chromatographic Chiral Separations* (M. Zief and L. J. Crane, eds.), Marcel Dekker, New York, pp. 15–52.

Wulff, G., and Minárik, M., 1990, Template imprinted polymers for h.p.l.c. separation of racemates, *J. Liq. Chromatogr.* **13**:2987–3000.

Wulff, G., and Sarhan, A., 1972, Use of polymers with enzyme-analogous structures for the resolution of racemates, *Angew. Chem. Int. Ed. Engl.* **11**:341.

Wulff, G., and Sarhan, A., 1982, Models of the receptor sites of enzymes, in: *Chemical Approaches to Understanding Enzyme Catalysis: Biomimetic Chemistry and Transition-State Analogs* (B. S. Green, Y. Ashani, and D. Chipman, eds.), Elsevier, Amsterdam, pp. 106–118.

Wulff, G., and Schauhoff, S., 1991, Racemic resolution of free sugars with macroporous polymers prepared by molecular imprinting. Selectivity dependence on the arrangement of functional groups *versus* spatial requirements, *J. Org. Chem.* **56**:395–399.

Wulff, G., and Vesper, W., 1978, Enzyme-analogue built polymers VIII. On the preparation of chromatographic sorbents with chiral cavities for racemic resolution, *J. Chromatogr.* **167**:171–186.

Wulff, G., and Vietmeier, J., 1989, Enzyme-analogue built polymers 26. Enantioselective synthesis of amino acids using polymers possessing chiral cavities obtained by an imprinting procedure with template molecules, *Makromol. Chem.* **190**:1727–1735.

Wulff, G., Sarhan, A., and Zabrocki, K., 1973, Enzyme-analogue built polymers and their use for the resolution of racemates, *Tetrahedron Lett.* **44**:4329–4332.

Wulff, G., Vesper, W., Grobe-Einsler, R., and Sarhan A., 1977, Enzyme-analogue built polymers, IV. On the synthesis of polymers containing chiral cavities, and their use for the resolution of racemates, *Makromol. Chem.* **178**:2799–2816.

Wulff, G., Schulze, I., Zabrocki, K., and Vesper, W., 1980, Bindungsstellen im Polymer mit unterschiedlicher Zahl der Haftgruppen, *Makromol. Chem.* **181**:531–544.

Wulff, G., Kemmerer, R., Vietmeier, J., and Poll, H.-G., 1982, Chirality of vinyl polymers. The preparation of chiral cavities in synthetic polymers, *Nouv. J. Chim.* **6**:681–687.

Wulff, G., Lauer, M., and Böhnke, H., 1984, Rapid proton transfer as cause of an unusually large neighboring effect, *Angew. Chem. Int. Ed. Engl.* **23**:741–742.

Wulff, G., Heide, B., and Helfmeier, G., 1986a, Molecular recognition through the exact placement of functional groups on rigid matrices *via* a template approach, *J. Am. Chem. Soc.* **108**:1089–1091.

Wulff, G., Oberkobusch, D., and Minárik, M., 1986b, On the dynamics of embedding in imprinted polymers, in: *Design and Synthesis of Organic Molecules Based on Molecular Recognition* (G. V. Binst, ed.), Springer-Verlag, Berlin, pp. 229–233.

Wulff, G., Poll, H.-G, and Minárik, M., 1986c, Racemic resolution on polymers containing chiral cavities, *J. Liq. Chromatogr.* **9**:385–405.

Wulff, G., Heide, B., and Helfmeier, G., 1987a, On the distance accuracy of functional groups in polymers and silicas introduced by a template approach, *React. Polym.* **6**:299–310.

Wulff, G., Kemmerer, R., and Vogt, B., 1987b, Optically active polymers with structural chirality in the main chain prepared through an asymmetric cyclopolymerization, *J. Am. Chem. Soc.* **109**:7449–7457.

Wulff, G., Vietmeier, J., and Poll, H.-G., 1987c, Influence of the nature of the crosslinking agent on the performance of imprinted polymers in racemic resolution, *Makromol. Chem.* **188**:731–740.

24

Bioseparation and Catalysis in Molecularly Imprinted Polymers

Lars I. Andersson, Björn Ekberg, and Klaus Mosbach

1. INTRODUCTION

The concept of molecular imprinting has been outlined in Chapter 23 of this volume. The preparation of molecular imprints in synthetic polymers is basically a three-step procedure (Fig. 1). The first step is the formation of specific and definable interactions between the monomer(s) and the print molecule. These interactions are subsequently responsible for the recognition of the print molecules by the imprinted polymer. The interactions can either be noncovalent bonds (Arshady and Mosbach, 1981, Ekberg and Mosbach, 1989), for example, ionic bonds and hydrogen bonds, or reversible covalent bonds, for example, boronic esters, (Wulff, 1986). In the first instance, the interactions are formed simply by mixing the print molecule with a suitable mixture of monomer(s) in an appropriate solvent prior to the polymerization. In the second instance, where covalent bonds are employed, an adduct composed of the print molecule and the monomer(s) is synthesized and is added to the polymerization mixture.

In the second step, the mixture of the print molecule and the monomer(s) is polymerized in the presence of a high percentage of cross-linking agent. The resulting polymer is rigid and insoluble. A high concentration of cross-linker is necessary to preserve the complementarity between the polymer and the print molecule that is created in the polymeric network during the polymerization reaction.

In the third step, the print molecule is removed from the polymer by extraction when noncovalent forces are utilized or, in the case of covalent bonds, by more drastic conditions such as acidic hydrolysis. "Imprints," which are complementary in both shape and chemical functionality to the print molecule, are now present within the polymeric network. Upon incubation of the polymer with substrates of similar structure, the print molecule is preferentially bound. For instance, if the print molecule is one of the enantiomers of a chiral compound, the polymer has the ability to selectively bind the imprinted enantiomer from a mixture of the optical antipodes (see Fig. 1, which shows the formation of imprints specific for the L-form of *N-tert*-butyloxycarbonyltryptophan).

Lars I. Andersson and Klaus Mosbach • Department of Pure and Applied Biochemistry, University of Lund, S-221 00 Lund, Sweden.   *Björn Ekberg* • Bio-Swede AB, IDEON, S-223 70 Lund, Sweden.

Molecular Interactions in Bioseparations, edited by That T. Ngo. Plenum Press, New York, 1993.

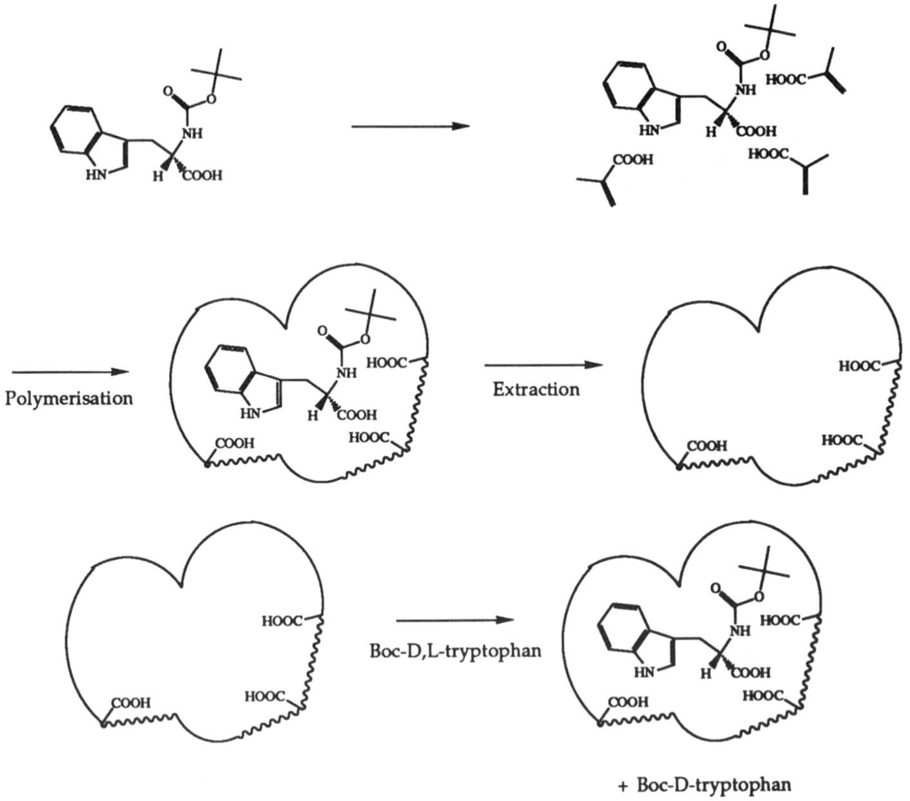

FIGURE 1. Schematic diagram depicting the preparation of molecularly imprinted polymers. The functional monomer, methacrylic acid, selected for its ability to form hydrogen bonds with a variety of chemical functionalities, is added to a solution of the print molecule, here Boc-L-tryptophan (Boc = *tert*-butyloxycarbonyl). The prearrangement step is followed by the polymerization step, using a high percentage of ethylene glycol dimethacrylate as cross-linking monomer. The print molecule is removed from the rigid insoluble polymer by extraction. The original enantiomer is preferentially bound to the resulting polymer upon incubation with the racemate of the print molecule.

2. ENANTIOMERIC SEPARATIONS

A growing need for enantiomerically pure compounds and analysis of enantiomers has led to the development of several chiral stationary phases (CSPs), a number of which are now commercially available. There is to date, however, no universal CSP; each CSP is able to resolve a group of compounds or a limited number of substance types. Therefore, there is a great demand for the development of new and more general CSPs. An alternate would be to develop a general, rapid, and straightforward approach to the design of tailor-made CSPs for each individual chiral separation of interest.

Over the past several years, considerable interest has been shown in the use of molecularly imprinted polymers for enantiomeric recognition and subsequent enantiomeric resolution. Although a great deal of the work has been performed in the batch mode, the true potential of molecularly imprinted polymers lies in their use as stationary phases in column chromatography. An inherent advantage of molecularly imprinted chiral stationary phases, in contrast to many other CSPs, is the predictable elution order of the enantiomers.

In order to use the polymers in the high-performance liquid chromatography (HPLC) mode, two requirements must be fulfilled. Firstly, the polymer must be able to withstand high pressure. Secondly, the polymer must be chemically inert to the eluent. The high content of cross-linking monomers in the polymerization mixture makes the polymers rigid enough to be used under HPLC conditions. In addition, we have not observed degradation of the polymer preparations by any of the solvents used. Polymer preparation and chromatographic elution are two separate events; consequently, the solvent of polymerization and the eluent are optimized independently. As in other chromatographic applications, it is advantageous to use small particles in long columns. At present, the polymers are prepared in bulk and then ground to particles of suitable size. Routinely, a fraction that passes through a 25-μm sieve is used, and the fines are removed by repeated sedimentation from acetonitrile. These particles can be dried under vacuum and stored for at least several years in the dry state without any detectable loss of enantiomeric separation capacity.

Several polymerization parameters have been investigated in order to determine the influence of these factors on the ability of the polymers to separate the enantiomers of the print molecule (O'Shannessy et al., 1989a,b). These include (i) the solvent employed during polymerization, (ii) the system for initiation of the polymerization, and (iii) the effect of molar ratio of functional monomer to print molecule. As a model system, the molecular imprinting of L-phenylalanine anilide using methacrylic acid as functional monomer was studied. In general, polymers prepared in a less polar solvent, for example, chloroform, are superior to those prepared in more polar solvents such as acetonitrile. Furthermore, polymers prepared at 0°C using a light-induced initiation system were superior to those prepared at 60°C using a heat-induced initiation system. An increase in the molar ratio of functional monomer to print molecule improved separation at the expense of peak broadening. The optimal ratio of functional monomer to print molecule was found to be approximately 4:1, in this system. This optimized polymer preparation protocol was used in the preparation of most of the CSPs presented in Table 1.

Molecular imprints were prepared against a number of chiral compounds (Table 1). These include amino acid esters, amino acid amides, N-blocked amino acids, N-blocked amino acid esters, amino alcohols, and carboxylic acid derivatives. Methacrylic acid was used as the functional monomer because the acid function of the monomer interacts ionically with the amino function of the print molecules. In addition, the carboxylic acid function on the functional monomers interacts via hydrogen bonding with a number of polar functionalities on the print molecule, including amides, carboxylic acids, carbamates, heteroatoms, and carboxylic esters. Other interactions between the print molecule and monomers may operate during imprinting and subsequent recognition, such as dipole–dipole interactions and hydrophobic forces. Molecular imprints can also be successfully prepared utilizing only nonionic interactions (Andersson and Mosbach, 1990). It may now be possible to prepare molecular imprints against a very large number of compounds.

Polymers were shown to efficiently resolve a racemate of the compound used as print molecule (Fig. 2). In addition, many of these polymer preparations were able to resolve enantiomers of substances very similar in structure to the print molecule (see below). However, it must be stressed that the best separation is always recorded for the enantiomers of the print molecule. It should also be noted that enantiomeric separation, at least for the print molecule, was observed for all of the stationary phases prepared. In all

TABLE 1

Separation and Resolution on a Selection of Molecularly Imprinted Polymers
Used as Chiral Stationary Phases in Chromatography

Print molecule[a,b]	α^c	R_s^d	Reference
L-Phenylalanine anilide	4.1	1.2	O'Shannessy et al., 1989
L-Leucine-β-naphthylamide	3.8	0.7	Andersson et al., 1990b
L-Phenylalanylglycine anilide	5.1	0.5	Andersson et al., 1990b
N,N-Dimethyl-L-phenylalanine anilide	3.7	1.4	Andersson et al., 1990b
L-proline anilide	4.5	1.0	Andersson et al., 1990b
N-Pyridylmethyl-L-phenylalanine anilide	8.4	1.1	Andersson et al., 1990b
Cbz-L-aspartic acid	2.2	1.7	Andersson and Mosbach, 1990
Cbz-L-glutamic acid	2.5	2.9	Andersson and Mosbach, 1990
Cbz-L-tryptophan	2.0	0.6	Andersson and Mosbach, 1990
Boc-L-tryptophan	1.9	0.8	Andersson and Mosbach, 1990
Boc-L-phenylalanine	1.8	1.4	Andersson and Mosbach, 1990
Cbz-L-tryptophan methyl ester	1.5	1.5	Andersson and Mosbach, 1990
Boc-L-proline-N-hydroxysuccinimide ester	1.2	0.8	Andersson and Mosbach, 1990
l-(−)-Mandelic acid	1.4	1.1	Andersson and Mosbach, 1990
(S)-(−)-Timolol	2.9	2.0	Fischer et al., 1991
(S)-(−)-Timolol[e]	2.5	1.9	Fischer et al., 1991

[a]Polymers were prepared with methacrylic acid (MAA) as the functional monemer and ethylene glycol
dimethacrylate (EDMA) as the cross-linking monomer.
[b]Abbreviations used: Cbz, benzyloxycarbonyl; Boc, tert-butyloxycarbonyl.
[c]Separation factor, α, is defined as the ratio of the capacity factor for the enantiomer used as print molecule to
the capacity factor for the other enantiomer. In cases where several values have been reported, only the best
value is given.
[d]R_s is the resolution factor calculated essentially as in Andersson et al. (1990b). In cases where several values
have been reported, only the best value is given.
[e]Itaconic acid was substituted for methacrylic acid.

instances it was shown that the L-form of a specific compound was eluted last on a column
packed with a material molecularly imprinted against this enantiomer. While on a stationary
phase made against the D-form, the D-enantiomer was the most strongly retained antipode.
In this context, a polymer imprinted against a racemate of the print molecule did not display
any enantiomeric separation.

The separation factors recorded on molecularly imprinted stationary phases are on the
average rather high; $\alpha = 4–8$ when ionic interactions are utilized, and $\alpha = 1.3–2.5$ when
only nonionic interactions are utilized. These figures compare well with those reported for
other chiral stationary phases. When only nonionic interactions are utilized, the interactions
between substrate and polymer are probably weaker compared with those for polymers
prepared with ionic interactions. This fact necessitates the use of less polar solvents as
eluents in the chromatographic analyses on polymers prepared with only nonionic interactions. As a result of the weaker (and faster) interactions, the peaks are narrowed and,
although the separation factors, α, are smaller (see above), the resolution, R_s, is increased
due to the improved peak shape (Fig. 2). Gradient elution schemes improve the peak shape,
increase the resolution, and shorten analysis times (Andersson et al., 1990b). An interesting
observation is that polymers prepared with only nonionic interactions have much higher
substrate selectivity for the print molecule in a mixture of similar substrates than those
polymer imprints prepared with ionic interactions.

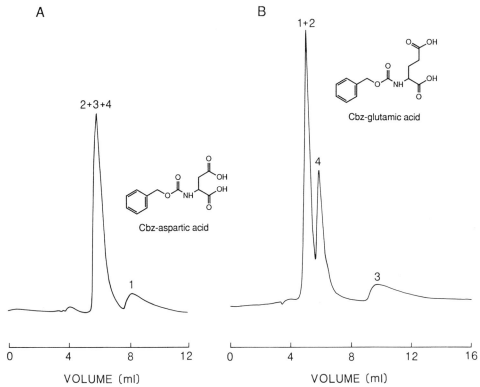

FIGURE 2. Separation of Cbz-L-aspartic acid (1), Cbz-D-aspartic acid (2), Cbz-L-glutamic acid (3), and Cbz-D-glutamic acid (4) on polymers prepared against Cbz-L-aspartic acid (A) and Cbz-L-glutamic acid (B). Particles (<25 μm) were packed into 200 × 4.5 mm i.d. columns. Analyses were performed isocratically using acetonitrile–chloroform–acetic acid (60:39.5:0.5, v/v/v) as the eluent at 0.1 ml/min at room temperature. Detection was at 250 mm. In both experiments, a mixture of 10 μg of each of the racemates of Cbz-aspartic acid and Cbz-glutamic acid was analyzed. (Cbz = benzyloxycarbonyl.)

We hypothesized that a polymer prepared using L-phenylalanine anilide as the print molecule may be able to separate also the enantiomers of other amide derivatives of amino acids, since the spatial positioning of functional groups within the polymer was defined (O'Shannessy et al., 1989c). We subsequently showed that efficient enantiomeric resolution of amide derivatives of amino acids, which included the anilides, p-nitroanilides, β-naphthylamides, and amides of various amino acids, was possible on the same polymer imprinted with L-phenylalanine anilide. In this context, it must be stressed that, whereas there is a certain freedom in the bulkiness of the various aliphatic and aromatic side groups, the substitution and positioning of the groups participating in the binding to the polymeric recognition sites are of extreme importance for enantiomeric separation. This concept may prove valuable in situations where the compound(s) needed to be separated is either difficult to prepare, available only in small quantities, or only available as a racemic mixture. In such situations, the design of a polymer with a readily available print molecule of similar structure could alleviate this limitation.

    In conclusion, the polymer preparations presented here were all shown to effect efficient separation of the enantiomers of the print molecule (enantioselectivity) in the HPLC mode (Fig. 2). The relatively high separation factors achieved are promising for future developments of polymers with practical applications in column chromatography. The ability of some polymers to resolve the enantiomers of compounds of similar structure to the print molecule may, in some instances, greatly simplify the design of molecularly imprinted CSPs. Furthermore, polymers, in particular those prepared utilizing only nonionic interactions, were also shown to separate the print molecule from other compounds very similar in structure (substrate selectivity) (Fig. 2). To summarize, molecular imprinting is an interesting and simple technique for the preparation of molecularly imprinted CSPs. The technique has now been developed to a stage where molecularly imprinted CSPs can potentially be prepared against a large number of compounds.

3. IMPRINTING OF COENZYMES

Imprints were made of NAD and two bis-NAD derivatives which differed in spacer length between the NAD moieties (Norrlöw *et al.*, 1987). The imprinting was accomplished by polymerizing a mixture of silanes containing various functionalities (e.g., amino, hydroxy, dodecyl, phenyl, and boronic) around the print molecule to form a thin layer on silica particles ("surface imprinting"). A correlation between the chromatographic elution times for NAD and the NAD derivatives and the respective print molecules was demonstrated. The retention time was always highest for the compound used as the print molecule. The imprinting of pyridoxal derivatives has also been described (see Section 7).

4. IMPRINTING OF METAL IONS

Molecular imprints against calcium ions were prepared in divinylbenzene-based polymers (Rosatzin *et al.*, 1991). A vinylic Ca^{2+}-selective, neutral ionophore, N,N'-dimethyl-N,N'-bis(4-vinylphenyl)-3-oxapentanediamide, was synthesized and used as the ion-complexing monomer (Fig. 3). 3-Oxapentanediamide derivatives have been used widely as electrically neutral carriers in calcium-ion-selective electrodes (ISE) (Ammann *et al.*, 1983). By analogy to other 3-oxapentanediamide derivatives, the ionophore is expected to form complexes with calcium ions with a 3:1 molar ratio. Therefore, the metal ion, added to the polymerization mixture in chloroform, was expected to act as a template for the ionophore during the polymerization (Fig. 3). The resulting polymers were analyzed for their ability to extract calcium ions from methanolic water. The polymers prepared against calcium ions were found to bind calcium ions with a K_{diss} value six times lower than that for reference polymers prepared in the absence of metal ions. We attribute the significant selectivity of our calcium-binding model system (for instance, binding of Mg^{2+} could not be detected) to the fact that ligand preorganization around the metal ion, followed by polymerization, results in recognition sites in which the coordinating oxygen atoms of the ionophoric residues are comfortably oriented (Fig. 3). In addition, the imprinting efficiency of this system is extremely good since the number of binding sites for calcium ions is very close (80%) to the theoretical value, taking into consideration the incorporation of the ionophore and the stoichiometry of complexation by calcium ions.

The results described further underline the great potential of molecular imprinting.

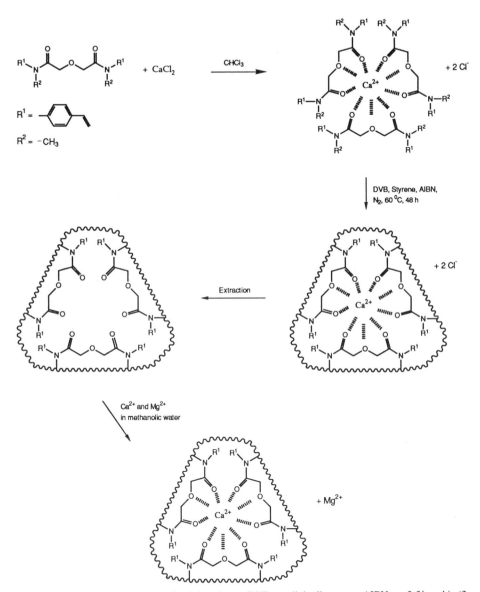

FIGURE 3. Molecular imprinting of calcium ions. (DVB = divinylbenzene; AIBN = 2,2'-azobis-(2-methylpropionitrile).)

Metal-imprinted polymers may find applications in metal-ion-selective sensors and, possibly, in the development of enzyme-like catalysts, since a metal ion often is a constituent of the active site of enzymes.

5. IMPRINTING OF PEPTIDES AND PROTEINS

As discussed above, the imprinting of, in particular, amino acid derivatives and low-molecular-weight compounds has been extensively studied by us using the noncovalent approach. The development of polymers possessing selectivity for peptides and proteins with potential as highly useful affinity phases for the isolation of high-value peptides and proteins will be the next step.

Using the noncovalent imprinting system described above (Section 2), involving the mixing of the print molecule with the functional monomer methacrylic acid and ethylene glycol dimethacrylate as cross-linker, imprints have been made against the dipeptide derivative L-phenylalanylglycine anilide (Andersson et al., 1990b). The resulting polymers were able to successfully resolve (separation factor 5.1) the racemic mixture D,L-phenylalanylglycine anilide.

In a preliminary study, following the approach described for NAD imprinting, imprints against proteins have been described for the first time (Glad et al., 1985). Employing a mixture of silanes, similar to the system described in connection with the imprinting of NAD coenzymes, these authors obtained polymers showing selectivity for the glycoprotein transferrin as compared to blank polymers without or with BSA as the print molecule. In particular, a specially synthesized boronate silane was added to interact with the glyco-moiety of the protein, which probably accounted for most of the recognition observed.

6. SUBSTRATE-SELECTIVE SENSORS

The potential of using molecularly imprinted polymers in analytical applications has been examined. The use of a flow-through column electrode for the separation and detection of small molecules in organic solvents has been described (Andersson et al., 1990a). Streaming-potential measurements are based on recording the potential across a packed bed of, for example, a chromatographic material placed in a continuous flow. Measurements of streaming potential can be regarded as a general method for the detection of binding reactions, as long as they involve a change in charge distribution on the surface at which they occur. The electrode in question consisted of a glass column in which the polymer, molecularly imprinted against L-phenylalanine anilide, was packed and where the end frits constituted the electrodes. The column resolved the enantiomers of phenylalanine anilide, and the recorded potentiometric signals could be correlated with the concentration of D- and L-phenylalanine anilide, respectively, in the sample. The results obtained clearly show the utility of potentiometric detection in organic solvents, as applied to separations on molecularly imprinted polymers. Moreover, such polymers may be useful in the separation and quantitative analysis of amino acid derivatives or other organic compounds.

The optical surface method ellipsometry has been applied to the detection of guest-selective recognition of the affinity binding of vitamin K_1 to an octadecylsilane-derivatized silicon surface (Andersson et al., 1988).

Recently, molecularly imprinted polymers have been utilized in conjunction with field effect electronic devices (Hedborg *et al.*, 1993). A sandwich polymerization procedure was developed for the fabrication of thin polymer membranes covalently attached to the silicon oxide surface capacitor structure. A sensor containing a phenylalanine anilide selective polymer measured this compound and could distinguish it from phenylalaninol.

7. SYNTHETIC ENZYME MODELS

In order to mimic enzymatic activity, we have prepared polymers possessing enzyme-like properties by the molecular imprinting technique. So far, both esterolytic and pyridoxal-dependent activities have been the focus of our studies.

Catalytic antibodies (catMABs) expressing enzymatic activity can be raised by employing transition-state analogs as haptens (Tramontano *et al.*, 1986; Pollack *et al.*, 1986). Analogously, molecular imprints were made against a transition-state analog, *p*-nitrophenyl methylphosphonate [see Fig. 4A (ii)], in poly[4(5)-vinylimidazole] by cross-linking with 1,4-dibromobutane (Robinson and Mosbach, 1989). The resulting polymer showed not only recognition for the original print molecule, but also an enhanced rate of hydrolysis of *p*-nitrophenyl acetate [Fig. 4A (i)]. The hydrolytic reaction was inhibited by the addition of the original print molecule, the transition state analog *p*-nitrophenyl methylphosphonate, to the reaction mixture.

The positioning of functional groups participating in the catalytic steps within the "active site" rather than mere stabilization of transition states is a further refinement of imprinted enzyme mimics. Employing suitable substrate analogs, this approach has been tried in the preparation of polymers possessing esterolytic activity (Leonhardt and Mosbach, 1987). These "esterase" polymers were prepared using 2-picolylamides of N-protected amino acids as print molecules [Fig. 4B (ii)]. Radical-induced copolymerization of 4(5)-vinylimidazole and divinylbenzene in the presence of metal ions (Co^{2+}) was performed. A complex is formed between the cobalt ions and vinylimidazole and the picolyl-amide derivatives. After polymerization and removal of the print molecule and the metal ion, the resulting polymers were allowed to react with the *p*-nitrophenyl ester of the N-protected amino acids [Fig. 4B (i)]. A clear preference for the hydrolysis of the *p*-nitrophenyl ester substrate corresponding to the original print molecule on the respective polymer was observed. In addition, some turnover was observed.

N-Pyridoxyl-L-phenylalanine anilide [Fig. 4C (ii)], a stable analog of the Schiff's base between pyridoxal and phenylalanine anilide, was synthesized (Andersson and Mosbach, 1989). The molecular imprinting of this coenzyme-substrate analog in methacrylate-based polymers resulted in the formation of recognition sites which could successfully separate the enantiomers of the coenzyme analog. The pyridoxal-catalyzed exchange of the α-proton on phenylalanine anilide [Fig. 4C (i)], which represents the first step in a number of enzymatic reactions involving pyridoxal, was studied. The presence of imprinted polymers in the reaction mixture resulted in a threefold increase in exchange rate compared with that observed for reference polymers, prepared with phenylalanine anilide as print molecule (or by substituting methyl methacrylate for methacrylic acid). The imprinting of substrate–coenzyme complexes leads to introduction of an additional "functional" group, that of the coenzyme, thereby increasing the number of catalytically active functionalities available.

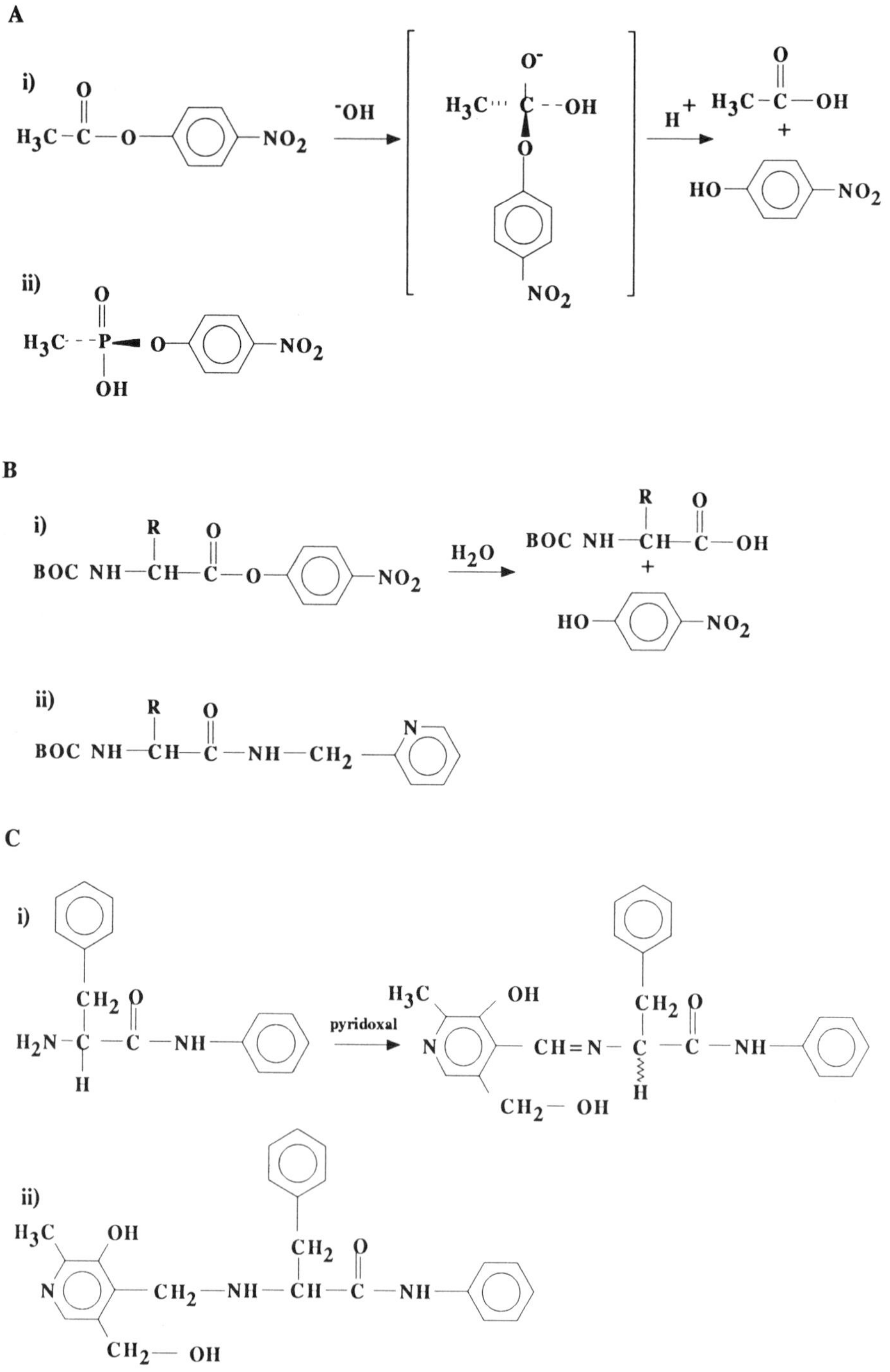

FIGURE 4. Schematics of the reactions of synthetic enzyme models, with the corresponding print molecules. See text for details. (BOC = *tert*-butyloxycarbonyl.)

8. CONCLUDING REMARKS

Molecularly imprinted polymers are potentially useful for the production of specialty separation media, in particular for applications in the resolution of chiral compounds.

The demonstration that molecularly imprinted polymers can be employed in analytical devices should spark interest in the use of these polymers in substrate-selective electrodes and biosensors. The advantages gained by using molecularly imprinted polymers instead of biomacromolecules are increased mechanical strength and increased temperature stability. Furthermore, a suitable biomacromolecule may not always be available.

The preparation of polymers with enzymelike properties may be, in the long run, a fruitful application of molecularly imprinted polymers. The use of catalytically active polymers has several advantages over the use of enzymes. First, the polymers are stable at high temperatures, withstand mechanical stress, and are compatible with most of the commonly used solvents. Second, there is the potential to prepare catalysts with a high number of active sites per unit weight, since the print molecule constitutes a significant part of the polymerization mixture. Finally, the possibility of tailor-making a catalyst for a specific reaction, for which it may be difficult to find a suitable "natural" enzyme, must not be overlooked.

In a recent publication, the use of imprinted polymers as antibody mimics have been described (Vlatakis *et al.*, 1993) Molecular imprints made against theophylline and diazepam showed strong binding (K_D-values as low as 10^{-8} M) and cross-reactivity profiles similar to those of antibodies. Such polymers could be used in a new radiolabelled ligand-binding assay, which measures drug levels in human serum with results comparable to those obtained using a well-established immunoassay technique. Thus, in the near future, we expect very promising developments in this particular area.

REFERENCES

Ammann, D., Morf, W. E., Anker, P., Meier, P. C., Pretsch, E., and Simon, W., 1983, Neutral carrier based ion-selective electrodes, *Ion Selective Electrode Rev.* **5:**3–92.

Andersson, L. I., and Mosbach, K., 1989, Molecular imprinting of the coenzyme-substrate analogue *N*-pyridoxyl-L-phenylalanine anilide, *Makromol. Chem. Rapid Commun.* **10:**491–495.

Andersson, L. I., and Mosbach, K., 1990, Enantiomeric resolution on molecularly imprinted polymers prepared with only non-covalent and non-ionic interactions, *J. Chromatogr.* **516:**313–322.

Andersson, L. I., Mandenius, C. F., and Mosbach, K., 1988, Studies on guest selective molecular recognition on an octadecyl silylated silicon surface using ellipsometry, *Tetrahedron Lett.* **1988:**5437–5440.

Andersson, L. I., Miyabayashi, A., O'Shannessy, D. J., and Mosbach, K., 1990a, Enantiomeric resolution of amino acid derivatives on molecularly imprinted polymers as monitored by potentiomeric measurements, *J. Chromatogr.* **516:**323–331.

Andersson, L. I., O'Shannessy, D. J., and Mosbach, K., 1990b, Molecular recognition in synthetic polymers. Preparation of chiral stationary phases by molecular imprinting of amino acid amides, *J. Chromatogr.* **513:**167–179.

Arshady, R., and Mosbach, K., 1981, Synthesis of substrate-selective polymers by host–guest polymerization, *Makromol. Chem.* **182:**687–692.

Ekberg, B., and Mosbach, K., 1989, Molecular imprinting: A technique for producing specific separation materials, *Trends Biotechnol.* **7:**92–96.

Fischer, L., Müller, R., Ekberg, B., and Mosbach, K., 1991, Direct enantioseparation of β-adrenergic blockers using a chiral stationary phase by molecular imprinting, *J. Am. Chem. Soc.* **113:**9358–9360.

Glad, M., Norrlöw, O., Sellergren, B., Siegbahn, N., and Mosbach, K., 1985, Use of silane monomers for molecularly imprinting and enzyme entrapment in poly-siloxane-coated porous silica, *J. Chromatogr.* **347:**11–23.

Hedborg, E., Andersson, L. I., Winquist, F., Lundström, I., and Mosbach, K., 1993, Some studies of molecularly imprinted polymer membranes in combination with field effect devices, *Sensors and Actuators A*, in press.

Leonhardt, A., and Mosbach, K., 1987, Enzyme-mimicking polymers exhibiting specific substrate binding and catalytic functions, *React. Polym.* **6:**285–290.

Norrlöw, O., Månsson, M.-O., and Mosbach, K., 1987, Improved chromatography: Prearranged distances between boronate groups by the molecular imprinting approach, *J. Chromatogr.* **396:**374–377.

O'Shannessy, D. J., Ekberg, B., and Mosbach, K., 1989a, Molecular imprinting of amino acid derivatives at low temperature (0°C) using photolytic homolysis of azobisnitriles, *Anal. Biochem.* **177:**144–149.

O'Shannessy, D. J., Ekberg, B., Andersson, L. I., and Mosbach, K., 1989b, Recent advances in the preparation and use of molecularly imprinted polymers for enantiomeric resolution of amino acid derivatives, *J. Chromatogr.* **470:**391–399.

O'Shannessy, D. J., Andersson, L. I., and Mosbach, K., 1989c, Moleular recognition in synthetic polymers. Enantiomeric resolution of amide derivatives of amino acids on molecularly imprinted polymers, *J. Mol. Recog.* **2:**1–5.

Pollack, S. J., Jacobs, J. W., and Schultz, P. C., 1986, Selective chemical catalysis by an antibody, *Science* **234:** 1570–1573.

Robinson, D. K., and Mosbach, K., 1989, Molecular imprinting of a transition state analogue leads to a polymer exhibiting esterolytic activity, *J. Chem. Soc., Chem. Commun.* **1989:**969–970.

Rosatzin, T., Andersson, L. I., Simon, W., and Mosbach, K., 1991, Preparation of Ca^{2+}-selective sorbents by molecular imprinting using polymerisable ionophores, *J. Chem. Soc. Perkin Trans. 2* **1991:**1261–1265.

Tramontano, A., Janda, K. D., and Lerner, R. A., 1986, Catalytic antibodies, *Science* **234:**1566–1570.

Vlatakis, G., Andersson, L. I., Müller, R., and Mosbach, K., 1993, Drug assay using antibody mimics made by molecular imprinting, *Nature* **361:**645–647.

Wulff, G., 1986, Molecular recognition in polymers prepared by imprinting with templates, in: Polymeric reagents and catalysts, ACS Symposium Series 308 (Ford, W. T., ed.), American Chemical Society, Washington, D.C., pp. 186–230.

25

Use of Heterobifunctional Ligands in Affinity Chromatographic Processes

Bo Mattiasson, Eva Linné, and Rajni Kaul

Affinity interactions in free solution occur without diffusional limitations and can be carried out in a large volume within a short time period. On the other hand, affinity binding in a heterogeneous phase, as in affinity chromatography, is limited due to exclusion effects and diffusional constraints of the solid supports employed. Diffusional restrictions have been shown to be reduced by decreasing the particle size. As purifications based on solid-phase affinity interactions are usually performed on a column, fairly clear samples need to be applied in small amounts.

The present chapter summarizes some efforts in the field of affinity interactions in free solution using heterobifunctional ligands in combination with chromatographic separation. The basic idea behind the use of heterobifunctional ligands is to have a soluble entity that can form affinity complexes through one of its functionalities, whereas the other is used to isolate the complex. Different kinds of molecules may be used as the second functionality. In Table 1 are listed some properties of the affinity complex that may be exploited in the separation process and also some general separation techniques that can be applied for its isolation and recovery.

1. FORMATION OF A HETEROBIFUNCTIONAL LIGAND

In most cases the heterobifunctional ligands have to be synthesized. The normal strategy is the covalent coupling of the ligand molecule to the structure representing the second functionality. A whole range of methods developed for solid-phase coupling may be made use of. Care has to be taken to avoid cross-linking during the coupling step as it may cause aggregation into insoluble aggregates, thereby reducing the yield of properly modified ligand. Even if coupling chemistry *per se* is well developed, more efforts are needed for the optimization of synthesis of soluble complexes.

There are two main routes followed so far for synthesis of heterobifunctional ligands. The first is activation of one of the components, with subsequent coupling to the other unit.

Bo Mattiasson, Eva Linné, and Rajni Kaul • Department of Biotechnology, Chemical Center, Lund University, S-221 00 Lund, Sweden.

Molecular Interactions in Bioseparations, edited by That T. Ngo. Plenum Press, New York, 1993.

TABLE 1
Properties of Affinity Complexes That Are Exploited
in Various Separation Methods

Property	Separation method
Size	Gel filtration
	Membrane filtration
Charge	Ion-exchange chromatography
Solubility	Precipitation/centrifugation
	Flotation
Surface properties	Partition in aqueous two-phase systems
	Extraction in microemulsions

The other approach has been to modify the ligand by covalent coupling of monomers that, in a subsequent polymerization step, are built into larger structures. This latter method has mainly been exploited in systems using acrylic polymers.

Gene technology offers interesting possibilities when it comes to the use of proteins as ligands. Fusion of affinity tails and formation of bifunctional proteins are both realistic today and are used to facilitate the purification of the proteins from recombinant organisms by affinity interactions between the affinity tail and an immobilized ligand. The only limitation in this latter effort is the need for cloning and gene technology work. For isolation of compounds from natural sources, it may be more realistic at present to utilize the synthetic approach.

2. MODE OF OPERATION

As shown in Table 1, several modes of separation may be applied for harvesting affinity complexes formed using heterobifunctional ligands. There are today a rapidly increasing number of applications in which a heterobifunctional ligand is involved in the primary interaction and the separation is subsequently based on the other functionality. Some of these applications are not necessarily performed in free solution. The heterobifunctionality of the ligands introduces new levels of freedom to systems using affinity interactions. Table 2 summarizes some systems studied. Most of the applications listed in the table are discussed in other chapters in this volume. This chapter will focus on the use of heterobifunctional ligands in chromatography.

There are two general modes of operating such systems. In the first approach, the heterobifunctional ligand is added to the homogenate to be fractionated. After the affinity complex is formed (in most cases this happens very quickly), the mixture is passed over a column with binding ability for the second component of the heterobifunctional molecule. From this column-bound heterobifunctional affinity complex, the target molecule is eluted, followed by a rinsing step in which the heterobifunctional ligand is recovered from the column. The second approach is to first add the heterobifunctional ligand to the column, thereby forming an affinity sorbent with the affinity ligand reversibly immobilized. When the column is established, conventional affinity chromatographic procedures may be carried

TABLE 2

Examples of Systems Based on the Use of Heterobifunctional Ligands[a]

Heterobifunctional ligand	Primary interaction	Secondary interaction	Target molecule	Application	Reference(s)
Serum albumin	Binding of anti-albumin antibodies	Binding between albumin and hydrophobic tails of the solid support	Antialbumin IgG	Sterilizable affinity column	Unpublished data
Octylglucoside	Con A-binding glucose residues	Octyl residue dissolved in organic phase in microemulsion	Con A	Efficient extraction using microemulsion	Hatton, 1987
Dinonylphenylethoxylate-STI	Trypsin binding to STI	Dinonylphenyl group binding to hydrophobic matrix	Trypsin	General hydrophobic affinity chromatography	Kaul et al., 1988
STI-dextran	Con A binding to dextran	Trypsin binding to STI-dextran	Con A	A general affinity chromatographic procedure	Mattiasson and Olsson, 1986; Olsson and Mattiasson, 1986
PEG-Cibacron blue	Lactate dehydrogenase binding to Cibacron blue	PEG partitioning to top phase in aqueous two-phase system	Lactate dehydrogenase	Affinity-facilitated extraction	Tjerneld et al., 1987
Colloidal silica–Cibacron blue	Alcohol dehydrogenase binding to Cibacron blue	Membrane separation of silica complexes from soluble proteins	Alcohol dehydrogenase	Membrane affinity purification	Ling and Mattiasson, 1989
STI-alginate	Trypsin binding to STI	Precipitation of alginate by means of Ca^{2+} ions	Trypsin	Affinity precipitation	Linné et al., 1992

[a]Abbreviations: STI, soybean trypsin inhibitor; PEG, poly(ethylene glycol); Con A, concanavalin A.

out. When the ligand denatures or whenever necessary, the column may be rinsed and recharged with a new ligand of the same or different specificity. Generally, it is desirable to utilize the same ligand.

3. APPLICATIONS

3.1. Use of Reversibly Immobilized Affinity Ligands to Design a Sterilizable Affinity Column

A crucial factor when dealing with affinity chromatographic beds is the operational lifetime of the column. Infections, proteases, and heavy-metal ions are three sources of problems when proteinaceous ligands are used. If a traditional affinity support gets infected, it is almost impossible to sterilize the column without losing the biologically active material. There may be exceptions in the case of extremely robust proteins.

Proteases and heavy metals may cause denaturation of the ligands, thereby ruining the column. In many cases it is not even possible to recover the solid support after the ligand has been destroyed.

A model study was set up to investigate the potential of sterilizable affinity columns (Linné and Mattiasson, unpublished results). As a first choice, a naturally occurring bifunctional structure, namely, human serum albumin, was used. Octyl-S-Sepharose, a hydrophobic gel that withstands sanitation, was used as the support material (Maisano et al., 1985). Defatted albumin is known to bind the octyl chains firmly. The reversibly bound albumin was utilized to isolate antialbumin immunoglobulin G (IgG). In order to hasten "infection" of the column, either bacteria or proteases were infused into the column. Rinsing of the column with 50% ethylene glycol helped to remove the albumin. The column material was then soaked in $0.2M$ NaOH according to the recommendations of the manufacturer. The albumin could be subjected to some separation steps, usually ultrafiltration, to remove cellular material. Proteases had to be removed through molecular sieving or ion-exchange chromatography. In this study, the albumin was bound and removed periodically in order to demonstrate the possibility of utilizing the concept of reversible immobilization of the ligand in combination with separate sterilization procedures for the ligand and the solid support.

Figure 1 shows the elution profiles for antiserum passing through the column with freshly added albumin. The first peak represents unretarded protein, and the peak after elution is the protein specifically retained on the affinity column. Elution profile I is prior to infection, II after infection, and III after stripping, sterilization, and reconditioning. There are some interactions between remaining hydrophobic groups on the support and some serum proteins. These may be minimized by blocking of the groups on the matrix by some inert protein. Anyhow, in the elution step, IgG is liberated in an almost symmetrical peak. The yield for elution step was 65%. The same procedure was repeated many times. "Infection" was introduced intermittently, and reconditioning was carried out according to the procedure described above. The repeated purification process showed a very similar elution pattern to that of the first one, whereas the system with albumin covalently bound to the support had lost its ability to a large extent.

The system utilized in this study has some limitations: we do not have control over possible IgG binding to denatured albumin, but the study anyway clearly demonstrates the

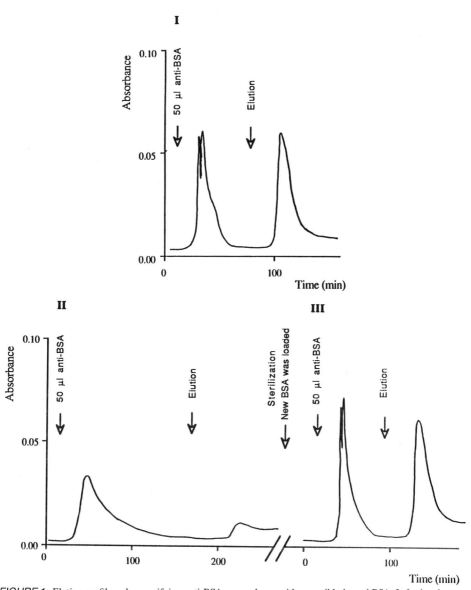

FIGURE 1. Elution profiles when purifying anti-BSA on a column with reversibly bound BSA: I, fresh column; II, after infection; III, after stripping, sterilization, and recharging the column. The first peak in the elution profile represents proteins with no affinity for the ligand, and the peak after elution represents the biospecifically retarded material, here anti-BSA antibodies.

possibility of utilizing reversibly bound affinity ligands in order to allow the sanitation of the column.

3.2.   Affinity Chromatographic Purification of Concanavalin A by Utilizing Trypsin–Dextran as a Heterobifunctional Ligand and Soybean Trypsin Inhibitor (STI)-Sepharose as Solid Support

To a particulate-free homogenate of *Canavalia ensiformis* seeds was added the heterobifunctional ligand trypsin–dextran with the idea that concanavalin A (Con A) in the homogenate would interact with the dextran part of the heterobifunctional ligand. The mixture was then passed over a soybean trypsin inhibitor (STI) column prior to displacement of Con A by a specific ligand, glucose (Olsson and Mattiasson, 1986). The yield in this process turned out to be very low, about 10–13%. A calculation based on the known affinity interactions for free reagents was carried out. It was assumed that the trypsin–STI interaction remains far stronger than the Con A–dextran interaction, so that it would be only a minor approximation to base the calculation of yield just on the Con A–dextran interaction. The calculations were carried out using two different estimations of the binding constant between chemically modified Con A and dextran. From these studies a good correlation between calulated and experimental yields was obtained for an association constant of 6×10^3. In a separate series of experiments using frontal analysis affinity chromatography, a value for the association constant of 6×10^3 was obtained, which supported the earlier observations and calculations.

3.3.   Use of Detergent-Modified Ligands in General Hydrophobic Affinity Chromatography

Dinonylphenylethoxylate epoxide (DNF 100), a generous gift from Dr. Krister Holmberg, Berol Nobel AB, Stenungsund, Sweden, was utilized for binding the affinity ligand, STI, via the epoxide group. By this chemical modification, the ligand becomes equipped with a hydrophobic handle that will make it utilizable in hydrophobic affinity chromatography (Kaul *et al.*, 1988) or even for extraction in microemulsions (Hatton, 1987). The target molecule in this study was trypsin from beef pancreas. A particulate-free homogenate was used. The heterobifunctional ligand was charged onto the hydrophobic column (Octyl-Sepharose), and, after proper equilibration with the running buffer, the column was ready for use. The sample was applied, and proteins not interacting with the affinity support were washed off the column. Trypsin was eluted by a pulse of $0.05M$ glycine–HCl buffer, pH 3.0. The ligand–detergent complex was removed from the column by washing with a 1% Tween 20 solution. The column was subsequently rinsed with 50% ethanol and eventually equilibrated with running buffer. The harvested ligand–detergent complex was separated from the Tween by a dialysis step. The STI–detergent complex was then ready for application on the column again. In Table 3 are shown some results from repeated runs using the same column that had been rinsed and recharged with the ligand.

4.   CONCLUSION

From the studies reported above, it is obvious that heterobifunctional ligands offer many new potentials in downstream processing. Besides the more straightforward applica-

TABLE 3
Capacity of Octyl-Sepharose DNF–STI Column
for Binding Trypsin[a]

Column	DNF:STI ratio	Loading step (mg)	Washing step (%)	Elution step (%)
1	5	20.0	1.4	78
		15.0	2.0	87
2[b]	10	15.0	ND[c]	90
		20.0	0	104

[a]From Kaul et al., 1988, with permission.
[b]In column 2, the DNF–STI complex was eluted out and reloaded prior to the enzyme purification step.
[c]ND, Not determined.

tions, such as affinity extraction and affinity precipitation, it is also possible to utilize these ligands in conventional systems, thereby introducing new degrees of freedom.

ACKNOWLEDGMENTS

This project was supported by The National Swedish Board for Technical Development (STU), The Forest and Agricultural Research Council (SJFR), and The Swedish Agency for Research Cooperation with Developing Countries (SAREC).

REFERENCES

Hatton, T. A., 1987, Extraction of proteins and amino acids using reversed micelles, in: *Ordered Media in Chemical Separations* (W. L. Hinze and D. W. Armstrong, eds.), ACS Symposium Series 342, American Chemical Society, Washington, D.C., pp. 170–183.

Kaul, R., Olsson, U., and Mattiasson, B., 1988, Affinity chromatography on a hydrophobic matrix using a heterobifunctional ligand, *J. Chromatogr.* **438:**339–346.

Ling, T. G. I., and Mattiasson, B., 1989, Membrane filtration affinity purification (MFAP). A continuous affinity purification method, *Biotechnol. Bioeng.* **34:**1321–1325.

Linné, E., Kaul, R., Garg, N., and Mattiasson, B., 1992, Evaluation of alginate as a ligand carrier in affinity precipitation procedures, *Biotechnol. Appl. Biochem.* **16:**48–56.

Maisano, F., Belew, M., and Porath, J., 1985, Synthesis of new hydrophobic adsorbents based on homologous series of uncharged alkyl sulphide agarose derivatives, *J. Chromatogr.* **321:**305–317.

Mattiasson, B., and Olsson, U., 1986, General chromatographic purification procedure based on the use of heterobifunctional affinity ligand, *J. Chromatogr.* **370:**21–28.

Olsson, U., and Mattiasson, B., 1986, Theoretical and experimental evaluation of the use of heterobifunctional affinity ligands in general chromatographic purification systems, *J. Chromatogr.* **370:**29–37.

Tjerneld, F., Johansson, F., and Joelsson, M., 1987, Affinity liquid–liquid extraction of lactate dehydrogenase on a large scale, *Biotechnol. Bioeng.* **30:**809–816.

26

Covalent Chromatography

S. Oscarsson and J. Porath

1. INTRODUCTION

Covalent chromatography is a method for separation of molecules, based on formation of reversible covalent bonds between functional groups in molecules and complementary structures on a stationary solid phase. Covalent chromatography thus involves a synthetic step by which a solute is covalently immobilized to a solid support—the chemisorbent—later followed by chemical cleavage and regeneration of the sorbent. Only a few methods have been explored so far, and we will here deal only with chromatography of proteins and peptides.

One of the first methods based on reversible covalent chromatography was presented as early as 1963 by Eldjarn and Jellum (1963), who were successful in isolating thiol-containing proteins such as hemoglobins, human albumin, urease, and glyceraldehyde phosphate dehydrogenase on organomercurial-derivatized Sephadex particles. This original work involved an immobilized metal ion on the chemisorbent. When thiol groups are present in the protein, they interact with the immobilized ion with the formation of covalent bonds (Fig. 1).

Covalent chromatography apparently may in its ideal form be used for selectively "fishing out" from a complex protein mixture only those proteins that can form covalent bonds with the ligands in the adsorbent. Ideally, this is a method directed toward a single species of surface-located side groups—in the Eldjarn–Jellum case, toward thiol groups. A method for methionine-containing proteins and peptides has been described by Shechter *et al.* (1977). The adsorbent contains reactive halogens such as —NH—CO—CH$_2$Cl. A sulfonium complex is formed, which in the desorption step can be cleaved by treatment with mercaptoethanol or, alternatively, by heating the solution at 110°C for 4 h (see Fig. 2). The former method may possibly be converted to a useful chromatographic procedure, but further developmental work is necessary, since the original adsorbent is not regenerated.

A method that, in principle at least, approaches ideal covalent chromatography is based on thiol–disulfide exchange. The method was introduced in 1973 by Brocklehurst and co-workers. In their version the adsorbent consists of pyridine-2-disulfide (PyS$_2$) coupled

S. Oscarsson • Biochemical Separation Centre, Uppsala University, S-751 23 Uppsala, Sweden.   *J. Porath* • Department of Biochemistry and the Division of Biotechnology, University of Arizona, Tucson, Arizona 85721.

Molecular Interactions in Bioseparations, edited by That T. Ngo. Plenum Press, New York, 1993.

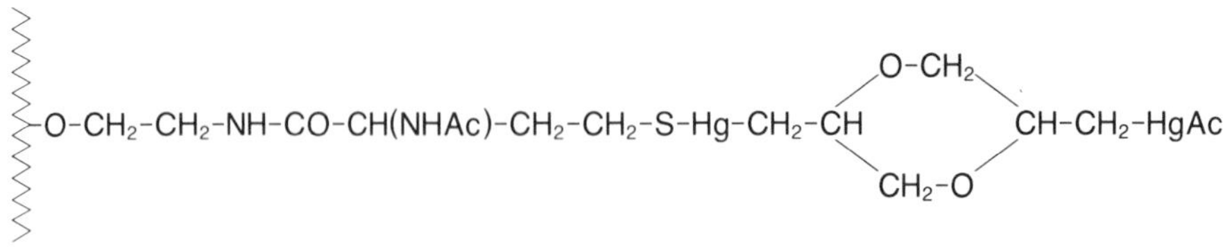

FIGURE 1. Schematic structure of the organomercurial polysaccharide matrix based on Sephadex that was introduced by Eldjarn and Jellum (1963).

to a solid support such as agarose that is derivatized with glutathione coupled via cyanogen bromide-activated agarose (Brocklehurst *et al.*, 1973). The thiol–disulfide exchange reaction between the disulfide on the agarose and the thiol groups in the proteins is favored by the formation of the highly resonance-stabilized thiopyridone during the reaction (see Fig. 3a). Pyridine-2-disulfide-modified polymer (Fig. 3b), a variant of Brocklehurst's adsorbent, was introduced by Axén *et al.* (1975), and alternative synthetic routes for preparation of the adsorbent were developed by Ngo (1986).

Certain forms of immobilized metal ion affinity chromatography (IMAC), where coordination is the main or only kind of interaction, may also be classified under the covalent chromatography concept (Porath and Olin, 1982). Recently developed alternative methods for covalent chromatography that have been presented by Carlsson and Batista-Viera (1991) are based on the use of disulfide oxides. Figure 4 shows the proposed structure of disulfide oxide-agarose and its reactions with thiol-containing molecules of low or high molecular weight. The original adsorbent can be regenerated, but the technique needs to be improved.

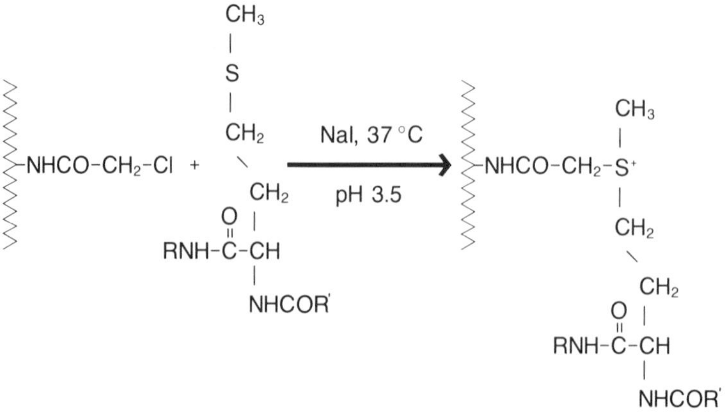

FIGURE 2. Schematic structure of chloroacetamidoethylpolyacrylamide and its reactivity with methionine-containing residues [according to Shechter *et al.* (1977)].

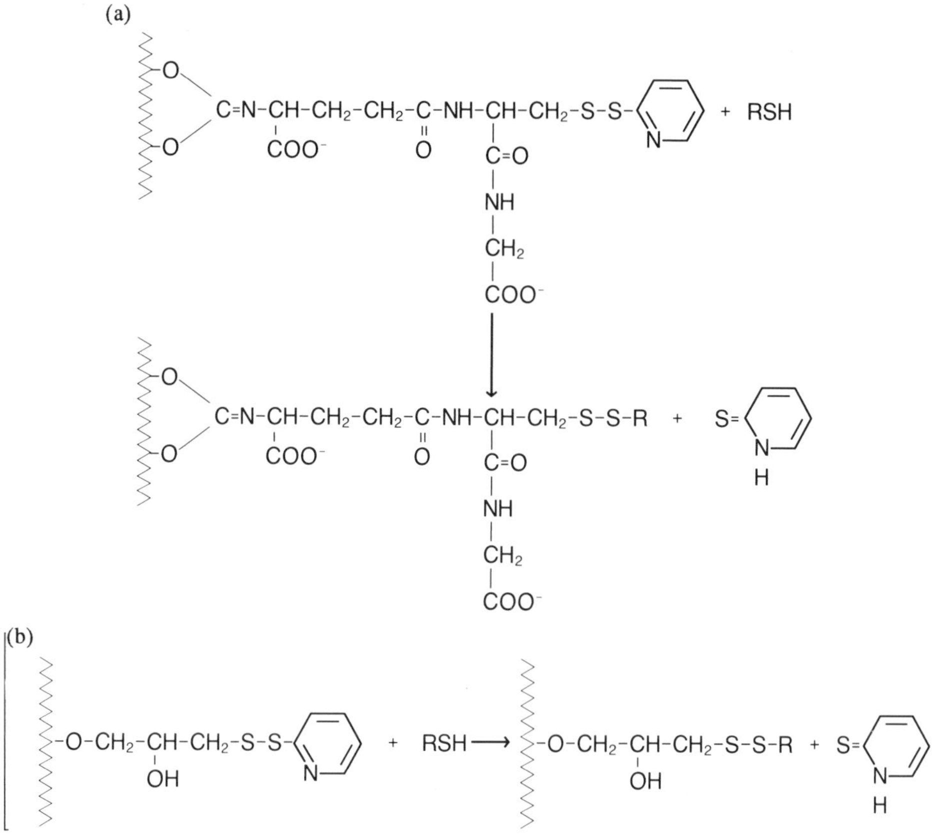

(a)

(b)

FIGURE 3. Schematic structures of glutathione-2-pyridyldisulfide agarose (a) [according to Brocklehurst *et al.* (1973)] and 3-(2-pyridyldisulfido)-2-hydroxypropylagarose (b), originally introduced by Axén *et al.* (1975), and their reactivities with thiol groups.

2. COVALENT CHROMATOGRAPHY ON PYRIDINE-2-DISULFIDE-DERIVATIZED AGAROSE

2.1. General Technique

The general conditions for covalent chromatography as well as the methods for preparation of the chromatographic support have been described by Brocklehurst *et al.* (1973) and Axén *et al.* (1975).

2.2. Salt Dependence

The covalent attachment of thiol proteins from serum to pyridine-2-disulfide-derivatized agarose was found to be very strongly salt-dependent (Oscarsson and Porath, 1989). In addition, a rather specific adsorption of immunoglobulins was found to occur,

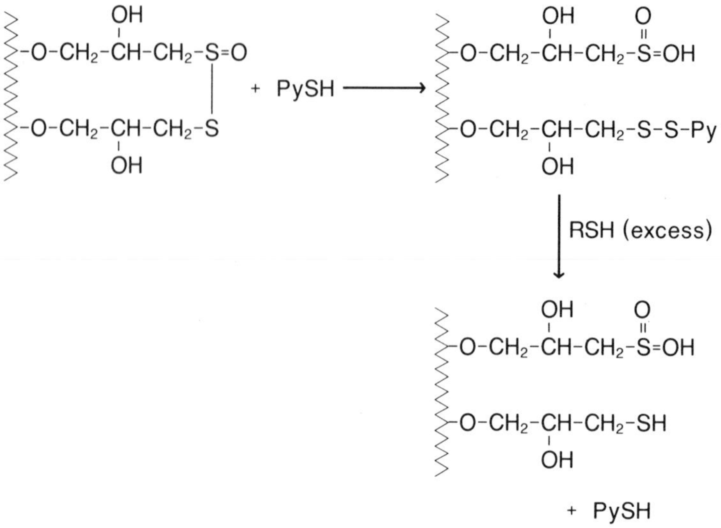

$$\begin{array}{c} \text{OH} \\ | \\ \text{>-O-CH}_2\text{-CH-CH}_2\text{-S=O} \\ | \\ \text{>-O-CH}_2\text{-CH-CH}_2\text{-S} \\ | \\ \text{OH} \end{array} \quad + \text{ PySH} \longrightarrow \begin{array}{c} \text{OH} \quad \text{O} \\ | \qquad \| \\ \text{>-O-CH}_2\text{-CH-CH}_2\text{-S=OH} \\ \\ \text{>-O-CH}_2\text{-CH-CH}_2\text{-S-S-Py} \\ | \\ \text{OH} \end{array}$$

$$\downarrow \text{ RSH (excess)}$$

$$\begin{array}{c} \text{OH} \quad \text{O} \\ | \qquad \| \\ \text{>-O-CH}_2\text{-CH-CH}_2\text{-S=OH} \\ \\ \text{>-O-CH}_2\text{-CH-CH}_2\text{-SH} \\ | \\ \text{OH} \end{array}$$

$$+ \text{ PySH}$$

FIGURE 4. Schematic structure of disulfide oxides on Sepharose and their reactivity with thiol-containing proteins or peptides according to Carlsson and Batista-Viera (1991).

which makes pyridine-2-disulfide-derivatized agarose even more versatile as a chromatographic adsorbent as such or in combination with the alternative thiophilic derivative of this adsorbent, 3-(2-pyridylthio)-2-hydroxypropyl agarose, described by Porath and Oscarsson (1988). Based on these observations, the original method has now been modified to make thiol–disulfide-based chromatography generally applicable for isolation of thiol-containing proteins.

The experimental situation for covalent chromatography becomes complicated if complex biological fluids, such as serum, are to be fractionated. The yield will be low in the absence of salt. With salt present, chemisorption increases, but there will be a concomitant increase in noncovalent adsorption of proteins (see Table 1).

TABLE 1

Protein Recoveries in Fractions Obtained after Application of Serum (0.5 ml)
on PyS$_2$ Gel (Bed Volume 4.6 ml) at Low and High Ionic Strengths

	Recovery (% of applied protein)	
Pooled fraction	0.1M Tris-HCl, pH 7.5	0.5M K$_2$SO$_4$ in 0.1M Tris, pH 7.5
Nonadsorbed protein	96.0	49.8
Protein desorbed with 0.1M Tris-HCl, pH 7.5	—	13.7
Protein desorbed with 40% ethylene glycol in 0.1M Tris-HCl, pH 7.5	2.85	1.4
Protein desorbed with 30% isopropanol in 0.1M Tris-HCl, pH 7.5	0	0.21
Protein desorbed with 0.1M Tris-HCl, pH 7.5, containing 5mM dithioerythritol	4.3	39.6

As shown in Table 1, with salt present, the covalently immobilized fraction of proteins increases from 4.3 to 39.6%. At the same time, the amount of proteins adsorbed and released from the adsorbent by deleting salt or by including ethylene glycol in the buffer increases from 2.85 to 15.1%.

2.3. Nature of the Proteins Adsorbed and Chemisorbed on Pyridine-2-disulfide-Derivatized Agarose

Qualitatively and quantitatively, adsorption of proteins strongly depends on the concentration of potassium sulfate used in the equilibrium buffer and present in the serum. In our studies, the 17 most common serum proteins were detected by the Ouchterlony technique. In buffered $0.5M$ potassium sulfate, IgA, IgM, IgG, C3 and C4, and α_2-macro-globulin were adsorbed and desorbed by salt exclusion.

The covalently attached proteins in the presence of $0.5M$ K_2SO_4 were albumin, prealbumin, IgG, IgA, C4, and α_2-macro-globulin as found by inclusion of reducing agents [such as dithiothreitol (DTT)] in the eluent. At a concentration of $0.65M$ potassium sulfate, the additional proteins adsorbed were albumin, ceruloplasmin, haptoglobin, transferrin, α_1-antitrypsin, and hemopexin. Thus, we found human serum albumin (HSA) to be adsorbed, although at an extremely low level of concentration: 40–50 μg of HSA out of a total of 25 mg was adsorbed.

After iodoacetamide treatment of these proteins in the presence of $0.5M$ K_2SO_4, all previously chemisorbed proteins except albumin, prealbumin, and α_1-antitrypsin became adsorbed to the PyS$_2$ gel in the presence of $0.5M$ K_2SO_4, which proves that among the major proteins in serum only these three can be covalently attached to the gel (see Table 2).

Other proteins may possibly contain thiol groups, but since they have thiophilic centra (Porath and Oscarsson, 1988), they become adsorbed both before and after iodoacetamide treatment. These proteins are desorbed after deletion of the salt from the buffer. Further proofs of covalent attachment of the proteins have been published (Oscarsson and Porath, 1989). Among the covalently attached proteins, the existence of a thiol group in α_1-antitryp-sin, IgA, and the κ-light chain of IgG has been shown (Laurell, 1979), and the existence of a thiol group in human serum albumin is well documented.

Chemisorption as a function of K_2SO_4 concentration in the buffer is illustrated in

TABLE 2
Results Obtained after Chromatography of Iodoacetamide-Treated Material
on PyS$_2$ in the Presence of $0.5M$ K_2SO_4[a]

Pooled fraction	Recovery[b] (%)	Components identified by Ouchterlony technique
Nonadsorbed protein	82.6	Albumin,[c] α_1-antitrypsin, prealbumin
Protein desorbed with $0.1M$ Tris-HCl, pH 7.5	19	IgA, IgG, C4, α_2-macroglobulin, albumin

[a]The total applied volume of iodoacetamide-treated material on the PyS$_2$ column was 3.4 ml with a total A_{280} of 9.18 units. To obtain this material, serum was chromatographed on a PyS$_2$ gel in the presence of $0.5M$ K_2SO_4–$0.1M$ Tris-HCl, pH 7.5, followed by desorption with $0.1M$ Tris-HCl, pH 7.5, with 40% (v/v) ethylene glycol, and with 30% (v/v) isopropanol in $0.1M$ Tris-HCl, pH 7.5. The material remaining on the column was desorbed with dithioerythritol and, after dialysis, treated with iodoacetamide.
[b]The recovery is estimated from absorbance values at 280 nm.
[c]The amounts of albumin in the two pooled fractions were 7.6 mg in the nonadsorbed fraction and 0.11 mg in the desorbed fraction.

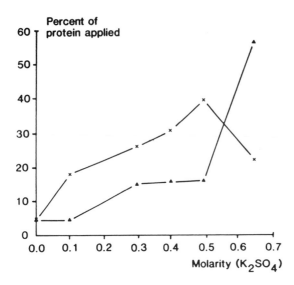

FIGURE 5. Adsorption and chemisorption of serum proteins on 3-(2-pyridyldisulfido)-2-hydroxypropylagarose as a function of the concentration of potassium sulfate in the buffer. Bed volume, 4.6–4.7 ml of gel; sample volume, 0.5 ml (containing 34 mg of protein); flow rate, 25 ml/cm² per hour; temperature, 25°C. ×, Percentage of total applied serum proteins that were covalently attached to 3-(2-pyridyldisulfido)-2-hydroxypropyl-agarose; ▲, percentage of total applied serum proteins that were noncovalently coupled to 3-(2-pyridyldisulfido)-2-hydroxypropyl-agarose. The plotted values were estimated from values obtained by amino acid analyses and absorbance measurements at 280 nm for the different fractions.

Fig. 5. Even a moderate potassium sulfate concentration increases the chemisorption, which reaches a maximum at about $0.5M$ K_2SO_4. The total amount of protein adsorbed due to covalent and noncovalent interaction increases drastically in the concentration range $0.5–0.65M$ K_2SO_4.

2.4. Kinetics

The study of chemisorption kinetics has been confined to HSA. Adsorption of mercaptalbumin, the SH-containing form of HSA, may be a special case or it may exemplify a more general aspect of importance for evaluation of covalent chromatography. In any case some comments are warranted.

Mercaptalbumin makes up about 60% of the serum HSA content, and it contains a single thiol group per molecule located in a 9.5-Å-deep pocket (Peters, 1984). Why is the adsorption facilitated by antichaotropic salts such as potassium sulfate? Among possible explanations we may consider the following:

a. The salting-out effect increases the solute concentration close to the phase boundary. The observed adsorption kinetic effect is explained by the mass-action law.
b. When brought in contact with a polymer matrix (or a ligand region in the matrix), a protein molecule may undergo changes in surface conformation, or it may be subjected to a wider range of conformational fluctuations. These changes may be strongly promoted by certain solutes such as K_2SO_4 and other water-structuring salts. As a consequence, the rate and extent of adsorption is increased.
c. The two above phenomena may be operating simultaneously, and, in addition, hydrophobic or/and thiophilic interaction may facilitate the chemisorption.

Reaction kinetics for interaction of native and thiolated HSA and pyridine-2-disulfide-substituted agarose and dextran have been studied, and the results obtained strongly support mechanism b (Oscarsson et al., 1992).

The kinetics of chemisorption and noncovalent adsorption of serum proteins on pyridine–2–disulfide substituted agarose in the presence and absence of salt is shown in Figs. 6a and 6b, which demonstrate that the kinetics are fast for both chemisorption and noncovalent adsorption.

In the absence of potassium sulfate, the chemisorption exhibits slower kinetics, and the reaction proceeds in a biphasic order with a higher initial rate. Details of these experiments and their interpretation have been reported by Oscarsson et al. (1992).

2.5. Proteins Isolated with Pyridine-2-disulfide-Derivatized Agarose

Pyridine-2-disulfide-derivatized agarose has been used mainly for isolation of thiol-containing proteins, but Tack et al. (1980) showed that thioester-containing proteins such as α_2-macroglobulin, C3, and C4 can be isolated by covalent chromatography. In the case of α_2-macroglobulin, there is a thioester between the sulfhydryl group of cysteine-948 and the α-carbonyl of glutamate-195 (Sottrup-Jensen et al., 1984). Probably, the thioester is hydrolytically unstable due to conformational changes in the protein as a consequence of adsorption to the agarose surface or simply owing to the reversibility of the thioester–thiol reaction (see Fig. 7). In both these two alternatives, a reactive thiol group is exposed that can react with the pyridine-2-disulfide-derivatized agarose.

3. PRACTICAL ASPECTS

A chromatographic experiment with pyridine-2-disulfide-derivatized agarose and potassium sulfate included in the buffer must proceed in at least four main steps. Potassium sulfate is added directly to the sample to reach a final concentration of $0.5M$ potassium sulfate. After application of the sample to the column, preequilibrated with buffer in $0.5M$ potassium sulfate, the adsorbed fraction is then desorbed from the column simply by deleting potassium sulfate from the eluent. The column is rinsed by equilibrating the bed with ethylene glycol in order to desorb the proteins still remaining on the column. Finally, desorption of the covalently attached proteins is accomplished by adding $5mM$ DDT to the buffer. The adsorbent is changed from a pyridine-2-disulfide-containing gel to an adsorbent containing a thiol group as the most important part. Regeneration of the adsorbent is possible. Various possible regeneration methods have been presented (Oscarsson and Porath, 1989). Before starting adsorbent regeneration, it is important to remove the tenaciously adsorbed macromolecules still on the column by using $0.1–0.5M$ NaOH. If such a treatment is suspected not to be efficient enough, we recommend analysis of the protein content on the gel. If proteins are found to be present by amino acid analysis, other regeneration methods may be tried (e.g., a mixture of 90% ethanol and 10% $1M$ NaOH if lipids or lipoproteins are the cause of incomplete regeneration; strong guanidinium chloride solution may also be tried).

4. APPLICATIONS

3-(2-Pyridyldisulfanyl)-2-hydroxypropylagarose (PyS$_2$-agarose) is commercially available under the trade name Thiopropyl Sepharose (Pharmacia Biotechnology Company,

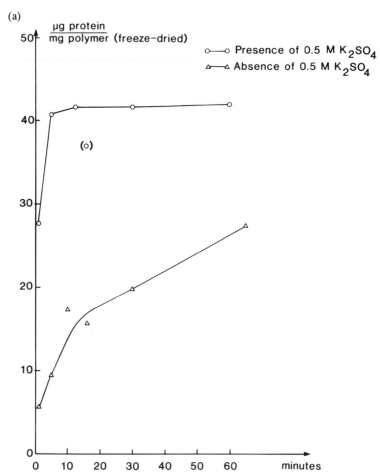

FIGURE 6. (a) Chemisorption of serum proteins as a function of time on pyridine–2–disulfide substituted agarose in the presence and absence of 0.5M K$_2$SO$_4$. Two hundred microliters of serum was added to six tubes, each containing 1.4 ml of suction-dried gel and 500 μl of 0.1M Tris-HCl, pH 7.5. The reaction was stopped after 1-min, 2-min, 10-min, 15-min, 30-min, and 60-min reaction, respectively, for the six vials by washing the gel on a sintered glass filter with 0.1M Tris-HCl, pH 7.5, followed by 40% (v/v) ethylene glycol in 0.1M Tris-HCl, pH 7.5, and finally distilled water. The amount of protein on the gel was determined by amino acid analysis. The same study was performed with 0.5M K$_2$SO$_4$ present in the buffer and by adding solid K$_2$SO$_4$ to the serum. (b) Adsorption of serum proteins as a function of time on 3-(2-pyridylthio)-2-hydroxypropylagarose in the presence of 0.5M K$_2$SO$_4$.

(b)

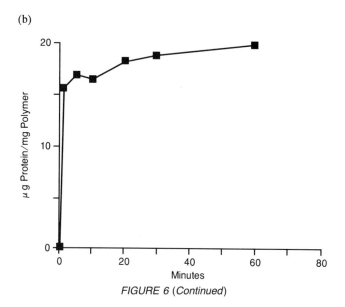

FIGURE 6 (Continued)

Uppsala, Sweden. It has been used for a large number of different applications during the past 15 years. In order to illustrate the great versatility of this and related techniques, some representative references are given below:

A. Purification and Isolation of Thiol-Containing Proteins and Peptides

A1. Human plasma lecithin-cholesterol acyltransferase purified >20,000-fold from plasma where thiopropyl-Sepharose is one chromatographic method used together with hydroxylapatite treatment; Holmqvist, L., 1987, *Biochem. Biophys. Methods* **14**(6):323–333.

A2. Bovine milk sulfhydryl oxidase prepared by covalent affinity chromatography; Janolino, V.G., and Swaisgood, H.E., 1990, *J. Dairy Sci.* **73**:308–313.

A3. Purification of penicillin acylase (EC 3.5.1.11) from *E. coli*; Boccu, E., Gianferrara, T., Gardossi, L., and Veronese, F. M., 1990, *Farmaco* **45**(2):203–214.

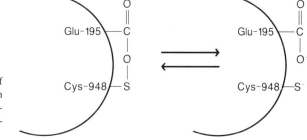

FIGURE 7. Schematic illustration of the thioester bond in α_2-macroglobulin between the sulfhydryl group in cysteine-948 and the α-carbonyl of glutamate-195.

A4. Metallothionein (MT) and its aggregates without or with Hg or Cd isolated by covalent affinity chromatography on 5,5′-dithiobis-(2-nitrobenzoate)-substituted Sepharose; Kabzinski, A. K. M., and Paryjczak, T., 1989, *Chromatographia* **27**(5–6):247–252.

A5. Isolation of reduced δ-(L-α-aminoadipyl)-L-cysteinyl-D-valine from culture broths; Orford, C. D., Adlard, M. W., and Perry, D., 1991, *J. Chem. Technol. Biotechnol.* **50**:523–533.

A6. Purification of procollagen type II by covalent chromatography with activated thiol-Sepharose 4B; Angermann, K., and Barrach, H. J., 1979, *Anal. Biochem.* **94**:253–258.

A7. Covalent chromatography as a means of isolating thiol peptides from large proteins. Application to human ceruloplasmin; Rydén, L., and Norder, H., 1981, *J. Chromatogr.* **215**:341–350.

A8. Purification of chymopapain A from the dried latex of papaya; Baines, B., Brocklehurst, K., Carey, P. R., Jarvis, M., and Salin, E., 1986, *Biochem. J.* **233**:119–129.

A9. Purification of actinidin from kiwifruit; Brocklehurst, K., Baines, B., and Malthouse, J. P. G., 1981, *Biochem. J.* **197**:739–746.

A10. Resolution of ox liver thiol-disulfide oxidoreductases by a new application of covalent chromatography; Hillson, D. A., and Freedman, B. K., 1979, *Biochem. Soc. Trans.* **7**:573–574.

A11. Covalent chromatography used for studying the role of sulfhydryl groups in pig brain purine nucleoside phosphorylase, PNP; Hakim, G., Solaini, G., and Rossi, C. A., 1980, *J. Solid-Phase Biochem.* **5**(4):185–192.

A12. Recent progress on the application of affinity and covalent chromatography to the purification of 1,25–dihydroxyvitamin D3 receptor from chick intestinal mucosa; Wecksler, W. R., Ross, F. P., Okamura, W. H., and Norman, A. W., 1979, in: *Proceedings of the 4th Workshop on Vitamin D (Vitamin D: Basic research and Its Clinical Applications)*, Walter de Gruyer and Co., Berlin, pp. 663–666.

A13. Purification of cytoplasmic aldehyde dehydrogenase by covalent chromatography on reduced thiopropyl-Sepharose 6B; Kitson, T. M., 1982, *J. Chromatogr.* **234**:181–186.

A14. Resolution of protein disulfide-isomerase and glutathione-insulin transhydrogenase activities by covalent chromatography; Hillson, D. A., and Freedman, R. B., 1980, *Biochem. J.* **191**:373–388.

A15. Resolution of thiol-containing proteins by sequential elution; Hillson, D. A., 1981, *J. Biochem. Biophys. Methods* **4**(2):101–111.

B. *Use of Covalent Chromatography for Sequential Analysis*

B1. Covalent chromatography for sequential analysis of membrane proteins; Ovchinnikov, Yu. A., and Abdulaev, N. G., 1986, in: *Methods of Protein Sequence Analysis* (K. A. Walsh, ed.), Humana, Clifton, New Jersey, pp. 189–209.

B2. Covalent chromatography for localization of essential cysteine residues during structural studies of proteins; Ustinnikova, T. B., Popov, V. O., and Egorov, Ts. A., 1988, *Bioorg. Khim.* **14**(7):905–909.

B3. Sequence homology between potato and rabbit muscle phosphorylase; Nakano, K., Fukui, T., and Matsubara, H., 1980, *J. Biochem.* **87**:919–927.

C. Use of Covalent Chromatography as a Tool in Immunoassay Techniques

C1. Application in immunoassay techniques where a specific antibody is fixed to the matrix via a disulfide bond. A sample mixture containing antigens was passed through the column. After rinsing the column, the formed immunocomplex was released from the column by splitting the disulfide bridge with a reducing agent such as DTE; Nakano, K., Fukui, T., and Matsubara, H., 1980, *J. Biochem.* **89:**223–229.

C2. Immunocomplex-immobilization technique; Oscarsson, S., and Carlsson, J., 1991, *Analyst* **116:**787–791.

REFERENCES

Axén, R., Drevin, H., and Carlsson, J., 1975, Preparation of modified agarose gels containing thiol groups, *Acta Chem. Scand., Ser. B* **29:**471–474.

Brocklehurst, K., Carlsson, J., Kierstan, M. P. J., and Crook, E. M., 1973, Covalent chromatography. Preparation of fully active papain from dried papaya latex, *Biochem. J.* **133:**573–584.

Carlsson, J., and Batista-Viera, F., 1991, Solid phase disulfides: A new approach to reversible immobilization and covalent chromatography on thiol compounds, *Biotechnol. Appl. Biochem.* **14:**114–120.

Eldjarn, L., and Jellum, E., 1963, Organomercurial-polysaccharide, a chromatographic material for the separation and isolation of SH-proteins, *Acta Chem. Scand.* **17:**2610–2621.

Millot, A. U., and Sebille, B., 1987, Rapid preparation of bovine mercaptalbumin by means of covalent chromatography on silica-based materials, *J. Chromatogr.* **408:**263–273.

Ngo, T. T., 1986, Facile activation of sepharose hydroxyl groups by 2-fluoro-1-methylpyridiniumtoluene-4-sulfonate: Preparation of affinity and covalent chromatographic matrices, *Biotechnology* **4**(2):134–137.

Oscarsson, S., and Porath, J., 1989, Covalent chromatography and salt-promoted thiophilic adsorption, *Anal. Biochem.* **176:**330–337.

Oscarsson, S., Medin, A., and Porath, J., 1992, Kinetic and conformational factors involved in chemisorption and adsorption of proteins on mercaptopyridine-derivatized agarose, *J. Colloid Interface Sci.* **152**(1):114–124.

Peters, T., Jr., 1984, Serum albumin, *Adv. Prot. Chem.* **37:**164–245.

Porath, J., and Olin, B., 1982, Immobilized metal ion affinity adsorption and immobilized metal ion affinity chromatography of biomaterials. Serum protein affinities for gel-immobilized iron and nickel ions, *Biochemistry* **22:**1621–1630.

Porath, J., and Oscarsson, S., 1988, A new kind of "thiophilic" electron-donor–acceptor adsorbent, *Makromol. Chem., Macromol. Symp.* **17:**359–371.

Shechter, Y., Rubinstein, M., and Patchornik, A., 1977, Selective covalent binding of methionyl-containing peptides and proteins to water insoluble polymeric reagent and their regeneration, *Biochemistry* **16:**1424–1430.

Sottrup-Jensen, L., Stepanik, T. M., Wierzbicki, D. M., Jones, C. M., Lonblad, P. B., Kristensen, T., Mortensen, S. B., Petersen, T. E., and Magnusson, S., The primary structure of α_2-macroglobulin and localization of a factor XIII cross-linking site, 1984, *Ann. N. Y. Acad. Sci.* **421:**41–61.

Tack, B. F., Harrison, R. A., Janatara, J., Thomas, M., and Prahl, J. W., 1980, Evidence for presence of internal thioester bond in third component of human complement, *Proc. Natl. Acad. Sci. USA* **77**(10):5764–5768.

27

Aza-Arenophilic Interaction
Novel Mode of Protein Adsorption and Applications in Immunoglobulin Purification

That T. Ngo and Neeta Khatter

1. INTRODUCTION

Protein adsorption to solid surfaces plays a crucial role in the organization of biological structures and the functioning of a living cell. Atkinson (1969) pointed out that the solvent capacity of water within a cell is not sufficient to solubilize all macromolecular poly-electrolytes and small molecules that coexist within the cell. The conservation of the solvent capacity of water can in part be achieved by adsorbing macromolecules onto the extensive membrane systems that permeate the cell. The adsorption of proteins (enzymes) to the membrane may keep them active by preventing their precipitation as biologically inactive agglomerates. A better understanding of the mechanism of protein adsorption to a well-defined solid surface can enhance our knowledge of the more complicated structure–function relationships in biological systems and allows us to further exploit this adsorption process for the purification of biomolecules by adsorption chromatography.

Specific adsorption of proteins to an affinity matrix involves a precise interaction of a specific region of the protein ligate with an equally specific area of an affinity ligand immobilized on the solid phase. A well-known example of such a specific interaction is the binding of CH2 and CH4 domains of the Fc fragment of an IgG to the Fc binding domain of protein A (Lindmark *et al.*, 1983). This specific interaction is exploited in the affinity purification of IgG using immobilized protein A gel.

Recently, a novel process of protein adsorption to immobilized nonionic sulfone-thioether ligands has been discovered by Porath *et al.* (1985). This interaction is termed "thiophilic." The proposed structure of the immobilized ligand can be presented as agarose-$CH_2CH_2SO_2CH_2CH_2SCH_2CH_2OH$. This immobilized ligand exhibits extraordinary selectivity in adsorbing IgG. The adsorption is promoted by "water-structuring" salts; however, the process is distinguishable from the conventional hydrophobic interactions. For

That T. Ngo and Neeta Khatter • BioProbe International Inc., Tustin, California 92680. All correspondence should be addressed to T. T. Ngo at Department of Developmental and Cell Biology, University of California, Irvine, California 92717.

Molecular Interactions in Bioseparations, edited by That T. Ngo. Plenum Press, New York, 1993.

example, serum albumin, a hydrophobic protein, is not adsorbed to the thiophilic gel under conditions under which immunoglobulins adsorb. Optimal conditions for selective and reversible adsorption of immunoglobulins involve the use of a loading buffer containing 7.5% ammonium sulfate and 0.5M NaCl (Hutchens and Porath, 1986). A one-step purification of monoclonal antibodies using a "thiophilic" adsorbent has been demonstrated (Belew et al., 1987). Neither the mechanism of thiophilic protein adsorption nor the thiophilic ligand acceptor sites on the protein are known in great detail presently. It was suggested that the permanent sulfone dipole and the free electron pair of the thioether can potentially act in concert as electron acceptor and donor sites, respectively, and that the two-carbon proximity of the thioether to the sulfone group may be a structural requirement for expression of the thiophilic character (Hutchens and Porath, 1987). Thus, it appears plausible that a ring structure can be formed between the thioether and sulfone moieties of the thiophilic ligands and the electron acceptor and donor site on the protein by an electron donor–acceptor or charge-transfer process (Hutchens and Porath, 1987).

We have recently discovered another interesting mode of protein adsorption to a surface having immobilized nitrogen-containing aromatic ligands (Ngo, 1991; Ngo and Khatter, 1990, 1991, 1992). The proposed structures of the ligands are depicted in Fig. 1. We provisionally term the adsorption as "aza-arenophilic" adsorption (aza for nitrogen and areno for aromatic). Aza-arenophilic gels show extraordinary protein adsorption selectivity and binding capacity (Ngo and Khatter, 1990). A version of an aza-arenophilic gel was able to purify immunoglobulins to a very high degree of purity in one step directly from a phosphate-buffered saline-diluted serum. The adsorption of proteins, in particular the immunoglobulins, from unfractionated serum takes place in phosphate-buffered saline (0.15M NaCl) only without the presence of "water-structuring" salts. The bound proteins can be eluted with acid buffers or with an "electron donor"-containing buffer at neutral pH (Ngo and Khatter, 1992). The selectivity of protein adsorption to aza-arenophilic gels can be varied by varying the composition and the salt concentration of the loading buffer. The selectivity of these gels can be further augmented by using selective elution buffers. It is also possible to alter the selectivity of the gel by changing the X and Y substituents of the immobilized ligands (see Fig. 1).

Here we report our studies on aza-arenophilic gels with respect to their preparation, the proposed structure of the immobilized ligands, their protein adsorption properties, their applications in immunoglobulin purification, the binding parameters of immunoglobulin and fragments therefrom, the selective enhancement of the specific radioactivity of ^{125}I-labeled proteins, and a proposed mechanism of protein adsorption.

2. PREPARATION OF AZA-ARENOPHILIC GELS

In general, aza-arenophilic gels were prepared by first reacting Sepharose CL-4B under anhydrous conditions with pentahalopyridine and 4-dimethylaminopyridine (DMAP) and then with a nucleophile solution such as mercaptoethanol, ethanolamine, ethylene glycol, or glycine. The preparation of gel no. 1 will be described in greater detail as an example of the preparative procedure. Sepharose CL-4B (100 ml) was washed with 5 × 100 ml distilled water and then suspended in 100 ml of distilled water in a large 2-liter beaker mounted on a shaker rotating at 100 rpm. To the gel suspension, 1 liter of dry acetone was

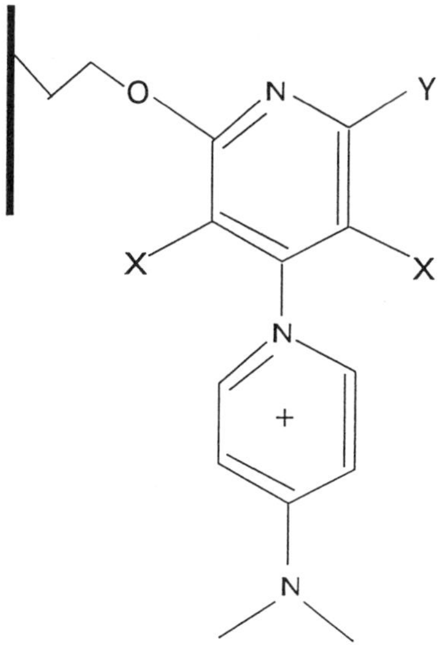

FIGURE 1. Proposed structure for aza-arenophilic gels.

GEL NUMBER	X	Y	GEL NUMBER	X	Y
(1)	F	$-O-CH_2-CH_2OH$	(9)	Cl	$-NH-CH_2-CH_2OH$
(2)	F	$-S-CH_2-CH_2OH$	(10)	Cl	$-OH$
(3)	F	$-NH-CH_2-CH_2OH$	(11)	Cl	$-NH-CH(COOH)-CH_2-CH_2-\overset{O}{\overset{\|}{C}}-OH$
(4)	F	$-OH$	(12)	Cl	$-NH-CH_2-\overset{O}{\overset{\|}{C}}-OH$
(5)	F	$-S-CH_2-CH_2-\overset{O}{\overset{\|}{C}}-OH$	(13)	Cl	$-S-CH_2-CH_2-\overset{O}{\overset{\|}{C}}-OH$
(6)	F	$-NH-CH_2-\overset{O}{\overset{\|}{C}}-OH$	(14)	Cl	$-S-CH_2-CH(OH)-CH(OH)-CH_2-SH$
(7)	Cl	$-O-CH_2-CH_2OH$	(15)	Cl	$-S-CH_2-CH(OH)-CH(OH)-CH_2-S-CH_2-\overset{O}{\overset{\|}{C}}-NH_2$
(8)	Cl	$-S-CH_2CH_2OH$	(16)	Cl	$-NH-CH_2-CH_2-NH_2$

added gradually over 30 min. The gel was filtered and resuspended in 1 liter of dry acetone and was agitated for 15 min. The gel was filtered and resuspended in 300 ml of dry N,N'-dimethylformamide (DMF), shaken for 5 min, filtered, and suspended in 100 ml of DMF containing 27.5 mmol of DMAP. To the gel suspension was immediately added 250 ml of DMF containing 25 mmol of pentafluoropyridine. The gel was tumbled at room temperature for 2 h. Then the gel was washed with 1 liter of DMF, 2 × 1 liter of acetone, 1 liter of distilled water, and 1 liter of 0.1M sodium bicarbonate. The washed activated gel was then suspended in an equal volume of 10% ethylene glycol in 0.1M sodium bicarbonate, pH 9, and tumbled at room temperature for 24 h. The gel was washed with 1 liter of 0.1M sodium bicarbonate and resuspended in 0.1M NaOH and tumbled for 14 h at room temperature. The gel was finally washed with 1 liter of distilled water, 1 liter of 1M NaCl, 1 liter of distilled water, and 1 liter of phosphate-buffered saline (PBS). The gel is stored in PBS at 4°C.

3. PROPOSED STRUCTURE OF THE IMMOBILIZED LIGANDS

It is well established that nucleophilic substitution reactions of monohalogenated pyridines occur most readily at 2- or 4-halogenated carbon by an addition–elimination mechanism (Katritzky and Lagowski, 1967; Gilchrist, 1985). The reactivity of halogen atoms in perhalogenated pyridines is greatly increased, and the two-step addition–elimination reactions with nucleophiles occur almost exclusively at carbon-4 (Chamber *et al.*, 1964; Banks *et al.*, 1965; Chamber and Sargent, 1981). Chamber *et al.* (1975) showed that nucleophilic attacks by pyridine on perhalogenated pyridine took place at position 4 of the perhalogenated pyridine to give a pyridinium adduct. By this analogy, the reaction of DMAP (I) and a perhalogenated pyridine (II) forms a pyridinium adduct (III) as shown in Fig. 2 [reaction (a)]. As mentioned above, the 2′-chloro group of the pyridine makes the 2′ carbon susceptible to a nucleophilic attack. The hydroxyl groups of Sepharose serve as nucleophiles and react with the pyridinium (III) via reaction (b) to form an activated gel (IV), which subsequently reacts with another nucleophile (Nu) at the 6′ position via reaction

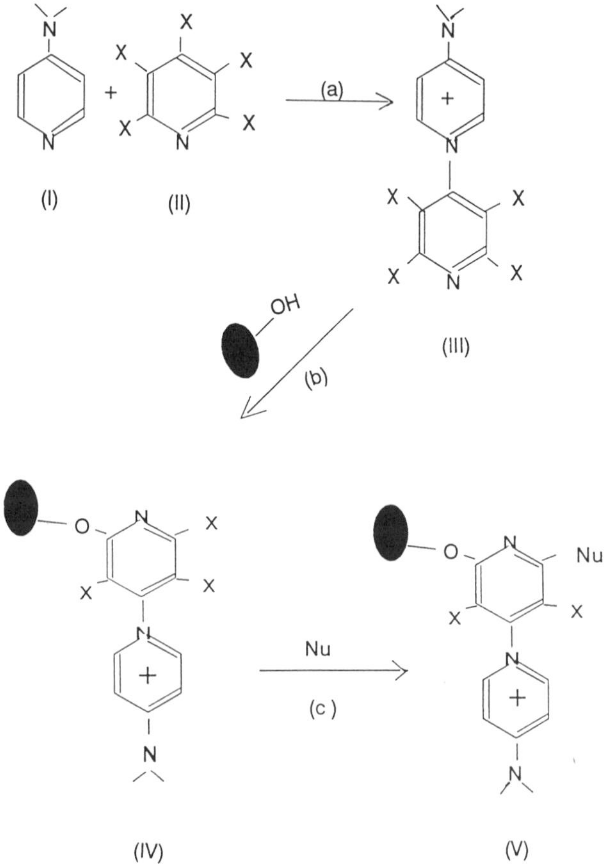

FIGURE 2. Reactions involved in the preparation of aza-arenophilic gel. Nu, Nucleophile; ●—OH, Sepharose CL-4B.

(c) to form the affinity gel (V). Reactions involved in the synthesis of aza-arenophilic gels are summarized in Fig. 2. Gels numbered 1 to 6 as listed in Fig. 1 were prepared by reacting Sepharose CL-4B with pentafluoropyridine and DMAP first and then with various nucleophiles listed under Y in Fig. 1. The gels numbered 7 to 16 were prepared by reacting Sepharose CL-4B with 3,5-dichloro-2,4,6-trifluoropyridine and DMAP and then with various nucleophiles listed under Y in Fig. 1. A total of 16 gels have been prepared, and the protein adsorption properties of these gels have been investigated in varying degrees of detail. Infrared spectroscopy of the aza-arenophilic gels showed the presence of a pyridinium ring in the structure of the immobilized ligand. The elemental analysis of the no. 9 gel showed the mole ratio of N:Cl:S to be 3:1.97:1.25, which is consistent with the theoretical mole ratio of 3:2:1 for this gel. The total concentration of the immobilized ligand calculated from the elemental analysis data was 1 to 1.3 mmol per gram of dry gel.

4. ADSORPTION OF PROTEINS TO AZA-ARENOPHILIC GELS

We have previously shown that in 20mM phosphate buffer, pH 7.4, almost all serum proteins were adsorbed to a difluorohydroxyethoxy-substituted aza-arenophilic gel (see Fig. 1, gel no. 1). The addition of 0.5M potassium phosphate caused a desorption of a large quantity of proteins; most of them were serum albumin. When the pH of the eluent was decreased to 5, 4, and 2.8, the desorbed proteins were mainly immunoglobulins (Ngo and Khatter, 1990). Similar results were obtained when diluted human serum was chromatographed on a dichloroethoxylated aza-arenophilic gel (gel no. 7 in Fig. 1). However, the adsorption of serum proteins appeared to be more complete, and proteins were eluted in a cleaner and more discrete manner (Fig. 3). The results of the analysis of eluted proteins by enzyme-linked immunosorbent assay (ELISA) for the distribution of serum albumin, IgG, IgA, and IgM are presented in Fig. 4. The ELISA revealed that (1) in 20mM phosphate, pH 7.4, virtually all serum proteins were bound, (2) the fraction of protein eluted by buffer containing 0.5M potassium sulfate (second pooled fraction) consisted mostly of albumin with a very small amount of the immunoglobulins, (3) the third pooled fraction desorbed by 0.1M glycine, pH 5, consisted mostly of IgG with only a small amount of albumin, and (4) the fourth and fifth pooled fractions desorbed at pH 4 and 2.8, respectively, consisted of all the immunoglobulin classes (IgG, IgA, and IgM) and some albumins. The amount of proteins contained in the pooled fourth and fifth fractions was much less than that in the second and third pooled fractions. The results of sodium dodecyl sulfate (SDS)-gradient (10–15%) polyacrylamide gel electrophoresis showed that proteins other than albumin and immunoglobulins were adsorbed and desorbed in different fractions (Ngo and Khatter, 1990). For example, appreciable amounts of proteins with the same molecular weight as transferrin were detected in the second pooled fraction and to a small degree in the third pooled fraction. The adsorption behavior of ten highly purified proteins on aza-arenophilic gel (gel no. 7) was also investigated. The proteins used have molecular weights ranging from 14,000 to 66,000 and isoelectric points ranging from 1 to 10.5. The adsorption experiments were carried with three different buffers: (1) 20mM phosphate, pH 7.4; (2) 20mM phosphate, pH 7.4, containing 10% ammonium sulfate; and (3) 20mM phosphate, pH 7.4, containing 20% ammonium sulfate. Proteins such as bovine serum albumin, human IgG, fetuin, papain, transferrin, pepsin, and trypsin adsorbed strongly to the gel without the presence of "water-structuring" salt, while other such as ribonuclease and

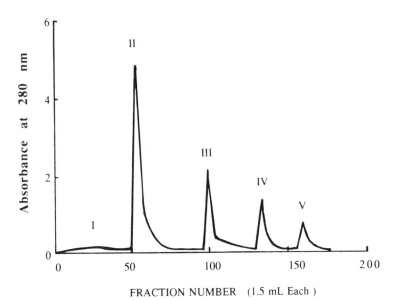

FIGURE 3. Chromatogram of human serum on aza-arenophilic gel no. 7. Human serum was diluted with 20mM phosphate buffer, pH 7.4. The diluted serum was applied at a flow rate of 1.25 ml/min at room temperature, and fractions of 1.5 ml were collected. Peak I, Nonadsorbed proteins; peak II, proteins desorbed by 10mM phosphate, pH 7.4, containing 0.5M K$_2$SO$_4$; peak III, proteins desorbed by 0.1M glycine, pH 5; peak IV, proteins desorbed by 0.1M glycine, pH 4; peak V, proteins desorbed by 0.1M glycine, pH 2.8.

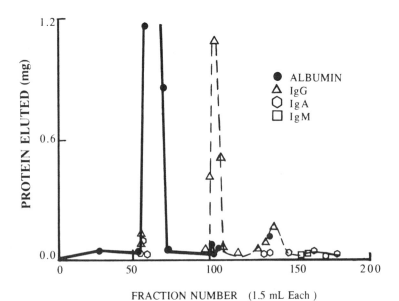

FIGURE 4. Analysis of the proteins in fractions shown in Fig. 3 by using specific sandwich ELISA technique.

lysozyme did not adsorb without the presence of "water-structuring" salt (Fig. 5). For bovine serum albumin and transferrin, which adsorbed strongly in 20mM phosphate, pH 7.4, the presence of 10% ammonium sulfate caused desorption from the gel. The adsorption of these proteins was increased by increasing the concentration of ammonium sulfate to 20%. The adsorption behavior of human IgG, however, was not altered by the addition of ammonium sulfate (Fig. 5). The adsorption of papain and myoglobin increased with increasing concentration of ammonium sulfate. The adsorption behavior of pepsin and trypsin is interesting. They have similar adsorption patterns in spite of the vast difference in their isoelectric points, being 1 and 10.5, respectively. Their adsorption decreased with increasing concentration of ammonium sulfate.

5. PURIFICATION OF IMMUNOGLOBULINS WITH AN AZA-ARENOPHILIC GEL: OPTIMIZATION

In view of the results presented in Fig. 3, which showed the selective adsorption and desorption of albumin and IgG by gel no. 7 with the use of various buffers, we attempted to simplify and optimize the conditions for IgG purification. When rabbit serum was diluted tenfold with 10mM sodium phosphate buffer, pH 7.4, containing 0.5M potassium phosphate, most of the serum proteins except IgG did not bind to aza-arenophilic gel no. 1. The bound IgG was eluted by using 0.1M glycine, pH 3.5. This purification procedure is identical to that using protein A affinity gel (Fig. 6). The isolated IgG still contains a number of non-IgG proteins. To further improve the purity of the isolated IgG and to

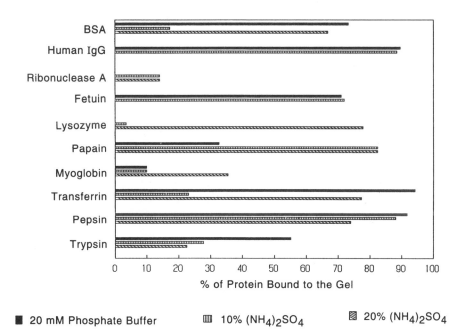

FIGURE 5. Aza-arenophilic adsorption behavior of selected proteins in 20mM phosphate, pH 7.4, alone or in the presence of either 10% or 20% $(NH_4)_2SO_4$.

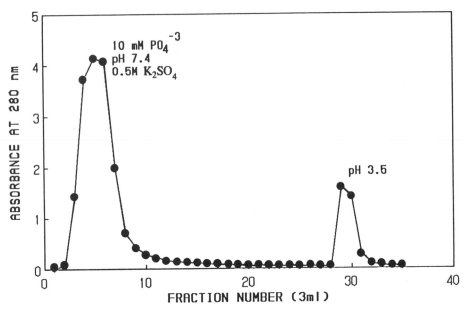

FIGURE 6. Chromatographic fractionation of rabbit serum on aza-arenophilic gel no. 1. Serum was diluted 10-fold with 10mM sodium phosphate, pH 7.4, containing 0.5M potassium phosphate. Elution was carried out with 0.1M glycine, pH 3.5.

optimize the operating conditions for the purification of IgG, a total of 16 different gels were synthesized and tested as affinity gels for IgG purification. The results from the screening of these gels showed that gel no. 8 (see Fig. 1 for the structure) with dichloro and hydroxy-ethylthio substituents appeared to be ideal for IgG purification. Using this gel, the serum needs only to be diluted fivefold with PBS. There is no need to add a high concentration of salt to the dilution buffer or to the serum. The bound IgG can be eluted by using 0.05M sodium acetate, with the pH adjusted to 3 with concentrated HCl not with glacial acetic acid. A typical chromatographic profile for IgG purification from goat serum using aza-arenophilic gel no. 8 is shown in Fig. 7. The purity of the isolated goat IgG was checked by using SDS-gradient polyacrylamide gel electrophoresis under both reducing and nonreducing conditions (Figs. 8a and 8b). The purity of IgG isolated from mouse, rabbit, and human serum was also examined with SDS-gradient electrophoresis (Figs. 8c–e). The electrophoretograms indicated a high degree of purity for the isolated IgG, with only a minor amount of other proteins. It must be emphasized that the use of a proper elution buffer is crucial for the successful purification of the antibody. For example, when the pH of the 0.05M sodium acetate buffer was adjusted to 3 with acetic acid (instead of concentrated HCl), the amount of IgG recovered was very low. It was also noticed that 0.05M sodium acetate was more effective than 0.1M sodium acetate in eluting the bound IgG. Chloride ion is more chaotropic than acetate ion and can thus promote the elution of IgG, whereas acetate ion is more effective than chloride ion in salting out proteins and therefore is expected to promote adsorption rather than desorption of IgG. For the elution of some more strongly adsorbed IgGs, the inclusion of 20% ethylene glycol or glycerol in the elution buffer may be necessary. Glycine buffer (0.1M), pH 3, can also elute the adsorbed

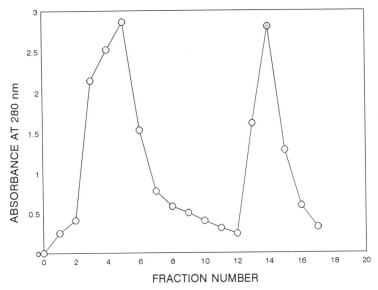

FIGURE 7. Purification of goat IgG using aza-arenophilic gel no. 8. Gel volume was 3 ml. One milliliter of serum was diluted 5-fold with PBS, pH 7.4, and was applied at room temperature. Fractions of 3 ml were collected. Elution was carried out with 0.1M sodium acetate, pH adjusted to 3 with concentrated HCl (not with glacial acetic acid).

antibodies. However, the antibody desorbed by using glycine buffer is less pure than that obtained by using acetate buffer. The amino group of glycine, with its lone-pair nitrogen electrons, may contribute additional eluting capability to desorb other non-IgG proteins. Indeed, many "electron-rich" compounds such as triethylamine, mercaptoglycerol, dithiothreitol, and acetonitrile were able to desorb bound IgG at near neutral pH (Ngo and Khatter, 1992). The ability to elute bound antibodies at neutral pH is an important advantage of aza-arenophilic gels in purifying acid-labile antibodies. It was also shown that neutral pH elution buffer consistently yielded antibodies with higher specific binding activity than those prepared by using acid elution buffer (Ngo and Khatter, 1992). We have used aza-arenophilic gel to purify IgG from the serum of 12 different animals. All IgGs tested so far adsorbed to the gel. The degree of purity of the isolated IgG varied from one animal species to another, the goat IgG having the highest degree of purity. The adsorption capacity of the gel for various IgGs is presented in Table 1. It should be mentioned that monoclonal IgM has also been purified with aza-arenophilic gel. However, the use of this gel for the purification of IgM from serum is not recommended because of the much higher concentration of IgG in the serum, which will compete with IgM for binding to the gel.

6. ADSORPTION OF IgG AND ITS PROTEOLYTIC FRAGMENTS TO AZA-ARENOPHILIC GEL

The adsorption of intact IgG molecules or their proteolytic fragments, that is, the Fc and Fab fragments, to aza-arenophilic gel no. 8 have been investigated under both

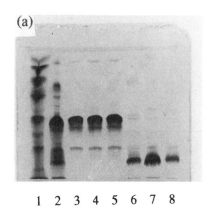

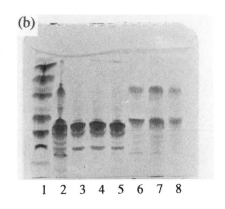

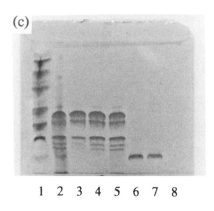

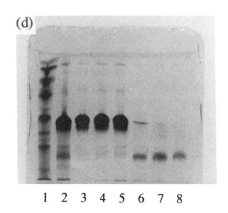

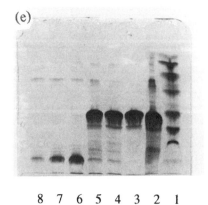

FIGURE 8. SDS-polyacrylamide gradient gel (10–15%) electrophoretic analysis of serum proteins fractionated on aza-arenophilic gel no. 8. For all electrophoretograms, lane 1 = molecular weight markers; lane 2 = unfractionated diluted serum; lanes 3–5 = unadsorbed, flow-through fractions; and lanes 6–8 = fraction eluted with acidic buffer (mainly IgG fractions). High-molecular-weight proteins remain close to the origin at the bottom, and low-molecular-weight proteins migrate to the upper portion of the gel. (a) Goat serum under nonreducing conditions; (b) goat serum under reducing conditions; (c) mouse serum under nonreducing conditions; (d) rabbit serum under nonreducing conditions; (e) human serum under nonreducing conditions.

TABLE 1

Antibody Adsorption to Aza-Arenophilic Gel No. 8

Source of antibody (serum)	Quantity adsorbed (mg antibody/ml gel)	Source of antibody (serum)	Quantity adsorbed (mg antibody/ml gel)
Bovine	6–7	Sheep	6–7
Chicken	7–8	Mouse	6–7
Donkey	4–8	Mouse	6–7
Goat	9–11	Mouse monoclonal	
Guinea pig	4–9	IgG2a	1.5–2.5
Horse	5–6	IgG2b	5–7
Human	9–11	IgG3	5–10
Pig	8–9	IgM (ascites)	2
Rabbit	9–10	IgM (cell culture)	>1.1
Rat	5–9		

equilibrium and pseudo-equilibrium conditions. The adsorption appeared to follow a simple Langmuir adsorption isotherm. Both types of experimental conditions resulted in similar adsorption parameters. The dissociation constants for IgG, Fab, and Fc were determined to be 17.4, 23.9, and 8.8 μM, respectively. The adsorption capacities were 14, 5.3, and 1.8 mg/ml of gel, respectively (Ngo and Khatter, 1992). The data showed that, in addition to binding IgG, the gel does bind both Fab and Fc fragments, albeit with different affinities and capacities. The Fc fragment has the highest affinity for the gel but the lowest adsorption capacity. The Fab fragment has the lowest adsorption affinity and medium adsorption capacity. The whole IgG, on the other hand, has medium affinity but the highest adsorption capacity. It thus appeared that the simultaneous presence of both Fab and Fc had a cooperative effect on adsorption to the gel.

7. SELECTIVE ENHANCEMENT OF SPECIFIC RADIOACTIVITY OF ^{125}I-LABELED PROTEINS

The chromatographic behavior of tyrosine, monoiodotyrosine, and diiodotyrosine on aza-arenophilic gel showed that diiodotyrosine was retarded most. This was followed by monoiodotyrosine in term of the retention time, and tyrosine was the least retarded. It is therefore possible to use aza-arenophilic gel to selectively enrich the specific radioactivity of ^{125}I-labeled proteins. The experiment was carried out as follows. Four milliliters of ^{125}I-labeled bovine serum albumin (BSA) having a specific radioactivity of 1000 cpm/mg of protein in a concentration of 20 mg/ml was added to 0.5 ml of aza-arenophilic gel at room temperature, and the gel suspension was tumbled end-to-end for 5 min. Then the gel was centrifuged briefly, and the supernatant was removed. The gel was washed 3 times in a buffer of 20mM sodium phosphate, pH 7.5. Then the ^{125}I-labeled BSA was eluted with two 4-ml portions of 0.1M glycine-HCl, pH 2.8. The eluted ^{125}I-labeled BSA has a specific radioactivity of 3550 cpm/mg of protein. This corresponds to a 350% increase in the specific activity of the protein. As increasing quantities of ^{125}I-labeled BSA were added to a fixed amount of aza-arenophilic gel, the amount of radioactivity bound to the gel increased almost linearly. However, the mass of the protein bound did not increase linearly; in fact, it

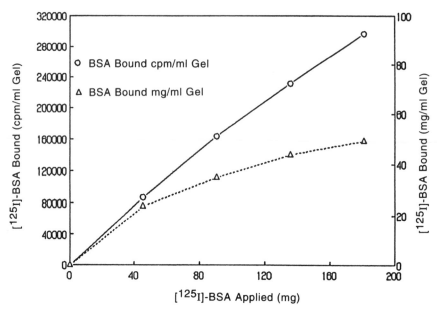

FIGURE 9. Enrichment of specific radioactivity of ^{125}I-labeled proteins by selective adsorption on aza-arenophilic gel. Four milliliters of ^{125}I-labeled bovine serum albumin (BSA) having a specific radioactivity of 1000 cpm/mg of protein in a concentration of 20 mg/ml was added to 0.5 ml of the aza-arenophilic gel no. 7 and was mixed at room temperature for 5 min. Then the gel was centrifuged briefly, and the supernatant was removed, The gel was washed 3 times with 20mM phosphate, pH 7.5, before the bound ^{125}I-labeled BSA was eluted with two 4-ml portions of 0.1M glycine-HCl, pH 2.8.

began to fall off (Fig. 9). It appeared that the preferential adsorption of ^{125}I-labeled proteins to the gel caused the discrepancy between the amount of bound radioactivity and the mass of the bound proteins. This discrepancy, in turn, led to an increase in the specific activity.

8. A PLAUSIBLE MECHANISM OF PROTEIN ADSORPTION TO AZA-ARENOPHILIC GELS

It should be obvious by now that the aza-arenophilic gels described here have unusual protein adsorption properties as shown by their protein adsorption selectivity and capacity. Early studies showed that at low ionic strength most serum proteins were adsorbed to all aza-arenophilic gels. Serum albumin, a "hydrophobic" protein, was in fact desorbed in the presence of a "water-structuring" salt such as potassium sulfate at 0.5M concentration (Ngo and Khatter, 1990). This phenomenon is contrary to the conventional hydrophobic interaction of serum albumin with long-chain alkyl- or aromatic-substituted gels, where the adsorption is promoted by "water-structuring" salts. The adsorption of IgG to aza-arenophilic gels takes place in either low- or high-ionic strength buffers with or without "water-structuring" salts. Therefore, the mode of adsorption of proteins to an aza-arenophilic gel cannot be totally attributed to either "pure" hydrophobic interactions or ionic interactions. Hydrophobic interactions may partially contribute to the overall mechanism of adsorption, because the adsorption of proteins to an aza-arenophilic gel increases slightly with increasing temperatures for the adsorption process (Ngo and Khatter, unpublished results). The adsorbed IgG can be selectively desorbed by a low-pH buffer or by a neutral pH buffer containing "electron-rich" compounds such as amines, nitriles, or thio compounds. Most of the aza-arenophilic gels do not contain element of sulfur, and yet they all are capable of adsorbing IgG with or without the presence of "water-structuring" salts. In this respect, the aza-arenophilic gels are distinguishable from thiophilic gels, which require the presence of sulfur as either a sulfone or a thioether in the ligand and the presence of "water-structuring" salt to promote the adsorption of IgG (Porath *et al.*, 1985; Porath, 1987; Hutchens and Porath, 1987). The pyridinium ring that is part of the ligand of aza-arenophilic gels can serve as an electron acceptor in charge-transfer interactions (Cilento and Giusti, 1959; Cilento and Tedeschi, 1961; Alivisatos *et al.*, 1960; Kosower, 1960; Ungar and Alivisatos, 1961; Alivisatos *et al.*, 1961). The relevant electron-rich ligates in a protein molecule are most likely the aromatic amino acid side chains. In this regard, tryptophan was estimated to be the strongest electron donor, followed by tyrosine and phenylalanine (Porath, 1989). Cilento and Giusti (1959) showed the transfer of an electron from the indole nucleus to a pyridine coenzyme. The oxidized pyridine coenzymes and their model, 1-benzyl-3-carboxamide-pyridinium chloride, form charge-transfer complexes in a 1:1 ratio with tryptophan and other indole derivatives in aqueous buffer systems (Cilento and Tedeschi, 1961; Alivisatos *et al.*, 1961). In addition to the role of substituents on the pyridinium ligand (Kosower, 1960), substituents on the indole nucleus of the ligate also play a decisive role in the formation of charge-transfer complexes. For example, indole with an electron-donating group, such as a methyl group, at position 2 is able to form a charge-transfer complex with nicotinamide adenine dinucleotide (NAD). On the other hand, an electron-withdrawing substituent, such as a phenyl group, at the 2 position of an indole ring renders the molecule incapable of forming a charge-transfer complex with NAD (Alivisatos *et al.*, 1961). The quaternary nitrogen of the pyridinium ring can also interact with the aromatic ring of an amino acid side chain via ion–dipole interaction. For example, recent studies have shown that highly hydrophobic molecules built up from etheno-anthracene units display a strong and fairly general affinity for quaternary ammonium compounds (Petti *et al.*, 1988). This complexing effect can be ascribed to an ion–dipole attraction between the positively charged quaternary ammonium ion and the electron-rich systems of the cyclophane. Furthermore, neutral molecules with electron-deficient systems are preferentially bound by the macrocyclic cyclophane. This binding suggests the operation of favorable donor–acceptor π-stacking interactions. A synthetic acetylcholine receptor comprising primarily aromatic rings has been prepared (Dougherty and Stauffer, 1990). This synthetic receptor provides an overall hydrophobic binding site capable of recognizing the positive charge of the quaternary ammonium group of acetylcholine through a stabilizing interaction with the electron-rich π systems of the aromatic rings (cation–π interaction). The biological receptors of cationic ligates such as acetylcholine and phosphocholine appear to comprise also primarily aromatic amino acid residues. A choline binding site of an antibody to phosphocholine comprises three aromatic amino acid residues in close proximity to the quaternary ammonium ion (Satow *et al.*, 1986; Getzoff *et al.*, 1988). The three aromatic amino acid residues that are within van der Waals contact distance are tryptophan (107H), tyrosine (33H), and tyrosine (107H). Two anionic residues, aspartate (97L) and glutamate (61H), also participate in the binding of the cationic ligate but are located further away than the aromatic residues. The hydrophobic environment created by the aromatic amino acid residues could result in a low-dielectric-constant space, which in turn could result in a higher effective local charge than might be predicted by considering the nearby acidic amino acid side chains. A recent study on the three-dimensional atomic structure of an acetylcholinesterase from *Torpedo californica* has shed light on the structure and chemical makeup of the choline binding site (Sussman *et al.*, 1991). The active site of this enzyme lies near the bottom of a deep and narrow gorge that reaches halfway into the protein. The quaternary ammonium ion appears to be bound not to a negatively charged "anionic" site, but rather to some of the 14 aromatic residues that line the gorge (Sussman *et al.*, 1991). These 14 aromatic amino acid residues comprise 5 tryptophan, 5 tyrosine, and 4 phenylalanine residues. Thus, the structural studies on the phosphocholine-binding antibody (Satow *et al.*, 1986; Davies and Metzger, 1983) and acetylcholinesterase (Sussman *et al.*, 1991) support the theory put forward by Dougherty and Stauffer (1990) that electrons of aromatic rings attract quaternary ammonium ions with greater affinity than isosteric uncharged ligands. Such attraction is thought to proceed via a stabilizing cation–π interaction. An analysis of 33 high-resolution protein crystal structures revealed a significant tendency for positively charged amino groups of lysine, arginine, asparagine, glutamine, and histidine to lie within 6Å of the π-electron clouds of aromatic amino acid side chains, where they make van der Waals contact with the $d(-)\pi$ electrons of the aromatic ring (Burley and Petsko, 1986). This interaction of a charged amino group with the face of an aromatic ring may be contrasted with the attraction of electron-rich oxygen or sulfur atoms to the electron-poor edge of such rings (Reid *et al.*, 1985).

From the foregoing discussion, we can postulate the underlying attractive forces that drive the adsorption of proteins to immobilized aza-arenophilic ligands. Four factors should be considered in relation to protein adsorption to the immobilized aza-arenophilic ligands: (1) the immobilized aza-arenophilic ligands with their extensive delocalized π-electron system allow the operation of a favorable π–π stacking type of interaction; (2) the pyridinium moiety of the ligand serves as an electron acceptor in the formation of a charge-transfer complex with an electron donor such as an aromatic amino acid residue; (3) hydrophobic interactions can occur subsequent to removal of bound water molecules, an entropy-driven process; and (4) the quaternary ammonium ion of the pyridinium group can interact with aromatic amino acid residues of a protein via a cation–π interaction. The presence of an N-dihalogenated pyridyl substituent on the pyridinium ring greatly enhances the electron-accepting character of the immobilized ligands. In the case of aza-arenophilic gel no. 8, which showed excellent utility for immunoglobulin purification, an additional factor should be considered. This gel, having a 2'-hydroxyethylthio substituent on the pyridine ring, may exhibit considerable "thiophilic" character as described by Porath and co-workers (Porath *et al.*, 1985; Porath, 1989; Porath and Belew, 1987). The coupling of mercaptopyridine to a divinylsulfone-activated gel gave an adsorbent that is more thiophilic than that obtained by coupling mercaptoethanol to a divinylsulfone-activated gel, the original thiophilic gel, or a gel obtained by coupling mercaptopyridine to an epoxide-activated gel (Porath, 1989). In an analysis of 36 high-resolution protein crystal structures, Reid *et al.* (1985) found that sulfur–aromatic interactions are commonly observed in the hydrophobic core of proteins and that about half of all electronegative sulfur atoms from cyst(e)ine and methionine residues are in contact with an aromatic ring (phenylalanine, tyrosine, and tryptophan). The sulfur atoms exhibit an affinity toward the edge of the aromatic ring. Thus, the electronegative sulfur atom is attracted by the positively charged aromatic ring hydrogens and avoid the region above the ring in the vicinity of the π electrons. It is clear that aromatic amino acid residues do interact intramolecularly with proximal methionine and cyst(e)ine residues. This sort of interaction between the electronegative sulfur of a thioether and aromatic amino acid residues may partially provide the driving force in thiophilic adsorption of proteins to a thiophilic gel. Porath and Belew (1987) proposed hypotheses on the mechanism of thiophilic adsorption. These hypotheses assume a simultaneous operation of both electron donor and acceptor sites in the ligand molecule that interact cooperatively with electron acceptor and donor sites, respectively, on the protein molecule. The electron donor site could be the sulfur atom of the thioether group, and the electron acceptor site could be either the methylene group next to the sulfone moiety or the $3d$ orbitals of the sulfur of the sulfone. Thus, the thioether sulfur of the immobilized ligand may interact with hydrogens on the edge of an aromatic ring. Such an interaction appears to be quite general and widespread in proteins. The electron acceptor of the ligand may interact with the π-electron clouds of the aromatic ring of a protein. A similar cooperative and multipoint interaction between the immobilized ligands of aza-arenophilic gels and protein molecules can also be envisaged. The pyridinium ring of the ligand is a good electron acceptor in the formation of a charge-transfer complex, and the "exo" ring heteroatom, such as O, N, or S, can serve as an electron donor. The formation of charge-transfer complexes between pyridinium rings and the aromatic rings of tryptophan, tyrosine, and phenylalanine is a well-known phenomenon (Kosower, 1960). The aza-arenophilic interaction described here is a form of charge-transfer adsorption, as discussed by Porath (1978, 1979).

We fully recognize that the plausible mechanism of protein adsorption to aza-arenophilic gels presented here is an oversimplification of what is, no doubt, a complicated adsorption process. It nevertheless has been a useful guide and has helped us to develop a neutral pH elution buffer using electron-rich compounds to compete with the electron donors of IgG. This model also helps to explain the preferential binding of ^{125}I-labeled proteins because of a favorable charge-transfer affinity between iodo compounds and the pyridinium ring (Kosower, 1960).

The successful development of aza-arenophilic gels as an effective charge-transfer adsorbent has been possible because of the excellent electron-accepting property of the pyridinium ring, which, in aza-arenophilic ligands, is augmented by (1) an N-dihalogenated pyridyl substituent; (2) the extensive delocalized π-electron system; and (3) the presence of a heteroatom, such as O, N, or S, next to the pyridine ring.

REFERENCES

Alivisatos, S. G. A., Mourkides, G. A., and Jibril, A., 1960, Non-enzymic reactions of indoles with coenzyme I, *Nature (London)* **186:**718–719.

Alivisatos, S. G. A., Ungar, F., Jibril, U. A., and Mourkides, G. A., 1961, Non-enzymic reactions of indole with pyridine coenzymes and related structures, *Biochim. Biophys. Acta* **51:**361–372.

Atkinson, D. E., 1969, Limitation of metabolite concentrations and the conservation of solvent capacity in the living cell, in: *Current Topics in Cellular Regulation*, Vol. 1 (B. L. Horecker and E. R. Stadtman, ed.), Academic Press, New York, pp. 29–43.

Banks, R. E., Burgess, J. E., Cheng, W. M., and Haszeldine, R. N., 1965, Heterocyclic polyfluoro-compounds. Part IV. Nucleophilic substitution in pentafluoropyridine: The preparation and properties of some 4-substituted 2,3,5,6-tetrafluoropyridines, *J. Chem. Soc.* **1965:**575–581.

Belew, M., Juntti, N., Larsson, A., and Porath, J., 1987, A one-step purification method for monoclonal antibodies based on salt-promoted adsorption chromatography on a "thiophilic" adsorbent, *J. Immunol. Methods* **102:** 173–182.

Burley, S. K., and Petsko, G. A., 1986, Amino–aromatic interactions in proteins, *FEBS Lett.* **203:**139–143.

Chamber, R. D., and Sargent, C. R., 1981, Polyfluoroaromatic compounds, in: *Advances in Heterocyclic Chemistry*, Vol. 28 (A. R. Katritzky and A. J. Boulton, eds.), Academic Press, New York, pp. 47–59.

Chamber, R. D., Hutchinson, J., and Musgrave, W. K. R., 1964, Polyfluoroheterocyclic compounds. Part II. Nucleophilic substitution in pentafluoropyridine, *J. Chem. Soc.* **1964**:3736–3739.

Chamber, R. D., Kenneth, W., Musgrave, R., and Urben, P. G., 1975, Pyridinium salts of halogenated heterocyclic compounds, *Chem. Ind.* **1975**:89.

Cilento, G., and Giusti, P., 1959, Electron transfer from the indole nucleus to the pyridine coenzyme, *J. Am. Chem. Soc.* **81**:3801–3802.

Cilento, G., and Tedeschi, P., 1961, Pyridine coenzymes. IV. Charge transfer interaction with the indole nucleus, *J. Biol. Chem.* **236**:907–910.

Davies, D. R., and Metzger, H., 1983, Structural basis of antibody function, *Annu. Rev. Immunol.* **1**:87–117.

Dougherty, D. A., and Stauffer, D. A., 1990, Acetylcholine binding by a synthetic receptor: Implications for biological recognition, *Science* **250**:1558–1560.

Getzoff, E. D., Tainer, J. A., Lerner R. A., and Geysen, H. M., 1988, The chemistry and mechanism of antibody binding to protein antigens, in: *Advances in Immunology*, Vol. 43 (F. J. Dixon, ed.), Academic Press, San Diego, pp. 1–98.

Gilchrist, T. L., 1985, *Heterocyclic Chemistry*, Pitman, London, pp. 247–269.

Hutchens, T. W., and Porath, J., 1986, Thiophilic adsorption of immunoglobulins—analysis of conditions optimal for selective immobilization and purification, *Anal. Biochem.* **159**:217–226.

Hutchens, T. W., and Porath, J., 1987, Thiophilic adsorption: A comparison of model protein behavior, *Biochemistry* **26**:7199–7204.

Katritzky, A. R., and Lagowski, J. M., 1967, *The Principles of Heterocyclic Chemistry*, Methuen & Co., London, pp. 41–76.

Kosower, E. M., 1960, Charge-transfer complexing of pyridinium rings, in: *The Enzymes*, Vol. 3 (P. D. Boyer, H. Lardy, and K. Myrback, eds.), Academic Press, New York, pp. 171–194.

Lindmark, R., Thoren-Tolling, K., and Sjoquist, J., 1983, Binding of immunoglobulins to Protein A and immunoglobulin levels in mammalian sera, *J. Immunol. Methods* **62**:1–13.

Ngo, T. T., 1991, Synthetic affinity ligand compositions and methods for purification and recovery of organic molecules, *U.S. Patent* 4,981,961.

Ngo, T. T., and Khatter, N., 1990, Chemistry and preparation of affinity ligands useful in immunoglobulin isolation and serum protein separation. *J. Chromatogr.* **510**:281–291.

Ngo, T. T., and Khatter, N., 1991, Rapid and simple isolation of multigram goat IgG from serum using Avid AL™ and radial flow column, *Appl. Biochem. Biotechnol.* **30**:111–119.

Ngo, T. T., and Khatter, N., 1992, Avid AL, a synthetic ligand affinity gel mimicking immobilized bacterial antibody receptor for purification of immunoglobulin G, *J. Chromatogr.* **597**:101–109.

Petti, M. A., Shepodd, T. J., Barrans, R. E., Jr., and Dougherty, D. A., 1988, "Hydrophobic" binding of water-soluble guests by high-symmetry, chiral hosts. An electron-rich receptor site with a general affinity for quaternary ammonium compounds and electron-deficient systems, *J. Am. Chem. Soc.* **110**:6825–6840.

Porath, J., 1978, Explorations into the field of charge-transfer adsorption. *J. Chromatogr.* **159**:13–24.

Porath, J., 1979. Charge-transfer adsorption in aqueous media, *Pure Appl. Chem.* **51**:1549–1559.

Porath, J., 1987, Salting-out adsorption techniques for protein purification, *Biopolymers* **26**:S193–S204.

Porath, J., 1989, Electron-donor–acceptor chromatography (EDAC) for biomolecules in aqueous solutions, in: *Protein Recognition of Immobilized Ligands* (T. W. Hutchens, ed.), Alan R. Liss, New York, pp. 101–122.

Porath, J., and Belew, M., 1987, "Thiophilic" interaction and the selective adsorption of proteins, *Trends Biotechnol.* **5**:225–229.

Porath, J., Maisano, F., and Belew, M., 1985, Thiophilic adsorption—a new method for protein fractionation, *FEBS Lett.* **185**:306–310.

Reid, K. S. C., Lindley, P. F., and Thornton, J. M., 1985, Sulphur–aromatic interactions in proteins, *FEBS Lett.* **190**:209–213.

Satow, Y., Cohen, G. H., Padlan, E. A., and Davies, D. R., 1986, Phosphocholine binding immunoglobulin Fab McPC603, an X-ray difraction study at 2.7 A, *J. Mol. Biol.* **190**:593–604.

Sussman, J. L., Harel, M., Frolow, F., Oefner, C., Goldman, A., Toker, L., and Silman, I., 1991, Atomic structure of acetylcholinesterase from *Torpedo californica*: A prototypic acetylcholine-binding protein, *Science* **253**:872–879.

Ungar, F., and Alivisatos, S. G. A., 1961, Spectrophotometric evidence of certain anion–DNP$^+$ interactions (including orthophosphates), *Biochim. Biophys. Acta* **46**:406–408.

VI

Affinity-Related Techniques

28

Model Systems Employing Affinity Chromatography for Extraction of Toxic Substances Directly from Whole Blood

Shlomo Margel and Leon Marcus

1. INTRODUCTION

Detoxification of harmful ions, molecules, and biocomplexes directly from the blood of patients without iatrogenic sequelae should be attainable with presently available technology. The techniques currently employed are nonspecific, and the sorbents for the most part are nonbiocompatible and expensive. Therefore, a major goal of research on blood detoxification should be the rational development of sorbents that are both specific and biocompatible. Such affinity sorbents are both possible and feasible. Eventually, bioengineering will produce large quantities of therapeutically useful affinity ligands.

Low-molecular-weight toxic molecules can filter across a concentration gradient (peritoneal dialysis or hemodialysis). Dialysis is nonspecific; only low-molecular-weight molecules below the cutoff limit of the membrane are removed. More sophisticated equipment is required to remove nondialyzable toxic molecules from blood. During plasmapheresis, for example, blood cells are separated from the blood by centrifugation. The soluble portion of the blood perfuses the sorbent, and the detoxified plasma is returned to the patient with his cells. Apheresis requires dedicated, highly sophisticated, expensive equipment and must be used because the sorbents disrupt blood cells. During plasma exchange, the entire plasma of a patient is exchanged with toxic molecule-free fluid. The rationale for this technique is that autoimmune mediators/activators, that is, antigens, antibodies, immune complexes, and others incompletely understood or as yet to be discovered, are removed en bloc from the circulation. Plasma exchange is nonselective and nonspecific. Furthermore, the plasma must be replaced with donor plasma, with the risk of transmission of hepatitis, AIDS, and so on. Alternatively, oncotic colloids may be infused, leaving the patient at risk until autologous blood components are formed. Clearly, either alternative carries risks that must be considered. However, for hemoperfusion, in which whole blood pervades a sorbent for removal of toxic components, only access for blood

Shlomo Margel • The Department of Chemistry, Bar Ilan University, Ramat Gan, Israel.   *Leon Marcus* • Pediatric Clinic, Kupat Holim Clalit Health Fund, Kiryat Moshe, Rehovot, Israel.

Molecular Interactions in Bioseparations, edited by That T. Ngo. Plenum Press, New York, 1993.

egress and return, a blood pump, the cartridge containing the sorbent, and standard blood inflow and outflow sets are required, all of which should be available in the emergency room or dialysis unit. Until recently, mainly nonspecific sorbents have been used for hemoperfusion. We will discuss the process of hemoperfusion in which specific or relatively specific ligands within agarose–polyacrolein microsphere beads (APAMB) chelate toxic molecules.

The ultimate goal for removal of pathological components from the blood must include the development of (1) specific chelators (affinity ligands) and (2) a biocompatible system such that we may perfuse whole blood through the ligands, thus obviating the preliminary step of separation of the formed elements of the blood from the plasma. The most selective sorbents developed to date are derived from the technique of affinity chromatography. Specific antigens or antibodies are coupled to a suitable matrix to bind complementary molecules. Either the backbone holding the ligands and the ligands themselves are biocompatible or the ligands and spacer arms may be packaged in a suitable biocompatible envelope.

The practical use of hemoperfusion therapy originated with the audacious clinical trials performed by Yatzidis (1964). Although the hemoperfusion therapy was somewhat efficacious, these clinical trials were fraught with difficulties: the charcoal adsorbent shed embolic particles, platelets and leukocytes (WBC) were depleted drastically, and erythrocytes (RBC) were disrupted, all with serious clinical sequelae. However, Yatzidis's pioneering trail was soon followed by Chang and others, who developed charcoal coated with various materials (Chang and Malave, 1970; Andrade et al., 1971; Chang et al., 1971), nonionic exchange resins (Pallota and Koppanyi, 1960; Rosenbaum et al., 1971), and other sorbents (Holloway et al., 1979; Seideman et al., 1984) to obviate these serious problems. Unfortunately, these sorbents remain nonspecific and remove essential biomolecules in addition to toxic ones (Winchester et al., 1977). Furthermore, many important toxic molecules are removed to a limited extent, if at all, by these sorbents.

2.   MATERIALS AND METHODS

2.2.   Cross-linked Agarose–Polyacrolein Microsphere Beads (APAMB)

We have designed, developed, and fabricated a novel adsorbent system, cross-linked APAMB, which consists of polyacrolein microspheres encapsulated in agarose. Covalently bound to the microspheres are affinity ligands (Margel et al., 1982; Margel and Offarim, 1983; Margel, 1982) or, if available, a specific nonantibody ligand, for example, deferoxamine for iron (Horowitz et al., 1984). Alternatively, the microspheres themselves may possess chelating properties, as in the case of polymercaptals (Margel and Hirsch, 1981; Margel, 1981). Any protein, antigen, antibody, or hapten attached to a carrier containing free amino groups may be bound covalently to the microspheres through the latter's aldehyde groups. Thus, the highly specific technique of affinity chromatography may be utilized. The APAMB are highly porous; protein molecules penetrate freely to interact with the ligand on the microspheres. High flow rates are readily obtained. For hemoperfusion, the optimal diameter of the APAMB is in the range 500–800 μm. Most importantly, the beads are compatible with blood cells and the soluble components of the blood so that whole blood directly perfuses the beads. Centrifugation and reconstitution steps are not required.

In some 100 *in vivo* hemoperfusion trials with rabbits, dogs, and humans, the safety and biocompatibility of our system have been validated by the following results:

1. Minimal loss of RBC, WBC, and platelets occurred (Fig. 1).
2. Routinely assayed soluble blood components—Cl^-, K^+, Na^+, Ca^{2+}, PO_4^{3-}, urea, uric acid, and creatinine—were not significantly affected (Fig. 2).
3. Likewise, C' (Margel and Marcus, 1986), total protein, albumin, bilirubin, and the enzymes serum glutamic oxalacetic transaminase (SGOT), lactate dehydrogenase (LDH), and alkaline phosphatase did not change significantly (Fig. 1).
4. Leakage of microspheres, agarose, and acrolein was below the limits of detection by nephelometry and/or spectrophotometry (Horowitz *et al.*, 1984).
5. Leakage of antibodies bound to the APAMB into the blood was insignificant as determined by radioimmunoassay (Margel and Marcus, 1986). Table 1 contains supporting data using a sensitive enzymatic assay method to detect leakage of ligands. When avidin was bound to the standard cyanogen-activated Sepharose 4B beads, the leakage was at least 50 times greater than found with APAMB.

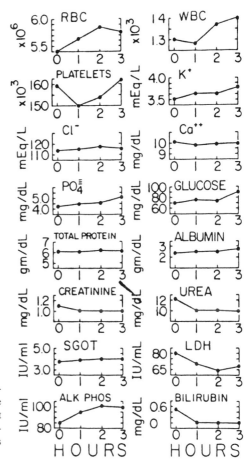

FIGURE 1. Various parameters of the formed elements and routinely assayed soluble components of the blood of 40 dogs during hemoperfusion trials. Weight of dogs, 15–25 kg. Column contained 17–38 g APAMB-antidigoxin. Blood flow rate, 80–120 ml/min. Values have been normalized about the mean value.

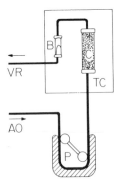

FIGURE 2. Schematic depiction of the hemoperfusion procedure. Abbreviations: AO, arterial outflow; P, peristaltic pump; C, column containing the APAMB; B, bubble trap; VR, venous return; TC, temperature-controlled environment.

6. The APAMB-specific ligands are routinely sterilized by irradiation with a cobalt source (2.5 Mrad). This procedure was found to result in almost no decrease in the ligand reactivity (Margel and Marcus, 1986).

2.2. Preparation of the APAMB Ligand

In brief, the microspheres are formed by cobalt irradiation of polyacrolein (Margel *et al.*, 1982) and are then encapsulated within molten agarose (Holloway *et al.*, 1979; Margel and Offarim, 1983). The agarose beads, in turn, are cross-linked (Porath *et al.*, 1975; Marcus *et al.*, 1984b). Amino ligands react with the APAMB in a single step carried out at physiological pH (Bayer *et al.*, 1986).

A diagram of the clinical procedure is shown in Fig. 2. The cannulas, tubing system, and column as well as the hemoperfusion procedure were described by us in detail for

TABLE 1

Leakage of Ligands Bound to APAMB
during Hemoperfusion of Dogs[a] and *In Vitro*[b]

	Ligand	Ligand/ml (ng)	Ligand/dog (μg)	Ligand released (%)
1	Antidigoxin	0.29[c]	0.20	0.16×10^{-3}
2	Antidigoxin	0.20[c]	0.17	0.14×10^{-3}
3	Antiparaquat	0.0[c]	0.0	0.0
4	Avidin	1.6[c]	—	0.9×10^{-3}
5	Avidin	0.0[d]	—	0.0

[a] Dogs were hemoperfused for 3 h at a blood flow rate of 120 ml/min and a body temperature of 98.6°F (37°C). Column contained 25 g of APAMB-antidigoxin or -antiparaquat, with 5 mg of antidigoxin or 18 mg of antiparaquat per gram of APAMB, respectively. A 60-ml sample of whole blood was drawn at $t = 0$ and $t = 3$ h. Antibody content of the plasma was quantified by radioimmunassay (^{131}I for antidigoxin; ^{14}C for antiparaquat).

[b] Plasma was perfused through 10 g of beads for 4 h at a flow rate of 35 ml/min. APAMB contained 17.8 mg avidin/g of beads. Avidin was quantified by an enzyme assay (Bayer *et al.*, 1986).

[c] Ligand in plasma.

[d] Ligand in saline.

rabbits (Mashiah *et al.*, 1984), dogs (Marcus *et al.*, 1985), and humans (Savin *et al.*, 1987). "Helsinki Committee" approval was received prior to initiation of human clinical trials. Protocols of safety and patient/patient family approval were adhered to strictly (Savin *et al.*, 1987).

Figure 3A is a photomicrograph of the APAMB. Figure 3B is a thin-section photomicrograph of the beads, showing the microspheres encapsulated within the agarose matrix.

3. RESULTS AND DISCUSSION

Several model systems which were examined during the course of our research are detailed below. These point to the utility of hemoperfusion therapy with biocompatible sorbents for various clinical problems. The uses are limited only by the imagination of the researcher and/or clinician.

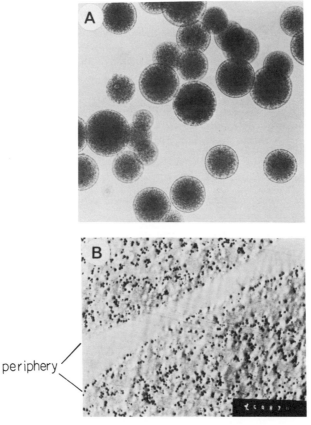

FIGURE 3. (A) Photomicrograph of APAMB; diameters range from 500 to 800 μm. (B) Thin-section photomicrograph of APAMB showing the microspheres encapsulated within the agarose matrix; mean diameter of the microspheres is 0.2 μm. Partial circumferences of two beads are depicted.

3.1. In Vivo *Hemoperfusive Removal of Specific Antigen*

3.1.1. Model System: Removal of Digoxin from the Blood of Intoxicated Animals with APAMB-Antidigoxin

Digoxin is widely used in the treatment of congestive heart failure and atrial fibrillation. Up to 35% of patients receiving this treatment suffer symptoms of digoxin intoxication, and up to 21% of these patients do not survive (Smith and Haber, 1973). The high rates of toxic sequelae result from the widespread use of digoxin in senior citizens with reduced renal elimination and increased myocardial sensitivity. Furthermore, therapeutic blood levels are close to that which produces toxicity, and thus digoxin intoxication is common. Additionally, digoxin has been used for self-poisoning, marked by an exceptionally high mortality rate. Current therapy, either hemodialysis (Ackerman *et al.*, 1967; Iisalo and Forrstrom, 1974) or hemoperfusion through coated charcoal (Carvallo *et al.*, 1975; Tobin *et al.*, 1977; Gibson *et al.*, 1978; Smiley *et al.*, 1978) or nonionic exchange resins (Winchester *et al.*, 1977), is not very effective. Recently, the eminently successful technique of infusion of antidigoxin antibodies or Fab antibody fragments to complex and inactivate digoxin was added to our armamentarium (Schmidt and Butler, 1971; Curd *et al.*, 1971; Smith *et al.*, 1976, 1982). The resultant immune complexes are excreted in the urine. Two theoretical drawbacks come to mind. First, activation of the immune system may preclude subsequent treatment by this modality. Second, some of these patients may not possess sufficient renal function to excrete the immune complexes. In addition, of course, the high cost is a factor to be considered. The therapy which we have developed and used successfully and which has proved successful in human clinical trials combines the advantages of hemoperfusion and antidigoxin antibodies. The sorbent, APAMB-antidigoxin, removes only digoxin from whole blood of intoxicated patients. Clinical manifestations of digoxin intoxication, namely, vomiting, diarrhea, tachycardia, and severe ECG changes, were ameliorated in the process. Patients and animals survived intoxications fatal to nonhemoperfused cohorts. Hemoperfusion was also effective in human patients with an impaired renal system.

3.1.1.1. Animal Hemoperfusion Trials.   Dogs were dosed with digoxin to cause severe arrhythmias which would have resulted in death without medical intervention. Table 2 summarizes the results from these clinical trials. Digoxin is rapidly removed from the serum by hemoperfusion through APAMB-antidigoxin. Digoxin bound to RBC receptors enters the serum and is removed at a rate slightly slower than that for digoxin already in the

TABLE 2
Summary of Digoxin Intoxication Clinical Trials in Dogs[a]

Trial no(s).	Therapy	Outcome
1, 2	None	Died
3–5	CPR/antiarrhythmics	Died
6	Hemoperfusion through APAMB lacking antibodies	Died
7–46	Hemoperfusion through APAMB-antidigoxin	Survived

[a]Cardiac disturbances were arrhythmias in all cases.

serum. Digoxin bound to heart and kidneys equilibrates with the serum component and is removed at an even slower rate (Gibson *et al.*, 1978). In any event, in chronic and acute intoxication, sufficient digoxin is removed within two blood volume perfusions through APAMB-antidigoxin to convert the arrhythmia from a potentially lethal state to a non-life-threatening one. Under our clinical trials, the ECG pattern returns to the preintoxicated state within 2 h. The dogs were well and survived indefinitely after hemoperfusion (Marcus *et al.*, 1985).

3.1.1.2. Human Hemoperfusion Trials. We also have used our technique in iatrogenically, acutely intoxicated humans. Thus far, 12 patients have been treated successfully (Savin *et al.*, 1987). These patients manifested toxicity as severe heart blocks, severe bradycardia, and in some cases severely abnormal arrhythmias. In all cases, life-threatening aspects of the intoxication were ameliorated by hemoperfusion through APAMB-antidigoxin. Of course, the underlying pathology that prompted digoxin therapy remained. In any event, hemoperfusion with affinity sorbents was extremely useful in bringing patients out of potentially lethal intoxications. A typical ECG series during therapy in humans is shown in Fig. 4; the severe heart block is decreased. Marked

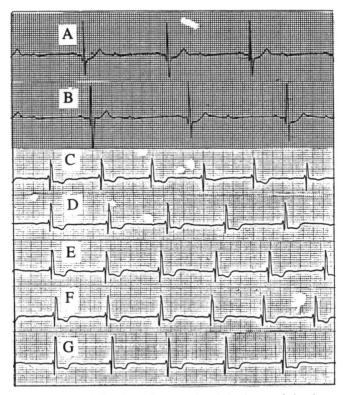

FIGURE 4. ECG strips of a digoxin-intoxicated human patient during hemoperfusion therapy with APAMB-antidigoxin. The patient was a 75-year-old female patient, weighing 65 kg. Column contained 30 g of APAMB-antidigoxin. Blood flow rate, 120 ml/min. ECGs were obtained at time = −2 h (A), at start of hemoperfusion (B), after 1 (C), 2 (D), 3 (E), and 4 (F) h of hemoperfusion, and 1 h post hemoperfusion (G).

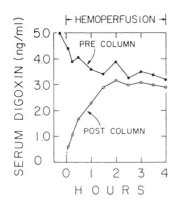

FIGURE 5. Rate of removal of digoxin from the blood of the patient in Fig. 4 during hemoperfusion with APAMB-antidigoxin.

improvement was noted within 30 min. The corresponding time course of removal of digoxin is shown in Fig. 5. All patients showed a change of heart rate toward normal, a decrease of nausea, and an interest in their surroundings and in food. The procedure was well tolerated. The formed elements and serum components of the blood were not altered significantly during therapy. There was little, if any, hemodynamic load, due to the small cartridges used. Neither hypotensive episodes nor bleeding diatheses occurred. Rebound or ECG manifestations due to digoxin toxicity were not seen in the subsequent days in which the patients were under observation. Hospital stay was greatly reduced.

3.1.2. Model System: Hemoperfusive Removal of Paraquat in Poisoned Animals

Paraquat is an effective herbicide and is safe when used with the usual precautions. However, accidental spray contamination, oral ingestion, or deliberate intake, that is, smoking "tobacco" treated with paraquat, causes severe morbidity and mortality reportedly ranging from 50% to 90% (Editorial, 1971; Tabei et al., 1982). Detoxification soon after contact is essential to obviate accumulation into tissue, notably into the pulmonary system (Sideman et al., 1984; Azhari et al., Okonek et al., 1976). Hemoperfusion of poisoned blood through activated charcoal or nonionic exchange resins is equivocally effective (Okonek et al., 1976; Sketris and Skoutakis, 1981; Winchester and Gelfand, 1978; Tabak et al., 1983; Lotan et al., 1983; Cavelli and Fletcher, 1977). Hemoperfusion through Fuller's earth is more effective than through encapsulated charcoal (Sideman et al., 1984); Tabak et al., 1983; Lotan et al., 1983). Currently permissible therapy includes peritoneal dialysis or hemodialysis (Eliahou et al., 1973; O'Brien, 1977). Unfortunately, these procedures are marginally effective at best.

In vitro experiments in our laboratory with APAMB-antiparaquat antibodies demonstrated the efficacy of the system (Azhari et al., 1986) The very rapid rate of removal of paraquat in an animal trial is shown in Fig. 6.

Charcoal and agarose-encapsulated Fuller's earth possess roughly the same capacity as our system (Sideman et al., 1984; Azhari et al., 1986) However, the APAMB-antiparaquat have much higher affinity and specificity for paraquat and remove the herbicide at a much more rapid rate than does charcoal. Infusion of acid citrate dextrose to obviate platelet depletion as in the Fuller's earth system is not required with the APAMB.

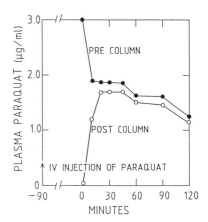

FIGURE 6. Rate of removal of paraquat from the blood of an intoxicated dog during therapy with APAMB-antiparaquat. Weight of dog wt, 10 kg. The dog received 2.2 of mg paraquat/kg wt by i.v. administration prior to hemoperfusion. Column contained 33 g of APAMB-antiparaquat. Blood flow rate, 80 ml/min.

3.2. Removal of Toxic Cations From Blood

3.2.1. Model System: Removal of Iron from Poisoned Animals with the Ligand Deferoxamine

The pathology, morbidity, and mortality of acute and chronic iron poisoning are well documented (O'Brien, 1977); a physiological route for the elimination of excess iron in humans is not known (McCance and Widdowson, 1937). Chronic iron overload occurs in patients treated with blood transfusions in thalassemia and aplastic anemia and in hemochromatosis. Furthermore, acute iron poisoning from accidental ingestion of iron tablets, mainly in the pediatric age group, is responsible for an estimated 2000 hospitalizations annually in the United States (Stein *et al.*, 1976).

The modern treatment of acute iron overload is slow infusion of deferoxamine in tandem with conventional life-supporting measures (Barry *et al.*, 1974; Modell 1979; Keberle 1964). The feroxamine formed is eliminated in the urine. Unfortunately, deferoxamine is expensive, has a short serum half-life, and, when given via the oral route, is both inefficient and somewhat toxic. Additional problems arise in cases of renal failure, since then the toxic feroxamine cannot be eliminated by urinary excretion.

Hemodialysis is ineffective; much of the excess iron is bound to specific protein carriers and nonspecifically to albumin and cannot cross the dialysis membrane. Nonspecific sorbents are likewise ineffective. Our affinity ligand hemoperfusion system, APAMB-deferoxamine, complexes iron and also is useful in patients with a nonfunctioning renal system. Excess iron, along with its nonspecific protein carrier, easily enters the APAMB-deferoxamine.

Hemoperfusive Removal of Iron through APAMB-Deferoxamine. Normal serum iron levels center narrowly about 1 ppm. Levels of 5 to 10 ppm are dangerous. Levels above 10 ppm are potentially lethal, requiring emergency, and sometimes heroic, measures. The rate of removal of iron during a hemoperfusion trial in a severely, acutely intoxicated dog is shown in Fig. 7. Up to 40% of an extremely high level of iron was removed in 3 h. Subsequent hemoperfusions would be required to bring the iron content below the danger level. However, this may still be preferred to the standard deferoxamine infusion therapy.

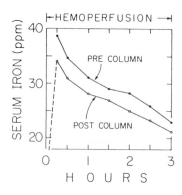

FIGURE 7. Rate of removal of iron from the blood of an "acutely" intoxicated dog during hemoperfusion with APAMB-deferoxamine. The dog received 60 mg of iron as ferric chloride dissolved in 100 ml of his own plasma by i.v. administration. Weight of dog, 17 kg. Column contained 50 g of APAMB-deferoxamine. Blood flow rate, 120 ml/min.

Hemoperfusion with APAMB-deferoxamine may be a useful method to remove iron in acute iron intoxication.

Since iron intoxication is potentially lethal in the toddler age group, it might be of value to explore the use of some of the bacterial iron-binding ligands which have a much greater affinity for iron (Neilands, 1974).

3.2.2. Model System: Removal of Mercury with Polymercaptal in Agarose–Polymercaptal Microsphere Beads (APMMB)

Poisoning due to organic or inorganic mercury compounds often results in nervous system involvement—at best, borderline mental retardation, and, at worst, brain damage and death (Aaronson, 1971; Grant, 1971). Accepted therapy includes intravenous (i.v.) infusion of the chelating drugs 2,3-dimercapto-1-propanol (BAL) and/or 2-amino-3-mercapto-3-methylbutanoic acid (penicillamine). BAL often causes renal damage and redistribution of the mercury in tissue; penicillamine is less efficacious and may cause nephrotic syndrome. Hemodialysis is somewhat effective (Kostyniak et al., 1977); hemoperfusion is more so. We have designed and used agarose–polymercaptal microsphere beads (APMMB) to take advantage of the mercury-binding capacity of the —SH groups (Margel, 1981). While these are less than specific, mercury intoxication may be so devastating that the need for an even partially effective therapy forces emergency procedures. The trials in rabbits, were somewhat difficult to perform technically since it was necessary to infuse large volumes of cysteine, via a third i.v. line, to displace mercury from tissue receptors. In some instances, we encountered problems of fluid overload in the rabbits. In addition, in these, our earliest experiences with animal hemoperfusions, reductions of 20–30% in the formed elements of the blood were not uncommon. Figure 8 depicts a trial in which a 50% reduction in serum mercury occurred in less than 1 h. The 67% decline over a period of 2 h indicates displacement of mercury from tissue (Margel et al., 1984a). We certainly will be able to overcome the technical difficulties in the future.

3.3. Removal of Specific Antibody and Immune Complexes; Autoimmunological Disorders

Numerous autoimmunological diseases have been described, many of which have been well studied with the result that an advanced understanding of the molecular biology of the

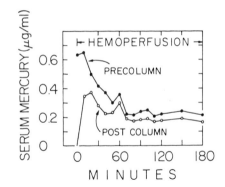

FIGURE 8. Rate of removal of mercury from the blood of a mercury-poisoned rabbit during hemoperfusion with APMMB. The rabbit received 4 mg of methylmercury by i.v. administration 90 min before initiation of therapy. Weight of rabbit, 4.0 kg. Blood flow rate, 10 ml/min. A 0.5*M* cysteine solution was infused at a rate of 1 ml/min.

disease processes has been achieved (Roitt, 1985). They are managed with steroids and cytotoxic drugs and recently by apheresis. Autoimmunological syndromes, the symptoms of which are caused by antibodies produced against the patient's own tissue, should be amenable to affinity-ligand hemoperfusion therapy.

Plasma exchange of patients with syndromes ascribed to autoantibodies has been employed for rheumatoid arthritis, myasthenia gravis, thyroiditis, glomerulonephritis, and other diseases with equivocal but encouraging results (Shumak and Rock, 1984). However, all soluble blood components are removed, requiring subsequent reinfusion of donor serum albumin or fresh frozen plasma with its attendant danger of transmission of hepatitis, AIDS, etc. Alternatively, saline may be reinfused, temporarily leaving the patient deficient in vital antibodies and coagulation and other factors. These dangers are obviated with hemoperfusion.

3.3.1. Model System: Removal of Anti-Bovine Serum Albumin (Anti-BSA) with APAMB-BSA

We developed a model system in which immunized rabbits contained elevated levels of anti-BSA (Marcus *et al.*, 1982) The ligand, BSA, was bound to the APAMB.

A composite of several typical rabbit hemoperfusion trials is shown in Fig. 9. Under our experimental conditions, 95% of the anti-BSA was removed in 120 min. In an extended series of rabbit trials, we demonstrated that the APAMB-BSA are indeed capable of removing specific antibodies while leaving the blood components unaffected (Marcus *et al.*, 1984). This is our model system presaging an attack on several autoimmune disorders.

Several serious syndromes which are potentially amenable to treatment by affinity chelating techniques and are in fact receiving urgent, aggressive attention are discussed below.

3.3.2. Model System: Removal of Circulating Immune Complexes with APAMB-C1q

It is generally accepted that the elicitation and subsequent deposition of immune complexes (ICs) in tissue play a key role in autoimmunological patholological processes.

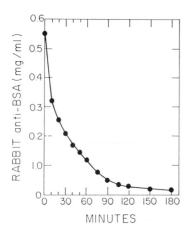

FIGURE 9. Rate of removal of anti-BSA from the blood of four immunized rabbits during hemoperfusion. Rabbit serum averaged 0.55 mg anti-BSA/ml. Weight of rabbits, 3.5–4.5 kg. Column contained 15 g of APAMB-BSA. Blood flow rate, 15 ml/min.

ICs interact with a number of blood cells—leukocytes, lymphocytes, platelets—as well as reticuloendothelial cells. Therapeutic approaches include inhibition of the inflammatory process by immunosuppression and/or cytotoxic chemicals as well as plasmapheresis. All carry the risks mentioned above.

ICs have been demonstrated in patients presenting with a variety of autoimmune, infectious, and malignant syndromes; a brief list would include DNA/anti-DNA IC in patients with systemic lupus erythematosus (Morimoto *et al.*, 1982), acetylcholine receptor/ anti-acetylcholine receptor IC in patients with myasthenia gravis (Barkas *et al.*, 1981), immunoglobulin G (IgG)/anti-IgG IC in patients with rheumatoid arthritis (Kunkel *et al.*, 1961), IC in patients with persistent hepatitis B virus (Wands and Zuraski, 1981; Brown *et al.*, 1984), and IC in patients with various malignant disorders (Gazitt *et al.*, 1982; Theophilopoulous and Dixon, 1980).

We have employed the affinity ligand C1q for removal of circulating ICs (Gazitt *et al.*, 1985). In *in vitro* affinity chromatography trials with APAMB-C1q, 80% of the BSA/anti-BSA IC was removed upon three perfusion flows of plasma containing the IC through the cartridge (Margel and Marcus, 1986). Similar hemoperfusions with plasma containing the hepatitis B virus surface antigen (HBVsAg) anti-HBVsAg IC removed a lesser degree of the IC (Gazitt *et al.*, 1985). In future work, the isolation of C1q will be scaled up for use in hemoperfusive therapeutic animal systems.

3.3.3. Removal of Anticholinesterase Receptor (AChR) Antibodies

Myasthenia gravis is a neurological disorder in which antibodies are formed against neurotransmitter receptor sites, inhibiting the passage of signal from nerves to muscles. The net result is general inhibition of activity. If respiratory muscles are affected, the disease is catastrophic. Heininger *et al.* (1985a, b), Yamazaki *et al.* (1982), and Sato *et al.* (1983) have developed a sorbent system in which a partially selective ligand, tryptophan, is linked to a poly (vinyl alcohol) gel. Plasma IgG, including up to 55% of the AChR antibodies, was removed in limited clinical trials. Plasmapheresis was employed. This does not meet the criteria listed as goals for removal of pathogenic molecules directly from whole blood but points to a system which may develop in that direction.

3.3.4. Tumor Therapy

Protein A stands out as the major (nonspecific) chelator for ICs. Protein A is a cell wall component of the bacterium *Staphylococcus aureus* which possesses a strong affinity for the Fc portion of IgG and thus has been coupled to various backbones and linkers for treatment of diverse autoimmune processes. Terman *et al*. (1980a, b) used treated whole cells of *S. aureus* C in dogs with tumors; a powerful necrotizing response was reported. A similar therapeutic response was confirmed by Messerschmidt *et al*. (1982). Serum blocking factors (SBF) play a role in the course of tumor development. Bansal *et al*. (1978) also utilized treated whole cells of *S. aureus* in a plasmapheresis system to treat one patient; tumor regression and decrease of SBF resulted. Plasma perfusion over a purified protein A–charcoal matrix was employed by Terman *et al*. (1981) in 5 patients with breast adenocarcinoma, with good responses. Bensinger *et al*. (1982) coupled the protein A to a silica matrix to treat 5 patients with breast adenocarcinoma; they reported a partial regression in 3 of the patients. MacKintosh *et al*. (1983) perfused plasma of 14 cancer patients over protein A coupled to Sepharose 4B with encouraging initial therapeutic results. The review by Jones *et al*. (1986) details a host of individually limited clinical trials reporting tantalizingly encouraging results. A practical device (the Prosorba column) in which protein A is coupled to a silica matrix was employed in studies conducted in 11 institutions. A cell/plasma separation step was required. In trials using well-established criteria for tumor evaluation, 45 of 83 patients responded to this therapy. Tumor shrinkage was noted in 17 patients, and tumor stabilization in 28 patients (Jones *et al*., 1986). Side effects were minimal and manageable.

3.4. Other Applications

3.4.1. Model System: Removal of LDL Cholesterol with anti-LDL Antibodies

Atherosclerotic cardiovascular disease is the leading cause of morbidity and mortality in the modern community (Castelli *et al*., 1977). It is now widely accepted that elevated plasma levels of low-density lipoprotein (LDL) correlate with an increased risk for the development of atherosclerosis. Lowering the LDL level in the blood via diet, drugs, and/or surgery is the current treatment of choice. Thompson *et al*. (1975) successfully lowered cholesterol levels in familial hypercholesterolemia patients via bimonthly plasma exchanges. Others reported similar success (Berger *et al*., 1978; King *et al*., 1980). However, large quantities of plasma are required as replacement, and the plasma itself may introduce viral, possibly fatal, infections. If fluids other than plasma replace the loss, there may be impaired clotting and/or immune response. To obviate some of these disadvantages, Lupein *et al*. (1976) and Graisely *et al*. (1980) used affinity chromatography, incubating the patient's blood extracorporeally with heparin linked to agarose beads. The levels of LDL-cholesterol were reduced. This procedure is somewhat nonspecific; other proteins, such as plasma coagulation factors (antithrombin III), components of the complement system, lipoproteins, and lipase, are known to bind with heparin. Recently, Yokohama *et al*. (1984) introduced a dextran sulfate sorbent covalently bound to cellulose beads that is relatively more specific for apolipoprotein B-containing lipoproteins. The system requires a cell/plasma separation step. A vastly improved technique introduced by Stoeffel and co-workers (Stoeffel and Demant, 1981; Stoeffel *et al*., 1981) also requires a preliminary cell/plasma separation step but uses a much more specific sorbent, that of anti-LDL antibodies. This has developed into a safe, economic, and efficient clinical procedure. Disadvantages of the system include (a) the requirement of a preliminary cell separation step, (b) the large bed volume of the affinity column (400 to 500 ml), and (c) the use of the cyanogen bromide coupling procedure, which leads to leakage of antibodies into the patient's bloodstream owing to the instability of the isourea bonds that are formed through this activation step (Margel and Offarim, 1983; Gray, 1980). To date, this therapy is the only practical method of removal of LDL-cholesterol other than the diet/drug/surgery triad.

We are developing an anti-LDL antibody-APAMB sorbent which has the advantages of (a) direct hemoperfusion without the cell/plasma separation step, (b) high capacity, (c) stable antibody–sorbent bond, and (d) high porosity (Margel et al., 1989).

3.4.2. Treatment of Endotoxic Shock

Fatal outcome in patients septicemic with gram-negative bacteria (endotoxic shock) remains high despite a new array of powerful antibiotics (Kreger et al., 1980a) and heroic efforts of improved life-support procedures (Kreger et al., 1980b). Most probably, the shock syndrome is triggered by reaction to bacterial cell wall–cell membrane components (lipopolycaccharide = endotoxin). Ziegler et al. (1982) reasoned that antiserum against core polysaccharide–lipid A complex could neutralize endotoxins and effect survival of patients in shock. Clinical trials were performed with 304 severely ill patients in eight participating hospitals. Along with the antiserum, antibiotics and treatment for hypotension, hypoxemia, acidosis, and other complications were given to the patients. Encouraging results were obtained in these patients compared with fatal outcomes in those receiving all of the usual life-support therapy without antiserum. Several international firms are marketing hyperimmune serum, with good results reported for any number of life-threatening syndromes.

One could envision several classes of hemoperfusion systems to treat endotoxic shock. Thus far, Hanasawa et al. (1989) have developed and taken to animal trials a Polymixin B-immobilized fiber (PMF-B) hemoperfusion system. It has done well in biocompatibility and animal survival trials. Nineteen of 20 control dogs expired within 18 h while 30 of 40 hemoperfused dogs survived 3 days, and 16 survived permanently.

3.4.3. Rescue Therapy after Administration of an Acutely Toxic Bolus of Chemotherapeutic Drug

The use of powerful antitumor drugs is limited by toxicity, although it has been established that higher doses might increase remission rates. With the so-called rescue technique, higher doses are presented to the patient, preferably by i.v. administration into the efferent blood vessel so that the drug need not traverse the entire body as well as the intended site. Then just distal to the organ to be treated, the drug is recaptured by perfusing the blood through a suitable sorbent. Several investigators (Leiter et al., 1966; Oberfeld et al., 1979; Ogata et al., 1974; Harada et al., 1981; Kihara et al., 1988) tested this technique using charcoal adsorbent to capture mitomycin C, 5-fluorouracil, and/or adriamycin. Initial studies indicate that selective arterial infusion, high-dose chemotherapy with recapture hemoperfusion may be a valuable treatment for advanced cancer patients. The system we have presented with suitable affinity sorbents would be very useful with this technique. Limitations in the use of this general technique are set only by the imagination of the clinician and/or investigator.

REFERENCES

Aaronson, I., 1971, Mercury in the environment, *Environment* **13:**16–27.

Ackerman, G. L., Doherty, J. E., and Flanagan, W. J., 1967, Peritoneal dialysis and hemodialysis of tritiated digoxin, *Ann. Int. Med.* **67:**718–729.

Andrade, J. D., Kunitomo, K., Van Wagenen, R., Kastigir, D., Gough, D., and Kolff, W. J., 1971, Coated absorbents for direct blood perfusion: HEMA/activated charcoal, *Trans. Am. Soc. Artif. Intern. Organs* **17:**222–228.

Azhari, R., Labes, A., Haviv, Y., and Margel, S., 1986, Extracorporeal specific removal of paraquat by hemoperfusion through antiparaquat conjugated to agarose–polyacrolein microsphere beads, *J. Biomed. Mater. Res.* **21:**25–41.

Bansal, S. C., Bensal, B. R. and Thomas, H. L., 1978, *Ex vivo* removal of serum IgG in a patient with colon carcinoma: Some biochemical, immunological and histological observations, *Cancer* **42:**1–18.

Barkas, T., Boyle, R. S., and Behan, P. O., 1981, Immune-complexes in myasthenia gravis, *J. Clin. Lab. Immunol.* **6:**27–36.

Barry, M., Flynn, D. M. Letsky, E. A., and Risdon, R. A., 1974, Long term chelation therapy in thalassemia major: Effect on liver concentration, liver histology and clinical progress, *Br. Med. J.* **2:**16–20.

Bayer, E. A., Ben-Hur, H., and Wilchek, M., 1986, A sensitive assay for biotin, avidin and streptavidin, *Anal. Biochem.* **154:**367–370.

Bensinger, W. I. Winet, J. P., Hennen, G., Franckenne, F., Schaus, C., Saint-Remy, M., Hoyoux, P., and Mahieu, P., 1982, Plasma perfusion over immobilized protein A for breast cancer, *N. Engl. J. Med.* **306:**935–936.

Berger, G. M. B., Miller, J. L., Bonnici, F., Jofe, H. S., and Oubovsky, D. W., 1978, Continuous flow plasma exchange in the treatment of homozygous familial hypercholesterolemia, *Am. J. Med.* **65:**243–253.

Brown, S., Howard, C. R., Steward, M. W., Ajdukiewiecz, H. C., and Whittle, A., 1984, Hepatitis B surface antigen containing immune complexes occur in seronegative hepatocellular carcinoma patients, *Clin. Exp. Immunol.* **55:**355–359.

Carvallo, A., Ramirez, B., Honig, H., Knepshield, J., Schreiner, G., and Gelfand, M. D., 1975, Treatment of digitalis intoxication by charcoal hemoperfusion, *Trans. Am. Soc. Artif. Intern. Organs* **22:**718–720.

Castelli, W. P., Doyle, J. T., Gordon, T., Hames, C., Hjortland, M. C., Hulley, F. B., Kagan, A., and Zukel, W. J., 1977, HDL cholesterol and other lipids in coronary heart disease, *Circulation* **55:**767–772.

Cavelli, R. D., and Fletcher, K., 1977, An effective treatment for paraquat poisoning, in: *Biochemical Mechanisms of Paraquat Poisoning* (A. P. Autor, ed.), Academic Press, London, pp. 213–229.

Chang, T. M. S., and Malave, N., 1970, The development and first clinical use of semipermeable microcapsules (artificial cells) formed from membrane activated charcoal, *Trans. Am. Soc. Artif. Intern. Organs* **16:**141–148.

Chang, T. M. S., Gonda, A., Dirks J. H., and Malave, N., 1971, Clinical evaluation of chronic, intermittent and short term hemoperfusion in patients with chronic renal failure using semipermeable microcapsules (artificial cells) formed from membrane coated activated charcoal, *Trans. Am. Soc. Artif. Intern. Organs* **17:**246–252.

Curd, J., Smith, T. W., Jaton, J. C., and Haber, E., 1971, *Proc. Natl. Acad. Sci. U.S.A.* **68:**2401–2406.

Editorial, 1971, Paraquat poisoning, *Lancet* **ii:**1018.

Eliahou, H. E., Almog, C., Gura, V., and Iaina, A., 1973, Treatment of paraquat poisoning by hemodialysis, *Isr. J. Med. Sci.* **9:**459–462.

Gazitt, Y., Klein, G., and Sulitzeanu, D., 1982, Reactivity with patient antibodies of partially purified gp40 antigen from immune complexes in Burkitt's lymphoma and nasopharyngeal carcinoma, *Int. J. Cancer* **29:**645–651.

Gazitt, Y., Margel, S., Lerner, A., Wands, J. R., and Shouval, D., 1985, Development of novel C1q immunoadsorbent for removal of circulating immunocomplexes: Quantitative isolation of hepatitis B virus surface antigen and immunocomplexes, *Immunol. Lett.* **11:**1–8.

Gibson, T. P., Lucas S. V., Nelson, H. A., Atkinson, A., Okida, J. T., and Ivanovich, P., 1978, Hemoperfusive removal of digoxin from dogs, *J. Lab. Clin. Med.* **240:**673–682.

Graisely, B., Cloarec, M., and Salmon, S., 1980, Extracorporeal plasma therapy of homozygous familial hypercholesterolemia, *Lancet* **ii:**1147–1150.

Grant, N., 1971, Mercury in man, *Environment* **13:**2–15.

Gray, G. R., 1980, Affinity chromatography, *Anal. Chem.* **52:**9R–15R.

Hanasawa, K., Aoki, H., Yoshioka, T., Matsuda, K., and Kodama, M., 1989, Novel mechanical assistance in the treatment of endotoxic and septicemic shock, *Trans. Am. Soc. Artif. Intern. Organs* **35:**341–343.

Harada, T., Ohmura, H., Nishizawa, O., and Tsuchida, S., 1981, Regional arterial infusion of an anticancer drug combined with direct hemoperfusion, *Tohoku J. Exp. Med.* **133:**423–429.

Heininger, K., Gaczkowski, A., Hartung, H. P., Toyka, K. V., and Borberg, H., 1985a, Plasma separation and immunoadsorption in myasthenia gravis, in: *5th Symposium on Therapeutic Plasmapheresis, Tokyo, Therapeutic Plasmapheresis (V)* (T. Oda, ed.), Schattuer Verlag, Stuttgart, pp. 50–55.

Heininger, K., Hendricks, M., and Toyka, K. V., 1985b, Myasthenia gravis: A new selective procedure to remove acetylcholine receptor autoantibodies from plasma, *Plasma Ther. Transfus. Technol.* **6:**771–775.

Holloway, C. J. K., Harstick, K., and Brunner, C., 1979, Agarose-encapsulated adsorbents, *Int. J. Artif. Organs* **2:**81–86.

Horowitz, D., Margel, S., and Shimoni, T., 1984, Iron detoxification by hemoperfusion through deferoxamine-conjugated agarose–polyacrolein microsphere beads, *Biomaterials* **6:**9–16.

Iisalo, E., and Forsstrom, J., 1974, Elimination of digoxin during maintenance hemodialysis, *Ann. Clin. Res.* **6:** 203–211.

Jones, F. R., Balint, J. P., and Snyder, H. W., 1986, Selective extracorporeal removal of immunoglobulin-G and circulating immune complexes, *Plasma Ther. Transfus. Technol.* **7:**333–349.

Keberle, H., 1964, The biochemistry of desferrioxamine and its relation to iron metabolism, *Ann. N.Y. Acad. Sci.* **119:**758–768.

Kihara, T., Nakazawa, H., Agishi, T., Honda, H., and Ota, K., 1988, Superiority of selective bolus infusion and simultaneous rapid removal of anticancer agents by charcoal hemoperfusion, *Trans. Am. Soc. Artif. Intern. Organs* **34:**581–584.

King, M. E., Breslow, J. L., and Lees, R. S., 1980, Plasma exchange therapy of homozygous familial hypercholesterolemia, *N. Engl. J. Med.* **302:**1457.

Kostyniak, P. J., Clarkson T. W., and Abbasi, A. H., 1977, An extracorporeal complexing hemodialysis system for the treatment of methylmercury poisoning. II. *In vivo* application in the dog, *J. Pharmacol. Exp. Ther.* **203:**253–263.

Kreger, B. E., Craven, D. E., Carling, P. C., and McCabe, W. R., 1980a, Gram-negative bacteremia, III. Reassessment of etiology, epidemiology and ecology in 612 patients, *Am. J. Med.* **68:**332–343.

Kreger, B. E., Craven, D. E., Carling, P. C., and McCabe, W. R., 1980b, Gram-negative bacteremia, IV. Reevaluation of clinical features and treatment in 612 patients, *Am. J. Med.* **68:**344–355.

Kunkel, H. G., Hberhard, H. J. M., Fudenberg, H. H., and Tomasi, T. B., 1961, Gamma globulin complexes in rheumatoid arthritis and certain other conditions, *J. Clin. Invest.* **40:**117–129.

Leiter, E., Edelman, S., and Brendler, H., 1966, Continuous preoperative intraarterial perfusion of renal tumors with chromatographic agents, *J. Urol.* **95:**169–175.

Lotan, N., Seideman, S., Tabak, A., Taitelman, U., Minich, H., and Lupovich, S., 1983, *In vivo* evaluation of a composite sorbent for the treatment of paraquat intoxication by hemoperfusion, *Int. J. Artif. Organs* **6:** 207–213.

Lupein, P. J., Moorjani, S., and Awad, J., 1976, A new approach to the management of familial hypercholesterolemia: Removal of plasma cholesterol based on the principle of affinity chromatography, *Lancet* **i:**1261–1265.

MacKintosh, F. R., Bennet, K., Schiff, S., Shield, J., and Hall, S. W., 1983, Treatment of advanced malignancy with plasma perfused over staphylococcal protein A, *Western J. Med.* **139:**36–45.

Marcus, L., Offarim, M., and Margel, S., 1982, A new immunoadsorbent for hemoperfusion: Agarose–polyacrolein microsphere beads, *In vitro* studies, *Med. Dev., Artif. Organs* **10:**157–171.

Marcus, L., Mashiah, A., Offarim, M., and Margel, S., 1984, Extracorporeal specific removal of antibodies by hemoperfusion through the immunoadsorbent agarose–polyacrolein microsphere beads, *In vitro* studies, *J. Biomed. Mater. Res.* **18:**1153–1167.

Marcus, L., Margel, S., Savin, H., Offarim, M., and Ravid, M., 1985, Therapy of digoxin intoxication in dogs by hemoperfusion through agarose–polyacrolein microsphere beads–antidigoxin, *Am. Heart J.* **110:**30–39.

Margel, S., 1981, A novel approach for metal poisoning treatment, A model. Mercury poisoning by means of chelating microspheres, hemoperfusion and oral administration, *J. Med. Chem.* **24:**1263–1266.

Margel, S., 1982, Agarose–polyacrolein microsphere beads: New effective immunoadsorbents, *FEBS Lett.* **145:**341–344.

Margel, S., and Hirsch, J., 1981, Hemoperfusion for detoxification of mercury. A model: Treatment of severe mercury poisoning by encapsulating chelating spheres, *Biomater. Med. Devices Artif. Organs* **9:**107–125.

Margel, S., and Marcus, L., 1986, Specific hemoperfusion through agarose acrobeads, *Appl. Biochem. Biotechnol.* **12:**37–66.

Margel, S., and Offarim, M., 1983, Novel effective immunoadsorbents based on agarose–polyaldehyde microsphere beads, *Anal. Biochem.* **128:**342–350.

Margel, S., Beitler, U., and Offarim, M. J., 1982, Polyacrolein microspheres: A new tool in cell biology, *J. Cell Sci.* **56:**157–175.

Margel, S., Marcus, L., Mashiah, A., Savin, H., and Dalit, M., 1984a, Extracorporeal removal of mercury by hemoperfusion through agarose–polymercaptal microsphere beads, *J. Biomed. Mater. Res.* **18:**617–629.

Margel, S., Marcus, L., Savin, H., Offarim, M., and Mashiah, A., 1984b, Specific removal of digoxin by hemoperfusion through agarose–polyacrolein microsphere beads–antidigoxin antibodies, *Biomater. Med. Devices Artif. Organs* **12:**25–36.

Margel, S., Fenaken, G., and Fainaru, M., 1991, Specific removal of LDL cholesterol by hemoperfusion, in: *Preventive Cardiology: Diagnostic and Metabolic Aspects of Cardiovascular Research* (W. Pelius, P. Lippert, and H. Weiland, eds.) (Urban and Schwarzenberg, Munchen) pp. 25–43.

Mashiah, A., Marcus, L., Savin, H., and Margel, S., 1984, The rabbit as an animal model for hemoperfusion: Surgical preparation and use, *Lab. Anim.* **18:**26–32.

McCance, R. A., and Widdowson, E. M., 1937, Absorption and excretion of iron, *Lancet* **ii:**680–694.

Messerschmidt, G. L., Bowles, C., and Alsaker, R., 1982, Long term follow up of dogs with spontaneous mammary tumors treated with *ex vivo* perfusion over *Staphylococcus aureus* Cowan I, *Proc. Am. Assoc. Cancer Res.* **23:**279–285.

Modell, B., 1979, Advances in the use of chelating agents for the treatment of iron overload, *Prog. Hematol.* **11:**267–312.

Morimoto, C., Hiroshi, S., Abe, T., and Steinberg, A. D., 1982, Correlation between clinical activity of systematic lupus erythematosus and the amounts of DNA/AntiDNA antibody immune complexes, *J. Immunol.* **129:**1960–1965.

Neilands, J. B. (ed.), 1974, *Microbial Iron Metabolism*, Academic Press, London.

Oberfeld, R. A., McCaffrey, J. A., Polio, J., Clouse, N. E., and Hamilton, T., 1979, Prolonged and continuous percutaneous intra-arterial hepatic infusion chemotherapy in advanced metastatic liver adenocarcinoma from colorectal primary cancer, *Cancer* **44:**414–423.

O'Brien, R. T., 1977, Iron overload—clinical and pathologic aspects in pediatrics, *Semin. Hematol.* **14:**115–126.

Ogata, J., Migata, N., and Nakamura, T., 1974, An experimental study of the chemotherapy of carcinoma of the urinary bladder, *Invest. Urol.* **11:**265–270.

Okonek, S., Hofmann, A., and Henningson, B., 1976, Efficiency of gut-lavage hemodialysis and hemoperfusion in the therapy of paraquat or diquat intoxication, *Arch. Toxicol.* **36:**43–51.

Pallota, A. J., and Koppanyi, T. J., 1960, The use of ion exchange resins in the treatment of phenobarbital intoxication in dogs, *J. Pharmacol. Exp. Ther.* **128:**318–327.

Porath, J., Loas, T., and Jansen, J., 1975, Agar derivatives for chromatography, electrophoresis and gel bound enzymes, *J. Chromatogr.* **103:**49–62.

Roitt, I., 1985, *Essential Immunology*, Blackwell Scientific Publications, Oxford.

Rosenbaum, J. L., Kramer, M. S., Raja, R., and Boreyko, C. N., 1971, Resin hemoperfusion: A new treatment for acute drug intoxication, *N. Engl. J. Med.* **284:**874–877.

Sato, T., Nishimiya, J., Arie, K., Anno, M., Yamawaki, N., Kuroda T., and Inagaki, K., 1983, Selective removal of anti-acetylcholine receptor antibodies in sera from patients with myasthenia gravis *in vitro* with a new immunoadsorbent, in: *3rd Symposium on Therapeutic Plasmapheresis, Tokyo, Therapeutic Plasmapheresis (III)* (T. Oda, ed.), Schattuer Verlag, Stuttgart, pp. 565–568.

Savin, H., Marcus, L., Margel, S., Ofarim, M., and Ravid, M., 1987, Treatment of digitalis intoxication in humans by hemoperfusion with agarose–polyacrolein microsphere beads–antidigoxin antibodies, *Am. Heart J.* **113:**1074–1084.

Schafer, A. I., Cheron, R. G., Dluhy, R., Cooper, B., Gleason, R. E., Soeldner, J. S., and Bunn, H. F., 1981, Clinical consequences of acquired transfusional iron overload in adults, *N. Engl. J. Med.* **304:**319–324.

Schmidt, D. H., and Butler, V. P., Jr, 1971, Reversal of digoxin toxicity with specific antibodies, *J. Clin. Invest.* **50:**1738–1744.

Shumak, K. H., and Rock, G. A., 1984, Therapeutic plasma exchange: Review, *N. Engl. J. Med.* **310:**762–771.

Sideman, S., Lotan N., Tabak, A., Manor, D., Mor, L., Taitelman, U., Brooks, J., and Tzipiniuk, A., 1984, Tailor made agarose based reactive beads for hemoperfusion and plasma perfusion, *Appl. Biochem. Biotechnol.* **10:**167–182.

Sketris, I. S., and Skoutakis, V. A., 1981, Dialysis and hemoperfusion of drugs and toxins, *Clin. Toxicol. Consult.* **3:**100–117.

Smiley, J. W., March, N. M., and DelGuerco, T., 1978, Hemoperfusion in the management of digoxin toxicity, *J. Am. Med. Assoc.* **240:**2736–2737.

Smith, T. W., and Haber, E., 1973, Digitalis, *N. Engl. J. Med.* **289:**1125–1129.

Smith, T. W., Haber, E., Yeatman, L., and Butler, V. P., Jr., 1976, Reversal of advanced digoxin intoxication with Fab fragments of digoxin specific antibodies, *N. Engl. J. Med.* **294:**797–800.

Smith, T. W., Butler, V. P., and Haber, E., 1982, Treatment of life threatening digitalis intoxication with digoxin specific Fab fragments: Experience in 26 cases, *N. Engl. J. Med.* **307:**1357–1362.

Stein, M., Blayney, D., and Feit, T., 1976, Acute iron poisoning in children, *Western J. Med.* **125:**289–294.

Stoeffel, W., and Demant, T., 1981, Selective removal of apolipoprotein B containing serum lipoproteins from blood plasma, *Proc. Natl. Acad. Sci. U.S.A.* **78:**611–615.

Stoeffel, W., Bomberg, H., and Greve, H., 1981, Application of specific extracorporeal removal of low density lipoprotein in familiar hypercholesterolemia, *Lancet* **ii:**1005–1007.

Tabak, A., Lotan, A., Sideman, S., and Taitelman, S., 1983, Bovel composite sorbent beads for paraquat removal by hemoperfusion, *Artif. Organs* **7:**176–185.

Tabei, K., Asano, Y., and Hosoda, S., 1982, Efficiency of charcoal hemoperfusion in paraquat poisoning, *Artif. Organs* **6:**37–42.

Terman, D. S., Yamamoto, T., and Mattioli, M., 1980a, Extensive necrosis of spontaneous canine mammary adenocarcinoma after extracorporeal perfusion over *Staphylococcus aureus* Cowans 1. Description of acute tumoricidal response: Morphologic, histologic, immunohistologic, immunologic and serologic findings, *J. Immunol.* **124:**795–805.

Terman, D. S., Yamamoto, T., and Tillquist, R. L., 1980b, Tumoricidal response induced by cytosine arabinoside after plasma perfusion over protein A, *Science* **209:**1257–1259.

Terman, D. S., *et al.*, 1981, Preliminary observations of the effects on breast adenocarcinoma of plasma perfused over immobilized protein A, *N. Engl. J. Med.* **305:** 1195–1200.

Theophilopoulous, A. N., and Dixon, F. J., 1980, The biology and detection of immune complexes, *Adv. Immunol.* **28:**89–220.

Thompson, G. R., Lowenthal, R., and Myant, M. B., 1975, Plasma exchange in the management of homozygous familial hypercholesterolemia, *Lancet* **i:**1208–1212.

Tobin, M., Steinbach, J., and Mookerjee, B., 1977, Hemoperfusion in digitalis intoxication: A comparative study of coated *vs.* uncoated charcoal, *Trans. Am. Soc. Artif. Intern. Organs* **23:**730–736.

Wands, J. R., and Zuraski, V. R., 1981, High affinity monoclonal antibodies to hepatitis B surface antigen produced by somatic-cell hydrids, *Gastroenterology* **80:**225–235.

Winchester, J. F., and Gelfand, M. C., 1978, Hemoperfusion in drug intoxication: Clinical and laboratory aspects, *Drug Metab. Rev.* **8:**69–104.

Winchester, J. F., Gelfand, M. C., Knepshield, J. H., and Schreiner, G. E., 1977, Dialysis and hemoperfusion of poisons and drugs-update, *Trans. Am. Soc. Artif. Intern. Organs* **23:**762–842.

Yamazaki, Z., Fujimori, Y., Takahama, T., Inoue, N., Wada, T., Kazama, M., Morioka, M., Abe, T., Yamawaki, N., and Inagaki, K., 1982, Efficiency and biocompatability of a new immunoadsorbent, *Trans. Am. Soc. Artif. Intern. Organs* **28:**318–323.

Yatzidis, H., 1964, A convenient hemoperfusion micro-apparatus over charcoal for the treatment of endogeneous and exogenous intoxications: Its use as an effect artificial kidney, *Proc. Eur. Dial. Transplant. Assoc.* **1:** 83–86.

Yokohama, S., Hayashi, R., Kikkawa, T., Tani, N., Takada, S., Hatanaka, K., and Yamamoto, A., 1984, Specific sorbent of apolipoprotein B containing lipoproteins for plasmapheresis: Characterization and experimental use in hypercholesterolemia rabbits, *Arteriosclerosis* **4:**276–285.

Ziegler, E. J., McCutchan, J. A., Fierer, J., Glauser, M. P., Sadoff, J. C., Douglas, H., and Braude, A. I., 1982, Treatment of gram-negative bacteremia and shock with human antiserum to a mutant *Escherichia coli*, *N. Engl. J. Med.* **307:**1225–1230.

29

Applications of Affinity Binding to the Development of Heparin Removing and Sensing Devices

Shu-Ching Ma and Victor C. Yang

1. INTRODUCTION

Affinity binding between a specific ligand and a receptor has been widely used in the development of immunoassays, purification of biological materials, and fabrication of biosensors. Because of the high specificity in the interaction between a receptor (e.g., antibody, binding protein, etc.) and its ligand, a sensing or purifying system thus developed would normally possess unique selectivity. In this chapter, we will discuss the applications of the affinity approach to the development of a heparin removal system and two heparin sensing devices (one is a disposable paper strip, and the other is a probe-type heparin electrode). These systems are established utilizing the specific binding of heparin either with its clinical antagonist (i.e., protamine) or with a strong heparin-complexing reagent, tridodecylmethylammonium chloride (TDMAC). These devices should be of significant value because they can improve the safety of current extracorporeal blood circulation (ECBC) procedures or provide a safeguard to heparin therapy.

1.1. Heparin Removal Following ECBC Procedures

Extracorporeal blood circulation has been widely employed in many clinical situations such as kidney dialysis, open-heart operations, organ transplantation, plasmapheresis, and blood oxygenations. It is estimated that in the United States alone approximately 15 million ECBC procedures are performed each year. In all these applications, blood is drawn from the patient and passed through an extracorporeal device. Upon contact with the synthetic materials forming the surfaces of such extracorporeal devices, the blood's precisely regulated hemostasis is disturbed and it tends to clot within the device. The thrombi formed occlude the perfusion channels in the machine. To prevent occlusion and maintain the fluidity of blood in the circulation, heparin, the most widely used anticoagulant, is

Shu-Ching Ma and Victor C. Yang • College of Pharmacy, The University of Michigan, Ann Arbor, Michigan 48109-1065. Correspondence should be addressed to Victor C. Yang.

Molecular Interactions in Bioseparations, edited by That T. Ngo. Plenum Press, New York, 1993.

systematically administered to the patient and the device prior to treatment. However, the high level of heparin required for this purpose poses a considerable hemorrhagic hazard to the patients (Swartz and Port, 1979; Hirsh, 1984; Kelton and Hirsh, 1984). The extent of the complications is enhanced for elderly patients, diabetic patients, patients with ulcers or other multiple traumata, and patients with current cardiac or vascular surgery. Indeed, heparin has been referred to as "the drug responsible for a majority of drug deaths in patients who are reasonably healthy" (Porter and Jick, 1977).

Because of the life-threatening nature of the hemorrhage associated with systemic heparinization, considerable amounts of effort have been directed at solving this problem. The approaches taken include (1) the administration of antiheparin compounds (e.g., protamine) to neutralize the anticoagulant activity of heparin (Jaques, 1973); (2) the development of low-molecular-weight heparins (Ljungberg et al., 1987); (3) the use of low-dose heparinization or regional heparin anticoagulation (i.e., by infusion of heparin into the blood entering the extracorporeal device and neutralization by infusion of protamine into the heparinized blood as it returns to the patient) (Swartz and Port, 1979); (4) the use of regional citrate anticoagulation [this approach is similar to that of the regional heparin anticoagulation as described in (3), except that citrate is used as the anticoagulant agent and calcium is used as the neutralizing agent] (Diederich et al., 1985); (5) the development of new blood-compatible materials, such as materials with surface-bound heparin, for construction of the extracorporeal devices (Nojiri et al., 1990); and (6) the development of an immobilized heparinase filter to degrade heparin at the termination of the ECBC procedure (Langer et al., 1982; Bernstein et al., 1987). Although these approaches have led to some improvement, they have not met much clinical success. Thus far, the approach which involves the administration of protamine at the conclusion of an ECBC procedure for heparin reversal still remains the only accepted method in clinical practice.

Protamine is composed of a group of heterogeneous polypeptides with an average molecular weight around 4500 daltons (Jaques, 1973). They are highly positively charged molecules due to the high content (about 67%) of arginine residues. The binding of protamine to negatively charged heparin causes the release of heparin from its binding to antithrombin III, rendering the anticoagulant function of heparin ineffective. In clinical practice, 1 mg of protamine is routinely used to neutralize 100 USP units of heparin (Jaques, 1973). Despite the approval of protamine for such clinical use by the Food and Drug Administration (FDA), it is nevertheless a toxic compound. Protamine-induced toxicity ranges from mild hypotension (Kirklin et al., 1986; Katz et al., 1987) to severe or ultimately fatal cardiovascular collapse (Olinger et al., 1980; Weiler et al., 1985). A recent study by Weiler et al. (1990) suggests that the administration of protamine sulfate during cardio-pulmonary bypass for heparin reversal may be a major cause of morbidity and mortality in patients undergoing these procedures.

A heparin-removing system which could potentially control both heparin- and protamine-induced complications is currently under development in our laboratory. The approach consists of placing a blood-compatible filter device containing immobilized protamine (defined as a protamine filter) at the distal end of the ECBC circuit (Yang, 1989; Yang et al., 1991). Such a protamine filter would bind and selectively remove heparin after heparin serves its anticoagulant purpose in the extracorporeal device and before it is returned to the patient. Meanwhile, the protamine filter would constrain protamine from entering the patient and allow for an external protamine treatment. Since the majority of protamine toxicity comes either from the circulating protamine–heparin complexes (Morel *et al.*, 1990) or from a direct interaction of protamine with certain organ tissue cells (e.g., mast cells) (Jaques, 1973), the use of an external protamine treatment would theoretically minimize the adverse effects of protamine.

1.2. Heparin Sensing Devices

Heparin is a polydisperse, highly negatively charged polysaccharide with an average molecular weight of about 15,000 daltons (Jaques, 1980). It is an anticoagulant widely used in the prophylaxis and treatment of venous thrombosis, disseminated intravascular coagulation, and peripheral arterial embolism (Jaques, 1980). The most important application of heparin is in medical procedures requiring ECBC, where plasma heparin levels are often maintained between 2 and 10 U/ml to ensure the blood fluidity. Although many heparin assays are available, most of these assays are based on blood clotting time estimation [e.g., the activated clotting time (ACT) and the activated partial thromboplastin time (aPTT)]. These methods lack defined biochemical mechanisms and are relatively inaccurate, slow, and cumbersome. In addition, they do not provide a direct heparin measurement. Furthermore, expensive specialized instruments are generally required for the measurements.

A non-clotting-based, simple strip-type method for heparin measurement has been developed in our laboratory (Chen and Yang, 1991). The system consists of a filter paper strip (as the solid phase) with covalently immobilized protamine on it. Theoretically, as a fixed volume of heparin sample migrates through the paper strip, heparin molecules will be attached to the immobilized protamine, and the area of paper to which the heparin binds (which can be detected using a thiazine dye that changes from blue to purple in the presence of heparin) will be proportional to the amount of heparin present in the sample. Since both the dimension of the paper strip and the sample volume are standardized, heparin concentration in the test sample can be determined from the length of the purple area using an appropriate, preconstructed standard plot.

Recently, based on the theory of ion-selective electrodes (ISE), we have developed an electrochemical sensor for heparin (Ma *et al.*, 1992). Polymeric membrane-type ion-selective electrodes have been used routinely within biomedical instruments to measure clinically important ions (e.g., Na^+, K^+, Li^+, H^+, Cl^-) in undiluted whole blood (Oesch *et al.*, 1986). Efforts to develop similar sensors (including immuno-based biosensors) suitable for the direct detection of larger biomolecules (e.g., drugs, specific proteins), however, have been less successful (Arnold and Meyerhoff, 1988) owing to the difficulty in identifying appropriate complexing agents and membrane chemistries that will yield a significant and specific electrochemical response toward the desired analytes. Tridodecylmethylammonium chloride (TDMAC), a quaternary ammonium salt, is known to possess strong ion association with heparin (Grode *et al.*, 1972). Indeed, TDMAC–heparin complexes have been incorporated into polymeric materials previously for the fabrication of biocompatible devices (Gott, 1972; Grode *et al.*, 1972). In addition, poly(vinyl chloride) (PVC) membranes doped with TDMAC or other quaternary ammonium salts have been suggested for fabricating conventional ISEs for small ions including chloride (Hartman *et al.*, 1978; Wegmann *et al.*, 1984). We have therefore selected TDMAC as the "heparin carrier" and incorporated it within a PVC polymer phase to fabricate a heparin-responsive electrode membrane. The membrane thus prepared exhibits significant potentiometric response to heparin in the clinically relevant concentration range.

In this chapter, we will provide a detailed review of the progress that has been made toward the development of the heparin removing and monitoring devices described above.

2. EXPERIMENTAL SECTION

2.1. Materials

Injectable heparin (sodium salt, from beef lung, 1,000 USP units/ml) was purchased from Elkins-Sinn (Cherry Hill, New Jersey), while solid heparin (sodium salt, from porcine internal mucosa, 169 USP units/mg) was purchased from Hepar Industries, Inc. (Franklin, Ohio). Protamine sulfate (10 mg/ml) was obtained from Eli Lilly & Co. (Indianapolis, Indiana). Actin-activated cephaloplastin reagent for the activated partial thromboplastin time (aPTT) assay was obtained from Dade (Miami, Florida). The Fibrometer (Fibrosystem BBL) was acquired from Becton-Dickinson Inc. (Hunt Valley, Maryland). Filter papers (grades 541 and 113) and chromatographic papers (grade 31ET Chr) were manufactured by Whatman Ltd. (Maidstone, England). Dioctylsebacate (DOS) and tridodecylmethyl-ammonium chloride (TDMAC) were purchased from Fluka Chemika-Biochemika (Ronkonkoma, New York). Freshly frozen citrated human plasma was obtained from the American Red Cross (Southfield, Michigan). Cellulose hollow fibers were obtained from the Travenol model 1500 CF Dialyzer. All other chemicals were reagent-grade, and distilled–deionized water was used for solution preparations.

2.2. Methods

2.2.1. Protamine Filters

2.2.1.1. Assays. Protamine concentration was assayed according to the method of Bradford (1976). Heparin concentration in solution was determined by the Azure A metachromasia assay (Jaques and Wollin, 1967). The biological activity of heparin in plasma was measured by the activated partial thromboplastin time (aPTT) clotting assay (Walenga *et al.*, 1986). The degree of activation of the hollow fibers was determined by measuring the cyanate ester concentrations on the activated polymers, according to the method of Kohn and Wilchek (1978).

2.2.1.2. Preparation of Fine Fiber Particles. Intact hollow fibers were chopped with a razor blade to small fragments of about 100–200 μm in size, immersed in liquid nitrogen until frozen, and then ground in a crucible. The freezing and grinding were repeated several times until particles that were approximately uniform in size (approximately 0.5-1.0 μm in diameter as measured under a light microscope) were obtained.

2.2.1.3. Fabrication of the Hollow Fiber Bundle. The Travenol model 1500 CF Dialyzer was disassembled, and the fibers were cut down to a length of 15 cm. Approximately 800–1000 of the cut fibers (about 3 g in dry weight) were collected and tied together at each end with a Teflon film to form a bundle. The bundle was then housed in a Tygon tubing which was fitted with a molded connector at each end (see Fig. 1). The junctions were sealed with epoxy, and the epoxy was cured at room temperature until it hardened.

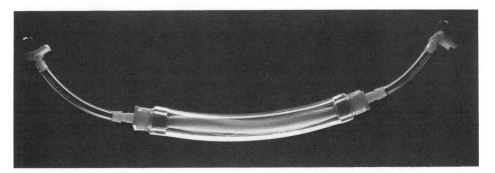

FIGURE 1. The protamine filter used for the *in vivo* experiments.

2.2.1.4.   Immobilization of Protamine onto the Hollow Fiber Bundle.   The fabricated bundle prepared above was circulated with 1*M* sodium carbonate solution for about 5 min. After the circulation, the bundle was placed in a fume hood and circulated for 5 min with CNBr solution. The CNBr-activated bundle was then washed with distilled water, 1m*M* HCl solution, and 0.1*M* NaHCO$_3$ buffer (pH 8.3 containing 0.5*M* NaCl). Immediately following the washing steps, protamine solution was circulated through the bundle for 2 h at room temperature. The bundle was then washed with high- and low-salt phosphate buffer to remove the nonspecifically adsorbed protamine. The hollow fiber bundle thus prepared that contained immobilized protamine was defined as the "protamine filter."

2.2.1.5.   Adsorption–Time curve of Heparin on the Protamine Filter.   Heparin solutions prepared in normal saline and at different concentrations were placed in a reservoir and recirculated through the protamine filter for 1 h. Samples were withdrawn from the reservoir at various time intervals and assayed for the residual heparin concentrations. Plain hollow fiber bundles (i.e., not treated with protamine) were used as the control to determine the amount of heparin that was adsorbed on the bundle through nonspecific, physical adsorption.

2.2.1.6.   Equilibrium Adsorption Isotherm.   Ground fiber particles containing immobilized protamine were used for this study. To each tube containing 0.5 g of the protamine-bound fiber particles, heparin solutions of different concentrations were added. The tubes were rotated for 1 h at room temperature. The fiber particles were then removed, and the equilibrium heparin concentration in the supernatant was measured by the Azure A assay. Plain ground fiber particles that were not treated with protamine were used as the control to assess the nonspecific heparin adsorption on the fibers.

2.2.1.7.   In vivo Experiments.   Healthy female mongrel dogs, weighing 20–30 kg, were selected for the study. Each animal was anesthetized with sodium pentobarbital (30 mg/kg), and the lungs were mechanically ventilated to maintain physiological blood gases. Hydration was maintained by administration of lactated Ringer solution (20-ml/kg bolus followed by a 10 ml/kg per hour infusion). The femoral artery and the femoral vein of the animal were cannulated, and to the cannulas a hollow fiber bundle was attached to allow blood to flow from the artery through the filter and into the vein. Blood flow through the fiber bundle was measured with an electromagnetic square-wave flow meter (Carolina Instruments, King, North Carolina) with the flow probe placed on the femoral vein. A venous thrombus trap (Baxter Healthcare Corporation, Round Lake, Illinois), consisting of an interior trap made of nylon with 200-μm meshes, was placed at the effluent site of the bundle to prevent clots from entering the animal. Heparin (1000 U/ml) was administered intravenously at a dose of 150 units/kg of body weight. Twenty minutes after heparin administration, the two valves on the hollow fiber bundle were opened to allow blood to circulate through the bundle. The time of opening the valves was defined as "time zero." Blood samples taken immediately before opening of the valves were used as the baseline samples. An oximetric catheter with an optical fiber (Oximetrix, Mountain View, California) was placed into the pulmonary artery for continuous determination of mixed venous oxygen saturations (SvO_2) and pulmonary artery systolic (PAS) and diastolic (PAD) pressures and for intermittent measurements of cardiac output using the thermodilution technique. Catheters were also inserted into the left carotid artery and jugular vein for monitoring systemic arterial blood pressure and blood withdrawal.

Twenty dogs were used for the entire *in vivo* studies. The animals were distributed into three groups (group I–III). Group I consisted of six dogs used as controls. In this group of dogs, a sham filter which contained no protamine was used, and the dogs were not given any protamine intravenously. Group II consisted of eight dogs used for the testing of the protamine filter. In this group the extracorporeal circuit was incorporated with a protamine filter which contained 60–70 mg of immobilized protamine, and no additional protamine was administered. Group III consisted of six dogs used for the testing of the response resulting from direct i.v. administration of protamine. In this group the extracorporeal circuit was incorporated with a sham filter which contained no protamine, and protamine was administered at time zero as a bolus dose (1 mg of protamine per 100 units of heparin) over a period of 5 s.

The blood activated clotting time (ACT) in all the animals was monitored periodically with a Hemochron (International Technidyne Corp., Metuchen, New Jersey). The plasma heparin activity was also determined using the aPTT and thrombin time (TT) assays. Hemodynamic changes in the animal were monitored every minute for 5 min and then at 10, 20, 30, and 60 min after opening of the fiber bundle. Blood samples were collected for all of the dogs from the arterial tubing of the bundle at time intervals of 3, 10, 20, 30, and 60 min after opening of the fiber bundle. Samples were drawn into VACUTAINER tubes containing EDTA (Becton Dickinson, Rockville, Maryland) for measurements of hematocrit, total blood hemoglobin, and complete blood cell counts and into VACUTAINER tubes containing sodium citrate for determinations of the heparin activity and the activation of the complement system.

2.2.2. Heparin Sensing Strips

2.2.2.1.   Selection of Support Material for the Fabrication of the Heparin Sensing Strip.   The support materials used for protamine immobilization were chosen to meet three practical requirements: (a) abundance of functional groups (e.g., hydroxyl) on the surface for chemical activation and subsequent protamine immobilization; (b) sufficient porosity for the rapid migration of liquid; and (c) adequate mechanical strength to withstand the vigorous conditions posed by the vigorous immobilization procedures. Whatman filter papers and chromatographic papers appear most appropriate because they are manufactured from cellulose and thus contain abundant hydroxyl groups on the surface, and they also possess relatively high porosity and mechanical strength.

2.2.2.2. Activation of the Support and Coupling of Protamine Sulfate to the Activated Paper Strip. Similar procedures to those described in Section 2.2.1 were used for paper activation and protamine immobilization. The amount of protamine immobilized on the paper strip was determined by measuring the protamine content in the solution prior to and following (i.e., in the drainage collected) the immobilization process. The protamine distribution density (PDD) value was obtained by dividing the amount of protamine immobilized by the total surface area of the paper strip.

2.2.2.3. Fabrication of the Heparin Sensing Device and Its Mode of Operation. A fibrin cup was slashed in the middle with a razor blade. A filter paper strip (standard dimension: 9 mm × 110 mm), with its top laminated with two slides of Parafilm, was inserted into the slot on the fibrin cup and then sealed at the contact edges with silicone rubber. The desired volume of the heparin sample was applied to the fibrin cup and allowed to migrate along the paper strip. Upon exhaustion of the entire sample volume, the device was transferred to a fume hood. Methylene Blue NNX dye solution thrusted by a CFC propellant was sprayed onto the paper strip immediately. The heparin migration distance was then detected by the length of the purple region on the strip.

2.2.2.4. The Standards Addition approach. A standards addition method for sample treatment was required for plasma specimens at low heparin concentrations (<5 U/ml). Aqueous heparin solution (10 U/ml) was mixed with an equal volume of the heparin specimens before assay.

2.2.3. Heparin Sensing Electrode

Polymer membranes were cast using the conventional method for ISE membrane preparation (Craggs *et al.*, 1974). Small disks (~5 mm i.d.) of polymer membranes were cut and incorporated into Phillips electrode bodies (IS-561, Glasblaserei Moller, Zurich). A $0.015M$ NaCl solution was used as the internal filling solution throughout the study. Electrodes were soaked in the same NaCl solution overnight prior to their use. The potentiometric response of the membrane electrode was measured relative to an external double-junction reference electrode at room temperature, with the sample solution being stirred constantly.

3. RESULTS AND DISCUSSION

3.1. Protamine Filter

3.1.1. Characteristics of Protamine Immobilization and Heparin Adsorption on the Protamine-Bound Cellulose Fiber Membrane

The protamine immobilization process consists of two steps: (i) activation of the hydroxyl groups on the cellulose membrane with CNBr to form reactive cyanate ester

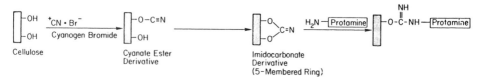

FIGURE 2. The mechanism describing the activation and coupling of protamine to cellulose hollow fibers.

derivatives; and (ii) coupling of protamine onto the activated cyanate ester groups through isourea linkages (see Fig. 2). A detailed investigation of the parameters affecting the immobilization process has been conducted (Kim *et al.*, 1992). Protamine immobilization onto the CNBr-activated fibers appeared to follow a bimolecular reaction mechanism and was dependent on both the cyanate ester concentration on the fibers and the protamine concentration in the coupling solution. Increasing the concentration of either of these two reactants (e.g., increasing the cyanate ester concentration by using a larger quantity of CNBr for fiber activation) resulted in an increase of protamine loading on the fibers. In addition, protamine loading on the fibers was also dependent on the duration of the coupling step, increasing with the prolongation of the incubation time. Since the CNBr concentration is normally fixed at 0.1 g/ml (above this concentration structural deterioration of the fibers may be introduced) and the incubation period is generally held to one hour (this is the time required to complete the coupling process), varying the protamine concentration in the coupling solution seems to be the easiest and most convenient means to control the protamine loading on the fibers.

A recirculation system consisting of a heparin reservoir and a pump was employed to characterize heparin adsorption for the purpose of simulating the *in vivo* blood circulation situation where the protamine filter would eventually be applied. All the heparin adsorption versus time curves at different heparin concentrations were found to follow a typical pattern: a rapid heparin adsorption occurred at the initial stage of the recirculation, followed by an asymptotic approach of the heparin concentration to a steady-state value. Results shown in Fig. 3 indicate that heparin adsorption on the protamine filter follows the Langmuir adsorption model:

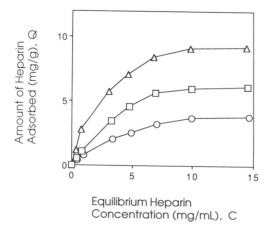

Equilibrium Heparin
Concentration (mg/mL), C

FIGURE 3. Equilibrium adsorption isotherm of heparin on the protamine-bound ground fiber particles. Protamine loading on the fiber was 6.5 ($\circ$), 12.7 ($\square$), and 20.7 mg/g ($\triangle$).

$$Q = \frac{KQ_sC}{1 + KC}$$

where Q, K, and Q_s are the specific amount of heparin adsorbed, the adsorption constant (ml/mg), and the saturation capacity (mg/g of fiber), respectively. In the low heparin concentration region (0.01–1.0 mg/ml), adsorption was first-order, the amount of heparin adsorbed being linearly proportional to the equilibrium heparin concentration. As the heparin concentration was increased from 1 to 10 mg/ml, adsorption became mixed-order. With further increase in the heparin concentration above 10 mg/ml, the amount of heparin adsorbed reached a plateau, which represented the saturation capacity (Q_s) of the protamine-bound fibers. In this case, the adsorption was zero-order (i.e., independent of the substrate concentration).

Figure 4 shows the double-reciprocal plots of the Langmuir adsorption isotherms given in Fig. 3. At three different protamine loadings, the adsorption constant K appeared to remain constant, whereas the saturation capacity Q_s varied with respect to the protamine loading. The average K value was estimated to be 0.30 ± 0.05 ml/mg.

3.1.2. *In Vivo* Function of the Protamine Filter in Removing Heparin and Preventing Protamine-Induced Complications

The rate of disappearance of heparin anticoagulant activity in the three animal groups is shown in Fig. 5. In animals receiving intravenous protamine administration, the blood heparin activity, as measured by the aPTT assay, rapidly decreased to the minimum. In contrast, heparin removal by the protamine filter appeared to follow a pattern consistent with a slower, regional-type, two-compartment kinetic model. A rapid removal, which occurred within the first 5 min of the experiment, was followed by a slower but relatively linear removal rate over a period of 60 min. At the 60-min mark, the heparin level had declined to a value close to that observed in animals receiving intravenous protamine.

Figure 6 shows that the use of the immobilized protamine filter attenuated protamine-induced hemodynamic responses. While i.v. injection of protamine in dogs (i.e., group III) produced typical and significant ($p < 0.05$) decreases in systemic arterial blood pressure

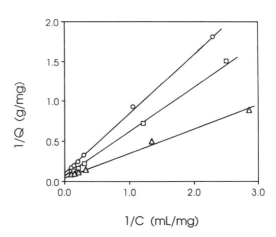

FIGURE 4. Double-reciprocal plots of the Langmuir adsorption isotherms in Fig. 3. Protamine loading on the fiber was 6.5 (O), 12.7 (□), and 20.7 mg/g (△).

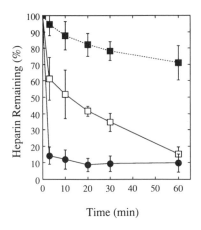

FIGURE 5. *In vivo* heparin removal. Heparin activity was determined by the aPTT heparin clotting assay. The data were normalized to 0% activity before heparin administration and 100% after heparin administration. ■, Group I animals (i.e., control dogs that received no protamine treatment); □, group II animals (i.e., dogs with the protamine filter); ●, group III animals (i.e., dogs that received intravenous administration of protamine). The data are presented as the means ± SD.

(-39.5 ± 9.2 mm Hg), cardiac output (-1.59 ± 0.23 l/min), and mixed venous oxygen saturation ($-7.5 \pm 1.3\%$) and increases in pulmonary artery systolic ($+12.7 \pm 4.4$ mm Hg) and diastolic pressures ($+10.0 \pm 3.6$ mm Hg), the use of the protamine filter (group II) did not elicit statistically significant change in any of the variables measured.

The use of the protamine filter also alleviated protamine-induced transient thrombocytopenic and granulocytopenic responses. The white blood cell counts and platelet counts reduced to 87.7 ± 7.5 and $83.3 \pm 5.0\%$ of baseline, respectively, in dogs with the protamine filter as compared to 35.5 ± 14.3 and $32.1 \pm 8.1\%$ of baseline in dogs receiving intravenous protamine.

The above results suggest that the protamine filter under development may provide a potential means to control both heparin- and protamine-induced complications simultaneously. This control would significantly improve the safety of current ECBC procedures. With nearly 15 million ECBC procedures performed each year, the benefits of the protamine filter could be extensive.

From a practical standpoint, the protamine filter possesses two major advantages. One advantage is the simplicity and flexibility of operating the device. The protamine filter can easily be interfaced with current ECBC devices without any additional apparatus or invasive procedures. The other advantage is the potential acceptance for clinical use. The filter is designed with a blood-compatible material (cellulose hollow fibers obtained from a clinical hemodialyzer) and a clinically accepted drug (protamine) and at the same time significantly minimizes the toxic effects of the drug. Its acceptance for clinical use, particularly in patients with increased risk of bleeding complications, would therefore be anticipated.

3.2. Heparin Sensing Strip

Among a group of 15 different filter papers and 8 chromatographic (Chr) papers manufactured by Whatman, we selected the grade 541 and grade 113 filter papers and the 31ET Chr paper to develop the heparin sensing strip. This is because of their superb stiffness and higher porosity compared with other types of papers.

Figure 7 shows the actual appearance of the paper strips (grade 541, PDD = 135.9

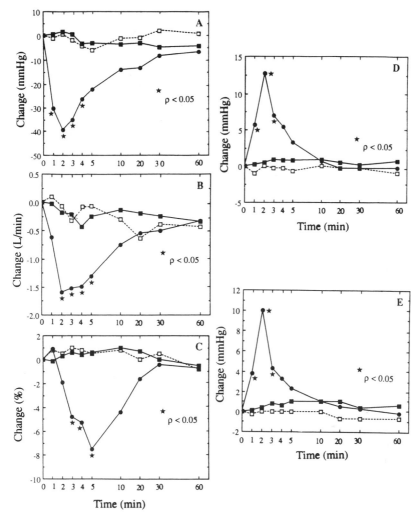

FIGURE 6. Hemodynamic changes in dogs. (A) Systemic arterial blood pressure (BP); (B) cardiac output (CO); (C) mixed venous oxygen saturation (SvO$_2$); (D) pulmonary artery systolic pressure (PAS); (E) pulmonary artery diastolic pressure (PAD). The experimental procedures were described in detail in Section 2.2.1.7. ■, Group I animals (i.e., control dogs that received no protamine treatment); □, group II animals (i.e., dogs with the protamine filter); ●, group III animals (i.e., dogs that received intravenous administration of protamine). To avoid clustering in the figure, data points are represented only by the mean values. *$p < 0.05$ from the control group.

μg/cm^2) following the completion of sample migration and treatment with Methylene Blue NNX dye. The purple area (the upper, darker portion) signifies the presence of heparin, whereas the blue area (the lower, lighter portion) indicates its absence. As illustrated, the heparin concentration was proportional to the length of the purple area on the paper strip.

A representative plot of the migration distance versus heparin concentration in plasma samples using either the grade 541 or grade 113 paper is shown in Fig. 8. A linear region is seen in each of the two plots, indicating the feasibility of using the strip to accurately detect

TOP

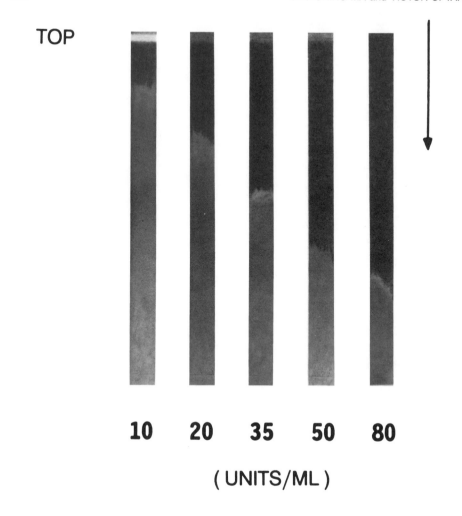

10   20   35   50   80

(UNITS/ML)

HEPARIN CONCENTRATION

FIGURE 7. Grade 541 paper strips (PDD = 135.9 μg/cm²) after sample migration and treatment with Methylene Blue NNX (a thiazine dye) solution (500 mg/liter). The concentration of heparin in the sample is indicated under each strip.

heparin concentrations in the range of 5–35 U/ml. At heparin concentrations of <5 U/ml and at particularly low PDD values (e.g., 22.9 μg/cm² on grade 541 paper), the heparin migration front exhibited a broad smearing (or diffuse zone), such that the exact locations of the fronts could not be identified. To solve this smearing problem, we used the standards addition method described in Section 2.2.2.4, adding a constant volume of aqueous heparin solution (10 U/ml) to an equal volume of plasma samples. This allowed adequate evaluation of the plasma samples with low heparin concentrations that were applied to paper strips with low PDD values. Calibration plots for the standards addition method had to be prepared separately.

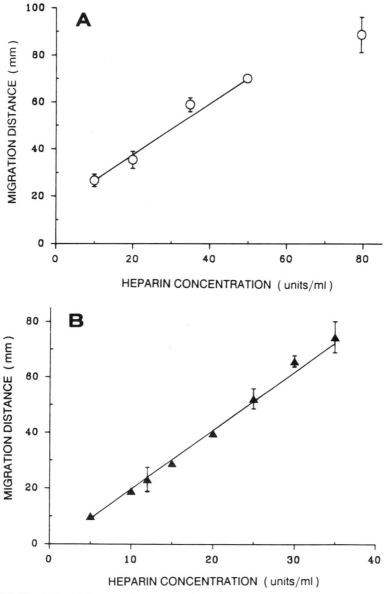

FIGURE 8. The relationship between migration distance and heparin concentration in plasma samples. (A) Grade 541 paper strips; sample volume, 300 μl; PDD, 135.9 μg/cm². (B) Grade 113 paper strips; sample volume, 300 μl; PDD, 219.8 μg/cm².

To validate the method, heparin concentrations in unknown plasma samples were determined by the strip method and the conventional aPTT heparin clotting assay. As shown in Fig. 9, the heparin levels established by the new method were in reasonably good agreement with those determined by the aPTT assay. A linear relationship ($r = 0.981$) with a slope of 0.995 and a y-intercept of 0.026 U/ml was observed in the least-square-fitted comparison plot, indicating that the proportional and constant errors of the new method

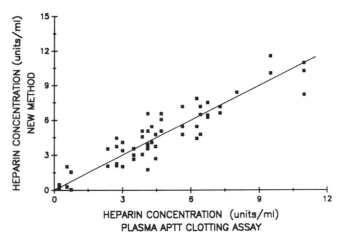

FIGURE 9. The correlation between the heparin concentrations determined by the strip assay and the conventional aPTT heparin clotting assay. Least-square-fitting was applied in the regression analysis.

were 0.5% and −0.026 U/ml, respectively. For a heparin level of 3 U/ml, the new method would yield a value of 2.985 ± 0.243 U/ml (confidence interval, 95%). The time required by the new heparin sensing method is, generally speaking, less than 10 min.

3.3. Heparin Sensing Electrode

To optimize the polymeric membrane's response toward the heparin molecule, various TDMAC-doped PVC membranes with different formulations of plasticizers and/or TDMAC were prepared and tested. As shown in Table 1, membranes prepared with a lower wt % plasticizer, relative to the PVC content, exhibited the greatest response to heparin. Indeed, the TDMAC-doped membrane-formulated with normal (high) levels of plasticizer (i.e., ~65 wt %) displayed significant response to heparin only at high heparin levels, whereas membranes formulated with lower wt % plasticizer exhibited much greater total heparin response and at lower concentrations of heparin. Membranes formulated with 65 wt % PVC, 33 wt % plasticizer, and 1.4–2.0 wt % TDMAC (i.e., membranes i and k) appeared to possess the optimum response to heparin and were thus used for further studies.

As shown in Fig. 10A, the electrode containing membrane i (see Table 1) was able to detect low levels of heparin (0.1–1.0 U/ml) in solution even in the presence of $0.12M$ chloride. Figure 10A also indicates that the response to chloride over its normal and quite narrow physiological concentration range is much less than the response to heparin. It should be noted that no response to added heparin was observed when a dialysis membrane (MW cutoff, 12,000 daltons) was placed over the surface of the sensing membrane to block the heparin–TDMAC interaction. Furthermore, when protamine was added to the heparin sample, the potential of the electrode immediately shifted toward a more positive value, as the activity of free heparin in solution decreased (data not shown). These observations indicate that the electrode is responding directly to the macromolecular heparin, and not to smaller ionic impurities within the heparin preparation used in these studies.

Linear potentiometric response was also observed in undiluted human plasma samples at heparin concentrations between 1.0 and 9.8 U/ml (Fig. 10B). The somewhat reduced

TABLE 1
Potentiometric Response of TDMAC-Based PVC
Membranes with Various Compositions to Heparin[a]

| Membrane | ΔE^c (mV) | Membrane composition[b] | | |
		PVC (wt %)	DOS (wt %)	TDMAC (wt %)
a	−11.6	33.0	66.0	1.0
b	−10.4	32.5	65.9	1.6
c	−2.1	32.7	65.3	2.0
d	−7.0	32.4	65.1	2.5
e	−5.0	32.4	64.6	3.0
f	−30.7	48.4	50.5	1.1
g	−41.8	65.8	33.3	0.9
h	−38.0	73.1	26.0	0.9
i	−44.4	65.5	33.1	1.4
j	−31.0	66.4	33.1	0.5
k	−45.0	65.5	32.5	2.0

[a]Heparin activity in the range 0–1.4 U/ml in 0.12M NaCl.
[b]PVC, Poly(vinyl chloride); DOS, dioctyl sebecate; TDMAC, tridodecylmethylammonium chloride.
[c]Average of duplicates obtained from three electrodes.

response in plasma versus saline solution may be due to the binding of heparin by endogenous proteins (e.g., antithrombin III) and/or the inhibition of heparin extraction into the membrane phase by surface-adsorbed plasma proteins. Nonetheless, given that heparin levels encountered in most surgical procedures are within the range 2–8 U/ml, the sensitivity of the electrode appears to be more than adequate to meet this clinical need. Preliminary studies using heparinized sheep whole-blood samples reveal similar electrode response characteristics.

The potentiometric response of the electrode toward various related compounds was examined. As shown in Table 2, the electrode displayed increased potentiometric response correlating to the degree of sulfate content of the heparin analogs and no measurable response to poly(vinyl sulfate) (PVS), a highly sulfated polymer, nor to sulfated or nonsulfated glucosamine residues, which are the major monosaccharide building blocks of heparin. These results indicate that heparin's high sulfate content and its specific polymeric structure are both essential for its interaction with the TDMAC-doped membrane.

The heparin sensor under development is the first known example of a polymer membrane electrode for direct monitoring of polyionic macromolecular species. The potentiometric response in saline solution and undiluted human plasma demonstrates that the sensor exhibits adequate selectivity and sensitivity to detect heparin in the clinically relevant concentration range (i.e., 1–10 U/ml). The response time of the electrode to heparin is sufficiently rapid to yield stable signals in 2–5 min at this range. Given the importance of and critical need for monitoring heparin levels, the sensor described here may ultimately offer attractive advantages over existing approaches based on clotting time measurements. Indeed, it is envisioned that the proposed membrane chemistry could be implemented in a disposable, single-use electrode format (similar to Kodak's Ektachem design) for rapid estimation of free heparin levels in undiluted blood. Work on demonstrating the clinical utility of this approach is currently in progress.

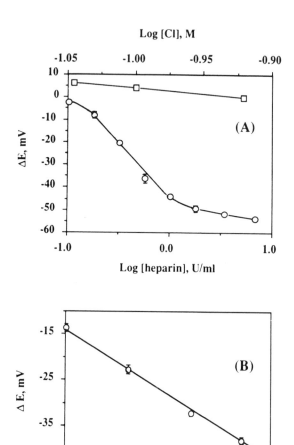

FIGURE 10. Response characteristics of heparin sensor. (A) Response toward Cl⁻ in the concentration range of 0.09–0.12M (□) and toward heparin in the presence of 0.12M NaCl (○). (B) Response toward heparin in citrated fresh human plasma sample. ΔE represents the potential change from the original cell potential (without added heparin or chloride) and is plotted against the logarithm of the concentration of the analytes. The results plotted in (A) and (B) are the means ± SD for duplicate measurements with two different heparin electrodes. Response slopes in the linear portions of the plots are 49.2 mV/dec and 28.2 mV/dec for 0.12M NaCl (heparin concentration between 0.2 and 1.0 U/ml) and undiluted human plasma (heparin concentration between 1.0 and 9.8 U/ml) samples, respectively.

TABLE 2
Potentiometric Response of Heparin Sensor
toward Various Compounds[a]

Compound[b]	ΔE^c (mV)	Sulfate content (wt %)
Glycosaminoglycans		
Heparin	−50	2.8
Dermatan sulfate	−25	1.1
Chondroitin sulfate A	−10	1.0
Hyaluronic acid	0	0
Poly(vinyl sulfate)[d]	0	62
Glucosamine residues		
Glucosamine	0	0
Glucosamine 2-sulfate	0	25
Glucosamine 3-sulfate (free acid)	0	27
Glucosamine 6-sulfate (free base)	0	27
Glucosamine 2,3-disulfate	0	42

[a]All compounds were prepared in 0.12*M* NaCl solution at the same concentration (i.e., 12 µg/ml). For heparin, this concentration is equivalent to an activity of 2 U/ml.
[b]Unless otherwise specified, all compounds are in their sodium salt form.
[c]Change in cell potential compared to solution of 0.12*M* NaCl; values are averages of duplicates obtained from three electrodes.
[d]Only the soluble part present in the supernatant is used; the insoluble part is filtered off.

ACKNOWLEDGMENTS

The authors thank Drs. Mark E. Meyerhoff, Jae-Seung Kim, Weiliam Chen, Ching-Leou C. Teng, Friedrich K. Port, Gerd O. Till, and Thomas W. Wakefield for their valuable advice and technical assistance. The work was partially supported by NIH Grant HL38353, a Whitaker Foundation Biomedical Engineering Grant, and a College of Pharmacy Upjohn Research Award.

REFERENCES

Arnold, M. A., and Meyerhoff, M. E., 1988, Recent advances in the development and analytical applications of biosensors, *CRC Crit. Rev. Anal. Chem.* **20**:149–196.

Bernstein, H., Yang, V. C., Lund, D., Randhawa, M., Harmon, W., and Langer, R., 1987, Extracorporeal enzymatic heparin removal: Use in a sheep dialysis model, *Kidney Int.* **32**:452–463.

Bradford, M. M., 1976, A rapid and sensitive method for the quantitation of microgram quantities of protein utilizing the principle of protein–dye binding. *Anal. Biochem.* **72**:248–254.

Chen, W., and Yang, V. C., 1991, Versatile, non-clotting-based heparin assay requiring no instrumentation, *Clin. Chem.* **37**:832–837.

Craggs, A., Moody, G. J., and Thomas, J. D. R., 1974, PVC matrix membrane ion-selective electrodes, *J. Chem. Educ.* **51**:541–544.

Diederich, D. A., Wiegmann, T. B., and Pinnick, R. V., 1985, Method for regional anticoagulation during extracorporeal dialysis. U.S. Patent 4,500,309.

Gott, V. L., 1972, Heparinized shunts for thoracic vascular operations, *Ann. Thoracic Surg.* **14**:219–220.

Grode, G. A., Falb, R. D., and Crowley, J. P., 1972, Biocompatible materials for use in the vascular system, *J. Biomed. Mater. Res. Symp.* **3:**77–84.

Hartman, K., Luterotti, S., Osswald, H. F., Oehme, M., Meier, P. C., Ammann, D., and Simon, W., 1978, Chloride-selective liquid-membrane electrodes based on lipophilic methyl-tri-*N*-alkylammonium compounds and their applicability to blood serum measurements, *Mikrochim. Acta* **II:**235–244.

Hirsh, J., 1984, Heparin induced bleeding, *Nouv. Rev. Fr. Hematol.* **26:**262–266.

Jaques, L. B., 1973, Protamine: Antagonist to heparin, *Can. Med. Assoc. J.* **108:**1291–1297.

Jaques, L. B., 1980, Heparins: Anionic polyelectrolyte drugs, *Pharmacol. Rev.* **31:**99–165.

Jaques, L. B., and Wollin, A., 1967, A modified method for the colorimetric determination of heparin, *Can. J. Physiol. Pharmacol.* **45:**787–794.

Katz, N. M., Kim, Y. D., Siegelman, R., Ved, S. A., Ahned, S. W., and Wallace, R. B., 1987, Hemodynamics of protamine administration, *J. Thoracic Cardiovasc. Surg.* **94:**881–886.

Kelton, J. G., and Hirsh, J., 1984, Bleeding associated with antithrombotic therapy, *Semin. Hematol.* **17:**375–379.

Kim, J., Yang, A. J., and Yang, V. C., 1992, Protamine immobilization and heparin adsorption on the protamine-bound cellulose fiber membrane, *Biotechnol. Bioeng.* **39:**450–456.

Kirklin, J. K., Chenoweth, D. E., Naftel, D. C., Blackstone, E. H., Kirklin, J. W., Bitran, D. D., Curd, J. G., Reves, J. G., and Samuelson, P. N., 1986, Effects of protamine administration after cardiopulmonary bypass on complement, blood components, and the hemodynamic state, *Ann. Thoracic Surg.* **41:**193–199.

Kohn, J., and Wilchek, M., 1978, A colorimetric method for monitoring activation of Sepharose by cyanogen bromide, *Biochem. Biophys. Res. Commun.* **84:**7–14.

Langer, R., Linhardt, R. L., Galliher, P. M., Flanagan, C. L., Cooney, C. L., and Klein, M. D., 1982, An enzymatic system for removing heparin in extracorporeal therapy, *Science* **217:**261–263.

Ljungberg, B., Blomback, M., Johnsson, H., and Lins, L. E., 1987, A single dose of low molecular weight heparin fragment for anticoagulation during hemodialysis, *Clin. Nephrol.* **27:**31–35.

Ma, S., Meyerhoff, M. E., and Yang, V. C., 1992, Heparin responsive electrochemical sensor, *Anal. Chem.* **64:**694–697.

Morel, D. R., Costabella, P. M. M., and Pittet, J. F., 1990, Adverse cardiopulmonary effects and increased plasma thromboxane concentrations following the neutralization of heparin with protamine in awake sheep are infusion rate-dependent, *Anesthesiology* **73:**415–424.

Nojiri, C., Park, K. D., and Grainger, D. W., 1990, *In vitro* nonthrombogenicity of heparin immobilized polymer surfaces, *Trans. Am. Soc. Artif. Intern. Organs* **36:**168–172.

Oesch, U., Ammann, D., Pham, H. V., Wuthier, U., Zund, R., and Simon, W., 1986, Design of anion-selective membranes for clinically relevant sensors, *J. Chem. Soc., Faraday Trans. 1* **1986:**1179–1186.

Olinger, G. N., Becker, R. M., and Bonchek, L. I., 1980, Noncardiogenic pulmonary edema and peripheral vascular collapse following cardiopulmonary bypass: Rare protamine reaction? *Ann. Thoracic Surg.* **29:** 20–25.

Porter, J., and Jick, J., 1977, Drug-related deaths among medical inpatients, *J. Am. Med. Assoc.* **237:**879–881.

Swartz, R. D., and Port, F. K., 1979, Preventing hemorrhage in high risk hemodialysis: Regional versus low total dose heparin, *Kidney Int.* **16:**513–518.

Walenga, J., Fareed, J., Hoppensteadt, D., and Emanuele, R. M., 1986, *In vitro* evaluation of heparin fractions: Old vs new methods. *CRC Crit. Rev. Clin. Lab. Sci.* **22:**361–389.

Wegmann, D., Weiss, H., Ammann, D., Morf, W. E., Pretsch, E., Sugahara, K., and Simon, W., 1984, Anion-selective liquid membrane electrodes based on lipophilic quaternary ammonium compounds, *Mikrochim. Acta* **III:**1–16.

Weiler, J. M., Freiman, P., Sharath, M. D., Metzger, W. J., Smith, J. M., Richerson, H. B., Ballas, Z. K., Halverson, P. C., Shulan, D. D. J., Matsuo, S., and Wilson, R. L., 1985, Serious adverse reactions to protamine sulfate: Are alternatives needed? *J. Allergy Clin. Immunol.* **75:**297–303.

Weiler, J. M., Cellhaus, M. A., Carter, J. G., Meng, R. L., Benson, P. M., Hottel, R. A., Schillig, K. B., Vegh, A. B., and Clarke, W. R., 1990. A prospective study of the risk of an immediate adverse reaction to protamine sulfate during cardiopulmonary bypass surgery, *J. Allergy Clin. Immunol.* **85:**713–719.

Yang, V. C., 1989, Extracorporeal blood deheparinization system, U.S. Patent 4,800,016.

Yang, V. C., Port, F. K., Kim, J. S., Teng, C. L. C., Till, G. O., and Wakefield, T. W., 1991, The use of immobilized protamine in removing heparin and preventing protamine-induced complications during extracorporeal blood circulation, *Anesthesiology* **75:**288–297.

30

Affinity Precipitation

Bo Mattiasson and Rajni Kaul

Precipitation and recovery of precipitates is one of the unit operations in chemical engineering that can be operated on a large scale. Experiences from the area of biotechnology have clearly shown that there is not enough selectivity in the precipitation methods used to be able to separate, for example, complex mixtures of proteins from cell homogenates. Moreover, methods to precipitate proteins with their biological functions remaining intact must be gentle. In recent years, much work has been done on improving precipitation processes, and there are, today, a few alternative strategies for the removal of cell debris by the use of efficient precipitation aids.

The separation technique that seems to hold most promise in terms of selectivity is that based on affinity interactions. Affinity interactions can be tailored to separate proteins from a complex mixture with a very high efficiency. Furthermore, the resolving power of the technique allows two proteins with only minor differences in molecular properties to be separated. The problem so far has been the limited capacity of affinity-based purification since it has almost exclusively been operated in a chromatographic mode.

With this background of precipitation as a high-capacity, low-selectivity technique and affinity chromatography as a high-selectivity, low-capacity technique, it was tempting to try to combine these techniques in such a manner as to exploit the advantages of each of them. The present chapter summarizes the current status of affinity precipitation and also discusses some possibilities and limitations for future developmental work.

1. METHODS TO INDUCE PRECIPITATION

There are quite a few methods to induce precipitation of polymeric molecules from an aqueous solution. Some of these are listed in Table 1.

As the technique of affinity precipitation is still young, many of the options for development of the technique still remain to be tested.

Bo Mattiasson and Rajni Kaul • Department of Biotechnology, Chemical Center, Lund University, S-221 00 Lund, Sweden.

Molecular Interactions in Bioseparations, edited by That T. Ngo. Plenum Press, New York, 1993.

TABLE 1
Methods Used to Induce Precipitation

- Addition of polyvalent ions to cross-link charged polymers
- Addition of charged polymers of the opposite charge
- pH shift to induce hydrophobic aggregation
- Temperature shift
- Addition of organic solvents
- Formation of large complexes through affinity interactions

2. PRINCIPLES OF SELECTION OF A PRECIPITATION METHOD

There are, in essence, two main strategies that may be employed in precipitating affinity complexes: one is to use the affinity interaction to create complexes that are insoluble because of their size, and the other is to induce precipitation by a group not directly involved in the affinity interaction. The former method was the first to be utilized, immunoprecipitation of antigens with divalent antibodies being a good illustration of this principle. There is an optimum ratio of the concentrations of the two entities involved in the affinity interaction, and if this optimum ratio is deviated from, the complexes formed start to dissolve. This means that affinity-mediated precipitation has a major drawback in that one needs to know and control the concentrations of the reactants to be precipitated. Furthermore, there is a need for at least divalent, and preferably multivalent, reactants. This latter requirement restricts the application of affinity-mediated precipitation to compounds with two or more binding sites for the ligand used. On the other hand, affinity-mediated precipitation is, at least in theory, very attractive since it is based strictly on affinity. This means that it would be possible to carry out precipitations with very specific affinity by exploiting this method.

Affinity precipitation of this kind has been mimicked by using "homobifunctional" ligands, such as nucleotides connected to a spacer (bis-NAD^+), which, on binding to the active sites on multisubunit proteins, cross-link them noncovalently, effecting precipitation (Mosbach and Larsson, 1979). A similar system was later set up using a bis-dye reagent (Hayet and Vijayalakshmi, 1986).

When there is a need for a slightly more robust precipitation method that can also deal with samples of unknown concentrations and/or with monovalent target molecules, then any of the other precipitation methods listed in Table 1 may be preferred. In such cases, the affinity ligand is bound to a polymer structure so that the product is soluble in the medium but may be precipitated by exploiting some of the properties of the polymer used as the backbone in synthesizing the "heterobifunctional" ligand. The method for inducing precipitation will very much depend on the polymer used, and also on the properties of the sample from which the precipitation will be carried out.

2.1. Ionic Cross-Linking

Ionic cross-linking of polymers through the addition of polyvalent counterions leading to the formation of large complexes that precipitate has been a very popular method of immobilization by entrapment. For example, alginate, often solubilized in the form of sodium alginate, may be precipitated through the addition of di- or trivalent metal ions. The efficiency of the precipitation depends on the metal ion. The quality of the alginate has also been shown to play an important role (Smidsrød and Skjåk-Braek, 1990).

2.2. Addition of Oppositely Charged Polymers

From the literature on entrapment, it is known that addition of a positively charged polymer to an alginate solution induces precipitation. In principle, a polyelectrolyte complex, consisting of a polycation and a polyanion may be precipitated by a change in pH or by an addition of salt. This phenomenon has been used in the context of affinity precipitation of lactate dehydrogenase employing polyethyleneimine bound cibacron blue complexed with polyacrylic acid (Dissing and Mattiasson, submitted for publication).

2.3. Addition of Hydrophobic Groups of Opposite Charge

It is possible to achieve precipitation by adding to a charged polymer in solution a detergent of opposite charge (Goddard, 1986). This technique has only recently started to be utilized and probably has a greater potential than can be judged from the present literature.

2.4. pH Shift to Induce Hydrophobic Precipitation

A polymer having both ionic and hydrophobic groups can behave as a charged polymer in a certain pH region and as a polymer with essentially hydrophobic properties in another pH region. Some polymers within this group are listed in Table 2.

The principle in using polymers like those listed above is then to operate in the pH range where the polymer is soluble during the affinity binding step and then to induce precipitation by a pH shift to facilitate the separation of the polymer-bound affinity complex.

2.5. Temperature-Induced Precipitation

Some polymers have interesting solubility curves as a function of temperature. Precipitation may occur when the temperature is increased or decreased. Since the temperature range in which the solubility of these polymers changes is small, it may be

TABLE 2
Polymers Used for pH-Induced Affinity Precipitation

Polymer	pH range in which polymer is soluble	pH range in which polymer is precipitated
Chitosan	<6.5	>6.5
Eudragit L	>5.0	<3.5
Eudragit S	>6.0	<4.5
Cellulose derivatives	>5.0	<3.5

possible to achieve dramatic precipitation effects by a temperature shift of only a few degrees Celsius.

3. AFFINITY PRECIPITATION USING HETEROBIFUNCTIONAL LIGANDS

Affinity precipitation employing heterobifunctional ligands results from affinity binding of the target molecule to the ligand entity and then a precipitation step induced by exploitation of the solubility properties of the polymer. A crucial point in this whole concept is then the formation of suitable heterobifunctional ligands. The ligand is coupled to the desired polymer by means of the same repertoire of coupling chemistries as used in affinity chromatography and enzyme immobilization. When designing an affinity precipitation procedure, it is important to ensure that the coupling of the ligand to the polymer results in a heterobifunctional ligand with retention of both the functionalities of affinity interaction and the ability to be precipitated. Hence, cross-linking of the polymer must be avoided during the coupling step, since it will influence both the availability of the ligand and the solubility of the complex.

A precipitation curve for a native polymer as a function of pH is shown in Fig. 1. After introduction of a ligand, a slight shift in the precipitation curve is seen, and when the affinity interaction has taken place, the curve is displaced even more.

The yield of the affinity precipitation step will be dependent upon the efficiency of both the affinity interaction and the precipitation behavior. If the precipitation is less efficient after affinity binding, it is always possible to add some native polymer to improve the precipitation behavior. This was clearly shown in one of the early studies in this field dealing with chitosan as the precipitating polymer (Senstad and Mattiasson, 1989a). The efficiency in precipitation trypsin from a solution increased from 80% to more than 95%.

Using synthetic polymers, for example, Eudragit or modified celluloses, it is possible to achieve very efficient precipitation. The amount of polymer left in solution is in most cases only a few percent. Provided the affinity interaction is good, a very efficient affinity precipitation procedure is achieved.

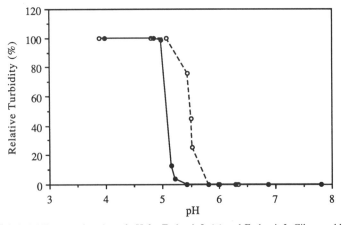

FIGURE 1. Solubility as a function of pH for Eudragit L (●) and Eudragit L-Cibacron blue (○).

3.1. Affinity Binding

The heterobifunctional ligand is added to the solution containing the target protein under conditions known to be favorable for binding, but still allowing the affinity complex to stay in solution. After binding has taken place, precipitation by any of the methods described above may be performed.

3.2. Washing

When forming an affinity precipitate from a crude solution, some nonspecific binding and also some entrapment of irrelevant material are likely to occur. Especially in cases using charged polymers, ion-exchange effects may cause some less desirable proteins to coprecipitate. Hence, the precipitate must be washed prior to elution of the target protein. This is done either by redissolving the precipitate in a suitable washing solution and then inducing floc formation and harvesting the precipitate again or by washing the precipitate thoroughly prior to the specific elution. It is very important to ensure that the target protein remains affinity-bound during these steps.

3.3. Dissociation

After the precipitate has been washed and harvested, dissociation of the bound protein is carried out. Ideally, this step is performed without solubilizing the polymer; alternatively, the complex is solubilized, dissociation takes place, and the polymer is precipitated while the target protein stays in solution. In most cases a shift in pH has been used to induce dissociation (Schneider *et al.*, 1981, Senstad and Mattiasson, 1989a), but chaotropic ions have also been used (Taniguchi *et al.*, 1989).

3.4. Separation of the Ligand–Polymer Complex from the Target Protein

It is essential to recover both the target protein and the ligand–polymer complex after the isolation procedure has been completed. The yield of protein of sufficient purity determines whether the procedure will be competitive or not, and in most cases the reuse of the ligand–polymer complex is a prerequisite to achieve an economical process.

Under ideal conditions, the target protein is liberated from the affinity complex without interfering with the solubility properties of the ligand–polymer complex. This would give the option of dissociating the protein while the affinity complex is in the precipitated form, or, alternatively, the affinity complex may first be dissolved and dissociated, and then the ligand–polymer complex is precipitated to separate it from the released protein. In cases in which pH-induced precipitation is used, recovery of the target protein and the ligand–polymer complex may turn out to be complicated, since pH is often used to induce dissociation. In such cases, the target molecule and the ligand–polymer complex must be separated under conditions under which they are dissociated, for example, by gel permeation chromatography or membrane filtration at pH values favoring dissociation (Senstad and Mattiasson, 1989a). This is not a favorable situation as it adds one more step to the purification process, and one should strive for systems where dissociation could be performed without interfering with the precipitation (Schneider *et al.*, 1981). One way to achieve this is to use biospecific displacement of the affinity-bound material.

3.5. Recycling of the Ligand–Polymer Complex

As stated above, recycling is a prerequisite for an economical process. For successful recycling, it is important to achieve complete precipitation and also to use an efficient washing and dissociation procedure. It should be borne in mind that the use of affinity precipitation comes very early in the purification scheme and that it may be realistic under such conditions not to operate with too sensitive ligands, but instead to use stable, low-molecular-weight group-specific ligands. This will then lead to a product in the elution step that is far from homogeneous, in contrast to the elution profiles obtained in affinity chromatography. However, affinity precipitation aims at isolation and enrichment of the target protein, and one can foresee that many different substances will bind to the ligand–polymer complex with varying degrees of interaction. The nonspecific interactions should be dealt with in the washing step, while the elution step may be more or less sophisticated. The easiest way is to elute everything that comes off and then to process this mixture in subsequent steps.

Recycling efficiencies may be improved by the addition of native polymer in the precipitation step. This strategy will reduce the losses of ligand, even if some of the polymer may be lost in each step.

3.6. Affinity Interactions Exploited in Affinity Precipitation

In principle, the whole range of affinity interactions used in chromatography are also available for affinity precipitation. There are, however, some constraints with regard to precipitation. In the chromatographic procedure, a large number of theoretical plates are used, and this means that there are many opportunities for the protein to interact with the ligand on its way down the chromatographic column. In the case of precipitation, no such possibilities exist as it is a batch adsorption process, and therefore one is restricted to using slightly stronger affinity interactions than are required in affinity chromatography. This is similar to the situation in affinity membrane filtration (Ling and Mattiasson, 1989).

4. EXAMPLES OF THE APPLICATION OF AFFINITY PRECIPITATION

Three examples of the application of the affinity precipitation technique are presented below. Our discussion of them is abbreviated since the introductory part of this chapter covers the general aspects. These and other examples of the application of affinity precipitation are summarized in Table 3.

4.1. Affinity Isolation of Wheat Germ Agglutinin (WGA)

The affinity isolation of wheat germ agglutinin (WGA) has been reported by Senstad and Mattiasson (1989b). WGA is a lectin with affinity for N-acetylglucosamine. Since chitosan is a copolymer of N-acetylglucosamine and glucosamine, one can predict that affinity interactions will take place. Chitosan is, furthermore, a polymer with suitable solubility properties in the sense that it is soluble below pH 6.5 and precipitates at higher pH values. Dissociation of the lectin from the ligand can be effected through the use of either monomeric sugars or low pH.

A flow chart for the purification procedure is presented in Fig. 2.

TABLE 3
Examples of Systems Utilized in Affinity Precipitation Studies

Target protein	Ligand–polymer[a]	Precipitation mode	Reference
Wheat germ agglutinin	Chitosan	pH	Senstad and Mattiasson, 1989b
Galactose dehydrogenase	Target protein with an affinity tail of polyhistidine	Metal affinity	Lilius et al., 1991
Trypsin	STI-chitosan	pH	Senstad and Mattiasson, 1989a
Trypsin	p-Aminobenzamidine-NIPAM– NASI copolymer	Temperature or ionic strength	Nguyen and Luong, 1989
Trypsin	p-Aminobenzamidine acrylate	pH	Schneider et al., 1981
Protein A	IgG-hydroxypropylmethyl- cellulose acetate, succinate	pH	Taniguchi et al., 1989
Antigen	mAb-NIPAM-glycidyl methacrylate	Temperature	Monji and Hoffman, 1987
Lactate dehydrogenase	Blue dextran	Con A	Senstad and Mattiasson, 1989c
Human IgG	Galactomannan-protein A	Tetraborate	Bradshaw and Sturgeon, 1990

[a]Abbreviations, STI, soybean trypsin inhibitor; NIPAM, N-isopropyl acrylamide; NASI, N-acryloxysuccionide; mAb, monoclonal antibody.

4.2. Isolation of Trypsin by Affinity Precipitation Using Chitosan–Soybean Trypsin Inhibitor Conjugates

Senstad and Mattiasson (1989a) have described the isolation of trypsin by affinity precipitation. Soybean trypsin inhibitor (STI) was covalently bound to chitosan. Mixing of the soluble heterobifunctional ligand with trypsin solution led to affinity interactions. A shift in pH from 5.5 to 8.5 induced floc formation, and the precipitate could be harvested. To improve the efficiency in the precipitation step, native chitosan was added after the affinity complex had been formed.

Dissociation was carried out by lowering the pH to 2.5, which also caused the precipitate to dissolve. The trypsin and the heterobifunctional ligand were separated through membrane filtration or, on a small scale, through gel permeation chromatography. When starting from a crude enzyme preparation, the purification factor was only 5.5; this is rather typical behavior for affinity precipitation procedures.

4.3. Utilization of Two Different Affinity Interactions in the Purification of Lactate Dehydrogenase (LDH) by Affinity Precipitation

Cibacron blue is a frequently used group-specific ligand for isolation of dehydrogen- ases and kinases. In a procedure reported by Senstad and Mattiasson (1989c) for the purification of lactate dehydrogenase (LDH) by affinity precipitation, the dye was bound to soluble dextran molecules. Affinity complexes were formed upon contacting the LDH molecules with the blue dextran. Since both reagents are polyvalent, there is a risk of spontaneous formation of affinity complexes large enough to induce precipitation. This happened in a few cases, but only to a very low degree. After the complexes were formed, they were precipitated by addition of the lectin concanavalin A (Con A), which binds the

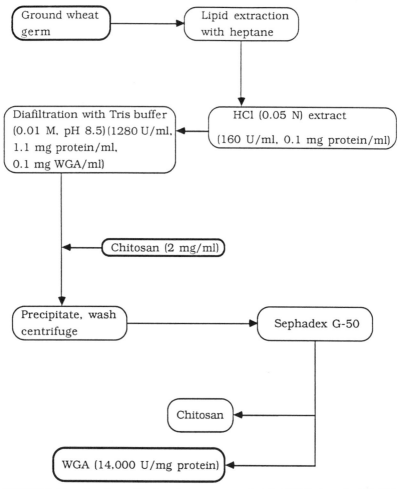

FIGURE 2. Flow chart for purification of wheat germ agglutinin (WGA). Overall yield, 70%.

dextran residues and induces precipitation. The efficiency of the precipitation step was approximately 85%. Dissociation of the affinity-bound LDH was carried out using KCl (1.5M KCl liberated all bound LDH). The Con A–blue dextran complexes were dissociated using glucose as competing ligand.

This example illustrates the possibility of utilizing two different binding reactions in order to generate a precipitate.

5. SIMILAR TECHNIQUES

Affinity precipitation is a good illustration of the application of principles established in the area of surface and colloid chemistry to problems of a biotechnological nature. The whole area of surface and colloid chemistry is now attracting much attention from biotechnologists, and this will lead to a revitalization of the work to develop new purification methods. In the area of affinity-mediated separation and secondary induced separation, there are several techniques with great similarities to affinity precipitation. Temperature-induced phase separation (Brodier, 1981) as well as phase separation induced by other physicochemical means represents this type of possibilities.

6. CONCLUDING REMARKS

Affinity precipitation is still at the starting point of its development. It is, however, already now obvious that the technique will offer new directions in downstream processing. Whether these will be generally accepted by industry remains to be seen.

ACKNOWLEDGMENTS

This work was supported by the National Swedish Board for Technical Development (STU) and the Swedish Agency for Research Cooperation with Developing Countries (SAREC).

REFERENCES

Bradshaw, A. P., and Sturgeon, R. J., 1990, The synthesis of soluble polymer–ligand complexes for affinity precipitation studies, *Biotechnol. Techniques* **4**:67–71.

Brodier, C., 1981, Phase separation of integral membrane proteins in Triton X-114 solution, *J. Biol. Chem.* **256**:1604–1607.

Goddard, E. D., 1986, Polymer–surfactant interaction. Part II. Polymer and surfactant of opposite charge, *Colloids Surf.* **19**:301–329.

Hayet, M., and Vijayalakshmi, M. A., 1986, Affinity precipitation of proteins using bis-dyes, *J. Chromatogr.* **376**:157–161.

Lilius, G., Persson, M., Bülow, L., and Mosbach, K., 1991, Metal affinity precipitation of proteins carrying genetically attached polyhistidine affinity tails, *Eur. J. Biochem.* **198**:499–504.

Ling, T. G. I., and Mattiasson, B., 1989, Membrane filtration affinity purification (MFAP). A continuous affinity purification method, *Biotechnol. Bioeng.* **34**:1321–1325.

Monji, N., and Hoffman, A. S., 1987, A novel immunoassay system and bioseparation process based on thermal phase separating polymers, *Appl. Biochem. Biotechnol.* **14**:107–120.

Mosbach, K., and Larsson, P.-O., 1979, Affinity precipitation of enzymes, *FEBS Lett.* **98**:333–338.

Nguyen, A. L., and Luong, J. H. T., 1989, Synthesis and applications of water-soluble reactive polymers for purification and immobilization of biomolecules, *Biotechnol. Bioeng.* **34**:1186–1190.

Schneider, M., Guillot, C., and Lamy, B., 1981, The affinity precipitation technique. Application to the isolation and purification of trypsin from bovine pancreas, *Ann. N.Y. Acad. Sci.* **369**:257–263.

Senstad, C., and Mattiasson, B., 1989a, Affinity-precipitation using chitosan as ligand carrier, *Biotechnol. Bioeng.* **33**:216–220.

Senstad, C., and Mattiasson, B., 1989b, Purification of wheat germ agglutinin using affinity flocculation with chitosan and a subsequent centrifugation or flotation step, *Biotechnol. Bioeng.* **34**:387–393.

Senstad, C., Mattiasson, B., 1989c, Precipitation of soluble affinity complexes by a second affinity interaction: A model study. *Biotechnol. Appl. Biochem.* **11**:41–48.

Smidsrød, O., and Skjåk-Braek, G., 1990, Alginate as immobilization matrix for cells. *Trends Biotechnol.* **8**:71–78.

Taniguchi, M., Kobayashi, M., Natsui, K., and Fujii, M., 1989, Purification of staphylococcal protein A by affinity precipitation using a reversibly soluble polymer with human IgG as a ligand, *J. Ferment. Bioeng.* **68**:32–36.

31

Affinity Separation of Nucleic Acids on Monosized Magnetic Beads

Mathias Uhlén, Ørjan Olsvik, and Erik Hornes

The introduction of solid-phase methods has had great impact on molecular biology in the area of synthesis and sequencing of proteins and DNA. High yield and reproducibility can be obtained, and automation is facilitated since reaction buffers can be rapidly and easily changed. There are many alternative solid supports, such as organic or inorganic polymer beads or the walls of test tubes or microtiter wells. An attractive alternative is the use of monosized magnetic particles, Dynabeads (Lea *et al.*, 1988; Ugelstad *et al.*, this volume, Chapter 16), which combines the convenience of magnetic motility with an adsorption and desorption of biomolecules at the surface. This provides reaction kinetics similar to those found in a free solution. Magnetic beads with covalently bound streptavidin can be used for directed immobilization of both double-stranded and single-stranded biotinylated DNA. The remarkable stability and strength of the biotin–streptavidin interaction ($K_d = 10^{-15}$ M) allow DNA manipulations, such as strand melting, elution, and hybridization (using alkali, temperature, or formamide) to be performed without interfering with the binding of the DNA to the beads.

Several alternative strategies exist for the selective introduction of biotin into one of the strands of the DNA. The most straightforward approach is to take advantage of the polymerase chain reaction (PCR) using a biotinylated oligonucleotide primer, as outlined in Fig. 1. The immobilized double-stranded DNA can subsequently be converted into single-stranded DNA by an appropriate melting procedure. Several applications have been developed for the use of the DNA bound to the magnetic beads (Fig. 1).

Mathias Uhlén • Department of Biochemistry and Biotechnology, The Royal Institute of Technology, S-100 44 Stockholm, Sweden.   *Ørjan Olsvik* • Enteric Diseases Branch, Division of Bacterial and Mycotic Diseases, Centers for Disease Control, Atlanta, Georgia 30333, and Department of Microbiology and Immunology, Norwegian College of Veterinary Medicine, Oslo 1, Norway.   *Erik Hornes* • Department of Microbiology and Immunology, Norwegian College of Veterinary Medicine, Oslo 1, Norway, and Dynal AS, 0212 Oslo, Norway.

Molecular Interactions in Bioseparations, edited by That T. Ngo. Plenum Press, New York, 1993.

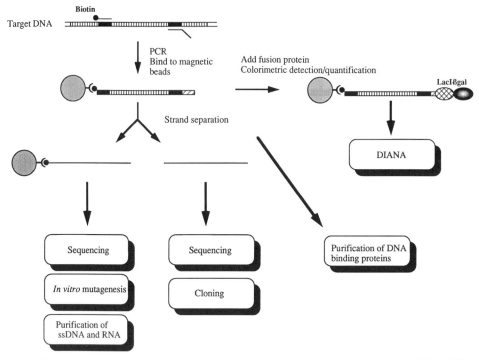

FIGURE 1. Schematic drawing of different strategies for the use of magnetic separation of DNA. PCR, Polymerase chain reaction; DIANA, detection of immobilized amplified nucleic acids.

1. PURIFICATION OF poly(A)⁺ MESSENGER RNA

Most strategies for mRNA isolation involve purification of total RNA as a first step and selection of poly(A)$^+$ RNA by affinity chromatography on oligo(dT) matrices as the second step. The conventional method using an oligo(dT)-cellulose column and gravity for elution is time-consuming and rather laborious. Newer variations of this strategy using oligo(dT) latex spheres in a spin column or in a batch procedure have made the process more user-friendly but still not ideal. The introduction of magnetic beads speeds up the process even more and makes it more flexible and at least as efficient as the nonmagnetic alternative with pure mRNA as a result. There are two main alternative methods to affinity-purify poly(A)$^+$ mRNA using oligo(dT) and magnetic beads. One utilizes a biotinylated oligo(dT) primer and streptavidin-coated magnetic beads, and the other uses oligo(dT) covalently coupled to magnetic beads [Dynabeads Oligo(dT)].

The first method utilizes a biotinylated oligo(dT) primer to hybridize to the 3′ poly(A)$^+$ tail in a first step and streptavidin-coated super paramagnetic beads (Dynabeads) to capture the hybrid in a second step. The system is not reusable.

The second system performs both steps simultaneously and has the greater potential for future applications. The use of magnetic oligo(dT) beads for purification of poly(A)$^+$ RNA works for crude RNA preparations or cell lysates that would clog standard oligo(dT)-cellulose column systems. By magnetic separation, followed by washing and elution steps, pure poly(A)$^+$ mRNA is obtained in the desired volume. The method for purification of cytoplasmic mRNA directly from tissue culture cells (Hornes and Korsnes, 1990) utilizes a gentle detergent such as Nonidet P-40 (NP-40) or Triton X-100 to lyse cells. The cells will disrupt almost immediately without disruption of the nuclei. The nuclei are removed by microcentrifugation, and mRNA is recovered directly from the nuclei-free cytoplasmic supernatant. Jakobsen *et al.* (1990) have shown that purification of mRNA directly from solid tissues takes only 15 minutes. With the use of this method, pure messenger RNA has been successfully prepared from several kinds of plant tissue such as embryo, aleurone, seedlings, roots, and leaves.

cDNA has been successfully synthesized onto Dynabeads directly after mRNA isolation (Wada *et al.*, 1991). Wada *et al.* used the commercially available Dynabeads Oligo(dT) to synthesize the first strand directly on the beads and then performed the PCR on the cDNA bound to beads for the typing of different leukemias. The Dynabeads bearing cDNA can be repeatedly used for other PCR analyses.

2. DNA SEQUENCING

DNA sequencing is one of the most central techniques in molecular biology and biotechnology. A protocol for manual and semiautomated sequencing of both plasmid and genomic DNA based on *in vitro* amplification by PCR and magnetic separation has been developed (Hultman *et al.*, 1991). First, a PCR is carried out using a pair of primers, one of which is biotinylated at the 5'-end. Note that for sequencing vector inserts, general primers can be used which are complementary to the sequence flanking the multilinker region of common cloning vectors. The PCR is in this case carried out directly on a single colony. In contrast, for direct genomic sequencing, a nested primer approach is normally carried out to minimize amplification of nonspecific sequences (Syvänen *et al.*, 1991). Second, the material from the PCR is bound to the magnetic beads with covalently bound streptavidin. The strand lacking the biotin is subsequently eluted and neutralized to yield a clean, single-stranded DNA molecule suitable as a sequencing template. The two strands are thereafter treated equally, and a standard dideoxy sequencing is performed. The reaction mixture is finally treated with formamide and directly applied to an automated or manual electrophoresis unit.

The method is well suited for automation. The use of magnetic beads facilitates liquid handling and ensures a flexible system where the quantification and strand separation can be performed under optimal conditions using a standard robotic work station equipped with a magnetic separator. The DNA to be analyzed is obtained directly by capturing the PCR product on a solid support. Thus, a large number of samples may be rapidly prepared for the sequencing system. With the use of a robotic work station with a heating/cooling device for the sequencing reactions, the complete method after the template preparation takes approximately 45 minutes.

Note that the direct sequencing method described here does not give rise to errors due to misincorporation during the PCR, which is a common problem for all strategies depending on *cloning* of the PCR products. The direct genomic sequencing methods have been used for *Chlamydia trachomatis* (Wahlberg *et al.*, 1990a), *Plasmodium falciparum* (Lundeberg *et al.*, 1990a), human immunodeficiency virus type 1 from clinical samples (Wahlberg *et al.*, 1991), human apolipoprotein E gene (Syvänen *et al.*, 1991), hypoxanthine phosphoribosyl transferase gene (Gibbs *et al.*, 1990), and cloned fragments with inserts of more than 70% GC (Hultman *et al.*, 1991).

3. SOLID PHASE CLONING

The ability to splice together gene or gene fragments by recombinant DNA techniques involving restriction enzymes and ligase is essential in molecular biology. However, the existing technology has several disadvantages. First, there is a need for convenient restriction sites, and when oligonucleotide linkers are used, additional bases are often introduced. In addition, the *in vitro* recombination process is rather inefficient and relatively cumbersome for screening of desired recombinants. Finally, the technology is rather time-consuming and not well suited for automation. There is therefore a need for alternative methods to create recombinant DNA molecules.

Recently, a solid-phase approach that creates recombinant DNA molecules without the use of restriction enzymes or ligase was described (Hornes *et al.*, 1990). The gene fragments can be combined independent of restriction sites, providing a flexible system suitable for automation. A single-stranded vector fragment is provided by selectively incorporating a biotin molecule into one of the strands of the vector DNA. This is achieved by restriction and fill-in using a biotinylated dNTP conjugate and DNA polymerase. The double-stranded DNA is bound to magnetic beads containing streptavidin, and the single-stranded vector is simply eluted with alkali. If more single-stranded vector fragment is needed, a run-off extension reaction with DNA polymerase can be carried out, one or several times, and the extended fragment is again eluted with alkali and collected.

The "insert" DNA to be cloned is obtained by PCR, using specific primers with handles consisting of vector regions. For genomic DNA, specific primers need to be synthesized, while for gene fragments inserted into vectors it is possible to use general PCR primers designed for the solid-phase cloning. In both cases, one of the primers contains a biotin molecule at the 5' end, allowing the *in vitro* amplified material to be captured by streptavidin-coated magnetic beads. A single-stranded insert fragment with flanking regions complementary to the vector fragment can subsequently be eluted with alkali. The two single-stranded fragments can then be mixed to form a gap-duplex molecule and transformed directly into *Escherichia coli*. The method allows cloning of any fragment, from any origin (chromosomal DNA, plasmid DNA, etc.), independent of restriction sites. No restriction enzymes or ligase is needed, and a very high fraction of recombinants is expected since the vector and insert alone theoretically are not inducing transformants. More than 90% correct recombinants were obtained when a human apolipoprotein E gene fragment was cloned from the human chromosome using this approach (Hornes *et al.*, 1990).

4. IN VITRO *MUTAGENESIS*

Many different approaches have been described for *in vitro* mutagenesis. A large number of these protocols involve oligonucleotide-mediated mutagenesis, which provides a precise approach to specific changes of the DNA. Recently, a method for *in vitro* mutagenesis using magnetic separation was described (Hultman *et al.*, 1990). The system is similar to the solid-phase cloning system described above. The vector and the insert fragment to be mutated are immobilized separately on magnetic beads through a biotin molecule incorporated into one of the strands.

For the fragment to be mutated, a single-stranded template suitable for primer-directed polymerase reactions is obtained by elution of the nonimmobilized fragment with alkali. Several variants of the system exist, but normally a double-primer system is employed in which a general primer, complementary to the vector part of the immobilized fragment, is combined with the specific mutagenesis primer with a mismatch in the region to the changed. An extension reaction using T4 DNA polymerase and T4 DNA ligase is thereafter performed to yield a mutated strand containing the desired mismatch. The produced strand is finally eluted by alkali and mixed with the single-stranded vector fragment. This yields a gap-duplex plasmid containing small double-stranded regions. The mutated region is single-stranded to avoid mismatch repair. The gap-duplex plasmid is finally transformed directly to *E. coli*, and clones are screened by conventional methods, most preferably by DNA sequencing, to identify mutated plasmids.

The mutagenesis has the advantage that the same result can be accomplished both with or without the use of PCR. This is important since accumulated polymerase errors are a major concern whenever PCR products are cloned (Hornes *et al.*, 1990). This makes it strongly desirable to sequence all PCR-produced fragments when they are used for cloning. Especially for large fragments, such as most cloning and expression vectors, this task is difficult and time-consuming. Therefore, protocols, such as the solid-phase method, where the vector can be produced with a non-PCR approach are attractive.

5. APPLICATIONS OF IMMUNOMAGNETIC SEPARATION (IMS) COMBINED WITH PCR

The development of PCR represents a new generation of DNA techniques available for clinical microbiological diagnostic use. By employing PCR, it is theoretically possible to identify genes encoding virulence factors directly in food or stool samples without precultivation of microorganisms. However, some disadvantages limit the technique for certain diagnostic uses. The template volume traditionally used in PCR has been from 5 to 20 µl. For several microbiological applications, for example, testing for *Salmonella*, requirements are one cultivatable organism per 100 g (or 100 ml) (10–25 g for some requirements) of sample. Using a 5–20-µl sample in PCR, this would restrict the sensitivity to a theoretical minimum of 50 to 200 organisms per milliliter. Additional factors preventing diagnostic use of PCR directly on clinical samples have been the sensitivity of the *Taq* polymerase to inhibitor elements in samples such as feces and blood. Several clinical samples such as feces, blood, and urine had to be diluted, or the DNA in the sample had to be purified for a successful test result. Different methods for purification of nucleic acids from the sample have been published, but they are all rather time-consuming.

Immunomagnetic separation (IMS) as a pre-PCR step has appeared to solve several of these problems. The bacteria in a large sample (1–10 ml) are concentrated in a suitable volume (10–100 µl) and simultaneously removed from the specific *Taq* polymerase inhibitors in the original test sample. After IMS, microwaving or boiling has been shown to be sufficient for preparing template DNA. Samples that have been frozen often contain nonviable cells. These bacterial cells can be extracted with IMS, and although they are not suitable for cultivation, PCR has appeared to be a good detection system for such organisms.

6. DETECTION OF IMMOBILIZED AMPLIFIED NUCLEIC ACIDS (DIANA)

The analysis of specific nucleotide sequences by the use of PCR has been used for detection of various pathogens, genetic disorders, and allelic variations (Innis *et al.*, 1990). The PCR is one way to overcome the limitations in sensitivity of conventional DNA detection methods (Innis *et al.*, 1990). Recently, a solid-phase assay for both qualitative or quantitative detection of DNA was described (Lundeberg *et al.*, 1991). The assay is used for *d*etection of *i*mmobilized *a*mplified *n*ucleic *a*cids (DIANA) and is based on selective introduction of biotin into the specific DNA fragment amplified by the PCR. The isotopic labeling used by Wahlberg *et al.* (1990b) for detection was later replaced by a colorimetric assay based on introduction of the *lac* operator sequence from *E. coli* during the amplification (Wahlberg *et al.*, 1990a; Lundeberg *et al.*, 1990b, 1991). In combination with a nested PCR procedure used to lower the background signal, and a colorimetric detection of bound material, the approach provides a rapid and reliable assay, suitable for both manual and automated clinical diagnosis. By using a solid-phase approach, electrophoresis, restriction mapping, and/or hybridization to discriminate the background from actual signal are avoided. Note that the positive samples can be directly sequenced if desired.

The first step is a nested PCR with outer and inner oligonucleotide-specific for the target sequence; during the inner amplification, biotin and a *lac* operator handle are selectively incorporated at the ends of the fragment. The fragment is immobilized on the streptavidin-coated magnetic beads and the nonbiotinylated fragment (due to unspecific annealing) is washed away. A recombinant fusion protein consisting of the *lac* repressor (LacI) and β-galactosidase is added to the beads; the bound enzyme conjugate is then detected by adding a chromogenic substrate specific for the β-galactosidase.

As a qualitative assay is not always sufficient for diagnosis, a quantitative assay has been developed. Lundeberg *et al.* (1991) have presented a quantitative assay based on a coamplification of an internal standard containing a cloned *lac* operator in the target region. The same primer site for both competitor and sample can be used, and the amplified fragment has the same length and G/C content. Since the protocol is similar to the enzyme-linked immunoassay (ELISA) protocol, extensively used in both clinical and research applications, the quantitative DIANA method is suitable for routine clinical and basic applications.

7. CONCLUDING REMARKS

Magnetic separation of DNA has proven to be useful for many applications in molecular biology. As described above, these procedures are well suited for sequencing, cloning, and *in vitro* mutagenesis. In addition, several qualitative and quantitative PCR methods for detection of RNA and DNA have been developed that take advantage of the efficient capture of the amplified material by the magnetic beads. These assays are rapid and eliminate the need for electrophoresis, restriction mapping, or hybridizations in the analysis of PCR-amplified materials. PCR-amplified materials combined with magnetic separation can also be used for rapid affinity purification of DNA-binding proteins from crude cell lysates (Gabrielsen *et al.*, 1989), for recovery of single-stranded DNA and RNA (Hornes and Korsnes, 1990), for genomic walking (Rosenthal and Jones, 1990), and for labeling of single-stranded DNA probes (Espelund *et al.*, 1990). Both manual and automated methods can be developed, and such procedures might have an impact on molecular biology in the near future.

REFERENCES

Espelund, M., Prentice, R. A., Jakobsen, S., and Jakobsen, K. S., 1990, A simple method for generating single-stranded DNA probes labeled to high activities, *Nucleic Acids Res.* **18:**6157–6158.

Gabrielsen, O. S., Hornes, E., Korsnes, L., Ruet, A., and Øyen, T. B., 1989, Magnetic DNA affinity purification of yeast transription factor—a new purification principle for the ultrarapid isolation of near homogenous factor, *Nucleic Acids Res.* **17:**6253–6267.

Gibbs, R. A., Nguyen, P. N., Edwards, A., Civitello, A. B., and Cashey, C. T., 1990. Multiplex DNA deletion, detection and exon sequencing of the HPRT gene in Lesch–Nyhan families. *Genomics* **7:**235–244.

Hornes, E., Hultman, T., Moks, T., and Uhlén, M., 1990, Direct cloning of the human genomic apolipoprotein E gene using magnetic separation of single-stranded DNA, *BioTechniques* **9:**730–737.

Hornes, E., and Korsnes, L., 1990, Magnetic DNA hybridization properties of oligonucleotide probes attached to superparamagnetic beads and their use in the isolation of poly(A) mRNA from eucaryotic cells, *Genet. Anal. Tech. Appl.* **7:**145–150.

Hultman, T., Bergh, S., Moks, T., and Uhlén, M., 1991, Bidirectional solid phase sequencing of *in vitro* amplified plasmid DNA, *BioTechniques* **10:**84–93.

Hultman, T., Murby, M., Ståhl, S., Hornes, E., and Uhlén, M., 1990, Solid phase *in vitro* mutagenesis using plasmid DNA template, *Nucleic Acids Res.* **18:**5107–5112.

Innis, M. A., Gelfand, D. H., Sninsky, J. J., and White T. J., 1990, *PCR Protocols*, Academic Press, San Diego, California.

Jakobsen, K. S., Breivold, E., Hornes, E., 1990, Purification of mRNA directly from crude plant tissues in 15 minutes using magnetic oligo dT microspheres. *Nucleic Acids Res.* **18:**3669.

Lea, T., Vartdal, F., and Nustad, K., 1988, Monosized, magnetic polymer particles: Their use in separation of cells and subcellular components, and in the study of lymphocyte function *in vitro*, *J. Mol. Recog.* **1:**9–17.

Lundeberg, J.,Wahlberg, J., and Uhlén, M., 1990a. Affinity purification of specific DNA fragments using a *lac* repressor fusion protein, *Genet. Anal. Tech. Appl.* **1990**(7):47–52.

Lundeberg, J., Wahlberg, J., Holmberg, M., Pettersson, U., and Uhlén, M., 1990b, Rapid colorimetric detection of *in vitro* amplified DNA sequences, *DNA Cell Biol.* **1990**(9):287–292.

Lundeberg, J., Wahlberg, J., and Uhlén, M., 1991, Rapid colorimetric quantification of PCR-amplified DNA, *BioTechniques* **10:**68–75.

Rosenthal, A., and Jones, D. S. C., 1990, Genomic walking and sequencing by oligocassette mediated polymerase chain reaction, *Nucleic Acids Res.* **18:**3095–3096.

Stegemann, J., Schwager, C., Erfle, H., Hewitt, N., Voss, H., Zimmermann, J., and Asorge, W., 1991, High speed on-line DNA sequencing on ultrathin slab gels. *Nucleic Acids Res.* **19:**675–676.

Syvänen, A.-C., Hultman, T., Söderlund, H., and Uhlén, M., 1991, Analyses of apolipoprotein E polymerhisms by solid phase sequencing with automatic fluorescent detection, *Genet. Anal. Tech. Appl.* **8:**117–123.

Wahlberg, J., Lundeberg, J., Hultman, T., and Uhlén, M., 1990a, General colorimetric method for DNA diagnostics allowing direct solid phase genomic sequencing of the positive samples. *Proc. Natl. Acad. Sci. U.S.A.* **87:**6569–6573.

Wahlberg, J., Lundeberg, J., Hultman, T., Holmberg, M., and Uhlén, M., 1990b, Rapid detection and sequencing of specific *in vitro* amplified DNA sequences using solid phase methods, *Mol. Cell. Probes* **1990**(4):285–297.

Wahlberg, J., Albert, J., Lundeberg, J., Fenyö, E.-M., and Uhlén, M., 1991, Analysis of the V3 domain in neutralization-resistant human immunodeficiency virus type 1 variants by a novel solid phase DNA sequencing method, *AIDS Res. Hum. Retroviruses* **7:**983–990.

32

Affinity Ultrafiltration for Protein Purification

Rajni Kaul and Bo Mattiasson

Purification of a biochemical product from fermentation broths or other complex biological mixtures generally involves a combination of techniques which resolve substances according to differences in their physicochemical properties. These techniques have traditionally been optimized individually rather than as a part of the integrated, continuous process. Moreover, a lot of the development and refinement of the procedures has taken place on a laboratory scale, with little consideration given to the economics of scaling up. As a result, downstream processing presently accounts for a major portion of the processing costs of producing a biotechnological product.

In order to cut down on costs and also to prevent the product loss that occurs during several purification stages, it is critical to design processes that possess high product recovery and resolution from impure feed streams.

1. AFFINITY CHROMATOGRAPHY

Among the available purification strategies, affinity chromatography is known to be the most powerful in terms of specificity. This has been amply demonstrated in the present volume. Affinity chromatography makes use of the natural affinity displayed between a biological macromolecule and a complementary ligand, for instance, a substrate analog or an inhibitor:

$$P + L \underset{k_{-1}}{\overset{k_1}{\rightleftharpoons}} PL$$

$$K_d = \frac{k_{-1}}{k_1} = \frac{[P][L]}{[PL]}$$

where K_d is the dissociation constant; k_1 and k_{-1} are forward and reverse rate constants, respectively; and [P], [L], and [PL] are the respective concentrations of the protein, the

Rajni Kaul and Bo Mattiasson • Department of Biotechnology, Chemical Center, Lund University, S-221 00 Lund, Sweden.

Molecular Interactions in Bioseparations, edited by That T. Ngo. Plenum Press, New York, 1993.

ligand, and the affinity complex. An extremely stable complex is represented by a low K_d value, and vice versa.

In affinity chromatography, the ligand is immobilized onto a solid matrix. After the protein–ligand affinity complex is formed and the impurities have been washed away, the bound protein molecule is dissociated from the ligand. This is achieved by a change in pH and/or ionic strength, by using chaotropic agents, or, in a more specific manner by using a solution of a free complementary ligand.

Despite the potential of this technique, its scaled-up application has been limited because of the high material costs, the requirement for sample pretreatment, and other problems characteristic of chromatographic separations. Currently, it is used only as a final step in the downstream processing of high-value products.

There have been attempts during the last decade to exploit affinity interactions in a mode which could be used early during the downstream processing of a protein. Such a configuration should be amenable to easy scale-up while at the same time overcoming the limitations of the chromatographic mode of operation.

2. ULTRAFILTRATION

Ultrafiltration is based on the ability of a membrane to separate molecules from other molecules or particles on the basis of size. Membrane filtration is an established technology in biochemical and pharmaceutical industries. In biotechnology, it has been integrated into downstream processing at various stages such as cell harvesting, separation of biomolecules, and concentration and diafiltration of products. It is simple, rapid, and economical and has low energy requirements.

Developments in polymer chemistry and module construction have resulted in the availability of an array of ultrafiltration membranes of high chemical and physical stability combined with high fluxes, for different scales of operation. Filtration is most effective when operated as a tangential flow (also termed cross-flow filtration) such that the flow of the feed is parallel to the membrane surface and is recirculated rapidly enough to prevent concentration polarization.

As a separation technique, ultrafiltration is limited by its low resolving power. At least a tenfold difference in molecular weight is necessary for separation of molecules.

3. AFFINITY ULTRAFILTRATION

In order to exploit the high-volume processing capability of membrane filtration, and at the same time make it extremely selective, the concept of affinity ultrafiltration has been developed. This technique works on the principle that when a protein to be purified is free in solution, it passes unimpeded through the pores of the ultrafiltration membrane, whereas when it is bound to a macromolecular affinity ligand placed on one side of the membrane, it is restricted to that side of the membrane (Mattiasson and Ramstorp, 1983; Mattiasson et al., 1983; Mattiasson and Ling, 1986; Luong et al., 1987). Compounds which do not bind the ligand pass freely through the membrane and are thus separated from the protein–ligand complex. After all the undesired components have been removed, the protein is desorbed from the ligand by the addition of an appropriate dissociating medium and is then recovered in the filtrate. The ligand is regenerated by removing the dissociating medium and recycled. The general procedure of affinity ultrafiltration is schematically represented in Fig. 1. This process can be run as a batch, semibatch, or continuous operation.

The respective sizes of the target molecule and the macromolecular ligand together with the cutoff of the filtration membrane determine the efficiency and quality of purification. The low selectivity of ultrafiltration membranes makes it imperative that the molecular weight of the macroligand be more than ten times that of the protein and the contaminant species.

The pores of the membrane should be of a size and uniformity which prevent the leakage of the ligand and, at the same time, should be highly permeable to the free protein and the impurities under the operating conditions. Therefore, prior to the affinity step, it is advantageous to study the behavior of the macromolecular ligand and the target protein individually on ultrafiltration, through the chosen membrane (Luong *et al.*, 1988a).

Adsorption of the macromolecular ligand to the membrane may lead to reduced filtration rates, and it may deposit as a layer on the membrane surface with its own rejection properties. It may be possible to minimize these interactions by introducing charges on the membrane surface. Protein–protein interactions such as complexation may also lead to enhanced rejection coefficients during protein ultrafiltration. The significance of the latter phenomena were well demonstrated in a study on the isolation of urokinase by affinity

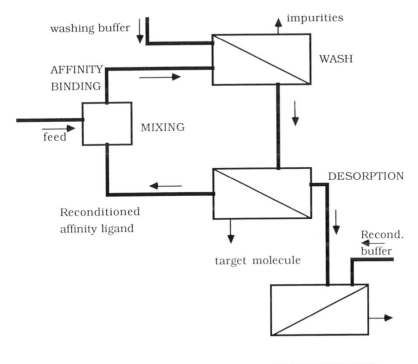

FIGURE 1. Schematic of the experimental setup for membrane affinity filtration.

filtration (Male *et al.*, 1990). When the enzyme, with a molecular weight of 31,000, was ultrafiltered in the absence of the affinity polymer using 10,000- or 100,000-MW-cutoff Millipore systems, it was not detected in either the filtrate or the retentate, suggesting that it was somehow binding to the polysulfone membrane. However, the enzyme was able to pass through a 100,000-MW-cutoff Amicon YM membrane (cellulose acetate) with greater than 95% recovery and was retained to a level of more than 95% by a 10,000-MW-cutoff Amicon membrane. Furthermore, during the purification of urokinase from human urine, it was noticed that the high-molecular-weight materials present in this crude source apparently formed a complex with the enzyme and prevented its passage through a 100,000-MW-cutoff membrane. The complex was dissociated by the addition of salt ($2M$ NaCl), resulting in 70% urokinase activity in the filtrate.

3.1. Affinity Binding

Binding between the ligand and the material to be purified can be achieved in one of the following ways:

1. by mixing together the crude solution with the ligand in the ultrafiltration unit or in a separate mixing chamber; or
2. by pumping the crude solution and the solution containing the ligand on either side of the filtration membrane and letting the binding take place when the protein passes over to the side where the ligand is located.

The first alternative is the method of choice for relatively clarified broths or when the solutions containing soluble and insoluble high-molecular-weight contaminants have been preultrafiltered by using a membrane with a suitable molecular weight cutoff. Most of the studies reported on affinity ultrafiltration have employed this procedure.

The second method is a more general approach and could be directly integrated with, for example, a fermentation broth or other complex mixtures of biological molecules. Here, the membrane forms a protective barrier for the ligand against particulate matter, infectious agents, and so on.

The purification of *Bacillus licheniformis* α-amylase by the latter approach with the use of a cross-linked starch slurry as the affinity adsorbent was studied by Somers and Van't Riet (1990). Recovery on the basis of enzyme activity was reported to be 95%.

The binding process during affinity ultrafiltration is influenced to a great extent by the concentration of the macromolecular ligand. The binding capacity initially increases with increasing ligand concentration but reaches a plateau at high concentrations as ligands become sterically inaccessible for protein binding. A similar phenomenon has been observed during affinity chromatography.

3.2. Washing

Washing of the affinity complex is usually performed in the diafiltration mode such that the volume of the retentate is maintained constant by adding the washing buffer at a rate equal to the filtration rate.

When affinity interactions are used in a chromatographic purification process, the binding strength only needs to be sufficient to provide effective separation so that the eventual dissociation of the complex to obtain the pure product is not difficult. However, in affinity ultrafiltration, washing is far more intense so that anything that is not firmly bound is washed away. The washing step begins with an equilibrium mixture of free and bound protein. As the washing proceeds, the free protein concentration decreases and the bound protein releases from the adsorbent in an effort to reestablish equilibrium. One strategy could be to use a high filtration rate to minimize loss of the bound component during the washing step. However, for very low affinity systems, this would result in a significant loss of the desired protein. According to Ling and Mattiasson (1989), if the binding between ligand and target protein is assumed to be in equilibrium, to get a calculated recovery of more than 90% when at least 99% of other material is washed out, the effective binding strength must be higher than $10^6 \, M^{-1}$. However, experimental and modeling work presented by Herak and Merrill (1989) using Cibacron blue-agarose and human serum albumin (HSA) indicated that the release of the target protein during the washing step is better represented by a model where no HSA is released from the gel. The observed slow kinetics of the release of the adsorbed protein during the washing step could be applicable to most other systems and may be exploited to minimize loss of the target protein.

3.3. Elution

For the elution of the bound protein, the buffer is changed so that the binding interaction is unfavorable and the complexed protein is released. Affinity between the protein and the ligand will change gradually as the elution buffer replaces the coupling buffer. The important aspect of this step is that the bound protein is readily released without an excessively large volume of buffer. Moreover, the conditions used must be favorable for the biomolecule, as well as the membrane. As in affinity chromatography, elution can be performed either by use of a competitive ligand or by altering pH, ionic strength, etc. The first alternative is attractive, since it is performed under conditions favorable for binding and preserves the biological activity of the molecule and also does not negatively influence the nature of the membrane. The drawbacks might be the high cost involved in using a competitive affinity ligand and the fact that the product is contaminated with the free ligand. Nonspecific elution, on the other hand, is cheap but may often be harmful either to the biomolecule to be purified or to the membrane. However, the choice of elution strategy will very much depend on the system concerned.

3.4. Reconditioning

After the elution step, the free macromolecular affinity ligand is reconditioned generally in a separate membrane filtration step where the dissociating buffer is replaced with the fresh binding buffer by diafiltration. It is then ready to be used for purification of subsequent batches.

The eluate containing the purified product is very dilute. Passage through another ultrafiltration membrane concentrates the product and at the same time removes the free competing ligands on the basis of molecular size.

4. MACROMOLECULAR AFFINITY LIGANDS USED FOR AFFINITY ULTRAFILTRATION

The choice of macromolecular affinity ligand could be considered to be the most important parameter determining the applicability of an affinity ultrafiltration system. It is important that the macroligands are either soluble or easily suspended in aqueous solutions so that they can be pumped through the filtration apparatus. The desirable characteristics of a macromolecular ligand are that it is relatively inexpensive, possesses a large specific surface area, and exhibits negligible nonspecific adsorption of undesired contaminants. Moreover, it should be stable under the required adsorption and elution conditions and show no attrition under high shear forces.

By and large, finding an affinity ligand possessing the molecular size desired in operating affinity ultrafiltration is difficult. This limitation is overcome by "tailoring" the macroligand; that is, a ligand molecule is coupled either to a water-soluble polymer or a water-insoluble carrier that does not biospecifically bind biomolecules. Table 1 lists various macroligands that have been employed for affinity ultrafiltration studies.

4.1. Water-Insoluble Macromolecular Affinity Ligands

One of the earlier studies on affinity ultrafiltration employed heat-killed cells of *Saccharomyces cerevisiae* (mean diameter, 5 μm) as a macroligand for the purification of concanavalin A (Con A; molecular weight, 102,000) from jack bean extract (Mattiasson and Ramstorp, 1984). Con A interacts with the carbohydrate residues exposed on the surface of the yeast cells. The solution containing the cells and the crude Con A extract was pumped through a hollow fiber unit (molecular weight cutoff; 1,000,000) at a flow rate of 500 ml/min. Glucose (150 g/liter) was used to dissociate the affinity complex, and the product stream from the dissociation step was passed through a membrane unit with a molecular weight cutoff of 35,000. The content of glucose was effectively reduced by a diafiltration of the solution. The yeast cell suspension was washed in the reconditioning step so that glucose was eliminated. Con A in a highly purified form was obtained by this procedure at an overall yield of 70%.

As seen in Table 1, carrier particles that are used in affinity chromatography, such as agarose and starch, could also be applicable for affinity ultrafiltration. Particles have been found to be relatively easy to pump through the cross-flow filtration apparatus with minimal fouling of the membranes (Herak and Merrill, 1989). However, particle degradation can be observed with time, depending on the durability of the particles. For example, Sepharose CL-6B, which has a more rigid gel structure consisting of a high cross-link density, was found to be fractured to a lesser extent (5%) after 24 h of circulation than Affi-Gel Blue, which was as much as 30–40% damaged in just 3 h (Herak and Merrill, 1989).

With the use of the commercially available affinity gel p-aminobenzyl-l-thio-β-D-galactopyranoside (PABTG) agarose, β-galactosidase was purified from *E. coli* homogenate by the continuous affinity recycle extraction (CARE) process (Pungor *et al.*, 1987).

From experience in classical affinity chromatography, it is known that the available surface area of the particles sets a limit on the amount of ligand and hence the amount of macromolecule that can be bound. The same principle applies to ultrafiltration affinity purification. To have a high binding capacity, the carrier particles must have a large accessible surface area, which means that they should be as small as possible. In this direction, Ling and Mattiasson (1989) attempted to use particles of fumed silica having a diameter of 12 nm and a surface area of 200 m^2/g with immobilized Cibacron blue for affinity ultrafiltration of alcohol dehydrogenase (ADH) and lactate dehydrogenase (LDH) from yeast and porcine heart, respectively. The ligand density was found to be 34 μmol/g (equivalent to 0.2 μmol/m^2), which was comparable to the density of 0.33 μmol/m^2 for

TABLE 1
Examples of Macromolecular Ligands Used for Protein Purification by Affinity Ultrafiltration

Carrier	Size	Ligand[a]	Target protein[b]	Reference(s)
Insoluble particles				
Saccharomyces cerevisiae cells	5 μm	Surface carbohydrate groups	Concanavalin A	Mattiasson and Ramstorp, 1984
Streptococcus cell groups	1 μm	Surface protein G	Bovine IgG	Chen, 1990
Starch granules	1–10 μm	Cibacron blue	ADH	Mattiasson and Ling, 1986
Silica nanoparticles	12 nm	Cibacron blue	ADH, LDH	Ling and Mattiasson, 1989
Agarose beads	45–165 μm	Cibacron blue	HSA	Herak and Merrill, 1989
Agarose beads	Not stated	PABTG	β-Galactosidase	Pungor et al., 1987
Sepharose beads	45–165 μm	Heparin	Lactoferrin	Chen, 1990
Sepharose beads	45–165 μm	Protein G	Bovine IgG	Chen, 1990
Liposomes	270 ± 5 Å	Biotin	Avidin	Powers et al., 1989a,b
Liposomes	600 Å	p-Aminobenzamidine	Trypsin	Powers et al., 1990
Soluble polymers				
Dextran	MW 2,000,000	Protein A	IgG	Nylén, 1980
Dextran	MW 2,000,000	Estradiol	$\Delta_{5\to4}$ 3-Oxosteroid isomerase	Hubert and Dellacherie, 1980
Dextran	MW 2,000,000	p-Aminobenzamidine	Trypsin	Adamski-Medda et al., 1981
Dextran	MW 2,000,000	STI	Trypsin	Choe et al., 1986
Polyacrylamide	MW > 100,000	m-Aminobenzamidine	Trypsin	Luong et al., 1988a,b; Male et al., 1987
Polyacrylamide	MW > 100,000	m-Aminobenzamidine	Urokinase	Male et al., 1990

[a]Abbreviations: PABTG, p-aminobenzyl-1-thio-β-D-galactopyranoside; STI, soybean trypsin inhibitor.
[b]Abbreviations: IgG, immunoglobulin G; ADH, alcohol dehydrogenase; LDH, lactate dehydrogenase; HSA, human serum albumin.

an NAD derivative coupled to silica beads used in high-performance liquid chromatography (HPLC) columns.

The silica nanoparticles formed a colloidal suspension and gave binding kinetics similar to those obtained for binding in free solution. This implied that binding and dissociation occurred almost instantaneously and that the process could be run continuously without the risk of incomplete binding. According to the authors, modification of the silica particles with methoxy–polyethelyne glycol (MPEG) to reduce the nonspecific adsorption caused a general decrease in binding capacity to an extent depending on the decrease in the number of available ligands.

The binding of the dehydrogenases with the macroligand was attempted after microfiltration of the respective homogenates to remove the particulate matter. The recoveries of both the enzymes on a preparative scale in a membrane unit were significantly lower than those obtained on an analytical scale in a test tube. This was attributed to the sensitivity of the enzymes to prolonged filtration. Losses due to adsorption of the enzymes to the filtration membrane in the elution step could also not be ruled out. A significant loss of LDH activity was observed during the washing step, which could be attributed to the low association constant of LDH and Cibacron blue.

The use of affinity-modified liposomes as macroligands for affinity cross-flow filtration systems seems to be quite an interesting approach (Powers *et al.*, 1989a,b, 1990). The unilamellar vesicles (average diameter of 200–700 Å) possess an extremely large surface area; for example, 300 Å-radius liposomes have a specific surface area of 100 m^2/ml of liposomes, which, in terms of area per volume of carrier, is approximately the same value as can be obtained with porous silica. The hydrophilic surface of the liposomes results in minimal nonspecific adsorption of undesired impurities, and, because the liposomes remain suspended in solution, membrane polarization and plugging during filtration steps are also avoided.

Unilamellar vesicles with affinity ligands can be prepared either by coupling the ligand covalently to the reactive functional groups on the outer surface of the preformed liposomes or by first coupling it to a reactive phospholipid in an appropriate organic solution, which is then mixed with unmodified phospholipid in aqueous solution and sonicated. The ligand density on the liposome surface can be carefully controlled by simply varying the amount of derivatized phospholipid used during liposome preparation. Because of the small size of liposomes, all interactions of liposomes with proteins are kinetically controlled, and, moreover, since the ligands are on the outer surface of the liposome, there are no diffusional limitations during either adsorption or desorption step.

The feasibility of using liposomes as ligand carriers for membrane-based affinity separations was first studied with avidin as the model protein and biotin as the affinity ligand (Powers *et al.*, 1989a). Avidin was found to bind stoichiometrically to liposome-bound biotin in a 1:1 ratio. Based on the measured binding rate constants, the equilibrium binding constant was determined to be $6 \times 10^7 \ M^{-1}$, as compared to $10^{15} \ M^{-1}$ with the free biotin. The decrease in binding constant was attributed to the steric hindrance between the avidin molecule and the liposomes during binding as the biotin was immobilized directly on the liposome surface; however, the binding constant was sufficient for purification purposes.

Using the biotinylated liposomes, avidin was specifically isolated from model protein mixtures consisting of avidin and cytochrome *c*, and avidin and lysozyme, and also from its naturally occurring source, hen egg white (Powers *et al.*, 1989a,b). In the latter case, hydrophobic impurities were first removed by centrifugation and ion-exchange chroma-tography, and subsequent affinity cross-flow filtration resulted in a recovery of 96% pure avidin with 80% of its original activity.

For the liposome-assisted affinity ultrafiltration process to be industrially feasible, the liposomes must be stable toward biological substances and have a relatively long shelf life. The susceptibility of liposomes to disruption by surfactants, hydrophobic molecules, and plasma proteins was overcome by the incorporation of large amounts of cholesterol (>33 mol %) into the lipid bilayer, leading to an enhanced stability and shelf life (Powers *et al.*, 1990).

The affinity liposomes with incorporated cholesterol were used for the purification of trypsin from a mixture of trypsin and chymotrypsin and from crude pancreatic extract (Powers *et al.*, 1990). The affinity ligand, *p*-aminobenzamidine (PAB), was coupled to the surface of the vesicles via a hydrophilic spacer arm, diglycolic acid, to prevent any steric hindrance during affinity binding. The equilibrium binding constant between trypsin and immobilized PAB, K_i, was shown to be dependent on the PAB density of the liposome surface. An increase in K_i was observed at high PAB loadings. Loss of trypsin during washing was thought to be due to the relatively low equilibrium constant of $1.7 \times 10^4 \ M^{-1}$ between trypsin and immobilized PAB. Elution of the bound trypsin was carried out with benzamidine as the eluent in a diafiltration mode to minimize the possibility of aggrega-tion of liposomes. From the crude extract, approximately 68% of the starting trypsin was recovered with a purity greater than 98%. It was observed that the same PAB-liposome solution could be used without any loss in trypsin binding capacity for at least five successive trypsin purifications from trypsin/chymotrypsin mixtures over a 30-day period.

4.2. Water-Soluble Macromolecular Affinity Ligands

The advantages of water-soluble biospecific polymers as ligand carriers lie in the high yields obtained from ligand coupling, the rapid control of the amount immobilized, and the easy complexation with the complementary macromolecule, which implies a high binding capacity. Protein binding takes place under equilibrium conditions, resulting in quantitative complexation.

For the purposes of affinity ultrafiltration, ligands bound to water-soluble polymers necessitate the use of ultrafiltration membranes that have much smaller pores and hence much lower flux rates than the microfiltration membranes; accordingly, a longer processing time or a larger membrane surface area is needed. Both reversible and irreversible fouling can be a significant problem in filtration of polymeric solutions. On the other hand, soluble polymer is not likely to be degraded by attrition.

As shown in Table 1, high-molecular-weight dextran was used as ligand carrier in a number of studies on affinity filtration (Nylén, 1980; Hubert and Dellacherie, 1980; Adamski-Medda *et al.*, 1981; Choe *et al.*, 1986). A number of established coupling mechanisms can be employed for attaching affinity ligands to the carbohydrate polymer. Unfortunately, a serious problem faced during purification of trypsin from a mixture of trypsin and chymotrypsin using dextran-bound trypsin inhibitors was the nonspecific adsorption of the proteins to the polymer backbone (Adamski-Medda *et al.*, 1981; Choe *et al.*, 1986).

Luong and co-workers synthesized a high-molecular-weight (100,000) soluble poly-mer having an acrylamide backbone with pendant aminobenzamidine groups for the purification of trypsin by affinity ultrafiltration (Luong *et al.*, 1988a; Male *et al.*, 1987). The polymer was found to be specific and effective in inhibiting trypsin.

The amount of inhibitor in the polymer could be varied by copolymerizing varying amounts of *N*-acryloyl-*m*-aminobenzamidine (referred to as the monomer) with acrylamide. The mechanism of trypsin inhibition by the affinity polymer was competitive, and the value of K_i appeared to depend on the acrylamide:monomer concentration ratio (Male *et al.*, 1987). At high monomer concentration, there was a decrease in the inhibition capacity of the polymer. An acrylamide:monomer ratio of 5:1 was found to be optimal for maintaining a balance between the number of available binding sites (high *M*-aminobenzamidine content) and the strength of the binding force (corresponding to low K_i).

The binding efficiency between the affinity polymer and trypsin was also dependent on the polymer concentration (Male *et al.*, 1987). Using an acrylamide:monomer ratio of 5:1, the amount of trypsin retained was about 90% at 0.75% polymer concentration. This level decreased to 54% when the concentration was reduced to 0.03%, even though the amount of enzyme introduced was also lowered by the same factor. This behavior was explained by the fact that decreasing the acrylamide concentration made the trypsin bound to the ligand more susceptible to the shear force resulting from the high flow rates through the membrane system. Though it may be possible to obtain a better binding efficiency using concentrated polymer solutions, that would most likely result in a very slow filtration rate due to the increased viscosity. Moreover, in the case of polymers exhibiting some nonspecific binding, these interactions may be modified in an unforeseeable manner with changes in concentrations of protein and macroligand.

The binding of the soluble affinity polymer with the enzyme was found to be instantaneous (Luong *et al.*, 1988b). This is an important observation as it suggests that the enzyme can be directly added to the polymer solution that is being passed through the ultrafiltration unit, instead of it being necessary to set up a holding tank for the binding to take place prior to ultrafiltration.

It was found important to maintain a neutral pH when the affinity polymer is used in conjunction with the ultrafiltration membrane plates. Both trypsin and chymotrypsin in a polymer solution passed through the ultrafilter very slowly at low and high pH values. Hence, acidic solution, which is often used to dissociate the trypsin–inhibitor complex during affinity chromatography, could not be used for elution from the soluble polymeric *M*-aminobenzamidine–enzyme complex.

The elution was found to be efficient when either arginine or benzamidine was used as the eluting agent. Trypsin with 98% purity and 90% yield was obtained in a batch process. The same polymer solution could be reused for more than one cycle of binding and elution. With the use of the same macroligand in a continuous process by recirculating the eluent as well as the ligand, trypsin was purified from crude porcine pancreatic extract with 97% purity and 77% yield. Based on the experimental data, a mathematical model has been developed to describe the dynamic behavior of the affinity cross-flow filtration process with this macromolecular ligand.

The polymer solution could be stored for several months without undergoing degradation or losing ligand molecules (Luong *et al.*, 1988c). The stored polymer exhibited the same binding capacity for trypsin as the freshly synthesized one. Also, the trypsin could be adsorbed by the polymer and stored until the enzyme was required for further processing.

The same soluble affinity polymer was used to purify urokinase from an artificial solution containing peroxidase and urokinase and from a crude urine source (Male *et al.*,

1990). In this case, the optimal acrylamide:monomer ratio was found to be 2:1. The process yields from the two sources were about 86% and 49%, respectively.

5. CONCLUDING REMARKS

In comparison to affinity chromatography, which is undoubtedly a superior analytical technique, affinity ultrafiltration is better suited for application on a large scale. Dilute feed streams can be handled in relatively short time periods. An affinity chromatography column is highly vulnerable to fouling and clogging, thereby reducing the system's productivity and limiting its use to fairly clear solutions. The ability of affinity ultrafiltration to process unclarified broths permits affinity interactions to be employed at the start of the downstream processing train such that the target protein is rapidly removed from a hostile environment.

The efficacy of the process is largely governed by the choice of the macromolecular affinity ligand–its selectivity, cost, and stability toward shear forces and recycling. It should be possible to extend the affinity ultrafiltration process to purify more than one biomolecule from a crude broth, either by using a group-specific ligand and manipulating the elution conditions or by employing different macroligands in a sequence.

ACKNOWLEDGMENTS

This work was supported by the National Swedish Board for Technical Development (STU) and the Swedish Agency for Research Cooperation with Developing Countries (SAREC).

REFERENCES

Adamski-Medda, D., Nguyen, Q. T., and Dellacherie, E., 1981, Biospecific ultrafiltration: A promising purification technique for proteins?, *J. Membr. Sci.* **9**:337.

Chen, J.-P., 1990, Novel affinity based processes for protein purification, *J. Ferment. Bioeng.* **70**:199.

Choe, T. B., Masse, P., and Verdier, A., 1986, Separation of trypsin from trypsin–chymotrypsin mixture by affinity-ultrafiltration, *Biotechnol. Lett.* **8**:163.

Herak, D. C., and Merrill, E. W., 1989, Affinity cross-flow filtration: Experimental and modeling work using the system of HSA and cibacron blue-agarose, *Biotechnol. Prog.* **5**:9.

Hubert, P., and Dellacherie, E., 1980, Use of water-soluble biospecific polymers for the purification of proteins, *J. Chromatogr.* **184**:325.

Ling, T. G. I., and Mattiasson, B., 1989, Membrane filtration affinity purification (MFAP) of dehydrogenases using cibacron blue, *Biotechnol. Bioeng.* **34**:1321.

Luong, J. H. T., Nguyen, A. L., and Male, K. B., 1987, Affinity cross-flow filtration for purifying biomolecules, *Bio/Technology* **5**:564.

Luong, J. H. T., Male, K. B., and Nguyen, A. L., 1988a, Synthesis and characterization of a water-soluble affinity polymer for trypsin purification, *Biotechnol. Bioeng.* **31**:439.

Luong, J. H. T., Male, K. B., and Nguyen, A. L., 1988b, A continuous affinity ultrafiltration process for trypsin purification, *Biotechnol. Bioeng.* **31**:516.

Luong, J. H. T., Male, K. B., Nguyen, A. L., and Mulchandani, A., 1988c, Affinity ultrafiltration for purifying specialty chemicals, in: *Biotechnology Research & Applications* (J. Gavora, D. F. Gerson, J. Luong, A. Storer, and J.H. Woodley, eds.), Elsevier, New York, pp. 79–93.

Male, K. B., Luong, J. H. T., and Nguyen, A. L., 1987, Studies on the application of a newly synthesized polymer for trypsin purification, *Enzyme Microb. Technol.* **9**:374.

Male, K. B., Nguyen, A. L., and Luong, J. H. T., 1990, Isolation of urokinase by affinity ultrafiltration, *Biotechnol. Bioeng.* **35:**87.

Mattiasson, B., and Ling, T. G. I., 1986, Ultrafiltration affinity purification. A process for large-scale biospecific separations, in: *Membrane Separations in Biotechnology* (W. C. McGregor, ed.), Marcel Dekker, New York, pp. 99–114.

Mattiasson, B., and Ramstorp, M., 1983, Ultrafiltration affinity purification, *Ann. N. Y. Acad. Sci.* **413:**307–310.

Mattiasson, B., and Ramstorp, M., 1984, Ultrafiltration affinity purification. Isolation of concanavalin A from seeds of *Canavalia ensiformis, J. Chromatogr.* **283:**323.

Mattiasson, B., Ling, T. G. I., and Nilsson, J. L., 1983, Ultrafiltration affinity purification, in: *Affinity Chromatography and Biological Recognition* (I. M. Chaiken, M. Wilchek, and I. Parikh, eds.), Academic Press, Orlando, pp. 223–227.

Nylén, U., 1980, Swedish Patent Application SE 8006102; European Patent Application, publication number 0 046 915.

Powers, J. D., Kilpatrick, P. K., and Carbonell, R. G., 1989a, Protein purification by affinity binding to unilamellar vesicles, *Biotechnol. Bioeng.* **33:**173.

Powers, J. D., Kilpatrick, P. K., and Carbonell, R. G., 1989b, Purification of avidin from egg whites using affinity-modified unilamellar vesicles, in: *Progress in Clinical and Biological Research* (D. A. Butterfield, ed.), Alan R. Liss, New York, pp. 237–246.

Powers, J. D., Kilpatrick, P. K., and Carbonell, R. G., 1990, Trypsin purification by affinity binding to small unilamellar liposomes, *Biotechnol. Bioeng.* **36:**506.

Pungor, E., Jr., Afeyan, N. B., Gordon, N. F., and Cooney, C. L., 1987, Continuous affinity-recycle extraction: A novel protein separation technique, *Bio/Technology* **5:**604.

Somers, W., and Van't Riet, K., 1990, Downstream processing of α-amylase by affinity interactions, in: *Proceedings from 5th European Congress on Biotechnology, Vol. 2, Copenhagen, July 8–13, 1990* (C. Christiansen, L. Munck, and J. Willadsen, eds.), Munksgaard International Publishers, Copenhagen, pp. 774–777.

33

Affinity Partitioning of Biomolecules in Aqueous Two-Phase Systems

Gerhard Kopperschläger and Gerd Birkenmeier

1. INTRODUCTION

Aqueous two-phase systems have been used for more than three decades in biochemical research as a tool for fractionation of proteins, nucleic acids, cell particles, and cells. The interest in applying these systems for large-scale processes in biotechnology has grown strongly during the last two decades. Two-phase systems form spontaneously upon mixing aqueous solutions of two hydrophilic polymers when certain threshold concentrations are exceeded.

The phenomenon of phase separation first discovered by Beijerinck in 1896 was rediscovered 60 years later by Albertsson (1956), who initially described two-phase systems composed of poly(ethylene glycol) (PEG) and dextran and PEG and salt. Albertsson immediately realized that, in addition to phase separation of aqueous solutions, partitioning of biomaterial takes place in such systems. Because aqueous polymer solutions constitute a gentle environment not only for biopolymers but also for cells and cell particles, partitioning in these systems can be exploited to effect separations otherwise difficult or impossible to achieve (Albertsson, 1971; Walter *et al.*, 1985).

The partitioning of proteins depends on the chemical nature of the phase-forming polymers and their molecular weight and concentration. The addition of salts and the pH also affect the distribution of the proteins. Furthermore, intrinsic properties of the proteins, such as size, conformation, and charge, are important for the protein partitioning behavior.

The partition coefficient, defined as the ratio of the protein concentrations in the upper and lower phases, is a characteristic property of a protein and can be viewed as the sum of individual terms (Albertsson, 1971):

$$\ln K = \ln K_{\text{charge}} + \ln K_{\text{hydrophobic}} + \\ \ln K_{\text{hydrophilic}} + \ln K_{\text{conformation}} + \ln K_{\text{ligand}} + \cdots \tag{1}$$

Many proteins have unique interactions with ligands that offer the possibility of altering partitioning in a rational way and achieving selectivity. Such an approach, called affinity partitioning, combines the partitioning behavior of biological macromolecules in aqueous

Gerhard Kopperschläger and Gerd Birkenmeier • Institute of Biochemistry, University of Leipzig, D–04103 Leipzig, Germany.

Molecular Interactions in Bioseparations, edited by That T. Ngo. Plenum Press, New York, 1993.

two-phase systems with the principle of biorecognition. Developments based on this approach range from the separation of single cell types using monoclonal antibodies to technical applications employing triazine dyes as cheap general ligands.

Small ligands are covalently coupled to one of the phase-forming polymers, most often to PEG. Therefore, in a PEG/dextran system the ligand will accumulate in the upper phase and the protein of interest will be transported into the upper phase by forming a biospecific complex.

Numerous ligands have been tested in affinity partitioning, including natural or pseudo-biospecific ligands of enzymes, antigens and antibodies, hormones and their corresponding receptors, and hydrophilic and hydrophobic ligands.

2. THEORETICAL CONSIDERATIONS

Flanagan and Barondes (1975) developed the first mathematical model to describe the linked equilibria of partition and complex formation in an affinity extraction process. The model implied that under saturating conditions the logarithm of the partition coefficient of the target protein is proportional to the number of binding sites (n) and the logarithm of the partition coefficient of the liganded polymer:

$$\Delta \log K_{max} = \log (K_c/K_p) = n \cdot \log K_L \qquad (2)$$

K_c, K_p, and K_L are the partition coefficients of the complex, the protein, and the ligand, respectively. Equation (2) implies that the binding constants for the ligand are independent and equal in both phases.

A more general expression which does not require saturation of the protein binding sites has been developed by Brooks et al. (1985):

$$K_c = K_p[1 + (C_L/k_d)_u]^n/[1 + (C_L/k_d)_l]^n \qquad (3)$$

where C_L is the concentration of the ligand and k_d the dissociation constant of the protein ligand complex in the upper (u) and lower (l) phases. This approach also assumed equal binding constants for all binding sites. At high ligand concentration with $(k_d)_u = (k_d)_l$, Eq. (3) reduces to Eq. (2).

The calculation of the number of binding sites, n, of a protein on the basis of Eq. (2) yields values that are inconsistent with many experimental data. This may be due to the assumption made in Eq. (2) that the dissociation constants in the upper and lower phases are identical, which is obviously not correct. In addition, it cannot necessarily be assumed that the binding constant for each ligand molecule is equal. An improved model has been derived by Cordes et al. (1987) assuming two, four, or six binding sites and taking into account distinct binding constants for the protein–ligand complex in the upper and lower phases and different classes of intrinsic dissociation constants.

3. AFFINITY LIGANDS

3.1. Mode of Attachment

Although all phase-forming polymers are capable of carrying covalently linked ligands, PEG has been preferentially used because of its availability in a large variety of molecular weights, its low price, and its stoichiometry of binding maximally two ligands per molecule. For a single derivatization, monoalkylated PEG can be used. Occasionally, dextran or modified starch has been applied as a carrier of the ligand (Johansson and Joelsson, 1987).

In order to increase the reactivity of the PEG, the terminal hydroxyl groups have to be substituted by more reactive electrophiles such as halides (Johansson, 1970; Bückmann et al., 1981).

The substitution of the hydroxyl groups of PEG by amines gives rise to the synthesis of a variety of derivatives under mild chemical conditions (Johansson et al., 1973).

3.2. Cofactors and Dyes as Ligands

Although cofactors of enzymes are suited to act biospecifically in affinity partitioning, their practical use is still not widespread. One reason might be that the procedure for coupling of the ligands to the polymer is laborious. Often, a spacer has to be introduced between the polymer and the cofactor in order to achieve biorecognition (Bückmann et al., 1987; Persson and Olde, 1988).

The discovery that reactive dyes are able to act as biomimetic ligands for many enzymes and other proteins has advanced affinity partitioning in different directions (Kopperschläger et al., 1982; Kopperschläger and Birkenmeier, 1990). The dyes can be easily bound to PEG under alkaline conditions with 30% yield. (Johansson and Joelsson, 1985a). Higher yields were obtained using amino-PEG (Cordes and Kula, 1986).

3.3. Other Ligands

Aliphatic esters of PEG have been synthesized as probes for hydrophobic affinity partitioning (Shanbhag and Johansson, 1974). Coupling of antibodies to PEG is usually performed by the reaction of activated PEG with available lysine residues of the protein (Karr et al., 1986; Sharp et al., 1986).

The construction of metal-chelating PEG by coupling of chloromonomethoxy-PEG to iminodiacetic acid (Wuenschell et al., 1990) or by reacting amino-PEG with bromoacetic acid (Birkenmeier et al., 1991) has initiated the development of a novel branch of affinity partitioning.

4. AFFINITY PARTITIONING AS AN ANALYTICAL APPROACH

4.1. Reactive Dyes as Biomimetic Ligands of Enzymes

Dye-ligand affinity partitioning has been elaborated as a sensitive method for studying enzyme–dye interactions. As a rule, the dye ligand is coupled to PEG because of the ease of steering the partition of bulk proteins into the opposite dextran-rich phase. The effect of affinity partitioning can be quantified by determining the difference of the partition coefficients of the target enzyme in the presence and absence of the dye-PEG, which is expressed as $\Delta \log K$. This measure, when plotted against the concentration of the dye-PEG, mostly followed a saturation function, as shown in Fig. 1. From a plot of this type, two parameters, namely, $\Delta \log K_{max}$ (maximum extraction power) and the half-saturation point $0.5 \times \Delta \log K_{max}$ can be calculated. The value of $\Delta \log K_{max}$ is somehow related to the

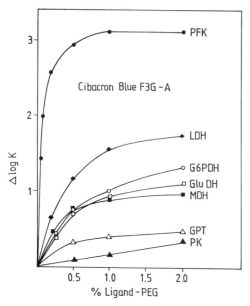

FIGURE 1. Affinity partitioning of enzymes as a function of increasing concentration of Cibacron Blue F3G-A-poly(ethylene glycol). The systems are composed of 5% PEG 6000, 7% dextran T500, and 50mM sodium phosphate buffer, pH 7.0. PFK, Phosphofructokinase; LDH, lactate dehydrogenase; G6PDH, glucose-6-phosphate dehydrogenase; Glu DH, glutamate dehydrogenase; MDH, malate dehydrogenase; GPT, glutamate pyruvate transaminase; PK, pyruvate kinase. (From unpublished results of the authors.)

number of binding sites (see Section 2) whereas $0.5 \times \Delta \log K_{max}$ is valid as a relative measure of the affinity of the ligand for the enzyme if values obtained under identical experimental conditions are compared (Kirchberger et al., 1989).

Affinity partitioning of enzymes exhibiting specific interaction with the dye ligand does not depend on the pH in the range of pH 6–9. The reason for this behavior is that the pK values of the dyes fall outside this pH range.

The protein–dye interaction can be influenced by ionic strength, temperature, competing ligands, and other factors. These effects are recognizable in affinity partitioning, even if small changes in the ligand–enzyme interaction do occur. Hence, the method turns out to be much more sensitive than other approaches (Kopperschläger and Birkenmeier, 1990). As an example, the influence of two effectors on affinity partitioning of lactate dehydrogenase (LDH) and pyruvate kinase (PK) is displayed in Fig. 2 (Kopperschläger et al., 1983). The interaction of Procion Navy HE-R with muscle pyruvate kinase was diminished when ATP was added to a two-phase system, as indicated by the decrease of $\Delta \log K$. When Mg^{2+} was supplemented, which gives rise to the formation of an Mg–ATP complex, the competitive effect was less pronounced, possibly due to the stabilization of the dye–enzyme complex by the excess of the metal ion. In the case of lactate dehydrogenase, there is a much stronger competition between Cibacron Blue F3G-A and NAD^+ since the dye more precisely mimics the natural ligand. As demonstrated, the competitive effect of NAD^+ was extremely amplified by addition of sodium sulfite owing to the formation of a highly stable ternary LDH–NAD^+–sulfite complex (Pfleiderer et al., 1958).

Similarly, the interaction of Procion Red HE-3B and dye analogs with LDH and the influence of NAD^+ and NADH were investigated by Kirchberger et al. (1989) in order to estimate the chemical requirement for the binding specificity.

Alkaline phosphatase was partitioned in two-phase systems containing Procion Red HE-3B-PEG (Kopperschläger et al., 1988). Besides inorganic phosphate, which is known

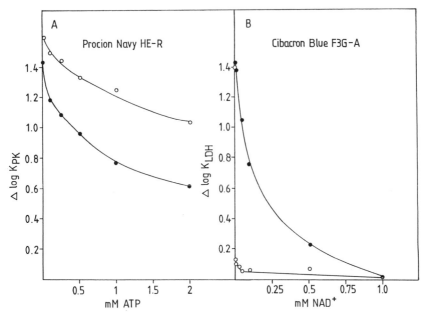

FIGURE 2. Affinity partitioning of pyruvate kinase (PK) and lactate dehydrogenase (LDH) as a function of increasing concentration of competing effectors. (A) Pyruvate kinase. System contains 5% poly(ethylene glycol) (PEG), 7% dextran, and 50mM sodium phosphate buffer, pH 7.0. Of the total PEG, 0.2% is Procion Navy HE-R-PEG. ●, ATP; ○, ATP plus 2mM magnesium chloride. (B) Lactate dehydrogenase. System composition without the dye-ligand as in (A). Of the total PEG, 0.5% is replaced by Cibacron Blue F3G-A-PEG. ●, NAD$^+$; ○, NAD$^+$ plus 2mM sodium sulfite. [From Kopperschläger *et al.* (1983).]

to inhibit the enzyme competitively, NADH, NADPH, 5'-AMP, and 5'-ADP were able to diminish the affinity partition of the enzyme but to different extents.

Affinity partitioning was found to be a simple and sensitive screening method in the selection of effective dye ligands for several affinity separation techniques. The determination of the maximum extraction power ($\Delta \log K_{max}$) makes it possible to decide whether an adequate interaction between an enzyme and a dye does or does not occur (Kopperschläger and Birkenmeier, 1990). Generally, the higher the $\Delta \log K_{max}$, the higher is the potential of the ligand when applied. This has been demonstrated, for example, for phosphofructokinase (Kopperschläger and Johansson, 1982), glucose-6-phosphate dehydrogenase (Johansson and Joelsson, 1985b), alkaline phosphatase (Kirchberger and Kopperschläger, 1990), and other enzymes. However, a linear relationship to predict the efficacy of all affinity ligands has not been found. Although a low $\Delta \log K_{max}$ is indicative of inability of the ligand to bind to the enzyme, a high value does not reflect *per se* usefulness in affinity separation practice (Naumann *et al.*, 1989).

4.2. Affinity Partitioning of Plasma Proteins

The first report demonstrating affinity partitioning of plasma proteins dealt with the extraction of an S-23 myeloma protein, with the use of dinitrophenyl-PEG (Flanagan and Barondes, 1975). Shanbhag and Johansson (1974, 1979) and Birkenmeier *et al.* (1986b) introduced hydrophobic acyl-PEGs of different chain lengths for the affinity partitioning of albumin and α-fetoprotein. Palmitate-PEG turned out to be a sensitive ligand for characterizing conformational changes in proteins which are accompanied by changes in hydrophobicity. For example, the conformational change of α_2-macroglobulin that occurs upon complex formation with proteinases or with primary amines is clearly recognized in affinity partitioning using palmitate-PEG (Birkenmeier et al., 1989).

Affinity partitioning of plasma proteins has also been performed with a view toward studying their interaction with reactive dyes. One example concerns the binding of Remazol Yellow GGL to prealbumin and albumin (Birkenmeier and Kopperschläger, 1987). It could be shown that albumin significantly increased the binding of prealbumin to the dye. The amplification was found to be inherent in albumin from human and from bovine plasma but not in egg albumin (Birkenmeier et al., 1984).

Two-phase partitioning has been applied also for studying the interaction of thyroxine-binding globulin with dyes (Birkenmeier et al., 1986a).

4.3. Metal-Chelate Affinity Partitioning

Metal-chelating groups coupled to agarose and other supports are able to form coordinative bonds with electron-donating N, S, and O atoms in peptides and proteins. This principle initiated a novel approach in affinity chromatography (Porath et al., 1975). The first data from the application of metal-chelating groups in affinity partitioning were reported by Wuenschell et al. (1990), Plunkett and Arnold (1990), Birkenmeier et al. (1991), and Arnold (1991). The chelating PEG when saturated with Cu^{2+} forms a 1:1 complex with the metal ion. Binding of the protein takes place by substitution of ligands in the inner coordination sphere of the metal. The binding to metal-chelate-PEG strongly depends on the number of available histidine residues in proteins, as these are the main targets for coordination.

The proteinase inhibitor α_2-macroglobulin was used to study the principle of metal-chelate–protein interaction in aqueous two-phase systems (Birkenmeier et al., 1991). As seen in Fig. 3, the protein partitions to different extents when added to metal iminodiacetate-PEG (IDA-PEG)-dextran systems containing different metals. Cu^{2+}-IDA-PEG causes a change in the partition coefficient of more than 1000-fold ($K_0 = 0.85$; $K_{aff} = 85$), yielding a maximum $\Delta \log K$ of 3. The complexed Zn^{2+} and Ni^{2+} give rise to lower efficacy, whereas the Fe^{3+}-IDA complex seems to have a negligible affinity for the protein. Dissociation of the metal-chelate–protein complexes was achieved by lowering the pH of the buffer or by adding competing electron donors, for example, free amino acids or chelators such as EDTA.

Using Cu^{2+}-IDA-PEG, Otto and Birkenmeier (1993) have probed the surface of lactate dehydrogenase isoenzymes from different species for the presence of metal chelate binding sites. It can be demonstrated that Cu^{2+}-IDA-PEG is able to recognize small differences in surface properties of different isoenzymes. Even cells (e.g., erythrocytes) have been found to display affinity to metal chelate-PEG and can be extracted in PEG/dextran aqueous two-phase systems (Walter et al., 1993).

4.4. Immunoligands in Affinity Partitioning

The most critical step in using immunoligands in affinity partitioning is the attachment of the antibodies to PEG. This modification often causes steric interferences with the

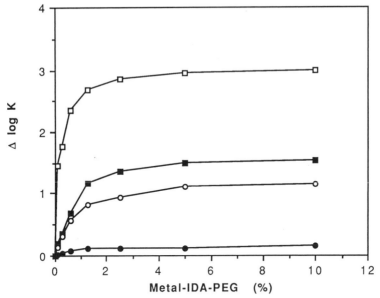

FIGURE 3. Effect of metal iminodiacetate-poly(ethylene glycol) (metal-IDA-PEG) on the partitioning of α_2-macroglobulin in PEG–dextran systems. The systems were composed of 5% PEG 6000, 7.5% dextran T70, $0.1M$ sodium sulfate, $10mM$ sodium phosphate buffer, pH 7.0, 20 μg α_2-macroglobulin, and increasing amounts of metal-IDA-PEG. The concentration of the liganded polymer is expressed as a percentage of the total PEG in the system. □, Cu^{2+}-IDA-PEG; ■, Zn^{2+}-IDA-PEG; ○, Ni^{2+}-IDA-PEG; ●, Fe^{3+}-IDA-PEG.

antibody binding reaction, and a compromise between both reactions must be found. Sharp *et al.* (1986) obtained an 18-fold increase in the partition coefficient of a rabbit anti-human red cell antibody when four to five molecules of PEG were coupled to one immunoglobulin molecule.

PEG-antibodies have been tested as immunologically specific affinity ligands for cell separation (Karr *et al.*, 1986, 1989). Mixtures of rabbit and human red cells were completely separable by sequential extraction steps owing to specific interaction of the surface antigens of cells with the derivatized antibody.

5. PURIFICATION OF PROTEINS BY AFFINITY PARTITIONING

Affinity partitioning has the potential to enrich proteins from crude material by single or multistep extraction. In comparison with other affinity separation techniques which are based on batch procedures, separations with liquid–liquid phase systems can be performed continuously and are easy to scale up.

Fig. 4 shows a general scheme for employing affinity partitioning in protein purification. In order to remove low-molecular-weight materials from the cell extract, as these often interfere with the protein–ligand complex formation, precipitation of the crude protein with PEG should be carried out as the first step. Alternatively, the extract can be partitioned at

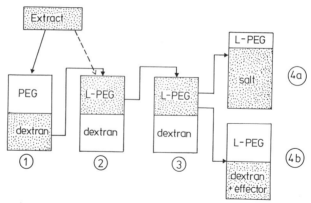

FIGURE 4. Scheme of affinity partitioning for enzyme purification. *Step 1*: Partition of an extract in a system without liganded poly(ethylene glycol) (PEG). *Step 2*: Partition of the dextran phase of step 1 or, alternatively, of the extract in the presence of liganded PEG. *Step 3*: Wash step of the top phase of step 2 with a fresh bottom phase. *Step 4a*: Formation of a PEG/salt two-phase system by adding solid salt to the top phase of step 3. *Step 4b*: Partition of the top phase of step 3 in the presence of a fresh bottom phase containing a competitive effector.

first in a system free of the affinity ligand in order to concentrate the bulk protein in one phase.

In most cases, only one of the phase-forming polymers is modified by a ligand. However, simultaneous substitution of the two polymers with different affinity ligands has also been reported (Johansson and Andersson, 1984; Pesliakas *et al.*, 1988).

A number of enzymes that have been purified by affinity partitioning are listed in Table 1.

TABLE 1
Purification of Enzymes by Means of Affinity Partitioning

Enzyme	Ligand	Purification (n-fold)	Reference
Phosphofructokinase (yeast)	Cibacron Blue F3G-A	10	Kopperschläger and Johansson, 1982
Glucose-6-phosphate dehydrogenase (yeast)	Procion Yellow HE-3G	22	Johansson and Joelsson, 1985b
Lactate dehydrogenase (muscle)	Procion Yellow HE-3G	30	Johansson and Joelsson, 1986
	Cibacron Blue F3G-A	25	
Formate dehydrogenase (*Candida boidinii*)	Procion Red HE-3B	6	Cordes and Kula, 1986
3-Oxosteroid isomerase	Estradiol	170	Hubert *et al.*, 1976
Alkaline phosphatase (calf intestine)	Procion Red HE-3B	13	J. Kirchberger, unpublished results
Alcohol dehydrogenase (yeast)	Light-resistant Yellow 2KT-Cu(II)	6	Pesliakas *et al.*, 1988

6. CONCLUDING REMARKS

Affinity partitioning offers numerous advantages in protein purification. Aqueous two-phase systems form a gentle environment that is highly suitable for maintaining proteins in their native structure. The binding capacity per unit volume seems to be considerably higher in affinity liquid–liquid phases, because of increased ligand density and free accessibility of the ligand, compared to liquid–solid interaction. Furthermore, the time requirement to reach equilibrium binding in solution is faster in comparison with that for solid affinity resins. For large-scale applications, an efficient recycling of ligands will, however, be necessary (Cordes and Kula, 1986). The technical development of aqueous two-phase systems is quite advanced, making use of commercially available equipment designed for solvent extraction.

In addition, affinity partitioning has been found to be a rather simple but sensitive method for the screening of ligands and for studying protein–ligand interaction.

REFERENCES

Albertsson, P.-A., 1956, Chromatography and partition of cells and cell fragments, *Nature (London)* **177**:771–774.

Albertsson, P.-A., 1971, *Partition of Cell Particles and Macromolecules*, 2nd ed., Wiley Interscience, New York.

Arnold, F. H., 1991, Metal affinity separations: A new dimension in protein bioprocessing, *Bio/Technology* **9**: 151–156.

Birkenmeier, G., and Kopperschläger, G., 1987, Interaction of the dye Remazol Yellow GGL to prealbumin and albumin studied by affinity phase partitioning, difference spectroscopy and equilibrium dialysis, *Mol. Cell. Biochem.* **73**:99–110.

Birkenmeier, G., Tschechonien, B., and Kopperschläger, G., 1984, Affinity chromatography and affinity partitioning of human serum prealbumin using immobilized Remazol Yellow GGL: Evidence that albumin increases binding of prealbumin to the dye, *FEBS Lett.* **174**:162–166.

Birkenmeier, G., Ehrlich U., Kopperschläger, G., and Appelt, G., 1986a, Partition of purified human thyroxine-binding globulin in aqueous two-phase systems in response to reactive dyes, *J. Chromatogr.* **360**:193–201.

Birkenmeier, G., Shanbhag, V. P., and Kopperschläger, G., 1986b, Comparison of the binding of poly(ethylene glycol)-bound fatty acids to human albumin and alpha-fetoprotein studied by affinity partitioning, *Biomed. Biochim. Acta* **45**:285–289.

Birkenmeier, G., Carlsson-Bostedt, L., Shanbhag, V. P., Kriegel, T., Kopperschläger, G., Sottrup-Jensen, L., and Stigbrand, T., 1989, Differences in hydrophobic properties of human alpha-2-macroglobulin and pregnancy zone protein as studied by affinity partitioning, *Eur. J. Biochem.* **183**:239–243.

Birkenmeier, G., Vijayalakshmi, M. A., Stigbrand, T., and Kopperschläger, G., 1991, Immobilized metal ion affinity partitioning, a method combining metal–protein interaction and partitioning of proteins in aqueous two-phase systems, *J. Chromatogr.* **539**:276–277.

Brooks, D. E., Sharp, K. A., and Fisher, D., 1985, Theoretical aspects of partitioning, in: *Partitioning in Aqueous Two-Phase Systems* (H. Walter, D. E. Brooks, and D. Fisher, eds.), Academic Press, London, pp. 11–84.

Bückmann, A. F., Kula, M.-R., Wichmann, R., and Wandrey, C., 1981, An efficient synthesis of high molecular weight NAD(H) derivatives suitable for continuous operation with coenzyme dependent enzyme systems, *J. Appl. Biochem.* **3**:301–315.

Bückmann, A. F., Morr, M., and Kula, M.-R., 1987, Preparation of technical grade poly(ethylene glycol) (PEG) (M_r 20,000)-N^6-(2-aminoethyl)-NADH by a procedure adaptable to large-scale synthesis, *Biotechnol. Appl. Biochem.* **9**:258–268.

Cordes, A., and Kula, M.-R., 1986, Process design for large scale purification of formate dehydrogenase from *Candida boidinii* by affinity partition, *J. Chromatogr. (Biomed. Appl.)* **376**:375–384.

Cordes, A., Flossdorf, J., and Kula, M.-R., 1987, Affinity partitioning: Development of mathematical model describing behavior of biomolecules in aqueous two-phase systems, *Biotechnol. Bioeng.* **30**:514–520.

Flanagan, S. D., and Barondes, S., 1975, Affinity partitioning—a method for purification of proteins using specific polymer-ligands in aqueous polymer two-phase systems, *J. Biol. Chem.* **250**:1484–1489.

Hubert, P., Dellacherie, E., Neel, J., and Ballieu, E.-E., 1976, Affinity partitioning of steroid binding proteins. The use of polyethyleneoxide-bound estradiol for purifying $\Delta_{5\rightarrow4}$ 3-oxosteroid isomerase, *FEBS Lett.* **65:** 169–174.

Johansson, G., 1970, Studies on aqueous dextran–PEG two-phase systems containing charged poly(ethylene glycol). Partition of albumin, *Biochim. Biophys. Acta* **222:**381–389.

Johansson, G., and Andersson, M., 1984, Parameters determining affinity partition of yeast enzymes using polymer-bound triazine dye ligands, *J. Chromatogr.* **303:**39–51.

Johansson, G., and Joelsson, M., 1985a, Preparation of Cibacron Blue F3G-A polyethylene glycol in large scale for use in affinity partitioning, *Biotechnol. Bioeng.* **27:**621–625.

Johansson, G., and Joelsson, M., 1985b, Partial purification of glucose-6-phosphate dehydrogenase from baker's yeast by affinity partitioning using triazine dyes, *Enzyme Microb. Technol.* **7:**629–634.

Johansson, G., and Joelsson, M., 1986, Liquid–liquid extraction of lactate dehydrogenase from muscle using polymer bound triazine dyes, *Appl. Biochem. Biotechnol.* **13:**15–27.

Johansson, G., Hartmann, A., and Albertsson, P.-A., 1973, Partition of protein in two-phase systems containing charged poly(ethylene glycol), *Eur. J. Biochem.* **33:**379–386.

Johansson, G., and Joelsson, M., 1987, Affinity partitioning of enzymes using dextran-bound Procion yellow HE–3G, *J. Chromatogr.* **393:**195–208.

Karr, L. J., Shafer, S. G., Harris, J. M., van Alstine, J. M., and Snyder, R. S., 1986, Immuno affinity partition of cells in aqueous polymer two-phase systems, *J. Chromatogr.* **354:**269–282.

Karr, L. J., van Alstine, J. M., Snuder, R. S., Shafter, S. G., and Harris, J. M., 1989, Cell separation by immunoaffinity partition in aqueous polymer two-phase systems: Application in cell biology and biotechnology, in: *Separation Using Aqueous Phase Systems* (D. Fisher, and I. A. Sutherland, eds.), Plenum Press, London, pp. 193–202.

Kirchberger, J., and Kopperschläger, G., 1990, An improved purification procedure of alkaline phosphatase from calf intestine by applying partitioning in aqueous two-phase systems and dye-ligand affinity chromatography, *Bioseparation* **1:**33–41.

Kirchberger, J., Cadelis, F., Kopperschläger, G., and Vijayalakshmi, M. A., 1989, Studies on the interaction of lactate dehydrogenase with structurally related triazine dyes using affinity partitioning and affinity chromatography, *J. Chromatogr.* **483:**289–299.

Kopperschläger, G., and Birkenmeier G., 1990, Affinity partitioning and extraction of proteins, *Bioseparation* **1:**235–254.

Kopperschläger, G., and Johansson, G., 1982, Affinity partitioning with polymer-bound Cibacron Blue F3G-A for rapid large-scale purification of phosphofructokinase from baker's yeast, *Anal. Biochem.* **124:**117–124.

Kopperschläger, G., Böhme, H.-J., and Hofmann, E., 1982, Cibacron Blue F3G-A and related dyes as ligands in affinity chromatography, in: *Advances in Biochemical Engineering* (A. Fiechter, ed.), Springer-Verlag, Berlin, pp. 101–138.

Kopperschläger, G., Lorenz, G., and Usbeck, E., 1983, Application of affinity partitioning in aqueous two-phase systems to the investigation of triazine dye–enzyme interaction, *J. Chromatogr.* **259:**97–105.

Kopperschläger, G., Kirchberger, J., and Kriegel, T., 1988, Studies on triazine dye–enzyme interaction by means of affinity partitioning, *Makromol. Chem., Macromol. Symp.* **17:**373–385.

Naumann, M., Reuter, R., Metz, P., and Kopperschläger, G., 1989, Affinity chromatography of bovine heart lactate dehydrogenase using dye ligands linked directly or spacer-mediated to bead cellulose, *J. Chromatogr.* **466:**319–329.

Otto, A., and Birkenmeier, G., 1993, Recognition and separation of isoenzymes by metal chelates. Immobilized metal ion affinity partitioning of lactate dehydrogenase isoenzymes, *J. Chromatogr.* **644:**25–33.

Persson, L. O., and Olde, B., 1988, Synthesis of ATP-polyethylene glycol and ATP-dextran and their use in the purification of phosphoglycerate kinase from spinach chloroplasts using affinity partition, *J. Chromatogr.* **457:**183–193.

Pesliakas, J.-H. J., Zutautas, V. D., and Glemza, A. A., 1988, Affinity partitiom of yeast and horse liver alcohol dehydrogenase in PEG-dyes/dextran two phase systems, *Chromatographia* **26:**85–90.

Pfleiderer, G., Jeckel, D., and Wieland, T., 1958, Über den Wirkungsmechanismus der Milchsäuredehydrogenasen, *Biochem. Z.* **330:**296–302.

Plunkett, S. D., and Arnold, F. H., 1990, Metal affinity extraction of human hemoglobin in aqueous polyethylene glycol–sodium sulfate two-phase systems, *Biotechnol. Tech.* **4:**45–48.

Porath, J., Carsson, J., Olsson, I., and Belfrage, G., 1975, Metal chelate affinity chromatography, a new approach to protein fractionation, *Nature (London)* **258:**598–599.

Shanbhag, V. P., and Johansson, G., 1974, Specific extraction of human serum albumin by partition in aqueous biphasic systems containing poly(ethylene glycol)-bound ligand, *Biochem. Biophys. Res. Commun.* **61:**1141–1146.

Shanbhag, V. P., and Johansson, G., 1979, Interaction of human serum albumin with fatty acids. Role of anionic group studied by affinity partition, *Eur. J. Biochem.* **93:**363–367.

Sharp, K. A., Yalpani, M., Howard, S. J., and Brooks, D. E., 1986, Synthesis and application of polyethylene glycol antibody affinity ligands for cell separation in aqueous polymer two-phase systems, *Anal. Biochem.* **154:**110–117.

Walter, H., Brooks, D. E., and Fisher, D., 1985, *Partitioning in Aqueous Two-Phase Systems, Theory, Methods, Uses, and Application to Biotechnology*, Academic Press, Orlando, Florida.

Walter, H., Widen, K. E., and Birkenmeier, G., 1993, Immobilized metal ion affinity partitioning of erythrocytes from different species in dextran-polyethylene glycol aqueous phase systems, *J. Chromatogr.* **641:**279–289.

Wuenschell, G. E., Naranjo, E., and Arnold, F. H., 1990, Aqueous two-phase metal affinity extraction of heme proteins, *Bioprocess Eng.* **5:**199–202.

34

Affinity Electrophoresis of Macromolecules

General Principles and Their Application to Nucleic Acids

Gabor L. Igloi

One can, by generalizing the basic idea behind affinity chromatography, suppose that any force which brings about the passage of one biomolecule past its matrix-bound interacting partner could be harnessed for separations based on biorecognition phenomena. The methods which have been devised in order to bring the interacting partners into mutual proximity reflect both the nature of the molecules and the nature of the subsequent analytical steps. In the field of nucleic acids, which will be considered in detail below, the Southern hydridization procedure, for example, requires the immobilization of a DNA fragment on a membrane and subsequent diffusion-mediated interaction with DNA of complementary sequence. One would, in this analytical case, not consider linking one of the two partners in this process to a column in order to bring about a biorecognition using hydrodynamic pressure as the driving force. For such experiments, the use of membranes and diffusion offers a much faster and more convenient route to specific interactions. Yet, for other purposes the passage of nucleic acids through a column containing a covalently bound complementary nucleic acid has, until recently, been the method of choice, as in the case of the selection of poly(A)-containing mRNAs by passage over poly(U) or poly(dT) columns. Although columns (hydrodynamic pressure) and membranes (diffusion) form the carriers of the bulk of affinity methods (as is evident from the contents of this volume), a third matrix, a polymeric gel (electric field), has also been demonstrated as being a useful "matchmaker" in the technique of affinity electrophoresis. This extension of the principle of affinity chromatography has been an established tool in immunoelectrophoresis for many years, and the nature of many enzyme–substrate or lectin–sugar complexes have been examined. However, just as the use of membranes was until recently a preserve of nucleic acids, so was affinity electrophoresis firmly in the domain of proteins.

The affinity electrophoretic theory and its application to the fractionation of proteins have been reviewed on a number of recent occasions (Horejsi, 1984; Takeo, 1984; Horejsi

Gabor L. Igloi • Institut für Biologie III der Universität Freiburg, D-79104 Freiburg, Germany.

Molecular Interactions in Bioseparations, edited by That T. Ngo. Plenum Press, New York, 1993.

and Ticha, 1986; Takeo, 1987; Heegaard and Bøg-Hansen, 1990; Bøg-Hansen, 1990) while its association with the analysis of certain structural features in nucleic acids has been largely ignored. This neglect is all the more remarkable since it is precisely the strong noncovalent but specific interactions characteristic of nucleic acid–nucleic acid and nucleic acid–protein recognition that lend themselves ideally to affinity electrophoretic analysis.

One should, perhaps, at this point define what types of interactions are to be considered in terms of affinity interactions, in general, and for the purposes of affinity electrophoresis, in particular. We are, clearly, concerned with the "biological" recognition between enzyme and substrate or between macromolecule and macromolecule on the basis of noncovalent, that is, electrostatic, hydrogen bond, and hydrophobic, forces. Despite the emphasis on biorecognition in this definition (Bøg-Hansen, 1990), one must also consider the specific "chemical" recognition between a reagent and a reversibly reactive group within a biomolecule, leading to a stable but easily dissociable new chemical entity [cf. covalent chromatography (Brockelhurst et al., 1973)]. Examples of such affinity matrices are the boronate derivatives, either for chromatographic (Schott et al., 1973) or electrophoretic (Igloi and Kössel, 1985) purposes. Techniques in which neither of the interacting partners is immobilized within the matrix may also fall within the definition of affinity-related separation methods [e.g., affinity elution (von der Haar, 1974) in the case of columns]. Indeed, the well-established methods of crossed electrophoresis or crossed immunoaffinity electrophoresis exemplify this type of separation. Similarly, the application of electrophoresis to the fractionation of complexes already present in solution, as in the case of the powerful gel retardation assay for detecting nucleic acid–protein complexes (Fried and Crothers, 1981; Garnier and Revzin, 1981; Carey, 1991) or of the more recently introduced affinophoresis (Shimura and Kasai, 1987), may be regarded as specialized forms of affinity electrophoresis, although it is, strictly speaking, more closely analogous to gel filtration than to affinity chromatography. In both these examples, complexation of interacting partners leads to, among other effects, a change in the net charge-to-size ratio, permitting a separation of complexed from uncomplexed biomolecules on the basis of a modified rate of migration: the matrix itself does not play a direct active role in the fractionation. One observes a reduction in mobility for protein–nucleic acid interactions and a stimulation in the mobility in the case of an interaction between a macromolecular polyelectrolyte bearing an affinity ligand (affinophore) and its target molecule.

1. THE MATRIX

The range of matrices available for electrophoresis, in general, together with their potential for further development has been reviewed by Righetti (1989).

The original concept of affinity electrophoresis, following earlier paper electrophoretic studies on the interactions between glycoproteins and a lectin (Nakamura et al., 1960) and incorporating the principles of crossed immunoelectrophoresis (Laurell, 1965), was described by Bøg-Hansen and co-workers (Bøg-Hansen, 1973; Bøg-Hansen et al., 1975) using a lectin added to an agarose matrix and glycoproteins as the mobile affinity partners. With the extension of the application to other systems, a number of methods of immobilization have been developed. At the present time, these can be classified into three general groups.

1.1. Molecular Entrapment

1.1.1. Agarose

In view of the apparent simplicity of their preparation, agarose gels have remained the most popular, in particular in routine clinical applications. For qualitative or diagnostic purposes, the absolute immobility of the matrix-bound component is not essential, but for quantitative interpretations (see below) it may sometimes facilitate the evaluation. Immobilization by entrapment can only be ensured if (a) the size of the ligand is sufficiently large, (b) the ligand is uncharged, or (c) electrophoresis is carried out at or near the pI of the ligand. In fact, total immobilization in agarose by entrapment is rarely achieved, but the notable and extensive literature on the application of entrapped concanavalin A (Con A) and related lectins (Bøg-Hansen and Takeo, 1980; Mackiewicz and Kushner, 1989; Heegaard et al., 1989; Hanham et al., 1986; Mackiewicz and Mackiewicz, 1986; Taketa et al., 1989; Taketa and Hirai, 1989; Breborowicz and Mackiewicz, 1989; and references therein) for the analysis of glycoproteins is evidence for the practical application of recognition processes in agarose gels between components of differing mobilities.

Since the efficiency of immobilization depends on the relationship between the size of the macromolecule and the porosity of the gel, prerequisite (a) is not easily satisfied in the highly porous agarose but is the main method of entrapment in the more tightly meshed polyacrylamide gels.

1.1.2. Polyacrylamide

The use of polacrylamide as a medium for affinity electrophoresis was developed in parallel to that of agarose, and this medium was also initially used for lectin–glycoprotein studies. In this case, however, the possibility of trapping a polysaccharide, by virtue of its size and lack of charge, in a polyacrylamide network could be utilized further for the study of sugar–enzyme interactions (see, e.g., Gerbrandy and Doorgeest, 1972; Takeo and Nakamura, 1972; Shimomura and Fukui, 1980; Maly et al., 1985; Matsumoto et al., 1990) and of the macromolecule–macromolecule interaction between entrapped lipopolysaccharide and an outer membrane protein (Borneleit et al., 1989a), between dextran and anti-dextran immunoglobulin G (IgG) (Tanaka et al., 1989), or between lactate dehydrogenase and immobilized Blue Dextran (Ticha et al., 1978). The effectiveness and simplicity with which biospecific interactions could be studied and quantified (see below) in this way was, nevertheless, limited by the nature of the available natural substrates having the required physical properties to enable immobilization. The idea of a macroligand created by coupling a small ligand of interest to an inert macromolecular carrier, dextran, was successfully introduced (Ticha et al., 1978; Cerovsky et al., 1980a,b) and extended to poly(ethylene glycol) (Müller et al., 1981). Other polymeric ligand adducts have been constructed from linear, water-soluble, polyacrylamide (Nakamura et al., 1980a,b) or methacrylamide derivatives (Ticha et al., 1980) giving rise to various nucleotide-bearing substrates for dehydrogenases or to substrates for trypsin (Cerovsky et al., 1980a,b). Copolymers of acrylamide have also been affinity electrophoretic substrates for galactosidases (Maly et al., 1985; Secova et al., 1988) and, with hydrophobic substituents, for albumin and β-lactoglobulin (Chen and Morawetz, 1981; Masson and Reybaud, 1988).

1.2. Particle Entrapment

Provided that a homogeneous distribution can be guaranteed, it is sometimes desirable to extend the ideas governing the entrapment of soluble ligands discussed above to the incorporation of derivatized particles into a gel (Horejsi *et al.*, 1982). Turkova *et al.* (1983) used periodate-oxidized Sephadex for the immobilization of aminophenylglycosides, followed by polymerization into a standard polyacrylamide gel. The advantage of this technique is that one can, in principle, use the same matrix for both affinity chromatography and affinity electrophoresis, provided that it does not interfere with the polymerization process or subsequent detection of the sample. In fact, the incorporation of the particles in a polymeric matrix may not always be required. A preparative form of affinity electrophoresis has been described (Bergenheim *et al.*, 1983) in which the separation takes place within an affinity chromatographic column.

1.3. Covalent Modification of the Electrophoretic Matrix

While the physical entrapment approach is widely applicable, it may suffer from the danger of microinhomogeneities in the gel. Moreover, some affinity electrophoretic investigations have required a covalent attachment of one of the partners to the matrix. These circumstances have created a need for procedures enabling matrix derivatization that are generally applicable yet sufficiently straightforward to be performed in biomedical environments.

1.3.1. Derivatives of Agarose

The large array of affinity chromatography media based on agarose is evidence for the feasibility of derivatizing agarose for affinity electrophoretic purposes as well. This potential was already recognized in 1982 by Horejsi *et al.*, who demonstrated the general utility of periodate-oxidized agarose for coupling ligands bearing primary amino groups such as aminophenyl sugars, aminobenzamidine, or acriflavine. After Schiff base formation, reduction with cyanoborohydride leads to stable morpholino derivatives. As an alternative, CNBr-activated Sepharose and epoxy-activated Sepharose, among others, could also be used to provide meltable agarose matrices (provided that the degree of cross-linking following activation permitted complete melting) although some difficulties with high background staining and/or gel distortion were encountered. To counteract the inherently poor resolution of proteins electrophoresed in agarose, these modified agaroses could also be used in conjunction with polyacrylamide in composite gels, as was shown for a Cibacron blue–agarose conjugate in its interaction with albumin (Johnson *et al.*, 1980). In the case of these agarose matrices would seem to be a rather easy way of obtaining the desired substitution; a limitation in their application remains the fact that the ligand must be stable under the conditions used to melt and cast the gels.

1.3.2. Derivatives of Acrylamide

Initial attempts using allylglycosides as direct copolymers of acrylamide to construct a homogeneous polymeric affinity electrophoretic matrix (Horejsi and Kocourek, 1974) were abandoned because of incomplete polymerization (Horejsi and Ticha, 1986), and, in any event, this approach seemed unattractive as a general procedure since each individual ligand required an initial chemical synthesis to provide the allylic double bond for polymerization. The direct chemical modification of acrylamide with ligands suitable for affinity electrophoresis is hampered, on the one hand, by the lack of suitably reactive functional groups in the acrylamide molecule and, on the other hand, by the requirement that the C=C bond be maintained intact to permit polymerization.

These difficulties were overcome by Schnaar and Lee (1975), who described the facile synthesis of N-succinimidyl and N-phthalimidyl esters of acrylic acid. These activated acrylamide derivatives react with primary amino groups under mild conditions, permitting the coupling of peptides (Brandley and Schnaar, 1988) and sugars (Weigel et al., 1979; Schnaar et al., 1978). In fact, these investigations involved the preformation of poly-acrylamide gels bearing the activated side chains prior to derivatization with the ligands and subsequent use as affinity surfaces. For affinity electrophoretic applications, one would couple the ligand to the activated ester monomer prior to polymerization. This has, as far as I am aware, yet to be tested. Direct acylation of an affinity ligand with acryloyl chloride following the method outlined by Whiteside et al. (1976) or Schnaar et al. (1978) is, chemically, a harsher procedure but can be applied to small ligands such as aminophenyl-boronic acid (Igloi and Kössel, 1985, 1987) or aminophenylmercuric acetate (Igloi, 1988) in a relatively straightforward manner, and the acryloyl monomers obtained can indeed be incorporated into polyacrylamide electrophoretic gels.

2. QUANTIFICATION

The observed dependence of the mobility of a substance in an affinity electrophoretic gel on the concentration of the immobilized ligand is a consequence of the equilibrium binding phenomenon taking place (Takeo, 1987). The relationship between these two parameters (migration distance and ligand concentration) was described quantitatively by Takeo and Nakamura (1972) and by Gerbrandy and Doorgeest (1972) with the assumption that the complex has zero mobility. Under these conditions, the dependence of the mobil-ity on the apparent binding constant K is given by

$$\frac{1}{r} = \frac{1}{R_0}\left(1 + \frac{c}{K}\right) \tag{1}$$

where r is the migration distance of one interacting partner in the presence of the immobilized ligand with a concentration c, and R_0 is the mobility at $c = 0$. Plotting $1/r$ versus c, therefore, results in a straight line whose intercept on the abscissa is $-K$. For the more frequently encountered situation in which the assumption of an immobile complex cannot be upheld, Bøg-Hansen and Takeo (1980) and Horejsi and Ticha (1986) developed this original affinity equation further, deriving what has become known as the general affinity equation:

$$\frac{1}{R_0 - r} = \frac{1}{R_0 - R_c}\left(1 + \frac{K}{c}\right) \tag{2}$$

where R_c is the mobility of the complex, and the assumption is made that c is much greater than the total concentration of substance being electrophoresed. This expression is clearly equivalent to Eq. (1) when $R_c = 0$.

A further useful feature of affinity electrophoresis is the disturbance of the simple binding equilibrium by the interaction of the complex with a competitor, for instance, an enzyme inhibitor (Secova et al., 1988) or a metal ion (Igloi, 1988), leading to a displacement of the mobile ligand from its immobilized complex. An analysis of the migration of the newly released ligand with respect to the concentration of the third species gives one the possibility of quantifying the interaction of, say, an enzyme with its inhibitor or a lectin with a sugar (Mackiewicz and Mackiewicz, 1986) according to the following equation (Takeo, 1984):

$$\frac{r}{R_0 - r} = \frac{K}{c}\left(1 + \frac{i}{K_i}\right) \tag{3}$$

where i is the concentration of the inhibitor and K_i is the apparent affinity constant of the enzyme for the inhibitor, and the assumption is made that the enzyme–inhibitor complex has the same migration as the free enzyme and that c as well as the concentration of the inhibitor is much higher than the enzyme concentration. More extensive theoretical considerations have been given by Horejsi and co-workers (Horejsi and Ticha, 1986; Matousek and Horejsi, 1982), but it is apparent from the practical application of these considerations that by using the simplified formulas in Eqs. (1)–(3), valuable quantitative information, suitable at least for comparative purposes, can be gained very rapidly by affinity electrophoresis. In contrast, affinity chromatography while offering the possibility of quantitative evaluation (Chaiken et al., 1992), would appear to suffer from limitations imposed by sample and column volumes as well as by currently available methods of detection.

3.   FACTORS AFFECTING THE MATRIX–LIGAND INTERACTION

Affinity electrophoresis, like any technique dealing with binding equilibria, is sensitive to changes in parameters that affect the binding process. Any change which shifts the equilibrium will influence the mobility of the migrating partner. Such alterations, can, under carefully controlled conditions and in combination with the quantitative approach described above, give useful, if only apparent, thermodynamic information concerning the binding process. In this way, the effects of pH (Ek et al., 1980), temperature (Takeo, 1987; Matsumoto et al., 1990), pressure (Masson and Reybaud, 1988), the involvement of other ligands (see above), and denaturation-induced conformational changes (Igloi, 1988) have been studied. On the other hand, one should be aware of possible unwanted perturbations of the interaction under investigation resulting from the use of inappropriate buffers (e.g., borate buffers when dealing with carbohydrates) or buffer components such as detergents or chaotropic agents. Similarly, the use of high concentrations of an immobilized but ionic ligand can, particularly in the case of polyacrylamide gels, lead to gel distortion through electroendosmotic effects. In such a case, the detection of weak complexes may be limited by the amount of immobilized ligand that can be incorporated without compromising the stability of the matrix.

4.   APPLICATIONS

4.1.   Proteins

Since the introduction of affinity electrophoresis in the mid-1970s, its use has expanded into all areas of protein research. It has become of particular value in routine clinical studies as exemplified by the work of Taketa and Hirai (1989), Breborowicz and Mackiewicz (1989), Rosalki and Foo (1989), and Heegaard *et al.* (1989). A comprehensive review of the extensive clinical applications does not fall within the scope of this chapter. Reviews covering many aspects have appeared from time to time; the most recent comprehensive review, listing about 70 different proteins which have been analyzed by this method, is that of Takeo (1987). In addition, Bøg-Hansen (1990) provided a historical overview of the development of affinity electrophoresis, while crossed affinity immuno-electrophoresis as a specific application has also been described (Heegaard and Bøg-Hansen, 1990). A series of reports in a paper symposium edited by Bøg-Hansen and Takeo describe the application of affinity electrophoresis to antihapten antibodies (Takeo *et al.*, 1989a,b), α-fetoprotein (Taketa *et al.*, 1989), human serum transferrin (Heegaard *et al.*, 1989), the regulation of plasma protein glycosylation (Mackiewicz and Kushner, 1989), and quantitative lipopolysaccharide–protein interactions (Borneleit *et al.*, 1989b) and illustrate recent progress in this field.

4.2. Nucleic Acids

As stated at the outset, the aim of this chapter is to consider in particular the advances which have been made in the affinity electrophoretic analysis of nucleic acids. These may be conveniently divided into those concerning analytical aspects (in terms of the presence or absence of a particular structural feature) and those giving information on the conformational state of the macromolecule.

4.2.1. Analysis of Unique Functional Groups

4.2.1.1. The Termini of RNA. A knowledge of the chemical nature of the 5′ or 3′ terminus of RNA can often reveal the origin or even function of a particular species. In particular, the presence or absence of a free ribose residue, as opposed to, say, the phosphorylated form, is a valuable indicator of the origin of the RNA under investigation. A group-specific boronate affinity chromatographic matrix for the binding of the *cis*-diol moiety found in free ribose groups was already developed in the early 1970s (Schott *et al.*, 1973), but the use of these columns has, with some exceptions (McCutchan *et al.*, 1975) and with any but the shortest oligonucleotides or merely nucleosides, been rather limited (but see also Beneš *et al.*, this volume, Chapter 20). In contrast to the problems apparently involved in the use of these matrices, aminophenylboronic acid linked directly to acrylamide [acryloylaminophenylboronic acid (APB)] (Igloi and Kössel, 1985) has proved to be a reliable electrophoretic medium for the analysis of all ribose modifications found at either end of naturally occurring RNAs (Igloi and Kössel, 1987). Table 1 summarizes the observed migration properties of terminally modified RNAs in the boronate affinity electrophoretic gels and gives an indication as to which additional treatment of the RNA, followed by another boronate gel analysis, can confirm the chemical nature of the RNA terminus. The principle of such an analysis is illustrated in Fig. 1 for each step of the chemical treatment of tRNA (approximately 80 nucleotides) that results in the removal of the terminal base. The free ribose is oxidized with periodate, giving rise to a faster migrating species. Amine-induced β elimination of the terminal base results in a shortened nucleotide chain with a 3′-terminal phosphate—exhibiting increased mobility—which can be removed with phosphatase to regain a free terminal ribose, which can once more interact

TABLE 1
Application of APB Gels to the Termini of RNA

Terminal structure	Occurrence	Specific test in conjunction with affinity electrophoresis	Mobility shift
3' Free ribose	Mature RNAs; tRNAs	Sensitive to IO_4^-	→
Phosphate	RNase degradation products; processing intermediates	Sensitive to phosphatase	←
5' Monophosphate	RNase degradation products; processing intermediates; mature RNAs	Adenylation with RNA ligase; shift eliminated by IO_4^-	←
Di/triphosphate	Primary transcripts	Guanylation with guanylyltransferase; shift eliminated by IO_4^-	←
Caplike	Eukaryotic mRNAs	Sensitive to IO_4^- and to tobacco acid pyrophosphatase	→

−APB **+APB**

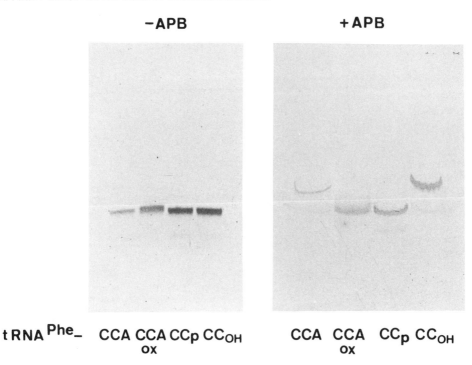

tRNA^{Phe}− CCA CCA CCp CC$_{OH}$ CCA CCA CCp CC$_{OH}$
           OX   OX

FIGURE 1. Influence of 3′-terminal modification on the migration of tRNA in an affinity electrophoretic gel (Igloi and Kössel, 1985). Electrophoresis patterns of, from left to right, tRNA-CCA, tRNA-CCA$_{ox}$ (the ribose of the terminal A destroyed by periodate oxidation), tRNA-CC$_p$ (the terminal oxidized A removed by β–elimination), and tRNA-CC$_{OH}$ (the terminal phosphate removed by phosphatase treatment) run on 10% polyacrylamide gels containing 7M urea in the absence (a) or presence (b) of 5% APB are shown. The samples were tRNA^{Phe} (yeast) (~20 ng/lane), and the bands were detected by silver staining (Igloi, 1983).

with the affinity gel and is thus retarded. The control gel, lacking the boronate component, can only resolve the starting material from the shortened product.

Of interest at the 5′ terminus of RNAs is analysis for the absence or presence of a 5′-5′-linked GTP or its derivative, found as the cap structure in eukaryotic mRNAs. Polyacrylamide may not be the ideal matrix for fractionating long mRNAs (>1000 nucleotides), but the electrophoresis of transcripts of over 600 nucleotides has confirmed the remarkable property of these gels in resolving species of this length differing from each other by merely one OH group (Igloi and Kössel, 1987) (Fig. 2). Using this method, Lien *et al.* (1987) showed convincingly that the 5′ end of duck hepatitis B virus plus-strand RNA primer is capped.

4.2.1.2. Modifications in vivo. Modified bases occur in DNA, in the form of methylated C or A residues, and, in particular, in RNA; in the case of tRNA, over 70 different modifications of both the sugar and the base have been recorded to date (Björk *et al.*, 1987). A number of these tRNA modifications involve the generation of chemical moieties that are sufficiently different from other reactive groups in the RNA molecule to permit group-

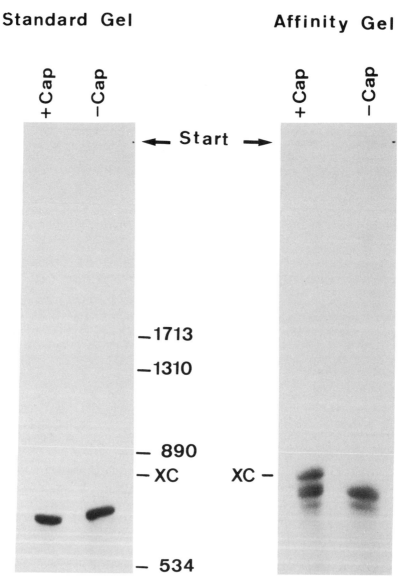

FIGURE 2. Detection of a 5'-capped product of SP6 RNA polymerase transcription (Igloi and Kössel, 1987). The cDNA for rabbit β-globin, cloned in the SP6 system, was transcribed with SP6 RNA polymerase in the presence or absence of m⁷GpppA and using [α-³²P]GTP. Aliquots were analyzed on horizontal 2% polyacrylamide gels containing 7M urea and either 0 or 25% APB, as shown. Note that the numbers alongside the pattern refer to DNA molecular weight markers (in bp) run simultaneously on the gel and detected by ethidium bromide staining.

specific binding by appropriate affinity electrophoretic gels. The modified bases which have been the subject of affinity electrophoretic analysis, so far, are summarized in Table 2.

a. Queosine. The modified base Q occurs at position 34 in several of those tRNAs which read XA(U/C) codons (Nishimura, 1978). Q is a derivative of deazaguanosine bearing a cyclopentendiol side chain, thereby offering a unique internal site within these tRNAs for an interaction between the diol function and an immobilized boronate (Fig. 3). The initial affinity chromatographic fractionation of this type (McCutchan et al., 1975) was followed by an affinity electrophoretic procedure using a boronate matrix (Igloi and Kössel, 1985).

b. Thiouridines. 4-Thiouridine (s^{4}U) is one of several thiouridine derivatives to be found in tRNAs. It has been detected in many bacterial tRNAs, where it is localized at position 8 in the polynucleotide chain. The detailed investigations of the interaction of s^{4}U as well as of s^{4}U-containing tRNAs with organomercurial compounds (Scheit and Faerber, 1973) formed the basis for the synthesis of acryloylaminophenylmercuric chloride (APM), a derivative of acrylamide which could be polymerized directly into an electrophoretic gel (Igloi, 1988). The specificity of the interaction with the matrix was demonstrated by targeted chemical or radiolytic modification of s^{4}U, which abolished the affinity retardation. With these affinity electrophoretic gels, it was possible to exploit the variation in the chemical reactivity of the sulfur atom, as reflected in the matrix–tRNA dissociation constants, with respect to its chemical environment. There was a clear difference in the affinity for s^{4}U compared with that for another thiouridine derivative, 5-methylamino-methyl-2-thiouridine (mnm^{5}s^{2}U), occurring in the anticodon of several tRNAs (Fig. 4).

Recently, Emilsson et al., (1992) were able to quantitate the growth rate dependent appearance of natural s^{4}U in the tRNAs of bacterial cultures by Northern hybridization following affinity electrophoresis in this system. Bezerra and Favre (1990), also using the immobilized mercury derivative of acrylamide, analyzed the incorporation of the exogenously added potential photoaffinity label 4-thiouridine into the total RNA population of Escherichia coli as a function of the number of exponentially growing generations undergone by the cells prior to contact with the thiolation medium.

A two-dimensional version of this type of fractionation (affinity matrix in the second

TABLE 2
Detection of Modified Bases in tRNA
by Affinity Electrophoresis

Functional group	Modified base[a]	Treatment	Matrix[c]	Reference[d]
Ribose	Q	None	APB	1
Thio	s^{4}U	None	APM	2
	mnm^{5}s^{2}U	None	APM	2
Amino	acp^{3}U	SPDP[b]	APM	3
	acp^{2}C	SPDP	APM	3

[a]For abbreviations, see Nishimura (1978).
[b]SPDP, N-Succinimidyl 3-(2-pyridyldithio)propionate.
[c]See text.
[d]1, Igloi and Kössel, 1985; 2, Igloi, 1988; 3, Igloi, 1992.

−APB   **+APB**

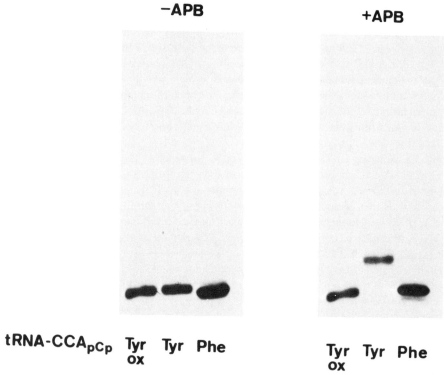

tRNA-CCA$_{pCp}$ Tyr   Tyr Phe   Tyr   Tyr Phe
                      ox   **ox**

FIGURE 3. Detection of the modified base Q in tRNA. tRNA^{Phe} (yeast) (no Q) and tRNA^{Tyr} (*E. coli*) (Q$_{34}$) were 3′-terminally radioactively labeled, and an aliquot of the tRNA^{Tyr} was subjected to periodate oxidation. Samples of each of the tRNA species were analyzed on a 5% polyacrylamide gel containing 7M urea with or without 10% APB as a copolymeric component, as indicated. [From Igloi and Kössel (1985).]

dimension) was used to detect thiolated modification(s) in total chloroplast tRNA (Igloi and Kössel, 1988).

c. Bases Bearing a Primary Amino Group.   Modifications of bases in tRNAs leading to the introduction of primary amino groups are not as common as thio substitutions. In fact, the only examples of this type are aminocarboxypropyluridine (acp^{3}U or X) and acp^{2}C (lysidine), which is an important element in the decoding function of tRNA^{Ile}(CAU) (Muramatsu *et al.*, 1988). One may, for the purposes of this subdivision, also regard aminoacylated tRNAs as bearing a modification with a free amino group, at the 3′ end of the polynucleotide chain.

Observations in our laboratory have shown that such primary amine side chains may be readily thiolated with the bifunctional reagent *N*-succinimidyl 3-(2-pyridyldithio)propion-ate (SPDP) (resulting in, among other things, a stabilization of the aminoacylester bond, in the case of the charged tRNA), and the corresponding tRNAs can subsequently be analyzed using the mercurial affinity electrophoretic gel system (Igloi, 1992).

d. Potential Applications to Other Modified Bases.   Of the other modified bases present in tRNAs, several possess structural features which are either analogous to ones already demonstrated to interact with either APB or APM or can be subjected to simple chemical

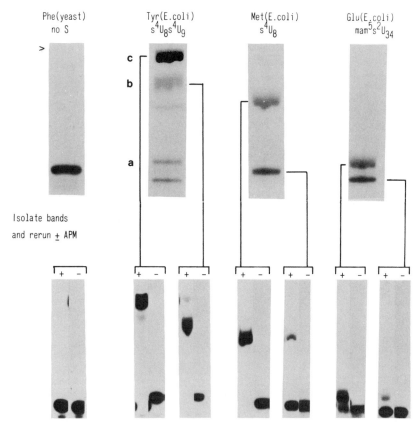

FIGURE 4. Affinity electrophoresis of thiolated tRNA species. Samples of each of the tRNAs indicated were 3'-terminally labeled and applied to a 5% polyacrylamide gel containing $7M$ urea and $25\mu M$ (10 μg/ml) APM. After recovery of the material from each band, it was rerun on an identical gel in the presence ($+$) or absence ($-$) of APM. Arrowheads indicate the position of sample application. [From Igloi (1988).]

modification in order to introduce a residue recognizable by these gels. Thiolation of amino groups as described above may be extended to free carboxyl moieties that have been previously linked to ethylenediamine (Ofengand et al., 1988). Taking these potential applications into consideration, in addition to those already described, 13 different modified bases become amenable to affinity electrophoretic analysis (Table 3).

4.2.1.3. Modifications In Vitro

a. Terminally. Modification of synthetic deoxyoligonucleotides at either terminus has achieved a considerable significance since the introduction of nonradioactive DNA sequencing and hybridization methods. Furthermore, the fields of application of phosphorothioate derivatives (Eckstein, 1985) of DNA are on the increase. The different reactivity of the sulfur in a wide range of different chemical and physical environments results in different mobilities in the mercury/acrylamide system described above. It is, therefore, not surprising that phosphorothioate monoesters behave differently from phosphorothioate diesters during affinity electrophoresis (Igloi, 1988). Indeed, the apparent dissociation

TABLE 3

Modified Bases in tRNA Potentially Amenable
to Affinity Electrophoretic Detection

Functional group	Modified base[a]	Treatment	Matrix[c]
Ribose	2'ribosylA	None	APB
	2'ribosylG	None	APB
Thio	mcm^{5}s^{2}U	None	APM
	s^{2}T	None	APM
	s^{2}C	None	APM
	s^{2}U	None	APM
	s^{2}Um	None	APM
Carboxy	o^{5}U	Cystamine[b]	APM
	cmo^{5}U	Cystamine	APM
	t^{6}A	Cystamine	APM
	mt^{6}A	Cystamine	APM
	ms^{2}T^{6}A	Cystamine	APM
	cmnm^{5}U	Cystamine	APM

[a]Documented in Edmonds et al. (1991).
[b]See Ofengand et al. (1988).
[c]APB, Acryloylaminophenylboranic acid; AMB, acryloyl-aminophenylmercuric chloride.

constant between phosphorothioate monoesters (as in the product of a transcription using GTPτS) and the mercury-substituted acrylamide of 0.8 μM is one of the strongest interactions measured in this system, while the diesters, whose sulfur atoms have a much reduced nucleophilic tendency, show no measurable affinity for the gel.

b. Internally. As mentioned above, the introduction of potentially photoactivatable modifications into certain positions within a polynucleotide chain has recently become of interest for studying specific protein–nucleic acid interactions (Connolly, 1991, Nikiforov and Connolly, 1992). In this case, a synthetic dodecanucleotide was synthesized containing s^{4}T (or s^{6}dG) at the recognition site of EcoRV. While these authors did not attempt an affinity electrophoretic purification of their synthetic product, Clivio et al. (1992) and Vitorino dos Santos et al. (1993a), describing the synthesis of oligonucleotides containing s^{4}U, were able to demonstrate clearly by affinity electrophoresis on APM that the crude reaction product could be fractionated into thiolated and nonthiolated species. Vitorino dos Santos et al., (1993b) used a phosphoramidite of deoxy-6-thioinosine to incorporate this photoreactive base analogue into oligonucleotides. Purification from non-thiolated products was achieved by APM electrophoresis, demonstrating the interaction of yet another thiolated species with the gel matrix. The random incorporation of s^{4}U into an RNA fragment by transcription in the presence of s^{4}UTP, as used for photoaffinity labeling of RNA polymerase (Khanna et al., 1991), was analyzed in detail by Dubreuil et al. (1991). By varying the concentration ratio of s^{4}UTP to (s^{4}UTP + UTP), the products of T7 RNA polymerase transcription, obtained as a homogeneous band on a denaturing polacrylamide gel, could be fractionated as a function of their s^{4}U content by affinity electrophoresis on the organomercurial matrix. Discrete bands containing zero to seven s^{4}U residues could be identified.

4.2.2. Conformation

The ligand-binding properties of macromolecules are not necessarily dependent on the maintenance of a biologically active conformation. In the case of an enzyme active site, perturbation of the tertiary structure usually leads to a reduction or loss of substrate recognition/binding. This is, however, not necessarily the case when a ligand is targeted at a specific chemical functional group as in the interaction between an organomercurial and a sulfur derivative. The sensitivity of affinity electrophoresis to conformational changes within nucleic acids has been demonstrated for tRNAs using the interaction of a specific sulfur-modified base with a mercurial affinity electrophoretic matrix as indicator. In the presence of denaturing agents (urea or formamide), quantitative changes in the affinity could be observed which were attributable to the conformational state of the tRNA in the vicinity of the thiolated residue, and hence the accessibility of the interacting base (Igloi, 1988, 1992).

The interpretation of the results is facilitated by considering the crystal structure (Fig. 5) as well as the numerous studies on the solution conformation (Kim, 1978). When considering $mnm^5s^2U_{34}$ at a position in the molecule which, under native conditions, is in a rather exposed environment, denaturation leads to a loss of binding which may be interpreted as being due to a reduced accessibility of the reacting sulfur atom in the more random conformation. Similarly, an electrophoretic comparison of a thiol residue introduced at acp^3U_{47} with the same modification at the 3' terminus gave rise to markedly

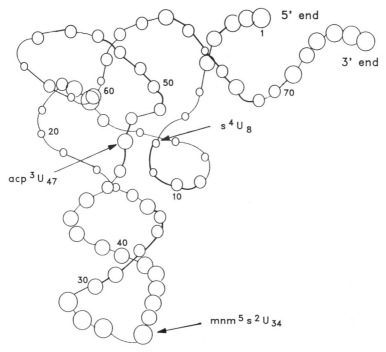

FIGURE 5. Schematic representation of the three-dimensional structure of tRNA. The positions of the modified bases detected by affinity electrophoresis are indicated. The flexibility of the polynucleotide chain is roughly proportional to the size of the circles denoting the individual bases (Kim, 1978).

different values for the binding constants to the mercurial matrix, leading to the interpretation that the more exposed and flexible terminal —SH is more available for interaction (Igloi, 1992). Although these differing affinities were measured in $7M$ urea, studies focusing on the modified base s^4U_8 had previously shown that under these conditions only partial unfolding of the tRNA conformation is achieved: 72% formamide is required for complete denaturation (Igloi, 1988). Interestingly, denaturation, which might at first sight be expected to bring this base into a more exposed situation, results (as was the case for mnm^5s^2U) in a weaker binding. Since competition experiments with Mg^{2+} had shown that the mercury binds to the region around s^4U_8 which is normally associated with a strongly bound Mg^{2+}, the effect of denaturation is in this case ascribed to the destruction of the architecture of a metal ion binding site.

4.2.3. Sequence-Specific Binding

4.2.3.1. Dyes. Studies on the sequence-specific interaction of organic cationic dyes with DNA led to the development of affinity electrophoretic gels with which a remarkable degree of resolution could be obtained (Müller *et al.*, 1981). The dyes, which could be classified as having a preference for GC- or AT-rich sequences, were linked to poly(ethylene glycol), which was then either immobilized by entrapment in polyacrylamide or used as a species migrating in agarose gels toward the cathode. In both systems, excellent fractionation was observed, in particular in two-dimensional gels, and could be correlated to the base composition of the DNA. For fragments with an AT content in the range 35–65%, a theoretical treatment of the results showed that a 0.5% difference in base composition between two fragments of equal length should be detectable. The direct coupling of acridine to periodate-oxidized agarose, giving a meltable derivative, has been reported (Horejsi *et al.*, 1982). A retardation of DNA by this matrix was observed.

Sequence-specific retardation has been observed in a procedure analogous to the mobility shift assay (see below) during DNA triple-helix formation between an appropriate target duplex (pyrimidine- or purine-rich) and a single strand designed to reside in the major groove by Hoogsteen base pairing (Cooney *et al.*, 1988).

4.2.3.2. Proteins. As macromolecular substrates for protein binding, large DNA or RNA molecules are not particularly well suited to immobilization in an electrophoretic matrix. In the first place, attachment to the gel should ideally be at one of the termini so that any potential protein binding site is not masked by itself being attached to the immobile phase. Even if this drawback can be overcome through immobilization by entrapment, it should be recalled that the incorporation of a polyelectrolyte into the gel matrix gives rise to endosmotic effects culminating in gel distortion. One approach to overcoming both problems was taken by Pitha (1975), who prepared acrylamide gels in the presence of electroneutral polyvinyl analogs of polynucleotides. It could be shown that poly(rA) was retarded by the poly(vinylU) gel. However, the behavior of other polyvinyl analogs was not entirely predictable, and these gels have not found general application. The apparent solution to the analysis of protein–DNA complexes by electrophoresis abandons any attempt at immobilization but relies instead on the changes in mobility caused by complex formation (a concept not altogether dissimilar to the original glycoprotein–lectin affinity electrophoresis). The gel retardation or, more accurately, mobility shift assay (Fried and

Crothers, 1981; Garnier and Revzin, 1981) now dominates virtually all investigations involving DNA binding proteins and is included in the standard repertoire of the molecular biologist (Ausubel *et al.*, 1988). Analogously, Konarska and Sharp (1986) were able to extend the gel retardation procedure to the analysis of RNA–protein complexes. The inclusion of nonspecific, nonradioactive DNA in the initial binding reaction enabled Strauss and Varshavsky (1984) to monitor crude protein extracts for binding to specific, radio-actively labeled DNA fragments. The anticipated problems arising from a kinetic instability of the complexes turned out to be unfounded. Complexes with half-lives of less than 1 min can be detected despite the fact that the actual electrophoresis time is considerably longer than this. This has been ascribed to a cage effect of the gel, which, while not affecting the dissociation rate of the complex, is thought to effectively increase the local concentration of the components and to enhance in this way the bimolecular reassociation reaction (Carey, 1991). In this respect, an active participation of the gel matrix in the maintenance of the complex is implied, and thus this procedure may be mechanistically distinct from a "pure" gel filtration analogy. Not only can complexes be detected in this fashion, but the stoichiometry and strength of the binding can be quantified (Fried, 1989), and the protein(s) involved in the binding can be characterized further by, for example, addition of a second dimension sodium dodecyl sulfate (SDS) gel. The possible evaluation of the data in terms of equilibrium binding constants, dissociation kinetics, binding competition (Fried, 1989; Carey, 1991), or conformational changes in the DNA induced by a biospecific interaction (Ceglarek and Revzin, 1989) has contributed to the power and popularity of the method.

As an alternative to the mobility shift method and with the aim of overcoming the potential limitations of this approach with respect to strong interactions, Lim *et al.* (1991) have applied the classical principle of affinity electrophoresis (gel entrapment of one component and electrophoretic migration of the interacting partner) to protein–DNA recognition. The procedure permits the analysis and quantitation of complex formation under equilibrium conditions while suffering the disadvantage of requiring larger amounts of protein (for incorporation into the gel) than is needed for mobility shift analysis, for example. Somewhat confusingly, since the distinction between this technique and affinity electrophoresis is not obvious, the term affinity coelectrophoresis has been introduced to describe this method.

5. CONCLUDING REMARKS

The initial step in the application of affinity electrophoresis to a wider range of bioseparations has certainly not been facilitated by the lack of a commercially available electrophoretic matrix suitable for direct derivatization, let alone an arsenal of prede-rivatized resins as is available to the affinity chromatographer. Fortunately, through the development of gel retardation methods, the requirement for specific affinity matrices for studying protein–DNA or protein–RNA interactions has been circumvented. Developments in other areas of affinity electrophoretic bioseparations, covering both proteins and nucleic acids, are evidence for the fact that the design of appropriate derivatives of the common electrophoretic media is the rate-limiting step in making this powerful technique even more widely applicable.

ACKNOWLEDGMENTS

I am grateful to Professor Hans Kössel for valuable discussions during the course of my work on affinity electrophores. This work was supported, in part, by the *Deutsche Forschungsgemeinschaft (SFB 206).*

REFERENCES

Ausubel, F. M., Brent, R., Kingston, R. E., Moore, D. D., Seidman, J. G., Smith, J. A., and Struhl, K. (eds.), 1988, *Current Protocols in Molecular Biology*, Vol. 2, Wiley Interscience, New York.

Bergenheim, N., Carlsson, U., and Hansson, C., 1983, Preparative affinity electrophoresis: Application to human erythrocyte carbonic anhydrase, *Anal. Biochem.* **134:**259–263.

Bezerra, R., and Favre, A., 1990, *In vivo* incorporation of the intrinsic photolabel 4-thiouridine into *Escherichia coli* RNAs, *Biochem. Biophys. Res. Commun.* **166:**29–37.

Björk, G. R., Ericson, J. U., Gustafsson, C. E. D., Hagervall, T. G., Jönsson, Y. H., and Wikström, P. M., 1987, Transfer RNA modification, *Annu. Rev. Biochem.* **56:**263–287.

Bøg-Hansen, T. C., 1973, Crossed immuno-affinoelectrophoresis. A method to predict the results of affinity chromatography, *Anal. Biochem.* **56:**480–488.

Bøg-Hansen, T. C., 1990, Affinity electrophoresis, *Phys.-Chem. Biol.* **34:**125–132.

Bøg-Hansen, T. C., and Takeo K., 1980, Determination of dissociation constants by affinity electrophoresis: Complexes between human serum proteins and concanavalin A, *Electrophoresis* **1:**67–71.

Bøg-Hansen, T. C., Bjerrum, O. J., and Ramlau, J., 1975, Detection of biospecific interaction during the first dimension electrophoresis in crossed immunoelectrophoresis, *Scand. J. Immunol.* **4**(Suppl. 2)**:**141–147.

Borneleit, P., Blechschmidt, B., and Kleber, H.-P., 1989a, Interactions between lipopolysaccharide and outer membrane proteins of *Acinetobacter calcoaceticus* studied by an affinity electrophoresis system, *Electrophoresis* **10:**234–237.

Borneleit, P., Blechschmidt, B., and Kleber, H.-P., 1989b, Lipopolysaccharide–protein interactions: Determination of dissociation constants by affinity electrophoresis, *Electrophoresis* **10:**848–852.

Brandley, B. K., and Schnaar, R. L., 1988, Covalent attachment of an Arg-Gly-Asp sequence peptide to derivatizable polyacrylamide surfaces: Support of fibroblast adhesion and long-term growth, *Anal. Biochem.* **172:**270–278.

Breborowicz, J., and Mackiewicz, A., 1989, Affinity electrophoresis for diagnosis of cancer and inflammatory conditions, *Electrophoresis* **10:**568–573.

Brockelhurst, K., Carlsson, J., Kierstan, M. P. J., and Crook, E. M., 1973, Covalent chromatography. Preparation of fully active papain from dried papaya latex, *Biochem. J.* **133:**573–584.

Carey, J., 1991, Gel retardation, *Methods Enzymol.* **208:**103–117.

Ceglarek, J. A., and Revzin, A., 1989, Studies of DNA–protein interactions by gel electrophoresis, *Electrophoresis* **10:**360–365.

Cerovsky, V., Ticha, M., Horejsi, V., and Kocourek, J., 1980a, Studies on lectins. XLIX. The use of glycosyl derivatives of dextran T-500 for affinity electrophoresis of lectins, *J. Biochem. Biophys. Methods* **3:**163–172.

Cerovsky, V., Ticha, M., Turkova, J., and Labsky, J., 1980b, Interaction of trypsin with immobilized *p*-aminobenzamidine derivatives studied by means of affinity electrophoresis, *J. Chromatogr.* **194:**175–181.

Chaiken, I., Rosén,S., and Karlsson, R., 1992, Analysis of macromolecular interactions using immobilized ligands, *Anal. Biochem.* **201:**197–210.

Chen, J.-L., and Morawetz, H., 1981, Affinity electrophoresis in gels containing hydrophobic substituents, *J. Biol. Chem.* **256:**9221–9223.

Clivio, P., Fourrey, J.-L., Gasche, J., Audic, A., Favre, A., Perrin, C., and Woisard, A., 1992, Synthesis and purification of oligonucleotides containing sulphur substituted nucleobases: 4-Thiouracil, 4-thiothymidine and 6-mercaptopurine, *Tetrahedron Lett.* **33:**65–68.

Connolly, B. A., 1991, Oligonucleotides containing modified bases, in: *Oligonucleotides and analogues. A practical approach* (F. Eckstein, ed.), IRL Press, Oxford, pp. 155–183.

Cooney, M., Czernuszewicz, G., Postel, E. H., Flint, S., and Hogan, M. E., 1988, Site-specific oligonucleotide binding represses transcription of the human c-*myc* gene *in vitro*, *Science* **241:**456–459.

Dubreuil, Y. L., Expert-Besancon, A., and Favre, A., 1991, Conformation and structural fluctuations of a 218 nucleotide long rRNA fragment: 4-Thiouridine as an intrinsic photolabelling probe, *Nucleic Acids Res.* **19**:3653–3660.

Eckstein, F., 1985, Nucleoside phosphorothioates, *Annu. Rev. Biochem.* **54**:367–402.

Edmonds, C. G. Crain, P. F., Gupta, R., Hashizume, T., Hocart, C. H., Kowalak, J. A., Pomerantz, S. C., Stetter, K. O., and McCloskey, J. A., 1991, Posttranscriptional modification of tRNA in thermophilic archaea (archaebacteria), *J. Bacteriol.* **173**:3138–3148.

Ek, K., Gianazza, E., and Righetti, P. G., 1980, Affinity titration curves. Determination of dissociation constants of lectin–sugar complexes and of their pH-dependence by isoelectric focusing electrophoresis, *Biochim. Biophys. Acta* **626**:356–365.

Emilsson, V., Näslund, A. K., and Kurland, C. G., 1992. Thiolation of transfer RNA in *E. coli* varies with growth rate, *Nucleic Acids Res.* **20**:4499–4505.

Fried, M. G., 1989, Measurement of protein–DNA interaction parameters by electrophoresis mobility shift assay, *Electrophoresis* **10**:366–376.

Fried, M., and Crothers, D. M., 1981, Equilibria and kinetics of *lac* repressor–operator interactions by polyacrylamide gel electrophoresis, *Nucleic Acids Res.* **9**:6505–6525.

Garnier, M., and Revzin, A., 1981, A gel electrophoretic method for quantifying the binding of proteins to specific DNA regions: Application to components of the *Escherichia coli* lactose operon regulatory system, *Nucleic Acids Res.* **9**:3047–3060.

Gerbrandy, S. J., and Doorgeest, A., 1972, Potato phosphorylase isoenzymes, *Phytochemistry* **11**:2403–2407.

Hanham, C. A., Chapman, A. J., Sheppard, M. C., Black, E. G., and Ramsden, D. B., 1986, Glycosylation of human thyroglobulin and characterization by lectin affinity electrophoresis, *Biochim. Biophys. Acta* **884**:158–165.

Heegaard, N. H. H., and Bøg-Hansen, T. C., 1990, Affinity electrophoresis in agarose gels. Theory and some applications, *Appl. Theor. Electrophor.* **1**:249–260.

Heegaard, N. H. H., Hagerup, M., Thomsen, A. C., and Heegaard, P. M. H., 1989, Concanavalin A crossed affinity immunoelectrophoresis and image analysis for semiquantitative evaluation of microheterogeneity profiles of human serum transferrin from alcoholics and normal individuals, *Electrophoresis* **10**:836–840.

Horejsi, V., 1984, Affinity electrophoresis, *Methods Enzymol.* **104**:275–281.

Horejsi, V., and Kocourek, J., 1974, Studies on phytohemagglutinin. XVIII. Affinity electrophoresis of phytohemagglutinins, *Biochim. Biophys. Acta* **336**:338–343.

Horejsi, V., and Ticha, M., 1986, Qualitative and quantitative applications of affinity electrophoresis for the study of protein–ligand interactions: A review, *J. Chromatogr.* **376**:49–67.

Horejsi, V., Ticha, M., Tichy, P., and Holy, A., 1982, Affinity electrophoresis: New simple and general methods of preparation of affinity gels, *Anal. Biochem.* **125**:358–369.

Igloi, G. L., 1983, A silver stain for the detection of nanogram amounts of RNA following two-dimensional electrophoresis, *Anal. Biochem.* **134**:184–188.

Igloi, G. L., 1988, Interaction of tRNAs and of phosphorothioate-substituted nucleic acids with an organomercurial. Probing the chemical environment of thiolated residues by affinity electrophoresis, *Biochemistry* **27**:3842–3849.

Igloi, G. L., 1992, Affinity electrophoretic detection of primary amino groups in nucleic acids: application to modified bases of tRNA and to aminoacylation, *Anal. Biochem.* **206**:363–368.

Igloi, G. L. and Kössel, H., 1985, Affinity electrophoresis for monitoring terminal phosphorylation and the presence of queosine in RNA. Application of polyacrylamide containing a covalently bound boronic acid, *Nucleic Acids Res.* **13**:6881–6898.

Igloi, G. L., and Kössel, H., 1987, Use of boronate-containing gels for electrophoretic analysis of both ends of RNA molecules, *Methods Enzymol.* **155**:433–448.

Igloi, G. L., and Kössel, H., 1988, Affinity electrophoresis for the separation of oligo and polynucleotides, *Nucleosides Nucleotides* **7**:639–643.

Johnson, S. J., Metcalf, E. C., and Dean, P. D. G., 1980, The determination of dissociation constants by affinity electrophoresis on Cibacron Blue F3G A-agarose-polyacrylamide gels, *Anal. Biochem.* **109**:63–66.

Khanna, N. C., Lakhani, S., and Tewari, K. K., 1991, Photoaffinity labelling of the pea chloroplast transcriptional complex by nascent RNA *in vitro*, *Nucleic Acids Res.* **19**:4849–4855.

Kim, S.-H., 1978, Crystal structure of yeast tRNA[Phe]: Its correlation to the solution structure and functional implications, in: *Transfer RNA* (S. Altman, ed.), MIT Press, Cambridge, Massachusetts, pp. 248–293.

Konarska, M. M., and Sharp, P. A., 1986, Electrophoretic separation of complexes involved in the splicing of precursors to mRNAs, *Cell* **46:**845–855.

Laurell, C.-B., 1965, Antibody–antibody crossed electrophoresis, *Anal. Biochem.* **10:**358–361.

Lien, J.-M., Petcu, D. J., Aldrich, C. E., and Mason, W. S., 1987, Initiation and termination of duck hepatitis B virus DNA synthesis during virus maturation, *J. Virol.* **61:**3832–3840.

Lim, W. A., Sauer, R. T., and Lander, A., 1991, Analysis of DNA–protein interactions by affinity coelectrophoresis, *Methods Enzymol.* **208:**196–210.

Mackiewicz, A., and Kushner, I., 1989, Affinity electrophoresis for studies of mechanisms regulating glycosylation of plasma proteins, *Electrophoresis* **10:**830–835.

Mackiewicz, A., and Mackiewicz, S., 1986, Determination of lectin–sugar dissociation constants by agarose affinity electrophoresis, *Anal. Biochem.* **156:**481–488.

Maly, P., Ticha, M., and Kocourek, J., 1985, Studies on lectins. LVII. Sugar-binding properties, as determined by affinity electrophoresis, of α-D-galactosidases from *Vicia faba* seeds possessing erythroagglutinating activity, *J. Chromatogr.* **347:**343–350.

Masson, P., and Reybaud, J., 1988, Hydrophobic interaction electrophoresis under high hydrostatic pressure: Study of the effects of pressure upon the interaction of serum albumin with a long-chain aliphatic ligand, *Electrophoresis* **9:**157–161.

Matousek, V., and Horejsi, V., 1982, Affinity electrophoresis: A theoretical study of the effects of the kinetics of protein–ligand complex formation and dissociation reactions, *J. Chromatogr.* **245:**271–290.

Matsumoto, A., Nakajima, T., and Matsuda, K., 1990, A kinetic study of the interaction between glycogen and *Neurospora crassa* branching enzyme, *J. Biochem.* **107:**123–126.

McCutchan, T. F., Gilham, P. T., and Söll, D., 1975, An improved method for the purification of tRNA by chromatography on dihydroxyboryl substituted cellulose, *Nucleic Acids Res.* **2:**853–864.

Müller, W., Hattesohl, I., Schuetz, H.-J., and Meyer, G., 1981, Polyethylene glycol derivatives of base and sequence specific DNA ligands: DNA interaction and application for base specific separation of DNA fragments by gel electrophoresis, *Nucleic Acids Res.* **9:**95–119.

Muramatsu, T., Nishikawa, K., Nemoto, F., Kuchino, Y., Nishimura, S., Miyazawa, T., and Yokoyama, S., 1988, Codon and amino-acid specificities of a transfer RNA are both converted by a single post-transcriptional modification, *Nature (London)* **336:**179–181.

Nakamura, S., Takeo, K., Tanaka, K., and Ueta, T., 1960, Kreuzungsverfahren bei der Papierelektrophorese, *Z. Physiol. Chem.* **318:**115.

Nakamura, S., Kuwahara, A., Ogata, H., and Takeo, K., 1980a, Study of the interaction between L-lactate dehydrogenase isoenzymes and immobilized 8-substituted adenosine 5′-monophosphate by means of affinity electrophoresis, *J. Chromatogr.* **192:**351–362.

Nakamura, S., Kuwahara, A., and Takeo, K., 1980b, Study on the interaction between NADP-dependent dehydrogenase and immobilized adenosine 2′-monophosphate by means of affinity electrophoresis, *J. Chromatogr.* **196:**85–99.

Nikiforov, T. T., and Connolly, B. A., 1992, Oligonucleotides containing 4-thiothymidine and 6-thiodeoxy-guanosine as affinity labels for the EcoRV restriction endonuclease and modification methylase, *Nucleic Acids Res.* **20:**1209–1214.

Nishimura, S., 1978, Modified nucleosides and isoaccepting tRNA, in: *Transfer RNA* (S. Altman, ed.), MIT Press, Cambridge, Massachusetts, pp. 168–195.

Ofengand, J., Denman, R., Nurse, K., Liebman, A., Malarek, D., Focella, A., and Zenchoff, G., 1988, Affinity labelling of tRNA-binding sites on ribosomes, *Methods Enzymol.* **164:**372–397.

Pitha, J., 1975, Affinity gel electrophoresis of polynucleotides, *Anal. Biochem.* **65:**422–426.

Righetti, P. G., 1989, Of matrices and men, *J. Biochem. Biophys. Methods* **19:**1–20.

Rosalki, S. B., and Foo, A. Y., 1989, Lectin affinity electrophoresis of alkaline phosphatase for the differentiation of bone and hepatobiliary disease, *Electrophoresis* **10:**604–611.

Scheit, K.-H., and Faerber, P., 1973, The interactions of 2-thiopyrimidine bases with hydroxymercurybenzene sulfonate, *Eur. J. Biochem.* **33:**545–550.

Schnaar, R. L., and Lee, Y. C., 1975, Polyacrylamide gels copolymerized with active esters. A new medium for affinity systems, *Biochemistry* **14:**1535–1541.

Schnaar, R. L., Weigel, P. H., Kuhlenschmidt, M. S., Lee, Y. C., and Roseman, S., 1978, Adhesion of chicken hepatocytes to polyacrylamide gels derivatized with *N*-acetylglucosamine, *J. Biol. Chem.* **253:**7940–7951.

Schott, H., Rudloff, E., Schmidt, P., Roychoudhury, R., and Kössel, H., 1973, A dihydroxyboryl-substituted methacrylic polymer for the column chromatographic separation of mononucleotides, oligonucleotides, and transfer ribonucleic acid, *Biochemistry* **12**:932–938.

Secova, E., Ticha, M., Kocourek, J., and Dey, P. M., 1988, Electrophoretic study of α-D-galactosidases from seeds of *Glycine soja* and *Vigna radiata* possessing erythroagglutinating activity, *J. Chromatogr.* **436**:59–66.

Shimomura, S., and Fukui, T., 1980, A comparative study on α-glucan phosphorylases from plant and animal: Interrelationship between the polysaccharide and pyridoxal phosphate binding sites by affinity electrophoresis, *Biochemistry* **19**:2287–2294.

Shimura, K., and Kasai, K., 1987, Affinophoresis in two-dimensional agarose gel electrophoresis: Specific separation of biomolecules by a moving affinity ligand, *Anal. Biochem.* **161**:200–206.

Strauss, F., and Varshavsky, A., 1984, A protein binds to a satellite DNA repeat at three specific sites that would be brought into mutual proximity by DNA folding in the nucleosome, *Cell* **37**:889–901.

Takeo, K., 1984, Affinity electrophoresis: Principles and applications, *Electrophoresis* **5**:187–195.

Takeo, K., 1987, Affinity electrophoresis, *Adv. Electrophoresis* **1**:229–279.

Takeo, K., and Nakamura, S., 1972, Dissociation constants of glucan phosphorylase of rabbit tissue studied by polyacrylamide disc gel electrophoresis, *Arch. Biochem. Biophys.* **153**:1–7.

Takeo, K., Suzuno, R., Tanaka, T., and Nakamura, K., 1989a, Complete separation of anti-hapten antibodies by two-dimensional affinity electrophoresis, *Electrophoresis* **10**:813–818.

Takeo, K., Tanaka, T., Nakamura, K., and Suzuno, R., 1989b, Studies on the heterogeneity of anti-hapten antibodies by means of two-dimensional affinity electrophoresis, *Electrophoresis* **10**:818–824.

Taketa, K., and Hirai, H., 1989, Lectin affinity electrophoresis of α-fetoprotein in cancer diagnosis, *Electrophoresis* **10**:562–567.

Taketa, K., Ichikawa, E., Sato, J., Taga, H., and Hirai, H., 1989, Two-dimensional lectin affinity electrophoresis of α-fetoprotein: Characterization of erythroagglutinating phytohemagglutinin-dependent microheterogeneity forms, *Electrophoresis* **10**:825–829

Tanaka, T., Nakamura, K., and Takeo, K., 1989, A simple method for determination of the concentration of anti-dextran IgG in antiserum by means of affinity electrophoresis, *Electrophoresis* **10**:178–182.

Ticha, M., Horejsi, V., and Barthova, J., 1978, Affinity electrophoresis of proteins interacting with Blue Dextran, *Biochim. Biophys. Acta* **534**:58–63.

Ticha, M., Barthova, J., Labsky, J., and Semansky, M., 1980, Determination of the interaction of lactate dehydrogenase with high-molecular-weight derivatives of AMP by affinity electrophoresis, *J. Chromatogr.* **194**:183–189.

Turkova, R., Ticha, M., and Kocourek, J., 1983, Studies on lectins. LIV. Affinity electrophoresis of lectins and trypsin on affinity gels prepared from Sephadex derivatives incorporated into a polyacrylamide gel matrix, *J. Chromatogr.* **257**:297–303.

Vitorino dos Santos, D., Vianna, A.-L., Fourrey, J.-L., and Favre, A., 1993a, Folding of DNA substrate hairpin ribozyme domains: use of deoxy 4-thiouridine as an intrinsic photolabel, *Nucleic Acids Res.* **21**:201–207.

Vitorino dos Santos, D., Fourrey, J.-L-., and Favre, A., 1993b, Flexibility of the bulge formed between a hairpin ribozyme and deoxy-substrate analogues, *Biochem. Biophys. Res. Commun.* **190**:377–385.

von der Haar, F., 1974, Affinity elution: Principles and application to purification of aminoacyl-tRNA synthetases, *Methods Enzymol.* **34**:163–171.

Weigel, P. H., Schnaar, R. L., Kuhlenschmidt, M. S., Schmell, E., Lee, R. T., Lee, Y. C., and Roseman, S., 1979, Adhesion of hepatocytes to immobilized sugars: A threshold phenomenon, *J. Biol. Chem.* **254**:10830–10838.

Whiteside, G. M., Lamotte, A., Adalsteinsson, O., and Colton, C. K., 1976, Covalent immobilization of adenylate kinase and acetate kinase in a polyacrylamide gel: Enzymes for ATP regeneration, *Methods Enzymol.* **44**:887–897.

35

Reversed Micelles for Protein Purification

Matthijs Dekker

1. GENERAL INTRODUCTION

Reversed micelles are aggregates of surfactant molecules containing an inner core of water molecules, dispersed in a continuous organic solvent medium. These systems are optically transparent and thermodynamically stable. The considerable biotechnological potential of these systems is derived principally from the ability of the water droplets to dissolve enzymes, without loss of activity, in much the same way as bulk water does. The use of these systems in the biotransformation of apolar compounds present in the bulk oil phase has been described in a number of publications (Luisi, 1985; Martinek and Semenov, 1981; Hilhorst *et al.*, 1983; Laane *et al.*, 1987; Fletcher and Robinson, 1985).

Proteins can be transported from one bulk aqueous phase to another via an intermediate reversed micellar phase, the transported protein molecules being transiently accommodated within the reversed micelles in this case (Luisi *et al.*, 1979; Van't Riet and Dekker, 1984; Göklen and Hatton, 1985). This phenomenon is a consequence of the reversibility of the phase transfer process.

In this chapter, recent advances in separating proteins and peptides by the use of reversed micelles are reviewed in Section 2. A case study on the straightforward application of these extractions in purifying specific peptides is given in Section 3.

2. RECOVERY OF PROTEINS FROM AQUEOUS SOLUTION

Specific extracellular enzymes are isolated from a fermentation broth using conventional processes by the stepwise removal of undesired compounds from the broth. Because these steps are usually not specific for the desired enzymes, several steps are necessary to obtain an enzyme preparation of the required purity. Consequently, new isolation techniques that are more selective for the required enzyme and easier to scale up are desirable. Liquid–liquid extraction procedures involving the use of reversed micellar systems have some very

Matthijs Dekker • Unilever Research Laboratorium, P.O. Box 114, 3130 AC Vlaardingen, The Netherlands.
Molecular Interactions in Bioseparations, edited by That T. Ngo. Plenum Press, New York, 1993.

promising features in this respect. The principle of this method is depicted schematically in Fig. 1. First, the enzyme is extracted from the aqueous phase to the reversed micellar phase under such conditions that the extent of phase transfer of the protein is maximal. Second, the enzyme is recovered from the organic phase, by extraction of the reversed micellar phase with a second aqueous phase, under such conditions that the transfer of the enzyme from the reversed micellar phase to the aqueous phase is maximal.

By appropriate manipulation of parameters such as solution pH and ionic strength, which determine the distribution behavior of the enzyme, it is possible to promote purification, and also concentration, of the required enzyme in a straightforward process.

2.1. The Distribution of Proteins

There is a large amount of evidence which indicates that electrostatic interactions play a very important role in determining the distribution coefficient of a protein between a reversed micellar phase and a conjugate aqueous phase. This influence is demonstrated clearly by the effect of aqueous phase pH, ionic strength, temperature, and surfactant type (anionic, cationic, nonionic, or affinity) on this distribution behavior.

The individual effects exerted by each of these parameters will now be discussed separately. A general overview of parameters influencing the distribution coefficient of a protein is given in Fig. 2.

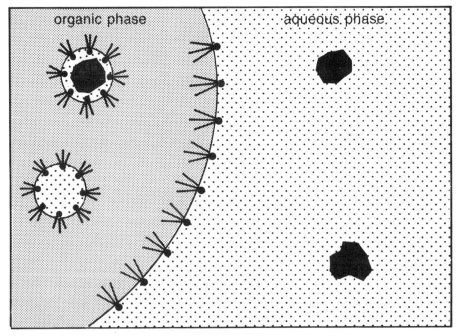

FIGURE 1. Schematic representation of the liquid–liquid extraction of enzymes between an aqueous phase and a reversed micellar phase.

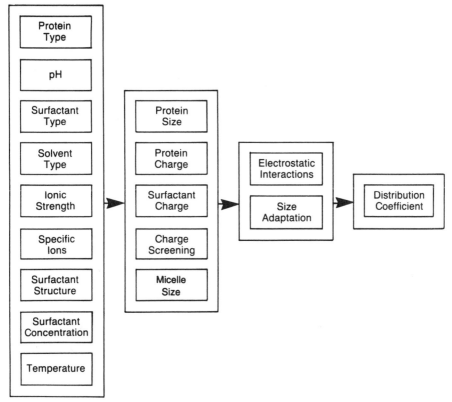

FIGURE 2. Parameters influencing the distribution coefficient of a protein in reversed micellar extraction.

2.1.1. Aqueous Phase pH

The aqueous phase pH determines the ionization state of the surface charged groups on the protein molecule. Attractive electrostatic interactions between the protein molecule and the surfactant head groups, which form the internal surface of the reversed micelle, will occur if the overall charge of the protein is opposite to the charge of the surfactant head groups. This implies that for cationic surfactants solubilization of the protein in reversed micelles is favored at pH values above the isoelectric point (pI) of the protein, whereas the opposite is true for anionic surfactants.

As the molecular mass of the protein increases, phase transfer can only be accomplished by increasing the absolute value of (pH-pI). The effect of pH on solubilization for proteins of different molecular mass might be explained by taking into account the fact that for larger proteins the size of the reversed micelle containing a protein molecule has to be significantly larger than the size of the empty reversed micelle. This energetically unfavorable transition of the micellar size has to be compensated for by more extensive electrostatic interactions in order to make the overall solubilization process feasible. This can be achieved by increasing the charge density of the protein by manipulation of the pH. For proteins whose size is smaller than the size of the water pool inside a reversed micelle, solubilization occurs as soon as the net protein charge is opposite to that of the reversed micellar interface.

This behavior is demonstrated clearly by the linear correlation attained between the molecular mass of a protein and the difference between the optimum pH of solubilization and the pI of the protein (Fig. 3; Wolbert *et al.*, 1989). The correlation holds for reversed micelles of both anionic and cationic surfactants.

2.1.2. Ionic Strength

The ionic strength of the aqueous phase determines the degree of shielding of the electrostatic potential imposed by a charged surface. This shielding causes at least two important effects in reversed micellar extraction. First, it decreases the electrostatic interaction between the charged protein molecule and the charged interface in the reversed micelle. Second, it reduces the electrostatic repulsion between the surfactant head groups, resulting in a decrease in the size of the reversed micelles at higher ionic strength (Luisi *et al.*, 1979; Göklen and Hatton, 1985).

An increase in the ionic strength results in a shift in the pH profile of the distribution behavior. Dekker *et al.* (1987b) studied the transfer of the enzyme α-amylase to a reversed micellar phase of trioctylmethylammonium chloride (TOMAC)/octanol in isooctane for different concentrations of sodium chloride as a function of the pH. As the salt concentration increased, a higher pH (resulting in a higher charge density on the protein) was required for maximal phase transfer. Thus, once again, electrostatic interactions are implicated as being of primary importance in the phase transfer mechanism.

The differences in the distribution behavior of proteins as a function of the pH and ionic strength of the aqueous phase have been used to advantage in the separation of a

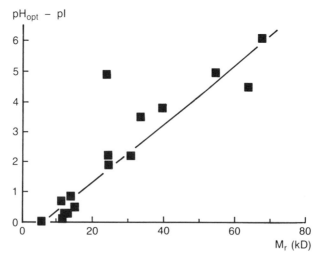

FIGURE 3. Relation between the molecular mass of a protein (M_r) and the difference between its pI and the optimal pH of extraction into a reversed micellar phase of TOMAC in isooctane. [Adapted from Wolbert *et al.* (1989).]

protein mixture (Göklen and Hatton, 1987). An aqueous solution of ribonuclease A, cytochrome *c*, and lysozyme was extracted with a reversed micellar phase of diethylhexyl sodium sulfosuccinate (AOT) in isooctane. The three proteins could be separated by a series of three extractions.

Rahaman *et al.* (1988) showed that the aqueous phase pH and ionic strength could be used to selectively extract an alkaline protease from a fermentation broth. By performing two series of extractions, 42% of the enzyme was recovered with a 2.2-fold increase in specific activity.

2.1.3. Temperature

Temperature is known to affect the phase behavior of microemulsions or reversed micellar solutions (Kahlweit and Strey, 1985). A recovery procedure using this temperature effect was developed by Dekker *et al.* (1991b). For a given composition of cationic surfactant in an organic solvent, a decrease in water solubilizing capacity was observed with increasing temperature. The process consisted of a forward extraction at low temperature (10°C). The temperature of the reversed micellar solution from this extraction, containing the enzyme, was subsequently increased, resulting in the formation, by the expulsion of water from the reversed micelles, of a separate aqueous phase, containing the enzyme in a very concentrated form. The recovery of the enzyme by means of increasing the temperature is a way of circumventing the low mass-transfer rates of back extraction as reported for some systems (Dekker *et al.*, 1990b).

2.1.4. Affinity Surfactants

Another way to promote transfer to a reversed micellar phase is by using biospecific interactions between an enzyme and a ligand (a substrate analog or product or an inhibitor). In principle, the ligand can be confined to the reversed micellar phase by conjugation to a suitable hydrophobic tail. In the reversed micelles, such a (polar) ligand would form a site-specific surfactant head group. The effect of such an affinity surfactant, octyl-β-D-glucopyranoside, has been shown for the solubilization of concanavalin A in AOT reversed micelles (Woll *et al.*, 1987, 1989). The protein was transferred to the reversed micellar phase at pH values at which no transfer was observed without the affinity surfactant. This transfer was inhibited by the addition of free ligand in the aqueous phase, indicating that an affinity interaction was required for the solubilization at these pH values.

Similar results were found for the affinity partitioning of the enzyme α-amylase with the use of surfactants of maltodextrin esterified with fatty acids (Dekker, 1990a).

2.2. Process Development

For the large-scale recovery of extracellular enzymes, an extraction with an organic solvent containing reversed micelles has interesting advantages over existing processes. A reversed micellar extraction combines the potential for concentration and purification of the enzyme in a single process. The liquid–liquid extraction technique in general is well known, and apparatus and scale-up rules have been established for numerous applications, including its use in processes for recovery of antibiotics and organic acids from fermentation broths.

The success of the type of extraction processes described here critically depends on the ability to direct the distribution coefficient of the desired enzyme between the aqueous phase and the reversed micellar phase, as discussed in Section 2.1.

2.2.1.   Forward and Back Extractions

Protein recovery from an aqueous solution by a solvent extraction using reversed micelles consists of a first step whereby the desired protein is transferred selectively from an aqueous solution to the reversed micellar solution (forward extraction). The protein is subsequently recovered from the reversed micelles by extraction with a second aqueous solution (back extraction) (Van't Riet and Dekker, 1984). These two extractions can be performed in a continuous mode, the reversed micellar phase circulating between the two extraction units. No specialized equipment is required; conventional solvent extraction equipment is, in principle, suitable for this application. The most important types of equipment are mixer/settler units, agitated columns, centrifugal extractors, and membrane extractors. For some reversed micellar systems, the use of mixer/settler units and agitated columns might be limited by emulsion formation between the aqueous phase and the reversed micellar phase because of the presence of surfactants that retard the coalescence.

2.2.2.   Mixer/Settler Extraction

The performance of a continuous forward and back extraction for the recovery of an enzyme was demonstrated by Dekker et al. (1986, 1989) with two mixer/settler units, one for each extraction step. The reversed micellar phase was circulated between the two units. The flow sheet for this extraction is shown in Fig. 4.

The distribution coefficient of the enzyme was manipulated by adjusting the pH and ionic strength of the aqueous phases. By this process, the enzyme α-amylase was concentrated 17-fold, with a recovery of enzymatic activity of 85%. The extraction performance could be predicted with sufficient accuracy by using a model which included

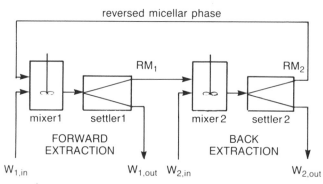

FIGURE 4. Flow sheet of the reversed micellar extraction process. The protein is extracted into the reversed micellar phase during a forward extraction. After phase separation, the reversed micellar phase is reextracted with a second aqueous phase to recover the product. In a continuous process, the reversed micellar phase is recirculated between the two extraction units.

distribution coefficients, mass-transfer rate coefficients, inactivation rate constants, phase ratios, and residence times during the forward and back extractions.

2.2.3. Membrane Extraction

A membrane can be used to immobilize an interface between two immiscible solvents. In this way a contacting area between two phases can be created without mechanical dispersion of one phase into the other. This type of extraction has been studied with hydrophobic membranes by several authors (Dekker *et al.*, 1987a; Dahuron and Cussler, 1988; Dekker *et al.*, 1991a). In these systems the membrane is wetted by the reversed micellar phase. The extraction rate is controlled by diffusion of the enzyme in the aqueous phase boundary layer (Dekker *et al.*, 1991a).

An interesting opportunity that arises from the use of membranes is the construction of a supported liquid membrane. In this type of operation, only the pores of the membrane are filled with the reversed micellar phase; the feed aqueous phase is located on one side of this membrane, while the strip aqueous phase is on the other side. However, the fouling of the interface between the liquids, and the difficulty of controlling the pressure to keep the interfaces stable, as found for membrane extraction (Dekker *et al.*, 1991a), will limit the application of this process.

2.2.4. Centrifugal Extraction

The use of centrifugal equipment to assist the phase separation in the extraction process has been studied for the enzyme α-amylase (Dekker *et al.*, 1991b). The temperature effect on the partitioning behavior (see Section 2.1.3) was used to recover the enzyme from the reversed micelles. In this way, addition of a second aqueous phase to the process was not required. Starting from a dilute aqueous enzyme solution, a concentration increase of 2000-fold could be accomplished, with an activity recovery of 73%. The forward extraction was done at 10°C, and, after phase separation by the first centrifuge, the reversed micellar phase containing the enzyme was heated to 35°C. This led to the formation of an excess aqueous phase containing the enzyme, which was subsequently removed in the second centrifuge.

3. CASE STUDY: APPLICATION TO PEPTIDE PURIFICATION

The objective of this work was to investigate the potential use of the reversed micellar extraction process to concentrate and purify casein phosphopeptides (CPP) from a tryptic casein digest. These peptides are of interest for various applications because of their capacity to form soluble complexes with metal ions.

3.1. Preparation of Peptides

In 1 liter of demineralized water, 100 g of sodium caseinate was dissolved, the pH was adjusted to 8.5 by the addition of 5*N* NaOH, and 0.1 g of trypsin (TPCK (L–(1–Tosylamide–2–phenyl) ethyl chloromethyl ketone) treated, 209 U/mg) was added. The solution was stirred at 30°C, with the pH maintained at 8.5. After 2 h no further pH change was measured, indicating that no further digestion took place. The digest obtained was stored at 4°C. Fast Protein Liquid Chromatography (FPLC) (anion-exchange column) analysis of the formed peptides is shown in Fig. 8A (Section 3.4). The CPP fractions eluted from the column between 20 and 27 min.

3.2. Forward Extraction Capacity

The extraction of CPP from casein digest was studied at different concentrations of digested casein. The composition of the reversed micellar phase was 8% (w/v) TOMAC and 4% (v/v) octanol in isooctane. The recovery and purity of the CPP in the reversed micellar solution were measured by performing a second extraction using an aqueous phase with a composition favoring transfer out of the reversed micelles ($0.5M$ NaCl, $0.1M$ NaAc, pH 5.5). The results of these measurements are shown in Fig. 5. Using equal volumes of digest and extractant, all the CPP was removed from the aqueous phase for up to 50 g of digested casein per liter. At 100 g/liter, the recovery drops considerably.

The CPP recovered in the second aqueous phase was purified considerably. In the starting casein digest, only some 4 to 5% of the protein is CPP. It should be noted that increasing the concentration of digested casein resulted in a more purified CPP in the second aqueous phase, indicating a competition between the different peptides for the reversed micelles. At these high concentrations of casein digest, the reversed micellar extraction resulted in a purification factor of 6–7, without optimizing the extractions with respect to pH and ionic strength.

3.3. Improving Purification

In order to improve the selectivity of the extraction process, a choice can be made among several options:

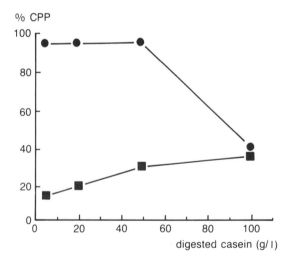

FIGURE 5. Effect of concentration of casein digest on the recovery (●) and purity (■) of casein phosphopeptides (CPP).

1. optimization of the forward extraction to prevent transfer of non-CPP peptides to the reversed micelles. This option was tested by performing the forward extraction at different NaCl concentrations. Unfortunately, the transfer of CPP was already completely inhibited at low concentrations of NaCl, while other peptides were still being transferred.

2. optimization of the back extraction to keep non-product peptides in the reversed micelles. The effect of NaCl concentration during the back extraction to an aqueous phase of 20mM Tris (pH 9) at an organic:water volume ratio of 3:1 is shown in Fig. 6. Up to 0.1M NaCl, no CPP was transferred to the second aqueous phase, while part of the other peptides were still being transferred. At 0.3M NaCl, the transfer of CPP was complete, giving a CPP purity of 50%. Further increase in the salt concentration resulted in the transfer of greater amounts of the other peptides, giving a decreased purity of the CPP.

3. removal of non-CPP peptides from the reversed micelles in a wash extraction and subsequent transfer of the CPP in an additional back extraction. Performing a wash extraction with 0.1M NaCl between the forward and back extractions should result in the removal of part of the other peptides that were also extracted during the forward extraction, giving a higher purity of the recovered CPP.

3.4. Performance of Combined Forward, Wash, and Back Extraction

A forward extraction of casein digest (50 g/liter) was performed with equal volumes of the aqueous and the reversed micellar phase. The reversed micellar phase from this forward extraction was subsequently extracted with washing buffer (0.1M NaCl) in a 3:1 volume ratio. To increase the concentration of purified CPP, the volume ratio of the reversed micellar phase to the back-extraction buffer was increased to 30:1. The recovery and purity of CPP in this back extraction as a function of the NaCl concentration are shown in Fig. 7. Much higher concentrations of NaCl were required for the recovery of CPP as compared to the

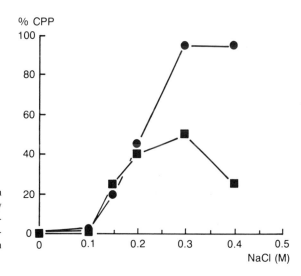

FIGURE 6. Effect of NaCl concentration during back extraction on the recovery (●) and purity (■) of casein phospho-peptides (CPP). Volume ratio of the reversed micellar phase to the back-extraction buffer was 3:1.

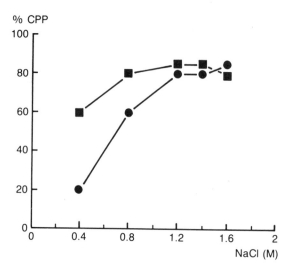

FIGURE 7. Effect of NaCl concentration during back extraction on the recovery (●) and purity (■) of casein phosphopeptides (CPP). Volume ratio of the reversed micellar phase to the back-extraction buffer was 30:1. The back extraction was preceded by a wash extraction at 0.1M NaCl; volume ratio of the reversed micellar phase to the wash extraction buffer was 3:1.

back extraction with a 3:1 volume ratio of the reversed micellar phase to the buffer (Fig. 6). This is due to the exchange of anions (Cl⁻) from the aqueous to the reversed micellar phase for the negatively charged CPP. From the amino acid sequence for CPP, it can be calculated that the concentration of negative charges on CPP is about 35mM in the starting casein digest. Since the extractions cause a 30-fold reduction in the volume in which the CPP is contained, about a 1M concentration of Cl⁻ ions is required in the final aqueous phase in order to release all the CPP out of the reversed micelles into this aqueous phase.

Up to 80–85% of CPP can be recovered in the back-extraction buffer, giving a 25-fold increase in the concentration of CPP as compared to that in the casein digest from which it was extracted. The FPLC chromatograms of the peptide mixture before and after the extraction procedure are shown in Fig. 8. The purity of the recovered CPP is increased from 4–5% to 80–85%, giving a purification factor of about 20. For reversed micellar extractions, such a high purification factor has never been published before. With protein mixtures, only purification factors of 2–6 have been obtained.

4. CONCLUSIONS

The recent developments in reversed micellar extractions clearly demonstrate that these processes have potential in biotechnology as a separation tool for proteins and peptides. Increasing the concentration and the purity of products has been demonstrated for several applications. The procedure requires only conventional extraction equipment with relatively simple modeling and scale-up rules. A drawback of these extractions might be the limitation on the surfactants and solvents that can be used for a particular application of a product. Further work is needed to establish for which bioproducts it is also economically feasible to apply the reversed micellar methodology in industrial processes.

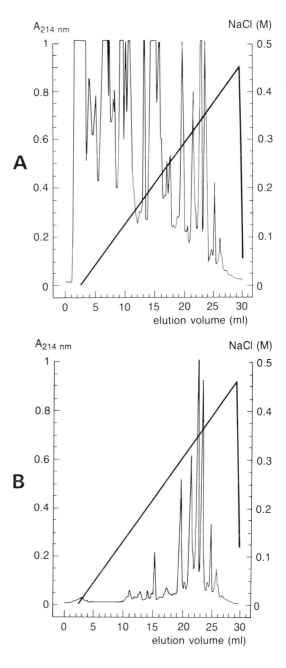

FIGURE 8. FPLC chromatograms of the casein digest before the forward extraction (A) and the aqueous phase obtained by the back extraction (B).

REFERENCES

Dahuron, L., and Cussler, E. L., 1988, Protein extraction with hollow fibers, *AIChE J.* **34:**130–136.

Dekker, M., Van't Riet, K., Weijers, S. R., Baltussen, J. W. A., Laane, C., and Bijsterbosch, B. H., 1986, Enzyme recovery by liquid–liquid extraction using reversed micelles, *Chem. Eng. J.* **33:**B27–B33.

Dekker, M., Van't Riet, K., Wijnans, J. M. G. M., Baltussen, J. W. A., Bijsterbosch, B. H., and Laane, C., 1987a, Membrane based liquid/liquid extractions of enzymes using reversed micelles, in: *Proceedings of the International Congress on Membranes and Membrane Processes (ICOM)*, Tokyo, pp. 793–794.

Dekker, M., Van't Riet, K., Baltussen, J. W. A., Bijsterbosch, B. H., Hilhorst, R., and Laane, C., 1987b, Reversed micellar extraction of enzymes: Investigations on the distribution behaviour and extraction efficiency of α-Amylase, in: *Proceedings of the 4th European Congress on Biotechnology*, Elsevier, Amsterdam, Vol. II, pp. 507–510.

Dekker, M., Van't Riet, K., Bijsterbosch, B. H., Wolbert, R. B. G., and Hilhorst, R., 1989, Modeling and optimization of the reversed micellar extractions of α-amylase, *AIChE J.* **35:**321–324.

Dekker, M., 1990a, Enzyme recovery using reversed micelles, Ph.D. Thesis, Agricultural University Wageningen, The Netherlands.

Dekker, M., Van't Riet, K., Bijsterbosch, B. H., Fijneman, P., and Hilhorst, R., 1990b, Mass transfer rate of protein extraction with reversed micelles, *Chem. Eng. Sci.* **45:**2949–2957.

Dekker, M., Koenen, P. H. M., and Van't Riet, K., 1991a, Reversed micellar—membrane—Extraction of Enzymes, Food Bioproduct. Process. *Trans. Inst. Chem. Eng.* **69**(C):54–58.

Dekker, M., Van't Riet, K., Van Der Pol, J. J., Baltussen, J. W. A., Hilhorst, R., and Bijsterbosch, B. H., 1991b, Temperature effect on the reversed micellar extraction of enzymes, *Chem. Eng. J.* **46:**B69–B74.

Fletcher, P. D. I., and Robinson, B., 1985, The partitioning of proteins between water-in-oil microemulsions and conjugate aqueous phases, *J. Chem. Soc. Faraday Trans. 1* **81:**2667–2679.

Göklen, K. E., and Hatton, T. A., 1985, Protein extraction using reversed micelles, *Biotechnol. Prog.* **1:**69–74.

Göklen, K. E., and Hatton, T. A., 1987, Liquid–liquid extraction of low molecular weight proteins by selective solubilization in reversed micelles, *Sep. Sci. Technol.* **22:**831–841.

Hilhorst, R., Laane, C., and Veeger, C., 1983, Enzymatic conversion of apolar compounds in organic media using an NADH-regenerating system and dihydrogen as reductant, *FEBS Lett.* **159:**225–228.

Kahlweit, M., and Strey, R., 1985, Phase behaviour of ternary systems of the type water-oil-nonionic amphiphiles (microemulsions), *Angew. Chem. Int. Ed. Engl.* **24:**655–669.

Laane, C., Hilhorst, R., and Veeger, C., 1987, Design of reversed micellar media for the enzymatic synthesis of apolar compounds, *Methods Enzymol.* **136:**316–229.

Luisi, P. L., 1985, Enzymes as guest molecules in inverse micelles, *Angew. Chem. Int. Ed. Engl.* **24:**439–460.

Luisi, P. L., Bonner, F. J., Pellegrini, A., Wiget, P., and Wolf, R., 1979, Micellar solubilization of proteins in aprotic solvents and their spectroscopic characterization, *Helv. Chim. Acta* **62:**740–753.

Martinek, K., and Semenov, A. N., 1981, Enzymes in organic synthesis: Physico-chemical means of increasing the yield of end product in biocatalysis, *J. Appl. Biochem.* **3:**93–126.

Rahaman, R. S., Chee, J. Y., Cabral, J. M. S., and Hatton, T. A., 1988, Recovery of an extracellular alkaline protease from whole fermentation broth using reversed micelles, *Biotechnol. Prog.* **4**(4):218–224.

Van't Riet, K., and Dekker, M., 1984, Preliminary investigations on an enzyme liquid–liquid extraction process, in: *Proceedings of the 3rd European Congress on Biotechnology*, Elsevier, Amsterdam, Vol. III, pp. 541–544.

Wolbert, R. B. G., Hilhorst, R., Voskuilen, G., Nachtegaal, H., Dekker, M., Van't Riet, K., and Bijsterbosch, B. H., 1989, Protein transfer from an aqueous phase into reversed micelles: The effect of protein size and charge distribution, *Eur. J. Biochem.* **184:**627–633.

Woll, J. M., Dillon, A. S., Rahaman, R. S., and Hatton, T. A., 1987, Protein separations using reversed micelles, in: *Protein Purification: Micro to Macro* (R. Burgess, ed.), Alan R. Liss, New York, pp. 117–130.

Woll, J. M., Hatton, T. A., and Yarmush, M. L., 1989, Bioaffinity separation using reversed micellar extractions, *Biotechnol. Prog.* **5:**57–62.

Index